Geometries and Transformations

Euclidean and other geometries are distinguished by the transformations that preserve their essential properties. Using linear algebra and transformation groups, this book provides a readable exposition of how these classical geometries are both differentiated and connected. Following Cayley and Klein, the book builds on projective and inversive geometry to construct "linear" and "circular" geometries, including classical real metric spaces like Euclidean, hyperbolic, elliptic, and spherical, as well as their unitary counterparts. The first part of the book deals with the foundations and general properties of the various kinds of geometries. The latter part studies discrete-geometric structures and their symmetries in various spaces. Written for graduate students, the book includes numerous exercises and covers both classical results and new research in the field. An understanding of analytic geometry, linear algebra, and elementary group theory is assumed.

Norman W. Johnson was Professor Emeritus of Mathematics at Wheaton College, Massachusetts. Johnson authored and coauthored numerous journal articles on geometry and algebra, and his 1966 paper 'Convey Polyhedra with Regular Faces' enumerated what have come to be called the Johnson solids. He was a frequent participant in international conferences and a member of the American Mathematical Society and the Mathematical Association of America.

Norman Johnson: An Appreciation
by Thomas Banchoff

Norman Johnson passed away on July 13, 2017, just a few months short of the scheduled appearance of his magnum opus, *Geometries and Transformations*. It was just over fifty years ago in 1966 that his most famous publication appeared in the *Canadian Journal of Mathematics*: "Convex Polyhedra with Regular Faces," presenting the complete list of ninety-two examples of what are now universally known as Johnson Solids. He not only described all of the symmetry groups of the solids; he also devised names for each of them, containing enough information to exhibit their constituent elements and the way they were arranged.

Now, as a result of this new volume, he will be further appreciated for his compendium of all the major geometric theories together with their full groups of transformations that preserve their essential characteristic features. In this process, he follows the spirit of Felix Klein and his Erlangen Program, first enunciated in 1870. Once again, a specific personal contribution to this achievement is his facility for providing descriptive names for each of the types of transformations that characterize a given geometry. His book will be an instant classic, giving an overview that will serve as an introduction to new subjects as well as an authoritative reference for established theories. In the last few chapters, he includes a similar treatment of finite symmetry groups in the tradition of his advisor and mentor, H. S. M. Coxeter.

It is fitting that the cover image and the formal announcement of the forthcoming book were first released by Cambridge University Press at the annual summer MathFest of the Mathematical Association of America. Norman Johnson was a regular participant in the MathFest meetings, and in 2016 he was honored as the person there who had been a member of the MAA for the longest time, at sixty-three years. His friends and colleagues are saddened to learn of his passing even as we look forward to his new geometrical masterwork.

GEOMETRIES AND TRANSFORMATIONS

Norman W. Johnson

CAMBRIDGE
UNIVERSITY PRESS

University Printing House, Cambridge CB2 8BS, United Kingdom

One Liberty Plaza, 20th Floor, New York, NY 10006, USA

477 Williamstown Road, Port Melbourne, VIC 3207, Australia

314–321, 3rd Floor, Plot 3, Splendor Forum, Jasola District Centre,
New Delhi – 110025, India

79 Anson Road, #06–04/06, Singapore 079906

Cambridge University Press is part of the University of Cambridge.

It furthers the University's mission by disseminating knowledge in the pursuit of
education, learning, and research at the highest international levels of excellence.

www.cambridge.org
Information on this title: www.cambridge.org/9781107103405

© Norman W. Johnson 2018

This publication is in copyright. Subject to statutory exception
and to the provisions of relevant collective licensing agreements,
no reproduction of any part may take place without the written
permission of Cambridge University Press.

First published 2018

Printed in the United States of America by Sheridan Books, Inc.

A catalogue record for this publication is available from the British Library.

Library of Congress Cataloging-in-Publication Data
Names: Johnson, Norman W., 1930–2017 author.
Title: Geometries and transformations / Norman W. Johnson,
Wheaton College.
Description: Cambridge, United Kingdom :
Cambridge University Press, 2017. |
Includes bibliographical references and index.
Identifiers: LCCN 2017009670 | ISBN 9781107103405 (hardback)
Subjects: LCSH: Geometry–Textbooks. | BISAC: MATHEMATICS / Topology.
Classification: LCC QA445 .J64 2017 | DDC 516–dc23
LC record available at https://lccn.loc.gov/2017009670

ISBN 978-1-107-10340-5 Hardback

Cambridge University Press has no responsibility for the persistence or accuracy
of URLs for external or third-party Internet Web sites referred to in this publication
and does not guarantee that any content on such Web sites is, or will remain,
accurate or appropriate.

To Eva

CONTENTS

PREFACE

THE PLANE AND SOLID GEOMETRY of Euclid, long thought to be the only kind there was or could be, eventually spawned both higher-dimensional Euclidean space and the classical non-Euclidean metric spaces—hyperbolic, elliptic, and spherical—as well as more loosely structured systems such as affine, projective, and inversive geometry. This book is concerned with how these various geometries are related, with particular attention paid to their transformation groups, both continuous and discrete. In the spirit of Cayley and Klein, all the systems to be considered will be presented as specializations of some projective space.

As first demonstrated by von Staudt in 1857, projective spaces have an intrinsic algebraic structure, which is manifested when they are suitably coordinatized. While synthetic methods can still be employed, geometric results can also be obtained via the underlying algebra. In a brief treatment of the foundations of projective geometry, I give a categorical set of axioms for the real projective plane. In developing the general theory of projective n-space and other spaces, however, I make free use of coordinates, mappings, and other algebraic concepts, including cross ratios.

Like Euclidean space, elliptic and hyperbolic spaces are both "linear," in the sense that two coplanar lines meet in at most one point. But spherical space is "circular": two great circles on a sphere always meet in a pair of antipodal points. Each of these spaces can be coordinatized over the real field, and each of them can be derived directly or indirectly from real projective space. For the three linear geometries, a metric can be induced by specializing an *absolute polarity* in projective n-space P^n or one of its hyperplanes. Fixing a *central inversion* in inversive n-space, realized as an oval n-quadric in P^{n+1}, similarly produces one or another kind of circular metric space.

Each of these real metric spaces has a *unitary* counterpart that can be derived analogously from complex projective space and coordinatized over the complex field. A unitary n-space may be represented isometrically by a real space of dimension $2n$ or $2n + 1$, and vice versa—a familiar example being the Argand diagram identifying complex numbers with points of the Euclidean plane. Many of the connections between real spaces and representations involving complex numbers or quaternions are well known, but new ground has been broken by Ahlfors (1985), Wilker (1993), Goldman (1999), and others.

From the point and hyperplane coordinates of a projective space we may construct coordinates for a metric space derived from it, in terms of which properties of the space may be described and distances and angles measured. Many of the formulas obtained exhibit an obvious symmetry, a consequence of the Principle of Duality that permits the interchange of projective subspaces of complementary dimensions. Along with the related concept of dual vector spaces, that principle also finds expression in convenient notations for coordinates and for corresponding bilinear and quadratic forms. The pseudo-symmetry of *real* and *imaginary* is likewise evident in inversive geometry as well as in the metric properties of hyperbolic and elliptic geometry.

With coordinates taken as vectors, transformations may be represented by matrices, facilitating the description and classification of geometric groups, as in the famous *Erlanger Programm* proposed by

Klein in 1872. Many of these occur among what Weyl called the "classical groups," and each geometric group is placed in an algebraic context, so that its relationship to other groups may be readily seen.

While there are still excellent treatments like that of Sossinsky (2012) that make little use of vectors and matrices, in recent years the study of geometries and transformations has benefited from the application of the methods of linear algebra. The treatment here has much in common with the approaches taken by such writers as Artzy (1965), Snapper & Troyer (1971), Burn (1985), and Neumann, Stoy & Thompson (1994). Conversely, as shown in some detail by Beardon (2005), many algebraic concepts can be illuminated by placing them in the framework of geometric transformations.

On the other hand, this book is not meant to be a text for a course in linear algebra, projective geometry, or non-Euclidean geometry, nor is it an attempt to synthesize some combination of those subjects in the manner of Baer (1952) or Onishchik & Sulanke (2006). My objective is rather to explain how all these things fit together.

The first ten chapters of the book essentially deal with transformations of geometries, and the last five with symmetries of geometric figures. Each continuous group of geometric transformations has various discrete subgroups, which are of interest both in their own right and as the symmetry groups of polytopes, tilings, or patterns. Every isometry of spherical, Euclidean, or hyperbolic space can be expressed as the product of reflections, and discrete groups generated by reflections—known as *Coxeter groups*—are of particular importance. Finite and infinite symmetry groups are subgroups or extensions of spherical and Euclidean Coxeter groups. Irreducible Euclidean Coxeter groups are closely related to simple Lie groups.

Certain numerical properties of hyperbolic Coxeter groups have been determined by Ruth Kellerhals and by John G. Ratcliffe and Steven Tschantz, as well as by the four of us together. A number of discrete groups operating in hyperbolic n-space ($2 \leq n \leq 5$) correspond to linear fractional transformations over rings of real, complex, or quaternionic integers. The associated *modular groups* are discussed in the final two chapters, which have for the most part been extracted from work done jointly with Asia Ivić Weiss.

One of the advantages of building a theory of geometries and transformations on the foundation of vector spaces and matrices is that many geometric propositions are easy to verify algebraically. Consequently, many proofs are only sketched or left as exercises. Where appropriate, however, I provide explanatory details or give references to places where proofs can be found. These include both original sources and standard texts like those by Veblen & Young and Coxeter, as well as relevant works by contemporary authors. In any case, results are not generally presented as formal theorems.

The reader is assumed to have a basic understanding of analytic geometry and linear algebra, as well as of elementary group theory. I also take for granted the reader's familiarity with the essentials of Euclidean plane geometry. Some prior knowledge of the properties of non-Euclidean spaces and of real projective, affine, and inversive geometry would be helpful but is not required. A few Preliminaries provide a short review. It is my hope that the book may serve not only as a possible text for an advanced geometry course but also as a readable exposition of how the classical geometries are both differentiated and connected by their groups of transformations.

Many of the ideas in this book are variations on themes pursued by my mentor, Donald Coxeter, who in the course of his long life did much to rescue geometry from oblivion. I am also grateful for the support and encouragement of my fellow Coxeter students Asia Weiss, Barry Monson, and the late John Wilker, as well as many stimulating exchanges with Egon Schulte and Peter McMullen. It was also a privilege to collaborate with Ruth Kellerhals, John Ratcliffe, and Steve Tschantz. Many details of the book itself benefited from the assistance of Tom Ruen, whose comments and suggestions were extremely helpful, along with Anton Sherwood and George Olshevsky. And I owe many thanks to my editors and the production staff at Cambridge, especially Katie Leach, Adam Kratoska, and Mark Fox.

PRELIMINARIES

THE INVENTION OF COORDINATES by Pierre de Fermat (1601–1665) and René Descartes (1596–1650) united what had been seen as the separate realms of geometry and algebra. Still deeper connections were revealed by the subsequent development of new kinds of geometry and the systematization of algebra. Before we proceed to examine some of those connections, it will be useful to set down a few basic facts about the geometries and algebraic systems themselves.

A. *Euclidean and other geometries.* Euclid's *Elements* deals with relations among points, lines, and planes and properties of geometric figures such as triangles, circles, and spheres. Among the fundamental concepts of Euclidean plane geometry are *collinearity*, *congruence*, *perpendicularity*, and *parallelism*. A rigorous treatment also involves *order* and *continuity*—relations not explicitly dealt with in the *Elements*. By omitting or modifying some of these concepts, a variety of other geometries can be constructed: the real affine and projective planes, the real inversive sphere, and the so-called non-Euclidean geometries. Like Euclidean geometry, all of these alternative systems have extensions to higher-dimensional spaces.

Any two points in the *Euclidean plane* E^2 are joined by a unique line, and lines are of infinite extent. Distances and areas can be measured with the aid of an arbitrarily chosen unit of length. Right angles provide a standard for angular measure. The Euclidean parallel postulate is equivalent to the assertion that through any point not on a given line there can be drawn *just one* line that does not intersect it. (The other postulates imply the existence of *at least one* such line.) From these assumptions it follows that the sum of the interior angles of every triangle is equal to two right angles, and that (the area of) the square on the hypotenuse of a right triangle is equal to the sum of (the areas of) the squares on the other two sides.

The real *affine plane* A^2 is the Euclidean plane without perpendicularity. There is no way to measure angles, and distances can be compared only for points on a line or on parallel lines. Nevertheless, areas can still be determined. Up to size, all triangles are equivalent, as are all parallelograms; there is no such thing as a right triangle or a square. Conics can be distinguished only as ellipses, parabolas, and hyperbolas—there are no circles.

By adopting the convention that all the affine lines parallel in a given direction meet in a unique "point at infinity" and that all such points lie on a single "line at infinity," we eliminate parallelism. When we admit the new points and the new line into the fold with the same rights and privileges as all the others, we have the real *projective plane* P^2. Incidences now exhibit a "principle of duality": any two points are joined by a unique line, and any two lines meet in a unique point. Angular measure, distance, and area are all undefined. Not only all triangles but all quadrilaterals are alike, and there is only one kind of nondegenerate conic. Congruence, perpendicularity, and parallelism have all disappeared; only the notion of collinearity remains.

Alternatively, the Euclidean plane can be given the topology of a sphere by adjoining a single "exceptional point" common to all lines. A line may then be regarded as a kind of circle. Extended lines and ordinary circles together form a set of "inversive circles" on the real *inversive sphere* I^2. Any three points lie on a unique inversive circle; points lying on the same circle are *concyclic*. Two circles may meet in two, one, or no real points. The distance between two points cannot be measured, but the angle between two intersecting circles can be. Thus collinearity has been replaced by concyclicity, and perpendicularity is still meaningful, but congruence and parallelism have been eliminated.

Though long suspected of being a theorem in disguise, the parallel postulate was eventually shown to be independent of the other assumptions governing the Euclidean plane. Replacing it with the contrary hypothesis—that through any point not on a given line there is *more than one* line not intersecting it—we obtain the *hyperbolic plane* of Bolyai and Lobachevsky. Moreover, if we do not assume that lines are of infinite length, we can construct a metrical geometry in which there are *no* nonintersecting lines: in the *elliptic plane* (the projective plane with a metric), any two lines meet in a point.

On the *elliptic sphere* (or simply "the sphere"), points come in antipodal pairs, and the role of lines is played by great circles; any two nonantipodal points lie on a unique great circle, and any two great circles meet in a pair of antipodal points. When antipodal points are identified, great circles of the elliptic sphere become lines of the elliptic plane (the two geometries are sometimes distinguished as "double elliptic" and "single elliptic" planes). Another possibility is the *hyperbolic sphere*, comprising two antipodal hemispheres separated by an "equatorial circle" of self-antipodal points. Two great circles either meet in a pair of antipodal points, are tangent at an equatorial point, or do not meet. Identification of antipodal points converts great circles of the hyperbolic sphere into lines of the hyperbolic plane.

The hyperbolic and elliptic planes and the elliptic sphere constitute the classical non-Euclidean geometries. Along with the hyperbolic sphere, they share with the Euclidean plane the notions of collinearity (or concyclicity), congruence, and perpendicularity. One notable difference is that the sum of the interior angles of a non-Euclidean triangle depends on its area, being proportionally greater than two right angles for an elliptic (spherical) triangle or proportionally less for a hyperbolic one.

Other properties of Euclidean and non-Euclidean geometries ("real metric spaces") and how they are related to real affine, projective, or inversive geometry will be described in greater detail beginning in Chapter 1. Although each geometry can be based on a selected set of postulates, a more instructive approach characterizes geometries by their transformation groups.

B. *Algebraic systems.* A *group* is a nonempty set G and a binary operation $G \times G \to G$, with $(a, b) \mapsto ab$, satisfying the associative law $(ab)c = a(bc)$, having an identity element, and with each element having a unique inverse. The group operation is commonly taken as multiplication, with the identity element denoted by 1 and the inverse of a by a^{-1}. The commutative law $ab = ba$ may or may not hold; when it does, the group is said to be *abelian*. Additive notation is sometimes used for abelian groups, with the identity denoted by 0 and the inverse of a by $-a$. The number of elements in a group G is its *order* $|G|$.

A subset of a group G that is itself a group with the same operation is a *subgroup*. For each element a of a (multiplicative) group G, the set of all distinct integral powers of a forms an abelian subgroup $\langle a \rangle$; the order of $\langle a \rangle$ is the *period* of a. The *center* $\mathrm{Z}(G)$ is the subgroup of elements that commute with every element of G.

A subgroup H of a group G is said to be *normal* (or "self-conjugate") if for any element $h \in H$ its conjugate ghg^{-1} by any element $g \in G$ is in H, i.e., if $gHg^{-1} \subseteq H$ for all $g \in G$; we write this as $H \lhd G$. If H is a normal subgroup of G, then for every $g \in G$, the left coset $gH = \{gh : h \in H\}$ is the same as the right coset $Hg = \{hg : h \in H\}$, and the set G/H of all such cosets is a group—the *quotient group* (or "factor group") of G by H—with $(g_1H)(g_2H) = (g_1g_2)H$. The number $|G : H|$ of cosets is the *index* of H in G. The center $\mathrm{Z}(G)$ of any group G is always normal, and $G/\mathrm{Z}(G)$ is the *central quotient group*.

If a and b are two elements of a group G, the element $a^{-1}b^{-1}ab$ is their *commutator*; this differs from the identity precisely when

$ab \neq ba$. The set of commutators generates a normal subgroup of G, the *commutator subgroup* (or "derived group") G'. The abelian group G/G' is the *commutator quotient group*.

If H and K are subgroups of a (multiplicative) group G having only the identity element 1 in common, if every element $g \in G$ is the product of some $h \in H$ and some $k \in K$, and if $hk = kh$ for every $h \in H$ and every $k \in K$, then G is the *direct product* of H and K, and we write $G = H \times K$. Necessarily both H and K are normal subgroups of G.

A group G may be presented in terms of a subset S of *generators*, and we write $G = \langle S \rangle$, if every element of G can be expressed as a product of (positive or negative) powers of elements of S. The generators satisfy certain *relations*, and G is the largest group for which the specified relations (but no others independent of them) hold. Thus the *cyclic* group C_p, of order p, is the group generated by a single element a satisfying the relation $a^p = 1$, while the *dihedral* group D_p, of order $2p$, is generated by elements a and b satisfying the relations $a^2 = b^2 = (ab)^p = 1$. For $p \geq 3$, these are, respectively, the rotation group and the full symmetry group (including reflections) of a regular p-gon.

A *transformation* is a permutation of the elements of an arbitrary set—e.g., some or all of the points of a space—or a mapping of one set into another. All the permutations of a given set S form the *symmetric* group $\mathrm{Sym}(S)$, any subgroup of which is a *transformation group* acting on S. A permutation of a finite set is *even* or *odd* according as it can be expressed as the product of an even or an odd number of transpositions interchanging two elements. The symmetric group on a set with n elements is denoted by S_n and has order $n!$. For $n \geq 2$, the even permutations constitute a subgroup of index 2 in S_n, the *alternating* group A_n, of order $\frac{1}{2}n!$.

When the points of a geometry are assigned suitable coordinates, each transformation preserving the fundamental properties of the geometry is represented by a particular type of invertible matrix, and groups of transformations correspond to multiplicative groups of matrices. Coordinates and matrix entries may be real numbers, or they may belong to more general number systems, e.g., *rings*. If we define a *semigroup* as a nonempty set with an associative binary operation but possibly lacking an identity element or inverses, then a ring R is a set whose elements form both an additive abelian group and a multiplicative semigroup, satisfying the distributive laws $a(b+c) = ab + ac$ and $(a+b)c = ac + bc$. The additive identity of R is its *zero*, and the multiplicative identity (if any) is its *unity*. The *trivial* ring 0 has only one element.

A ring in which multiplication is commutative is a *commutative ring*. An *integral domain* is a nontrivial commutative ring with unity without zero divisors, i.e., such that $ab = 0$ implies that either $a = 0$ or $b = 0$. An integral domain in which every element $a \neq 0$ has an inverse a^{-1}, so that the nonzero elements form a multiplicative group, is a *field*. Among the systems of interest are the integral domain $\mathbb{Z}$ of integers and the fields $\mathbb{Q}$, $\mathbb{R}$, and $\mathbb{C}$ of rational, real, and complex numbers. For each prime or prime power q there is a finite field $\mathbb{F}_q$ with q elements; such systems, also called *Galois fields* and denoted $\mathrm{GF}(q)$, were first investigated by Évariste Galois (1811–1832).

A ring R is *ordered* if it has a nonempty subset R^+ of *positive* elements, closed under both ring operations, such that for each element a in R just one of three cases holds: $a \in \mathsf{R}^+$, $a = 0$, or $-a \in \mathsf{R}^+$; in the last case, a is said to be *negative*. Then $a < b$ if and only if $b - a$ is positive. The ring $\mathbb{Z}$ and the fields $\mathbb{Q}$ and $\mathbb{R}$ are ordered, but the field $\mathbb{C}$ is not. (Either i or $-$i would have to be positive, but their squares both equal the negative number -1.) No finite ring can be ordered. The real field $\mathbb{R}$, which has additional properties of continuity, is a *complete* ordered field.

A transformation T mapping the elements of a group, ring, or other algebraic system U to a similar system V, written $T : U \to V$, is a *homomorphism* if it preserves the system operation(s), carrying sums or products in the *domain* U into sums or products in the *codomain* V. The *kernel* Ker T is the set of elements in U that are mapped into the identity element of V (the zero element in the case of a ring homomorphism). The *image* (or "range") Img T is the set of elements in V to which elements of U are mapped. When the systems are groups, Ker T is a normal subgroup of U and Img T is a subgroup of V.

The mapping $T : U \to V$ is a *monomorphism* or "one-to-one" transformation if Ker T contains only the identity element of U; it is an *epimorphism* or "onto" transformation if Img T is the entirety of V. A homomorphism taking U to V that is both one to one and onto has an inverse taking V to U that is also a homomorphism. Such a mapping is called an *isomorphism* (we write $U \cong V$) or, if U and V are the same system, an *automorphism*. If a is a fixed element of a group G, the mapping $x \mapsto axa^{-1}$ ("conjugation by a") is an *inner automorphism* of G.

When transformations of geometric points are expressed as algebraic homomorphisms, successive operations are normally carried out from left to right, as in the diagram

$$U \xrightarrow{T_1} V \xrightarrow{T_2} W$$

The *product* (T_1 followed by T_2) of the homomorphisms $T_1 : U \to V$ and $T_2 : V \to W$ is then the homomorphism $T_1T_2 : U \to W$, with $x(T_1T_2)$ defined as $(xT_1)T_2$.* Multiplication of homomorphisms is always associative: $(T_1T_2)T_3 = T_1(T_2T_3)$. Any algebraic system has at least the identity automorphism $x \mapsto x$, and every automorphism has an inverse. Thus the set of all automorphisms of an algebraic system forms a group.

C. *Linear algebra.* Of primary importance in our study of geometries and transformations are *vector spaces*, additive abelian groups

* Homomorphisms may be distinguished from ordinary functions, which typically precede their arguments and so are normally composed from right to left. Besides allowing them to be carried out in the order they are written, left-to-right composition of point mappings is compatible with transformations of dual systems (e.g., left and right vector spaces) as well as systems in which multiplication is noncommutative.

whose elements ("vectors") can be multiplied by the elements of a field ("scalars"). The set of vectors is closed under such "scalar multiplication"; i.e., if $\boldsymbol{V}$ is a (left) vector space over a field F, then for all scalars λ in F and all vectors $\mathbf{x}$ in $\boldsymbol{V}$, $\lambda\mathbf{x}$ is also in $\boldsymbol{V}$. Scalar multiplication also has the properties

$$\lambda(\mathbf{x}+\mathbf{y}) = \lambda\mathbf{x}+\lambda\mathbf{y}, \quad (\kappa+\lambda)\mathbf{x} = \kappa\mathbf{x}+\lambda\mathbf{x}, \quad (\kappa\lambda)\mathbf{x} = \kappa(\lambda\mathbf{x}), \quad 1\mathbf{x} = \mathbf{x}.$$

For each positive integer n, the canonical vector space F^n comprises all lists $(x_1, \ldots, x_n)$ of n elements of F (the "entries" of the list), with element-by-element addition and scalar multiplication. The vector space F^1 is the field F itself.

Given an ordered set $[\mathbf{x}_1, \ldots, \mathbf{x}_k]$ of vectors in a vector space $\boldsymbol{V}$, for any ordered set of scalars $(\lambda_1, \ldots, \lambda_k)$ the vector $\lambda_1\mathbf{x}_1 + \cdots + \lambda_k\mathbf{x}_k$ is a *linear combination* of the vectors. If S is a subset of $\boldsymbol{V}$ and if every vector $\mathbf{x}$ in $\boldsymbol{V}$ can be expressed as a linear combination of vectors in S, the set S *spans* $\boldsymbol{V}$. If a linear combination of distinct vectors $\mathbf{x}_i$ in a set S is the zero vector only when all the scalar coefficients λ_i are zero, the vectors in S are *linearly independent*. If the vectors in an ordered set S spanning a vector space $\boldsymbol{V}$ are linearly independent, S is a *basis* for $\boldsymbol{V}$, and the expression for each $\mathbf{x}$ in $\boldsymbol{V}$ is unique. Every vector space has a basis, and the number of basis vectors, called the *dimension* of the vector space, is the same for any basis. (The empty set is a basis for the zero-dimensional vector space $\mathbf{0}$.)

A vector-space homomorphism, preserving vector sums and scalar multiples, is a *linear transformation*. (When scalars are written on the left, linear transformations go on the right, and vice versa.) If $\boldsymbol{V}$ is a vector space over F, a linear transformation $\boldsymbol{V} \to \mathsf{F}$ is a *linear form* on $\boldsymbol{V}$. In our treatment, coordinates of points and hyperplanes function as row and column vectors, and geometric operations—expressed algebraically as linear transformations of coordinates—are represented by matrices. Basic geometric properties, such as distances and angles, are defined by means of *bilinear forms*, functions $\boldsymbol{V} \times \boldsymbol{V} \to \mathsf{F}$ that map pairs of vectors into scalars, preserving linear combinations. The relevant theory of finite-dimensional vector spaces will be developed beginning in Chapter 4.

Many algebraic systems can be dualized. In particular, corresponding to each linear form on a given vector space V is a *covector* of the *dual* vector space $\check{V}$. The *annihilator* of a vector $\mathbf{x} \in V$ is the set of covectors $\check{\mathbf{u}} \in \check{V}$ for which the corresponding linear form maps $\mathbf{x}$ to 0. If V is a left vector space, its dual $\check{V}$ is a right vector space, and vice versa. If the elements of V are rows, the elements of $\check{V}$ are columns. When V is finite-dimensional, the dual of $\check{V}$ is isomorphic to V, so that V and $\check{V}$ are mutually dual vector spaces.

If V is an n-dimensional vector space over a field F, we may express the fact that a covector $\check{\mathbf{u}} \in \check{V}$ belongs to the annihilator of a vector $\mathbf{x} \in V$ (and vice versa) by writing $\mathbf{x} \lozenge \check{\mathbf{u}}$. A one-to-one linear cotransformation $V \to \check{V}$ mapping each vector $\mathbf{x}$ to a covector $\check{\mathbf{x}}$ is a *polarity* provided that $\mathbf{x} \lozenge \check{\mathbf{y}}$ whenever $\mathbf{y} \lozenge \check{\mathbf{x}}$, and vectors $\mathbf{x}$ and $\mathbf{y}$ are said to be *conjugate* in the polarity. These concepts can be extended to the $(n - 1)$-dimensional projective space $\mathrm{P}V$ whose "points" are one-dimensional subspaces $\langle \mathbf{x} \rangle$ spanned by nonzero vectors $\mathbf{x} \in V$.

A *module* has the structure of a vector space except that scalars are only required to belong to a ring. An *algebra* $\mathbf{A}$ is a module over a ring R in which there is also defined a multiplication of module elements, distributive over addition and such that

$$\lambda(\mathbf{xy}) = (\lambda\mathbf{x})\mathbf{y} = \mathbf{x}(\lambda\mathbf{y})$$

for all λ in R and all $\mathbf{x}$ and $\mathbf{y}$ in $\mathbf{A}$. If each nonzero element has a multiplicative inverse, $\mathbf{A}$ is a *division algebra*.

D. *Analysis.* The assignment of coordinates establishes a correspondence between points of a geometric line and elements of some number system. When this system is an ordered field (e.g., $\mathbb{Q}$ or $\mathbb{R}$), sets of collinear points have a definite linear or cyclic *order*, which can be described, following Moritz Pasch (1843–1930), in terms of one point lying between two others or, following Giovanni Vailati (1863–1909), one pair of points separating another pair. The order relation can be used to define line segments, rays, and the like.

The property that sets real and complex geometries apart from others is *continuity*, which essentially means that no points are "missing" from a line. The notion of continuity is implicit in the theory of proportion developed by Eudoxus (fourth century BC) and presented in Book V of Euclid's *Elements*. As we shall see in Chapter 2, a formal definition can be based on the theory of rational "cuts" invented by Richard Dedekind (1831–1916), so that each point but one of a "chain" corresponds to a unique real number. (If the definition of continuity were modified to allow infinitesimal quantities, one could even identify the points of a line with the "hyperreal" numbers of nonstandard analysis.)

E. *Arithmetic.* A nonzero integer b is a *divisor* of an integer a if there is an integer c such that $a = bc$. A positive integer p greater than 1 whose only positive divisors are 1 and p itself is a *prime*. If a and b are integers with $b > 0$, then there exist unique integers q (the "quotient") and r (the "remainder"), such that $a = bq + r$ with $0 \leq r < b$. The process of determining q and r is called the *division algorithm*.

The *greatest common divisor* of two nonzero integers a and b is the largest integer that is a divisor of both; we denote it by $\gcd(a, b)$ or simply (a, b). When $(a, b) = 1$, a and b are said to be *relatively prime*. Given two nonzero integers a and b, we can repeatedly apply the division algorithm to obtain a decreasing sequence of positive integers $r_1 > r_2 > \cdots > r_k$, where

$$a = bq_1 + r_1,\ b = r_1q_2 + r_2,\ r_1 = r_2q_3 + r_3,\ \ldots,\ r_{k-1} = r_kq_{k+1} + 0.$$

Then r_k, the last nonzero remainder, is the greatest common divisor (a, b). This process, described in Book VII of the *Elements*, is called the *Euclidean algorithm*.

The Euclidean algorithm can be used to prove the Fundamental Theorem of Arithmetic: every positive integer can be uniquely factored into the product of powers of primes. These ideas can be extended to other integral domains and number systems, but not every system allows a version of the Euclidean algorithm or has the unique factorization property.

F. *Geometric transformations.* Each of the geometries described above has certain sets of transformations that permute the points while preserving such properties as collinearity and congruence. All the transformations that leave a particular property invariant form a group, and in later chapters we shall exhibit the relevant transformation groups for the different geometries.

For each geometry to be considered coordinates can be assigned so that points correspond to vectors and geometric transformations to linear transformations of a vector space. Linear transformations in turn can be represented by matrices. When coordinate vectors $(x) = (x_1, \ldots, x_n)$ are regarded as rows ($1 \times n$ matrices), a linear transformation $\cdot A$, determined by an $n \times n$ matrix A, maps the point X with coordinates (x) to the point X' with coordinates $(x)A$, and geometric transformation groups correspond to multiplicative groups of matrices. As a rule, the groups in question turn out to be subgroups of the *general linear* group GL_n of invertible $n \times n$ matrices or its central quotient group, the *projective general linear* group PGL_n.

In a variation of this procedure, each point of a projective line FP^1 over a field F can be identified uniquely either with some element $x \in \mathsf{F}$ or with the extended value ∞ (an extra element that behaves like the reciprocal of 0). A "projectivity" of FP^1 can be expressed as a *linear fractional* transformation of the extended field $\mathsf{F} \cup \{\infty\}$, defined for given field elements a, b, c, d ($ad - bc \neq 0$) by

$$x \mapsto \frac{ax + c}{bx + d}, \quad x \in \mathsf{F} \backslash \{-d/b\},$$

with $-d/b \mapsto \infty$ and $\infty \mapsto a/b$ if $b \neq 0$ and with $\infty \mapsto \infty$ if $b = 0$. Such mappings can also be represented by 2×2 invertible matrices over F, constituting the projective general linear group $\mathrm{PGL}_2(\mathsf{F})$.

This book is primarily concerned with geometries that can be coordinatized over the real field $\mathbb{R}$, but many results can be generalized to the complex field $\mathbb{C}$ and other systems, including the noncommutative division ring ("skew-field") of quaternions. Also, real spaces may be represented by complex or quaternionic models. For example, the real inversive sphere I^2 can be modeled by the complex projective line, with circle-preserving "homographies" of I^2 corresponding to linear fractional transformations of $\mathbb{C} \cup \{\infty\}$.

Besides the mostly continuous groups of transformations of whole spaces, we shall also investigate the discrete symmetry groups of certain geometric figures, such as polytopes and honeycombs. In the latter part of the book we shall rely heavily on H. S. M. Coxeter's theory of groups generated by reflections. Discrete "modular" groups that operate in hyperbolic n-space for $2 \leq n \leq 5$ are defined by linear fractional transformations (or 2×2 matrices) over rings of real, complex, or quaternionic integers.

1

HOMOGENEOUS SPACES

EUCLIDEAN GEOMETRY is the prototype of a real *metric space*, in which the distance between any pair of points is a nonnegative real number, so that lengths and angles can be measured in a consistent way. In a *homogeneous* space, any point can be taken to any other point by a distance-preserving transformation, or *isometry*. Euclidean and other homogeneous spaces can be distinguished by how their subspaces are related. Every isometry can be expressed as the product of reflections, a fact that enables us to classify the different isometries of n-dimensional space. Complex conjugation allows a real-valued distance function to be defined on spaces coordinatized over the complex field, and each real metric space has a *unitary* counterpart.

1.1 REAL METRIC SPACES

The classical real metric spaces are what Riemann (1854) called "spaces of constant curvature." For each $n \geq 0$ we have *spherical, Euclidean, hyperbolic,* and *elliptic* n-space, denoted respectively by S^n, E^n, H^n, and eP^n. For $n \geq 1$ such geometries are n-manifolds (locally Euclidean topological spaces); for $n \geq 2$ the spaces S^n, E^n, and H^n are simply connected—any two points can be linked by a path, and every simple closed curve can be shrunk to a point. Having constant curvature means that each of these spaces is *homogeneous* (any two points are related by a distance-preserving transformation) and *isotropic* (all directions are alike). In later chapters we shall see how the different geometries can be derived from projective, affine, or inversive spaces. Here we note a few of their distinctive properties.

Each n-dimensional real metric space contains various lower-dimensional subspaces. A k-dimensional subspace of Euclidean, hyperbolic, or elliptic n-space ($k < n$) is a *k-plane* E^k, H^k, or eP^k—a *point* if $k = 0$, a *line* if $k = 1$, a *plane* if $k = 2$, a *hyperplane* if $k = n-1$. Spherical k-space can be embedded in any metric n-space ($k < n$) as a *k-sphere* S^k—a *pair* of points if $k = 0$, a *circle* if $k = 1$, a *sphere* if $k = 2$, a *hypersphere* if $k = n - 1$. The points of S^n occur in antipodal pairs; a k-sphere in S^n that contains such a pair is a *great* k-sphere, a k-dimensional subspace of spherical n-space. The empty set $\varnothing$ may be regarded as a universal subspace of dimension -1.

Subspaces of the different geometries are related in different ways. Any two points of E^n, H^n, or eP^n are joined by a unique line, and any two nonantipodal points of S^n lie on a unique great circle. Any two great circles on a sphere S^2 intersect in a pair of antipodal points (see Figure 1.1a), while two lines in the elliptic plane eP^2 intersect in a single point; in each case, the great circles or lines have a unique common perpendicular. Two lines in the Euclidean plane E^2 either intersect or are *parallel*, and parallel lines in E^2 are everywhere equidistant (Figure 1.1b). In the hyperbolic plane H^2, two nonintersecting lines are either *parallel* or *diverging* ("ultraparallel"). Parallel lines in H^2 are asymptotic, converging in one direction (Figure 1.1c), while two diverging lines have a unique common perpendicular, away from which the distance between them increases without bound (Figure 1.1d).

As its center recedes along a given line, the limit of a circle through a fixed point in E^2 is a straight line, but in H^2 the limit is a *horocycle*—a "circle" with its center at infinity. Similarly, the locus of points at a constant distance from a given line in E^2 is a pair of parallel lines, but in H^2 it is a two-branched "equidistant curve" or *pseudocycle*. In eP^2 the locus is a circle (or a single point). In S^2 the locus of points

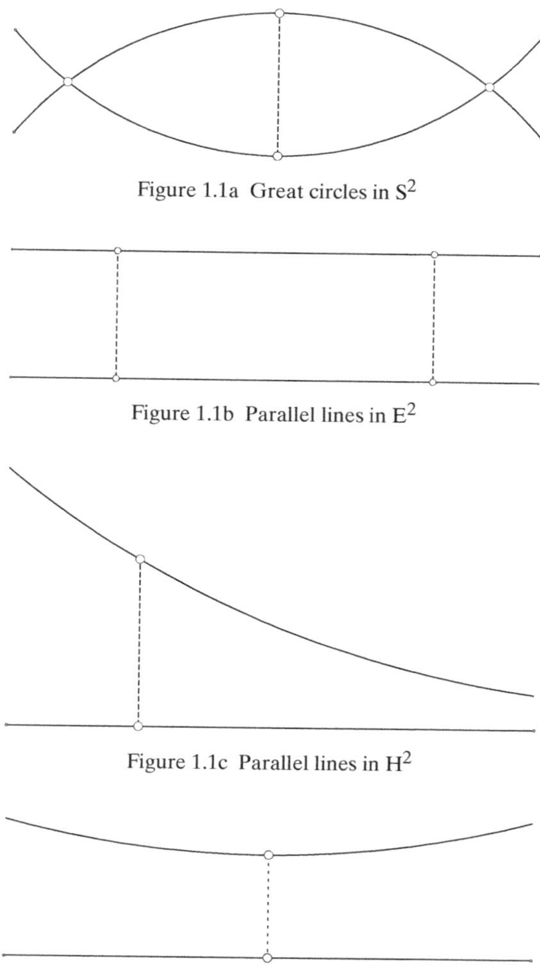

Figure 1.1a Great circles in S^2

Figure 1.1b Parallel lines in E^2

Figure 1.1c Parallel lines in H^2

Figure 1.1d Diverging lines in H^2

at a constant distance from a great circle is an antipodal pair of small circles (or points).

Even before the seminal work of Lobachevsky (1829–30) and Bolyai (1832), Gauss's student F. L. Wachter (1792–1817) observed that, independently of the parallel postulate, the geometry of a sphere whose radius increases without bound becomes Euclidean in the limit

(Stäckel 1901; cf. Bonola 1906, § 30; Sommerville 1914, p. 15; Coxeter 1942, pp. 7, 220). In Euclidean n-space the limit of a k-sphere S^k ($k < n$) is a k-plane E^k, while in hyperbolic n-space it is a *k-horosphere*, isometric to Euclidean k-space.

Hyperbolic k-space can be embedded in hyperbolic n-space ($k < n$) not only as a k-plane H^k but also as one of the two sheets of an "equidistant k-surface" or *k-pseudosphere*. A portion of the hyperbolic plane (a horocyclic sector) can be embedded in Euclidean 3-space as a "pseudospherical surface" (Beltrami 1868a; cf. Bonola 1906, § 67; Sommerville 1914, pp. 168–170; Coxeter 1942, pp. 256–258; Stillwell 1996, pp. 1–5), but Hilbert (1901) showed that there is no way to embed all of H^2 in E^3. Still, H^k can be isometrically immersed in E^n — possibly with self-intersections—if n is sufficiently large, as when $n \geq 6k - 5$ (Blanuša 1955; cf. Alekseevsky, Lychagin & Vinogradov 1988, chap. 8, §§ 6–7).

Ordinary Euclidean n-space E^n ($n \geq 1$) may be extended by adjoining to the ordinary points of each line one *absolute point* (or "point at infinity"), the resulting *extended Euclidean* space being denoted $\bar{E}^n$. All the absolute points of $\bar{E}^n$ lie in an *absolute hyperplane*, and each ordinary k-plane, together with all the k-planes parallel to it, meets the absolute hyperplane in an *absolute* $(k-1)$*-plane*. The geometry of the absolute hyperplane of $\bar{E}^n$ is elliptic.

Likewise, ordinary hyperbolic n-space H^n ($n \geq 1$) may be extended by adjoining to the ordinary points of each line an *absolute pair* of points—i.e., two "points at infinity," one in each direction—the resulting *extended hyperbolic* space being denoted $\bar{H}^n$. All the absolute points of $\bar{H}^n$ lie on an *absolute hypersphere*, and each ordinary k-plane meets the absolute hypersphere in a unique *absolute* $(k-1)$*-sphere*. Each ordinary plane is thus bounded by an *absolute circle*. As noted by Liebmann (1905, pp. 54–58), the geometry of the absolute hypersphere of $\bar{H}^n$ is inversive.

Extended hyperbolic n-space can be completed by embedding it in *projective* n-space P^n, thereby adjoining to the ordinary and absolute

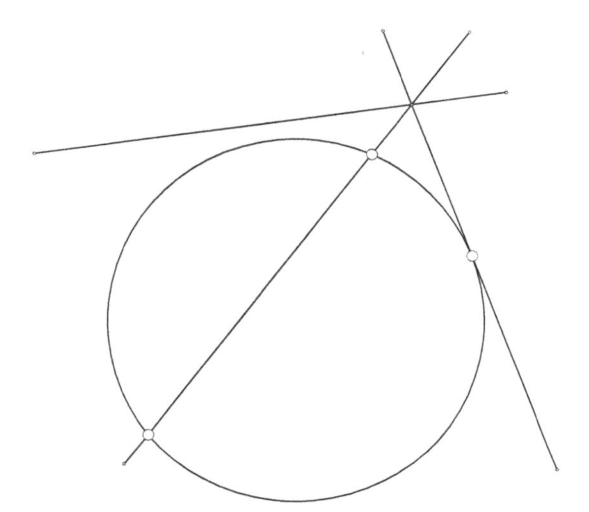

Figure 1.1e Ordinary, absolute, and ideal lines

points a set $\check{\mathrm{H}}^n$ of *ideal* (or "ultra-infinite") points, exterior to the absolute hypersphere (Bonola 1906, app. IV; Sommerville 1914, pp. 47–50). In *complete hyperbolic n-space* hP^n, any $k+1$ points of general position $(1 \le k \le n-1)$ lie in a unique k-plane, which is *ordinary*, *absolute*, or *ideal* according as it meets the absolute hypersphere in an absolute $(k-1)$-sphere, a single absolute point, or not at all.* The two-dimensional case is illustrated in Figure 1.1e. The absolute hypersphere is the locus of self-conjugate points in an *absolute polarity* that pairs each ordinary, absolute, or ideal point with an ideal, absolute, or ordinary hyperplane.

Indeed, as shown by Cayley (1859) and Klein (1871, 1873), each of the non-Euclidean n-spaces eP^n and H^n (as contained in hP^n) results from fixing a polarity in projective n-space (an *involution* if $n = 1$),

* The geometry of an ordinary, absolute, or ideal k-plane is respectively hyperbolic, co-Euclidean, or elliptic. The points and subspaces of an ordinary k-plane hP^k may themselves be ordinary, absolute, or ideal. An absolute k-plane $\check{\mathrm{E}}^k$ (dual to Euclidean k-space) has a bundle of absolute subspaces passing through its one absolute point; all other points and subspaces are ideal. All the points and subspaces of an ideal k-plane eP^k are ideal.

and Euclidean n-space E^n can be obtained from *affine* n-space A^n by fixing a polarity in the (projective) hyperplane at infinity. For both eP^n and E^n, the polarity is *elliptic*; i.e., there are no self-conjugate points. For hP^n, the polarity is *hyperbolic*: the locus of self-conjugate points—the absolute points of $\bar{H}^n$—is an *oval* $(n-1)$-*quadric* in P^n (a *conic* in P^2, a pair of points in P^1). Fixing a hyperbolic polarity in the hyperplane at infinity of A^n produces *Lorentzian* n-space (sometimes called "Minkowski space"), which is homogeneous but not isotropic.

In projective $(n+1)$-space P^{n+1}, an oval n-quadric Φ is a conformal (angle-preserving) model for *inversive* n-space I^n, alias the *Möbius n-sphere*. When we specialize an *inversion* on I^n, induced by a collineation of P^{n+1} that takes Φ into itself but interchanges "antipodal" points, we obtain a *central* n-sphere, with a metric. If the *central inversion* so defined has no fixed points, this is the *elliptic* n-sphere S^n. Otherwise, we have the *hyperbolic* n-sphere $\ddot{S}^n$, with an "equatorial" $(n-1)$-sphere of fixed ("self-antipodal") points. When antipodal points are identified, S^n is converted to elliptic n-space eP^n, and $\ddot{S}^n$ to extended hyperbolic n-space $\bar{H}^n$ (Johnson 1981, pp. 452–454).

1.2 ISOMETRIES

According to the *Erlanger Programm* of Klein, a geometry is characterized by the group of transformations that preserve its essential properties. Each of the metric spaces to be considered here has a group of *isometries* (or "rigid motions"), transformations that preserve distances. A Euclidean space also has a larger group of *similarities*, preserving ratios of distances.

A. *Isometry groups.* An isometry of Euclidean n-space E^n taking points X and Y respectively to points X' and Y' keeps the distance between them the same: $|X'Y'| = |XY|$. A similarity of scale $\mu > 0$ has $|X'Y'| = \mu|XY|$ for every pair of points. The isometry group of E^n is the *Euclidean* group E_n, a normal subgroup of the Euclidean similarity, or *euclopetic*, group $\tilde{\mathrm{E}}_n$ (i.e., if $R \in \mathrm{E}_n$ and $S \in \tilde{\mathrm{E}}_n$, then $SRS^{-1} \in \mathrm{E}_n$).*

A *translation* in E^n is a direction-preserving isometry, with each point moving the same distance, so that $|XX'|$ is the same for all points X. The *translation* group T^n is a normal subgroup of E_n. Moreover, as the additive group of the vector space $\mathbb{R}^n$, T^n is abelian; i.e., the translations form a commutative subsystem of isometries. Likewise, a *dilatation* is a similarity that preserves or reverses directions, so that $X'Y'$ is parallel to XY, and the dilatations (including translations) constitute a normal subgroup of $\tilde{\mathrm{E}}_n$. Detailed treatments of Euclidean isometries and similarities, especially in E^2, are readily available (e.g., see Yaglom 1955, Martin 1982, or Barker & Howe 2007).

The isometry group of the elliptic $(n-1)$-sphere S^{n-1} is the *orthogonal* group $\mathrm{O}_n \cong \mathrm{E}_n/\mathrm{T}^n$, induced by the isometries of E^n that leave a given point fixed. The center of O_n—the subgroup of isometries of S^{n-1} that commute with every isometry—is the *unit scalar* (or "central") group $\bar{\mathrm{S}}\mathrm{Z}$, the group of order 2 generated by the central inversion, sometimes denoted simply by '2'. Elliptic $(n-1)$-space $\mathrm{eP}^{n-1} = \mathrm{S}^{n-1}/2$ results from the identification of antipodal points of S^{n-1}, and its isometry group is the *projective orthogonal* group $\mathrm{PO}_n \cong \mathrm{O}_n/\bar{\mathrm{S}}\mathrm{Z}$. The isometry group of H^{n-1} is the *projective pseudo-orthogonal* group $\mathrm{PO}_{n-1,1}$, the central quotient group of the isometry group $\mathrm{O}_{n-1,1}$ of the hyperbolic $(n-1)$-sphere $\ddot{\mathrm{S}}^{n-1}$ (Johnson 1981, pp. 446–447).

Euclidean, hyperbolic, and spherical spaces are orientable, with two distinguishable *senses*, as are odd-dimensional elliptic spaces. An isometry of an orientable space is either *direct* (sense-preserving) or *opposite* (sense-reversing), the direct isometries forming a subgroup of index 2 in the full isometry group. In a Euclidean space, the same is true of similarities.

* Appended to a suitable stem, the suffix "-petic" (from the Greek *anapeteia* expansion) indicates the result of combining a measure-preserving transformation and a scaling operation.

The set of images of a given point of a metric space under the operations of a group of isometries is its *orbit*. If every point has a neighborhood that includes no other points in its orbit, the group is said to be *discrete*.

B. *Products of reflections.* The transformation that leaves every point of a geometry invariant is the *identity* 1. An isometry of a real metric space, other than the identity, that leaves invariant every point of an ordinary hyperplane or great hypersphere is a *reflection*, and the invariant hyperplane or hypersphere is its *mirror*. In elliptic n-space, extended Euclidean n-space, or complete hyperbolic n-space, a reflection leaves one other point invariant—the *pole* of the mirror. In any n-space, a reflection interchanges noninvariant points in pairs and thus is *involutory*, i.e., of period 2.

Every isometry of n-space can be expressed as the product of reflections. Let us say that an isometry of n-space is of *type* l if it can be expressed as the product of l, but no fewer, reflections (Coxeter 1988, p. 43). Thus the identity is of type 0 and a reflection is of type 1. By a famous theorem of Élie Cartan (1869–1951) and Jean Dieudonné (1906–1992), the type of an isometry of n-space is at most $n + 1$; that is, every isometry can be expressed as the product of $n + 1$ or fewer reflections. In elliptic n-space, the maximum type is n or $n + 1$, whichever is even (Coxeter 1942, pp. 114, 130). In an orientable space, an isometry of type l is direct if l is even, opposite if l is odd.

In any n-space, the product of two reflections whose $(n - 1)$-dimensional mirrors intersect in an ordinary $(n - 2)$-plane or a great $(n - 2)$-sphere is a *rotation* about the intersection, the *pivot* of the rotation, through twice the angle between the mirrors. If the angle is commensurable with π, the period of the rotation is finite; in any case, points in the orbit of a point not on the pivot lie on a circle. When the mirrors are coincident, the rotation reduces to the identity.

In hyperbolic n-space ($n \geq 2$), two ordinary hyperplanes are parallel when their absolute $(n-2)$-spheres are tangent. The product of two reflections in parallel mirrors is a *striation*, essentially a "pararotation" about the common absolute $(n-2)$-plane of the two mirrors. When the mirrors coincide, the transformation is just the identity. Except for this, every striation is of infinite period. Points in the orbit of any ordinary point lie on a horocycle.

In Euclidean or hyperbolic n-space, the product of two reflections in mirrors that are both orthogonal to some ordinary line is a *translation* along the line, the *axis* of the translation, through twice the distance between the mirrors. When the mirrors are coincident, the translation reduces to the identity. Otherwise, in E^n ($n \geq 2$) the mirrors are parallel, and any line parallel to the given line can serve as the axis of the translation. In H^n ($n \geq 2$), the mirrors are diverging, and the axis is unique. Except for the identity, a translation is of infinite (or imaginary) period. In the Euclidean case, repeated translates of any point are collinear. In the hyperbolic case, a point on the axis translates to other points on the axis, while points in the orbit of a point not on the axis lie on one branch of a pseudocycle. A hyperbolic translation can be thought of as a "hyperrotation" about the ideal $(n-2)$-plane absolutely polar to the axis.

Every isometry of even type $2k$ (other than the identity) can be expressed as the commutative product of $k-1$ rotations and a translation, of $k-1$ rotations and a striation, or of k rotations. Every isometry of odd type $2k+1$ can be expressed as the commutative product of an isometry of type $2k$ and a reflection. The *rank* of an isometry is the smallest integer m such that it can be effected in E^n or H^n (or eP^n if of odd type) for any $n \geq m$ or in S^n (or eP^n if of even type) for any $n \geq m-1$. Table 1.2 lists the type and rank of every nontrivial isometry of S^n (or eP^n), E^n, or H^n.

The prefixes "trans-" and "stria-" and the combining term "rotary" are used in naming products of isometries. A *transflection* (or "glide reflection") in E^n or H^n ($n \geq 2$) is the product of a translation and an

Table 1.2 *Isometries of Real Metric Spaces*

Isometry	Spaces	Type	Rank
Reflection	S, E, H	1	1
Translation	E, H	2	1
Striation	H	2	2
Rotation	S, E, H	2	2
Transflection	E, H	3	2
Striaflection	H	3	3
Rotary reflection	S, E, H	3	3
Rotary translation	E, H	4	3
Rotary striation	H	4	4
Double rotation	S, E, H	4	4
Rotary transflection	E, H	5	4
Rotary striaflection	H	5	5
Double rotary reflection	S, E, H	5	5
$(k-1)$-Tuple rotary translation	E, H	$2k$	$2k-1$
$(k-1)$-Tuple rotary striation	H	$2k$	$2k$
k-Tuple rotation	S, E, H	$2k$	$2k$
$(k-1)$-tuple rotary transflection	E, H	$2k+1$	$2k$
$(k-1)$-tuple rotary striaflection	H	$2k+1$	$2k+1$
k-tuple rotary reflection	S, E, H	$2k+1$	$2k+1$

orthogonal reflection. A *striaflection* (or "parallel reflection") in H^n ($n \geq 3$) is the product of a striation and a reflection. A *rotary translation*, the product in E^n or H^n ($n \geq 3$) of a rotation and an orthogonal translation, may be called more simply a *twist* (Veblen & Young 1918, p. 320; Coxeter 1974, p. 7).

Euclidean and non-Euclidean isometries can also be classified according to other criteria, such as the nature of their invariant subspaces, if any, or the critical values of their displacement functions, measuring the distance between a given point and its image. These approaches are usefully employed by Alekseevsky, Vinberg & Solodovnikov (1988, chap. 5), whose terminology differs in some respects from that used here. In particular, their "parallel displacement along a line" is our *translation*, and their "parabolic translation" (Coxeter's "parallel displacement") is our *striation*.

C. *Special cases.* The product of reflections in l mutually orthogonal hyperplanes or great hyperspheres of n-space $(1 \le l \le n)$ is an involutory isometry, an *inversion* in the $(n - l)$-plane or great $(n - l)$-sphere common to the l mirrors. When $l = 1$, this is just a reflection. For $l = 2$, it is a *half-turn*; for $l = 3$, a *flip reflection*; for $l = 2k \ge 4$, a *k-tuple half-turn*; for $l = 2k + 1 \ge 5$, a *k-tuple flip reflection*. The product of reflections in $n + 1$ mutually orthogonal great hyperspheres of S^n is the central inversion. The product of reflections in $n + 1$ mutually orthogonal hyperplanes of eP^n is the identity. Some writers call all involutory isometries "reflections."

The term "inversion" is also used for certain involutory transformations of inversive n-space I^n. While this duplication of terminology might seem unfortunate, the two meanings of the word are not totally unrelated. In the derivation of S^n as the elliptic n-sphere, a fixed "elliptic" inversion of I^n becomes the central inversion of S^n, which is an inversion in a point—the center of an n-sphere—when S^n is embedded in some $(n + 1)$-dimensional metric space. A fixed "hyperbolic" inversion of I^n defining the hyperbolic n-sphere $\ddot{S}^n$ likewise becomes an inversion in a hyperplane, i.e., a reflection.

In odd-dimensional elliptic or spherical space eP^{2k-1} or S^{2k-1}, the product of rotations through equal angles in k mutually orthogonal planes or great spheres is a special k-tuple rotation called a *Clifford translation* (Clifford 1873; Bonola 1906, app. II; Coxeter 1942, pp. 135–141). This transformation has the property, shared with Euclidean translations, that the distance between any point and its image is constant. A Clifford translation in elliptic 3-space leaves invariant one of two sets of "parallel" (actually skew) lines lying in a ruled quadric—a *Clifford surface*—which has zero curvature everywhere and thus a Euclidean metric.

The orbit of a point in Euclidean n-space under a discrete group of translations is called a *lattice*; a line through two points of a lattice passes through an infinite number of lattice points, equally spaced. A lattice can also be realized in hyperbolic n-space as the orbit of a point under a discrete group of striations that leave some absolute point $\dot{O}$ invariant; lattice points are equally spaced on horocycles with center $\dot{O}$.

EXERCISES 1.2

NOTE: The symbol '$\vdash$' preceding a statement indicates that it is a theorem to be proved.

1. $\vdash$ The product of reflections in two perpendicular lines of E^2 or H^2 or two orthogonal great circles of S^2 is a half-turn.

2. $\vdash$ The product of two half-turns in E^2 is a translation.

3. $\vdash$ The set of all translations and half-turns in E^2 forms a group.

4. $\vdash$ Every isometry in S^2 is a reflection, a rotation, or the commutative product of a reflection and a rotation.

5. $\vdash$ Every isometry in E^3 is either a reflection, a rotation, or a translation or the commutative product of two of these.

1.3 UNITARY SPACES

Each of the real metric spaces eP^n, E^n, H^n, and S^n has a complex analogue—namely, *elliptic unitary* n-space PU^n, *Euclidean unitary* n-space EU^n, *hyperbolic unitary* n-space HU^n, and *spherical unitary* n-space SU^n—each of which can be derived from a complex projective space of dimension n or $n+1$. The basic structure of these spaces is described in this section; a more detailed treatment will be given starting in Chapter 7.

Self-conjugate points of a polarity correspond to solutions of a quadratic equation. By the Fundamental Theorem of Algebra, the complex field $\mathbb{C}$ is algebraically closed, so that there are no irreducible quadratic polynomials over $\mathbb{C}$. Hence there are no elliptic polarities in complex projective n-space $\mathbb{C}\mathrm{P}^n$. However, unlike the real field $\mathbb{R}$, which has no nontrivial automorphisms, $\mathbb{C}$ has an involutory automorphism $* : \mathbb{C} \to \mathbb{C}$ of complex conjugation, with $z = x + yi$ mapped to $z^* = \bar{z} = x - yi$. Thus $(z+w)^* = z^* + w^*$ and $(zw)^* = z^*w^*$ for all z and w. The nonnegative real number $zz^* = |z|^2 = x^2 + y^2$ is the *norm* of z. This automorphism can be used to define an *antipolarity* in $\mathbb{C}\mathrm{P}^n$ (an *anti-involution* if $n = 1$), pairing each point with a unique hyperplane, and complex antipolarities (like real polarities) may or may not have self-conjugate points.

Each of the unitary metric spaces $\mathrm{P}\mathrm{U}^n$, $\mathrm{E}\mathrm{U}^n$, and $\mathrm{H}\mathrm{U}^n$ can be obtained by fixing an *absolute antipolarity* in $\mathbb{C}\mathrm{P}^n$ or, in the Euclidean unitary case, by first fixing a hyperplane $\mathbb{C}\mathrm{P}^{n-1}$ and then (if $n > 1$) fixing an absolute antipolarity in $\mathbb{C}\mathrm{P}^{n-1}$ (cf. Coolidge 1924, pp. 133–154; Mostow 1980, pp. 180–183; Goldman 1999, pp. 11–25). For both $\mathrm{P}\mathrm{U}^n$ and $\mathrm{E}\mathrm{U}^n$, the absolute antipolarity is *elliptic*; i.e., there are no self-conjugate points. For $\mathrm{H}\mathrm{U}^n$ the absolute antipolarity is *hyperbolic*: the locus of self-conjugate points is an *oval $(n-1)$-antiquadric* in $\mathbb{C}\mathrm{P}^n$ (an *anticonic* in $\mathbb{C}\mathrm{P}^2$, a *chain* in $\mathbb{C}\mathrm{P}^1$).

The complex projective line $\mathbb{C}\mathrm{P}^1$ with one point fixed is the Euclidean unitary line $\mathrm{E}\mathrm{U}^1$. As evidenced by the well-known *Argand diagram* (after J. R. Argand, 1768–1822),* $\mathrm{E}\mathrm{U}^1$ has the same metric as the real Euclidean plane E^2. Likewise, $\mathbb{C}\mathrm{P}^1$ with an absolute anti-involution is, according as the anti-involution is elliptic (no fixed points) or hyperbolic (a chain of fixed points), either the elliptic unitary line $\mathrm{P}\mathrm{U}^1$ (which has the same metric as the real elliptic sphere S^2) or the hyperbolic unitary line $\mathrm{H}\mathrm{U}^1$ (which has the same metric as the

* Though he was not the first to identify $\mathbb{C}$ with $\mathbb{R}^2$, it was Argand's 1806 pamphlet that first gained the attention of the mathematical community, as the insightful 1797 paper of Caspar Wessel (1745–1818) went virtually unnoticed for nearly a century. The two-dimensional real representation of complex numbers was also fully understood by Gauss.

real hyperbolic sphere $\ddot{\mathrm{S}}^2$). In general, a unitary n-space EU^n, PU^n, or HU^n has a representation as a real Euclidean $2n$-space E^{2n} or as a real elliptic or hyperbolic $2n$-sphere S^{2n} or $\ddot{\mathrm{S}}^{2n}$.

In complex projective $(n+1)$-space $\mathbb{C}\mathrm{P}^{n+1}$, an oval n-antiquadric serves as a conformal model for *inversive unitary* n-space IU^n—the *Möbius n-antisphere.* This is to be distinguished from the *complex inversive* (or "complex conformal") n-space $\mathbb{C}\mathrm{I}^n$ described for $n = 2$ by Veblen & Young (1918, pp. 264–267) and more generally by Scherk (1960). (That geometry, in which complex conjugation plays no role, is modeled in $\mathbb{C}\mathrm{P}^{n+1}$ not by an n-antiquadric but by an n-quadric, defined by a polarity rather than an antipolarity.) Fixing a family of "unitary inversions" converts IU^n into spherical or pseudospherical unitary n-space SU^n or $\breve{\mathrm{S}}\mathrm{U}^n$, which has a representation as a real elliptic or hyperbolic $(2n+1)$-sphere.

EXERCISES 1.3

1. Show that the equation $x_1{}^2 + x_1x_2 + x_2{}^2 = 0$ has no nontrivial solutions in real numbers x_1 and x_2.
2. Find a nontrivial solution of the equation $z_1{}^2 + z_1z_2 + z_2{}^2 = 0$ in complex numbers z_1 and z_2.
3. Show that the equation $z_1\bar{z}_1 + z_2\bar{z}_2 = 0$ has no nontrivial solutions in complex numbers z_1 and z_2.

2

LINEAR GEOMETRIES

Each of the spaces described in Chapter 1 can be derived by introducing an appropriate metric into projective, affine, or inversive geometry. Affine geometry is Euclidean geometry without perpendicularity, and projective geometry is affine geometry without parallelism. Both of these are *linear* geometries—two points lie on a unique line, and two lines meet in at most one point. The real projective plane is based on a few elementary axioms, but for higher dimensions we make use of coordinates. In this chapter we shall see how the selection of an *absolute polarity* converts projective or affine n-space into a linear metric space—elliptic, hyperbolic, or Euclidean n-space. The elliptic and hyperbolic n-spheres on the other hand are *circular* metric spaces; in the next chapter we shall show how each of them can be obtained from inversive n-space.

2.1 PROJECTIVE PLANES

Following the path originally marked out by Arthur Cayley (1821–1895) and more fully delineated by Felix Klein (1849–1925), our starting point for defining and coordinatizing various real and complex spaces is projective geometry. The two-dimensional theory can be based on the primitive terms *point* and *line* and a relationship of *incidence*, together with certain properties of *harmonic* and *quadrangular* sets and the notions of *order* and *continuity*.

A. *Incidence.* A *projective plane* is a set of primitive objects, called *points* and *lines*, with an undefined relation of *incidence*. If a point P is incident with a line $\check{Q}$, we write $P \Diamond \check{Q}$ or $\check{Q} \Diamond P$ and say that P "lies on" $\check{Q}$ or that $\check{Q}$ "passes through" P. If two points are incident with a given line, the line is said to *join* the points, while two lines incident with a given point are said to *meet* (or "intersect") in the point. Three or more points lying on one line are *collinear*, and three or more lines passing through one point are *concurrent*. In any projective plane, we make the following assumptions of incidence (cf. Veblen & Young 1910, pp. 16–18; Hall 1959, p. 346; Artzy 1965, p. 201; Onishchik & Sulanke 2006, ¶1.1.2):

Axiom I. *Any two distinct points are incident with just one line.*

Axiom J. *Any two lines are incident with at least one point.*

Axiom K. *There exist four points no three of which are collinear.*

Definitions, descriptions, and statements can be dualized by interchanging the words "point" and "line," "join" and "meet," "collinear" and "concurrent," etc. The Principle of Duality says that the dual of any valid statement is also valid, a consequence of the fact that the axioms imply their own duals. The following propositions are easily verified.

Theorem Ǐ. *Any two distinct lines are incident with just one point.*

Theorem J̌. *Any two points are incident with at least one line.*

Theorem Ǩ. *There exist four lines no three of which are concurrent.*

A *triangle* PQR or $\check{P}\check{Q}\check{R}$ is a set of three noncollinear points P, Q, R and the three lines $\check{P} = QR$, $\check{Q} = RP$, $\check{R} = PQ$ joining them in pairs or, dually, a set of three nonconcurrent lines $\check{P}$, $\check{Q}$, $\check{R}$ and the three points $P = \check{Q}{\cdot}\check{R}$, $Q = \check{R}{\cdot}\check{P}$, $R = \check{P}{\cdot}\check{Q}$ in which pairs of them meet. The three points are the triangle's *vertices*, and the three lines are its *sides*. For $n > 3$, an *n-gon* is a set of n points and n lines, its *vertices* and *sides*, with the former joined cyclically in pairs by the latter. Triangles and n-gons are self-dual figures.

A *complete quadrangle PQRS* is a set of four points P, Q, R, S, no three collinear, and the six lines QR and PS, RP and QS, PQ and RS joining them in pairs. The four points are the quadrangle's *vertices*, and the six lines are its *sides*. The three points $QR{\cdot}PS$, $RP{\cdot}QS$, and $PQ{\cdot}RS$ in which pairs of opposite sides meet are the quadrangle's *diagonal points*. A complete quadrangle is a configuration $(4_3, 6_2)$ of four points and six lines, with three lines through each point and two points on each line.

Dually, a *complete quadrilateral* $\check{P}\check{Q}\check{R}\check{S}$ is a set of four lines $\check{P}$, $\check{Q}$, $\check{R}$, $\check{S}$, no three concurrent, and the six points $\check{Q}.\check{R}$ and $\check{P}.\check{S}$, $\check{R}.\check{P}$ and $\check{Q}.\check{S}$, $\check{P}.\check{Q}$ and $\check{R}.\check{S}$ in which pairs of them meet. The four lines are the quadrilateral's *sides*, and the six points are its *vertices*. The three lines $\check{Q}.\check{R} \mid \check{P}.\check{S}$, $\check{R}.\check{P} \mid \check{Q}.\check{S}$, and $\check{P}.\check{Q} \mid \check{R}.\check{S}$ joining pairs of opposite vertices are the quadrilateral's *diagonal lines*. A complete quadrilateral is a configuration $(6_2, 4_3)$ of six points and four lines, with two lines through each point and three points on each line.

B. *Harmonic and quadrangular sets.* Of fundamental importance is the *harmonic construction*, first employed by Philippe de La Hire in 1685. If A, B, and C are three distinct points on a line $\check{S}$ (see Figure 2.1a), let S be any point not lying on $\check{S}$, and let R be any point on the line $\check{C} = CS$ other than C or S. Let the lines $\check{A} = AS$ and $\check{Q} = BR$ meet in the point P, and let the lines $\check{B} = BS$ and $\check{P} = AR$ meet in the point Q. Finally, let the line $\check{R} = PQ$ meet $\check{S}$ in the point D. Then we call the ordered tetrad (AB, CD) a *harmonic set*, and points C and D are said to be *harmonic conjugates* with respect to points A and B, written $AB \mathrel{/\!\!H\!\!/} CD$.

We see that $PQRS$ is a complete quadrangle and that $A = QR \cdot PS$ and $B = RP \cdot QS$ are two of its diagonal points. If the third diagonal point $PQ \cdot RS$ also lies on $\check{S}$, then it coincides with C and D, and we have $AB \mathrel{/\!\!H\!\!/} CC$. In this case, the seven points A, B, C, P, Q, R, S and

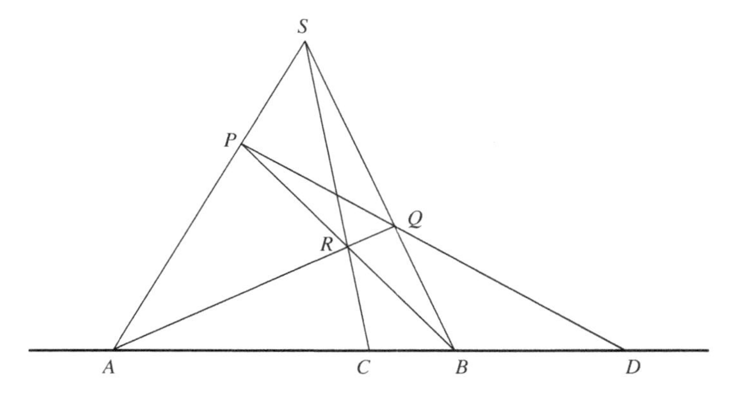

Figure 2.1a The harmonic construction

the seven lines $\check{A}$, $\check{B}$, $\check{C}$, $\check{P}$, $\check{Q}$, $\check{R}$, $\check{S}$ form a self-dual *Fano configuration* 7_3, with three lines through each point and three points on each line. Such finite geometries were investigated by Gino Fano (1871–1952), and the configuration $7_3 = \mathbb{F}_2\mathrm{P}^2$ is the simplest finite projective plane.

To avoid this and other anomalies that can occur when points C and D coincide, we may make a further assumption, known as "Fano's Axiom":

Axiom L. *The three diagonal points of a complete quadrangle are never collinear.*

From Axioms I, J, K, and L we can easily prove the dual of Axiom L:

Theorem $\check{\mathrm{L}}$. *The three diagonal lines of a complete quadrilateral are never concurrent.*

Fano's Axiom guarantees that C and D are distinct points but raises the question of whether the location of the fourth harmonic point D is independent of the choice of the auxiliary points S and R in the above construction. An affirmative answer requires still another assumption:

Axiom M. *Each point of a harmonic set is uniquely determined by the remaining points.*

Two triangles are said to be *perspective from a point* if the lines joining pairs of corresponding vertices are concurrent, and *perspective from a line* if the points in which corresponding sides meet are collinear. Given Axioms I, J, K, L, and M, one can prove the following result, a special case of the celebrated two-triangle theorem of Girard Desargues (1593–1662):

Theorem 2.1.1 (Minor Theorem of Desargues). *If two triangles are perspective from a point, and if two of the three points in which corresponding sides meet lie on a line through this point, then so does the third.*

A projective plane in which the Minor Theorem holds is called a *Moufang plane*, after Ruth Moufang (1905–1977), who first investigated such planes. For more details, see Bruck 1955 (pp. 12–16), Hall 1959 (pp. 366–372), or Artzy 1965 (pp. 220–226).

Given a complete quadrangle $PQRS$, let $\check{O}$ be a line not passing through any vertex (see Figure 2.1b). Let the concurrent sides PS, QS, RS meet $\check{O}$ in the respective points A, B, C, and let the nonconcurrent sides QR, RP, PQ meet $\check{O}$ in the respective points D, E, F. Then we call the ordered hexad (ABC, DEF) a *quadrangular set* and write $ABC \mathbin{\text{Q}} DEF$. The points of the set need not all be distinct. If $A = D$ and $B = E$, so that $\check{O}$ passes through two of the three diagonal points, then the tetrad (AB, CF) is a harmonic set. A stronger assumption than Axiom M is

Axiom N. *Each point of a quadrangular set is uniquely determined by the remaining points.*

Since a harmonic set is a special case of a quadrangular set, Axiom N implies Axiom M. Given Axioms I, J, K, and N (but without

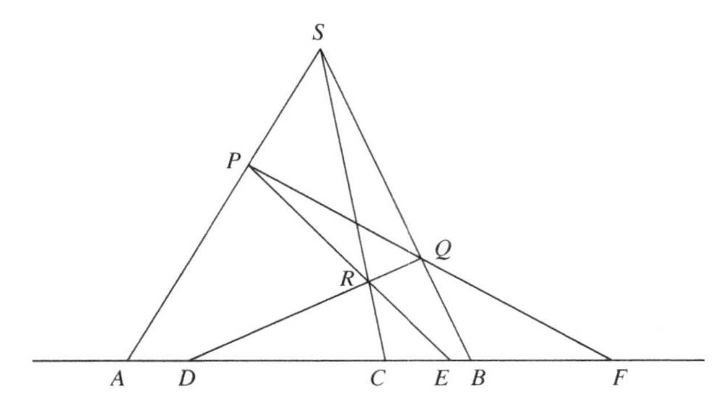

Figure 2.1b Quadrangular set of points

necessarily assuming Axiom L), the restriction of the Minor Theorem can be removed, and the plane is fully Desarguesian:

THEOREM 2.1.2 (Desargues). *If two triangles are perspective from a point, they are perspective from a line.*

The converse of Desargues's Theorem is its dual, which is automatically true by the Principle of Duality. The vertices and sides of the two triangles, the lines joining pairs of corresponding vertices, the points in which corresponding sides meet, and the center and axis of perspectivity form a *Desargues configuration* 10_3 of ten points and ten lines, with three lines through each point and three points on each line (see Figure 2.1c). When the center lies on the axis, as in the Minor Theorem, the ten points and ten lines form a *Moufang superfiguration* (cf. Grünbaum 2009, pp. 86–87). The finite plane $\mathbb{F}_2P^2$, too small to contain a pair of perspective triangles or a Desargues configuration, is trivially Desarguesian.

If the nonrestricted Desargues's Theorem or its equivalent is taken as an axiom, then Axiom N can be proved as a theorem (cf. Veblen & Young 1910, pp. 16–18, 47–50). It is evident that every Desarguesian plane is a Moufang plane, though there are Moufang planes that are not Desarguesian.

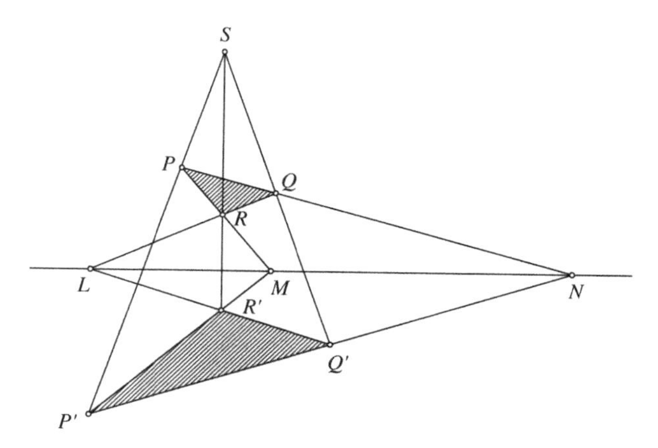

Figure 2.1c The Desargues configuration

The roles of the different sides of a quadrangle used in the construction of a harmonic or quadrangular set are to some extent interchangeable. Hence if one of the ordered tetrads

$$(AB,\ CD),\ (AB,\ DC),\ (BA,\ CD),\ (BA,\ DC)$$

is a harmonic set, then so are the others. Given Axioms I, J, K, L, and N, one can also show that the harmonic relation is symmetric (Veblen & Young 1910, p. 81):

THEOREM O. *If (AB, CD) is a harmonic set, then so is (CD, AB).*

The two triples of a quadrangular set can be subjected to the same permutation; thus $ABC \mathrel{\text{Q}\!\!\!\wedge} DEF$ implies $ACB \mathrel{\text{Q}\!\!\!\wedge} DFE$ and $BCA \mathrel{\text{Q}\!\!\!\wedge} EFD$. Likewise, if one of the ordered hexads

$$(ABC,\ DEF),\ (AEF,\ DBC),\ (DBF,\ AEC),\ (DEC,\ ABF)$$

is a quadrangular set, then so are the others. However, the symmetry of the quadrangular relation cannot be proved from the above axioms but must be postulated. We may omit Axiom L and replace Axiom N by a stronger assumption:

AXIOM P. *If (ABC, DEF) is a quadrangular set, then so is (DEF, ABC).*

This makes corresponding points of the two triples pairs of an involution $(AD)(BE)(CF)$, so that Axiom P implies Axiom N. Given Axioms I, J, K, and P, one can derive the theorem first proved by Pappus of Alexandria in the fourth century:

THEOREM 2.1.3 (Pappus). *If the six vertices of a hexagon lie alternately on two lines, then the three points in which pairs of opposite sides meet lie on a third line.*

The vertices and sides of the hexagon, together with the three points and the three lines, form a *Pappus configuration* 9_3 of nine points and nine lines, with three lines through each point and three points on each line (see Figure 2.1d).

If Pappus's Theorem or its equivalent is taken as an axiom, then Axiom P can be proved as a theorem (cf. Veblen & Young 1910, pp. 95, 100–101). Gerhard Hessenberg showed in 1905 that in any projective plane, Pappus's Theorem implies Desargues's Theorem, although the converse does not hold (cf. Coxeter 1949/1993, § 4.3; Artzy 1965, pp. 235–237). Thus every Pappian plane is Desarguesian, but there exist Desarguesian planes that are not Pappian.

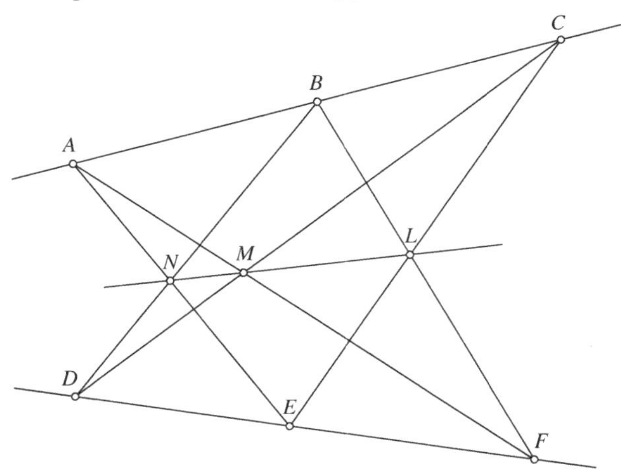

Figure 2.1d The Pappus configuration

C. *Order and continuity.* In a plane satisfying Axioms I, J, K, and M, a point P on a line $\check{O}$ is said to be *harmonically related* to three distinct points A, B, and C on $\check{O}$ if it is one of a sequence of points A, B, C, $D, \ldots$, each of which, after the first three, is the harmonic conjugate of one of the preceding points with respect to two others. All points harmonically related to A, B, and C form the *net of rationality* (or "harmonic net") $\mathsf{Q}(ABC)$, described in the real case by A. F. Möbius in 1827. If L, M, and N are three distinct points of $\mathsf{Q}(ABC)$, then $\mathsf{Q}(LMN) = \mathsf{Q}(ABC)$ (Veblen & Young 1910, pp. 84–85).

The number of points in a net of rationality may be either finite or infinite but is the same for any net on any line. If finite, it must be of the form $p+1$, where p is a prime. There may be more than one net of rationality on a line, possibly infinitely many. In a Desarguesian plane with $p+1$ points in a net of rationality, the number of points on a line may be either finite, of the form p^m+1, or infinite.

To ensure that there are not only infinitely many points on every line but infinitely many in any net of rationality, we make the following assumption:

Axiom Q. *No net of rationality is finite.*

This axiom (also due to Fano) is considerably stronger than Axiom L, which requires only that a net of rationality have more than three points.

In a projective plane satisfying Axioms I, J, K, M, and Q, a *circuit* is an infinite set of collinear points, including the entire net of rationality of any three points in the set, with a relation of *separation*, written $AB//CD$ ("A and B separate C and D"), having the following properties (cf. Veblen & Young 1918, pp. 44–45; Coxeter 1949/1993, § 3.1; Artzy 1965, pp. 239–240):

(1) If $AB//CD$, then the points A, B, C, and D are distinct.

(2) The relation $AB//CD$ implies $AB//DC$ and $CD//AB$.

(3) If A, B, C, and D are four distinct points, one and only one of the relations $AB//CD$, $BC//AD$, and $CA//BD$ holds.

(4) If $AB//CD$ and $BC//DE$, then $CD//EA$.

(5) If $AB \not\parallel CD$, then $AB//CD$.

If $AB//CD$, still other permutations are valid, so long as we have 'A' and 'B' together on one side and 'C' and 'D' together on the other. Note that the above properties are not axioms applying to all sets of collinear points but are part of the definition of a circuit, which may or may not include all the points of a line.

If A, B, and C are distinct points of a circuit, the *segment* AB/C ("AB without C") is the set of all points X such that $AB//CX$. If $AB//CD$, the two points A and B decompose the circuit into just two segments, AB/C and AB/D, with AB/C containing D and AB/D containing C.

Separation, as a way of characterizing *order* in a projective plane, was first defined by Giovanni Vailati in 1895. The following axiom of *continuity* is essentially the "cut" devised by Richard Dedekind in 1872 (see Figure 2.1e):

Axiom R. *Any three collinear points A, B, C belong to a unique circuit* $\mathbb{R}(ABC)$ *with the following property: If the points of the net of rationality* $\mathsf{Q}(ABC)$ *belonging to the segment* AB/C *are divided into two nonempty sets such that* $AY//BX$ *for any point X of the first set and any point Y of the second, then there exists a point M in the segment (not necessarily in the net of rationality) such that* $AM//BX$ *for every point* $X \neq M$ *in the first set and* $AY//BM$ *for every point* $Y \neq M$ *in the second set.*

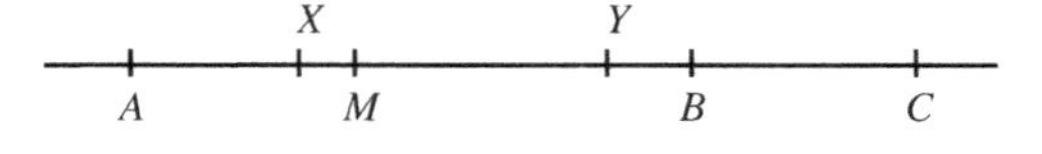

Figure 2.1e The chain $\mathbb{R}(ABC)$

The circuit $\mathbb{R}(ABC)$ is called a *chain*. Axioms Q and R imply that a chain is a compact continuum of points, having the topology of a real circle S^1. To obtain the real projective plane P^2, we may combine Axioms I, J, K, P, Q, and R with the following axiom of completion (cf. Coxeter 1949/1993, app. 1):

Axiom S. *There is not more than one chain on a line.*

All ten of the foregoing axioms hold in P^2, since Q implies L and P implies both M and N. While other combinations of the axioms generally have a variety of models, the set {I, J, K, P, Q, R, S} is categorical.

D. *Coordinates.* The points and lines of a projective plane can generally be coordinatized by some algebraic system, with operations such as addition and multiplication. One such system is a *division ring*; the elements of a division ring K form an additive abelian group T(K), the nonzero elements form a multiplicative group GL(K), and multiplication is distributive over addition. If multiplication is commutative, K is a *field*; if not, K is a *skew-field*.*

In 1899 David Hilbert famously showed how certain geometric properties of a projective plane correspond to algebraic properties of its coordinate ring. In particular, Hilbert found that Pappus's Theorem is equivalent to the commutative law of multiplication $ab = ba$, while Desargues's Theorem is equivalent to the associative law $(ab)c = a(bc)$.

In 1907 Oswald Veblen and J. H. M. Wedderburn described a class of non-Desarguesian projective planes that can be coordinatized over what have come to be known as *Veblen-Wedderburn systems*. In 1933 Ruth Moufang showed that at least the Minor Theorem of Desargues holds so long as multiplication obeys the alternative laws $(aa)b = a(ab)$ and $(ab)b = a(bb)$. A Veblen-Wedderburn system having the properties of a division ring except that multiplication is alternative

* As used here, the terms *field* and *skew-field* are mutually exclusive. The group GL(K) is often denoted by K^* or $K^\times$.

but not necessarily associative is an *alternative division ring*. In 1943 Marshall Hall unified these theories by means of more primitive systems called *ternary rings*, in which addition and multiplication are bound together in a single operation combining three elements at a time. Various properties of projective and affine planes depend on the structure of their coordinatizing ternary rings (Hall 1959, pp. 356–366; Artzy 1965, pp. 203–219).

Wedderburn proved in 1905 that every finite associative division ring is commutative and hence a finite field $\mathbb{F}_q$, where q is a prime or a prime power. Emil Artin and Max Zorn later showed that the same is true of a finite alternative division ring. Thus every finite Desarguesian (or Moufang) plane is Pappian. Finite projective planes of order q, i.e., with $q+1$ points on a line, exist for every prime power $q = p^m$; one always has at least the Desarguesian plane $\mathbb{F}_q\mathrm{P}^2$. Whether q can have other values is not known, though certain values (e.g., 6 and 10) can definitely be excluded.

The *characteristic* of a ring R with unity 1 is the smallest positive integer p (necessarily a prime if R is a division ring) for which the sum of p 1's equals the zero element; if no such integer exists, the characteristic is 0. For a projective plane satisfying Axiom M, the number of points in a finite net of rationality is one more than the characteristic of the coordinatizing ring. Axiom L excludes projective planes coordinatized by rings of characteristic 2, and Axiom Q allows only rings of characteristic 0.

The fields $\mathbb{Q}$, $\mathbb{R}$, and $\mathbb{C}$ of rational, real, and complex numbers all coordinatize Pappian projective planes. Hamilton's skew-field $\mathbb{H}$ of quaternions, associative but not commutative, coordinatizes a Desarguesian plane $\mathbb{H}\mathrm{P}^2$ that is not Pappian. The Cayley-Graves algebra $\mathbb{O}$ of octonions, alternative but not associative, coordinatizes a Moufang plane $\mathbb{O}\mathrm{P}^2$ that is not Desarguesian.

The axiomatic and algebraic theory of projective planes can be extended to spaces of three or more dimensions, but with one major qualification. In any projective 3-space it is possible to prove Desargues's Theorem from the three-dimensional axioms of incidence alone (Veblen & Young 1910, p. 41; Robinson 1940, pp. 14–16; Artzy 1965, pp. 232–233). This means that for $n > 2$ every projective n-space is coordinatizable by a division ring. Thus no non-Desarguesian plane can be embedded in any higher-dimensional projective space.

EXERCISES 2.1

1. Assuming Axioms I, J, and K, prove Theorems Ǐ, J̌, and Ǩ.
2. Prove that in any projective plane there are at least three points on every line and at least three lines through every point.
3. ⊢ The four vertices of a complete quadrangle and any two of its diagonal points are the vertices of a complete quadrilateral.
4. Construct an incidence table for the Fano configuration 7_3, showing which of the seven points lie on each of the seven lines.
5. Assuming Axioms I, J, K, and L, prove Theorem Ľ.
6. Choose three points A, B, and C on a line in P^2, and find the point D such that $AB \,\mathcal{H}\, CD$. Then find the point E such that $CD \,\mathcal{H}\, AE$. Are B and E the same point?

2.2 PROJECTIVE n-SPACE

For $n \geq 1$ we may regard *projective n-space* as a partially ordered set of projective k-planes ($0 \leq k \leq n - 1$), its *subspaces*, satisfying certain axioms of incidence or coordinatized by a particular number system—e.g., a field or a division ring. A 0-plane is a *point*, a 1-plane is a *line*, a 2-plane is just a *plane*, an $(n - 2)$-plane is a *hyperline*, and an $(n - 1)$-plane is a *hyperplane*. Two subspaces are *incident* if one is contained in the other. Any two distinct points are incident with a unique line, and any two distinct hyperplanes are incident with a unique hyperline.

A. *Homogeneous coordinates.* In a projective n-space FP^n over a field F—even a skew-field would suffice—each point X may be assigned coordinates

$$(x) = (x_1, x_2, \ldots, x_{n+1}), \tag{2.2.1}$$

and each hyperplane $\check{U}$ ("co-U") may be assigned coordinates

$$[u] = [u_1, u_2, \ldots, u_{n+1}]. \tag{2.2.2}$$

Such coordinates are not unique but *homogeneous*, meaning that the x's or u's are not all zero and that any coordinate vector may be multiplied by an arbitrary nonzero scalar. (That point and hyperplane coordinates can always be consistently assigned will be shown in Chapter 4.) It is convenient to regard (x) as a *row* and $[u]$ as a *column*, belonging to the dual (left and right) vector spaces F^{n+1} and $\check{\mathsf{F}}^{n+1}$.* Each row (x) or column $[u]$ spans a one-dimensional subspace $\langle(x)\rangle = \{\lambda(x) : \lambda \in \mathsf{F}\}$ or $\langle[u]\rangle = \{[u]\rho : \rho \in \mathsf{F}\}$. The point X and the hyperplane $\check{U}$ are incident, written $X \Diamond \check{U}$, if and only if $\langle[u]\rangle$ is contained in the annihilator of $\langle(x)\rangle$ (and vice versa), i.e., if and only if $(x)[u] = 0$, where $(x)[u]$ is the scalar (1×1 matrix) $x_1u_1 + \cdots + x_{n+1}u_{n+1}$.

Given $n+1$ points in FP^n, let A be the $(n+1) \times (n+1)$ matrix having the coordinates of the ith point as its ith row (a_i), with jth entry a_{ij}. Then the points are all incident with a common hyperplane if and only if $\det A = 0$. Any n points of general position (meaning that their coordinate vectors are linearly independent) lie in a unique hyperplane, coordinates for which are the cofactors of a row of a matrix whose other rows are the coordinates of the n points. For instance, when $n = 2$, two points X and Y with coordinates $(x) = (x_1, x_2, x_3)$ and $(y) = (y_1, y_2, y_3)$ determine a line XY with coordinates

$$(x) \vee (y) = \left[\begin{vmatrix} x_2 & x_3 \\ y_2 & y_3 \end{vmatrix}, \begin{vmatrix} x_3 & x_1 \\ y_3 & y_1 \end{vmatrix}, \begin{vmatrix} x_1 & x_2 \\ y_1 & y_2 \end{vmatrix} \right]. \tag{2.2.3}$$

* Some writers, e.g., Onishchik & Sulanke (2006), distinguish point and hyperplane coordinates through the duality of subscripts and superscripts, following the conventions of tensor notation.

Dually, given $n + 1$ hyperplanes, let A be the $(n + 1) \times (n + 1)$ matrix whose jth column $[a_j]$ is the coordinate vector of the jth hyperplane. Then the hyperplanes are all incident with a common point if and only if $\det A = 0$. Any n hyperplanes of general position pass through a unique point, coordinates for which are the cofactors of a column of a matrix whose other columns are the coordinates of the n hyperplanes. For instance, when $n = 2$, two lines $\check{U}$ and $\check{V}$ with coordinates $[u] = [u_1, u_2, u_3]$ and $[v] = [v_1, v_2, v_3]$ meet in a point $\check{U}\cdot\check{V}$ with coordinates

$$[u] \wedge [v] = \left(\begin{vmatrix} u_2 & v_2 \\ u_3 & v_3 \end{vmatrix}, \begin{vmatrix} u_3 & v_3 \\ u_1 & v_1 \end{vmatrix}, \begin{vmatrix} u_1 & v_1 \\ u_2 & v_2 \end{vmatrix} \right). \tag{2.2.4}$$

Although neither distances nor ratios of distances are defined, FP^n admits a "ratio of ratios." Let X and Y be two points, and $\check{U}$ and $\check{V}$ two hyperplanes, with respective coordinate vectors (x) and (y), $[u]$ and $[v]$. Provided either that $X \not\Diamond \check{U}$ and $Y \not\Diamond \check{V}$ or that $X \not\Diamond \check{V}$ and $Y \not\Diamond \check{U}$, the ordered tetrad $(XY,\ \check{U}\check{V})$ has a *cross ratio* $\|XY,\ \check{U}\check{V}\|$, whose value when both conditions hold is given by

$$\|XY,\ \check{U}\check{V}\| = \frac{(x)[u]}{(x)[v]} \Big/ \frac{(y)[u]}{(y)[v]}. \tag{2.2.5}$$

If either $X \Diamond \check{U}$ or $Y \Diamond \check{V}$, then $\|XY,\ \check{U}\check{V}\| = 0$. If either $X \Diamond \check{V}$ or $Y \Diamond \check{U}$, then we write $\|XY,\ \check{U}\check{V}\| = \infty$. Thus the cross ratio (like the result of a linear fractional transformation) belongs to the extended field $\mathsf{F} \cup \{\infty\}$.* The classical cross ratio $\|XY,\ UV\|$ of four collinear points, which goes back to Pappus, is now defined as $\|XY,\ \check{U}\check{V}\|$, where $\check{U}$ and $\check{V}$ are any hyperplanes such that $U \Diamond \check{U}$ and $V \Diamond \check{V}$, subject to the conditions given above (Veblen & Young 1910, pp. 160–161; Coxeter 1942, pp. 84–85, 90).

* In a projective n-space over a skew-field K, the cross ratio $\|XY,\ \check{U}\check{V}\|$ is defined as the class of elements of $\mathsf{K} \cup \{\infty\}$ conjugate to

$$\{(x)[v]\}^{-1}\{(x)[u]\}\{(y)[u]\}^{-1}\{(y)[v]\}$$

(cf. Baer 1952, pp. 71–73; Onishchik & Sulanke 2006, ¶1.4.2).

If F is of characteristic 0, the cross ratio $\|AB,\ CD\|$ of four points of a circuit, not necessarily distinct but with $A \neq D$ and $B \neq C$, is an element of some ordered subfield of F, with $\|AB,\ CD\| < 0$ if and only if $AB//CD$ (Coxeter 1949/1993, § 12.4). If $BC//AD$, then $\|AB,\ CD\|$ is between 0 and 1; if $CA//BD$, then $\|AB,\ CD\| > 1$. A tetrad (AB, CD) with $\|AB,\ CD\| = -1$ is a harmonic set, with $AB \not{H}\ CD$.

B. *Analytic proofs.* Incidence relations between points and hyperplanes of projective n-space can be determined from their respective coordinates, providing an alternative means of proving theorems. As an example, recall that Theorem 2.1.2 (Desargues) holds only in projective planes satisfying Axiom N and Theorem 2.1.3 (Pappus) only in planes satisfying Axiom P. But Hessenberg showed that every Pappian plane is Desarguesian. A synthetic proof of this fact is not difficult but requires the consideration of various special cases (cf. Artzy 1965, pp. 235–236). However, in view of Hilbert's discovery of the connection between Pappus's Theorem and the commutativity of multiplication, any Pappian plane can be coordinatized over a field. This enables us to prove Desargues's Theorem in a projective plane FP^2 as follows:

Given two triangles PQR and $P'Q'R'$ perspective from a point S, we may without loss of generality suppose that coordinates have been assigned so that (assuming Axiom L) their respective vertices are the points

$$P = (0,\ 1,\ 1),\quad Q = (1,\ 0,\ 1),\quad R = (1,\ 1,\ 0),$$
$$P' = (x,\ 1,\ 1),\quad Q' = (1,\ y,\ 1),\quad R' = (1,\ 1,\ z),$$

where x, y, and z are arbitrary elements, other than 0 or 1, of the coordinatizing field F. Then the lines

$$PP' = [0,\ 1,\ -1],\quad QQ' = [-1,\ 0,\ 1],\quad RR' = [1,\ -1,\ 0]$$

are concurrent in the point $S = (1,\ 1,\ 1)$. The sides

$$QR = [-1,\ 1,\ 1],\qquad RP = [1,\ -1,\ 1],\qquad PQ = [1,\ 1,\ -1]$$

of the first triangle meet the corresponding sides

$$Q'R' = [yz - 1, 1 - z, 1 - y],$$
$$R'P' = [1 - z, zx - 1, 1 - x],$$
$$P'Q' = [1 - y, 1 - x, xy - 1]$$

of the second triangle in the points

$$L = (z - y, yz - y, z - yz),$$
$$M = (x - zx, x - z, zx - z),$$
$$N = (xy - x, y - xy, y - x).$$

Since their coordinate vectors are linearly dependent, the points L, M, and N are collinear.

A Pappian plane in which Axiom L is false has a coordinate field of characteristic 2. In such a plane, points P, Q, and R as given above would all lie on the line [1, 1, 1]. However, a different choice of coordinates, say

$$P = (1,\ 0,\ 0), \qquad Q = (0,\ 1,\ 0), \qquad R = (0,\ 0,\ 1),$$

makes it possible to construct a valid proof for this case also.

C. *Transformations.* A one-to-one incidence-preserving transformation of FP^n taking k-planes into k-planes is a *collineation*; one that relates subspaces to their duals, taking k-planes into $(n-k-1)$-planes, is a *correlation*. Of necessity if $n > 1$, and as a further requirement if $n = 1$, harmonic sets go to harmonic sets (or their duals). A collineation or correlation that preserves cross ratios is said to be *projective*; if $n = 1$, it is a *projectivity*.

A *perspective* collineation leaves fixed every hyperplane through some point O, the *center*, and every point on some *median hyperplane* $\check{O}$ (the *axis* if $n = 2$). It is an *elation* if $O \lozenge \check{O}$, a *homology* if $O \not\lozenge \check{O}$. If P is a noninvariant point of a homology $O\check{O}P \mapsto O\check{O}P'$, and if $\dot{O}$ is then the point in which the line OP meets $\check{O}$, then the cross ratio $\|O\dot{O},\ PP'\|$ is the same for every choice of P (Robinson 1940, p. 135).

Over a field F not of characteristic 2, a homology with cross ratio -1 is a *harmonic homology*, of period 2, with $O\dot{O} \mathrel{H} PP'$. A projective correlation of period 2 is a *polarity*, which pairs each point with a unique hyperplane. A projectivity of period 2 is an *involution*, which interchanges points on FP^1; an involution has either two fixed points or none if char $\mathsf{F} \neq 2$.*

Just as the points and hyperplanes of a projective space may be represented by row and column vectors, so collineations and correlations may be represented by matrices. If A is an $m \times n$ matrix over a field or other commutative ring, with $(i, j) \Rightarrow a_{ij}$, the *transpose* of A is the $n \times m$ matrix $A^{\vee}$ with $(i, j) \mapsto a_{ji}$.† An $n \times n$ matrix A is *symmetric* if $A^{\vee} = A$, *skew-symmetric* if $A^{\vee} = -A$. With $(A + B)^{\vee} = A^{\vee} + B^{\vee}$ and $(AB)^{\vee} = B^{\vee}A^{\vee}$, transposition is a "co-automorphism" of the ring of $n \times n$ matrices.

Any invertible symmetric matrix A of order $n+1$ defines a polarity $\langle \cdot A \rangle^{\vee}$ on FP^n (an involution if $n = 1$). The *polar* of a point X with coordinates (x) is the hyperplane $\check{X}'$ with coordinates $[x'] = A(x)^{\vee}$, and the *pole* of a hyperplane $\check{U}$ with coordinates $[u]$ is the point U' with coordinates $(u') = [u]^{\vee}A^{-1}$. Thus $(x'') = (x)$ and $[u''] = [u]$. Two points X and Y or two hyperplanes $\check{U}$ and $\check{V}$ are *conjugate* in the polarity, i.e., each is incident with the polar or pole of the other, if and only if

$$(x)A(y)^{\vee} = 0 \quad \text{or} \quad [u]^{\vee}A^{-1}[v] = 0.$$

In other words, conjugacy in the polarity is defined by means of a symmetric *bilinear form* $\cdot A \cdot : \mathsf{F}^{n+1} \times \mathsf{F}^{n+1} \to \mathsf{F}$, with $[(x), (y)] \mapsto$

* By what is often called the Fundamental Theorem of Projective Geometry, on a projective line over a field, two triples (ABC) and $(A'B'C')$ of distinct points define a unique projectivity $ABC \Rightarrow A'B'C'$. The projectivity is an involution $(AA')(BB')(CC')$ if and only if $(ABC,\ A'B'C')$ is a quadrangular set (Veblen & Young 1910, p. 103; Coxeter 1949/1993, § 4.7). Thus, given Axiom P, the statement $ABC \mathrel{Q} A'B'C'$ may be interpreted to mean that the projectivity determined by the two triples is an involution.

† Curiously, there is no standard notation for the transpose of a matrix. Since matrix transposition is a dualizing operation, I employ the inverted circumflex or *caron*—the mark commonly used for duals—as an appropriate symbol for the transpose.

$\sum\sum x_i a_{ij} y_j$. Values of such bilinear forms (or the forms themselves) may be denoted by expressions like $(x\ y)$ (cf. Sommerville 1914, p. 130; Coxeter 1942, p. 224; Coxeter 1949/1993, § 12.6). Bilinearity of a form $(x\ y)$ means that it is "linear" as a function $\mathsf{F}^{n+1} \to \mathsf{F}$ of either (x) or (y) individually, i.e., that

$$(w+x\ \ y) = (w\ \ y) + (x\ \ y) \quad \text{and} \quad (x\ \ y+z) = (x\ \ y) + (x\ \ z),$$
$$(\lambda x\ \ y) = \lambda(x\ \ y) \quad \text{and} \quad (x\ \ \mu y) = (x\ \ y)\mu, \tag{2.2.6}$$

for all vectors (w) or (z) and for all scalars λ or μ. The form is *symmetric* if $(x\ y) = (y\ x)$ for all (x) and (y), *skew-symmetric* if $(x\ y) = -(y\ x)$. The *dual* of a bilinear form $(x\ y) = (x)A(y)^\vee$ is the form $[u\ v] = [u]^\vee A^{-1}[v]$.

The locus $(x\ x) = 0$ of self-conjugate points or the envelope $[u\ u] = 0$ of self-conjugate hyperplanes (if any) of a given polarity is an $(n-1)$-*quadric*, a *conic* if $n = 2$ (von Staudt 1847, p. 137). There is only one kind of projective conic, as opposed to the familiar Euclidean (or affine) trichotomy of ellipses, parabolas, and hyperbolas. Higher-dimensional quadrics may be of different types (e.g., "oval" or "ruled").

When the field F is not of characteristic 2, a quadric is a self-dual figure: the polars of the points of a quadric locus are the hyperplanes of a quadric envelope, and vice versa. When char $\mathsf{F} = 2$, the polars of a quadric locus do not form a quadric envelope but all pass through a point, and the poles of a quadric envelope do not form a quadric locus but all lie in a hyperplane.

Other geometries can be obtained from projective geometry through various specializations. Affine n-space FA^n is projective n-space with one hyperplane and all the subspaces incident with it removed; two subspaces of FP^n whose intersection lies on this "hyperplane at infinity" become *parallel* subspaces of FA^n. Following Cayley (1859, 1872) and Klein (1871, 1873), we may derive elliptic, Euclidean, or hyperbolic geometry by fixing a certain *absolute polarity* in real projective n-space $\mathrm{P}^n = \mathbb{R}\mathrm{P}^n$. Many projective properties, usually including some form of duality, carry over to these other spaces.

EXERCISES 2.2

1. Given the points $I = (1, 0, 0)$ and $E = (1, 1, 1)$, find the coordinates of the line $\check{L} = IE$.
2. Given the lines $\check{I} = [1, 0, 0]$ and $\check{E} = [1, 1, 1]$ and the result of Exercise 1, find the coordinates of the points $P = \check{L}\cdot\check{I}$ and $Q = \check{L}\cdot\check{E}$.
3. Use Formula (2.2.5) and the results of the preceding exercises to find the cross ratio $\|IE, \check{I}\check{E}\|$ and hence the cross ratio $\|IE, PQ\|$.
4. $\vdash$ The twenty-four permutations of four collinear points A, B, C, D yield at most six different cross ratios, which for distinct points take the values $x, 1/x, 1-x, 1/(1-x), x/(x-1), (x-1)/x$.
5. Show that if A, B, and C are three distinct points on a line, then $\|AB, CA\| = \infty$, $\|AB, CB\| = 0$, and $\|AB, CC\| = 1$.
6. How does the proof of Desargues's Theorem given above depend on the commutativity of multiplication?
7. Use suitably chosen coordinates to prove Desargues's Theorem for a projective plane over a field of characteristic 2.

2.3 ELLIPTIC AND EUCLIDEAN GEOMETRY

When a polarity on P^n is defined by a matrix E that is congruent to the identity matrix $I = I_{n+1} = \backslash 1, 1, \ldots, 1\backslash$ (i.e., $PEP^{\check{}} = I$ for some invertible matrix P), the quadratic forms $(x) \mapsto (x\,x)$ and $[u] \mapsto [u\,u]$ on $\mathbb{R}^{n+1}$ and $\check{\mathbb{R}}^{n+1}$ corresponding to the dual bilinear forms

$$(x\,y) = (x)E(y)^{\check{}} \quad \text{and} \quad [u\,v] = [u]^{\check{}}E^{-1}[v]$$

are *positive definite*. That is, $(x\,x) > 0$ and $[u\,u] > 0$ for all nonzero (x) and $[u]$; consequently there are no real self-conjugate points or hyperplanes. Fixing such an *elliptic* polarity converts projective

n-space P^n into *elliptic* n-space eP^n. The distance between two points X and Y in eP^n is given by

$$[XY] = \cos^{-1} \frac{|(x\,y)|}{\sqrt{(x\,x)}\sqrt{(y\,y)}}, \tag{2.3.1}$$

and the angle between two hyperplanes $\check{U}$ and $\check{V}$ by

$$(\check{U}\check{V}) = \cos^{-1} \frac{|[u\,v]|}{\sqrt{[u\,u]}\sqrt{[v\,v]}}. \tag{2.3.2}$$

The distance from a point X to a hyperplane $\check{U}$ is given by

$$[X\check{U}) = \sin^{-1} \frac{|(x)[u]|}{\sqrt{(x\,x)}\sqrt{[u\,u]}}. \tag{2.3.3}$$

Distances and angles both range from 0 to $\pi/2$.

A *circle* in the elliptic plane eP^2 is the locus of points at a fixed distance (the *radius*) from a given point (the *center*). As an "equidistant curve," a circle is also the locus of points at a fixed distance (the *coradius*) from a given line (the *axis*), the polar of the center in the absolute polarity. For a real, nondegenerate circle, each of these distances must be between 0 and $\pi/2$.

Alternatively, we may first convert projective n-space P^n into extended affine n-space by singling out a hyperplane P^{n-1} "at infinity," say the hyperplane $x_{n+1} = 0$, in which we then fix an elliptic polarity, determined by a bilinear form on $\mathbb{R}^n$. (We may regard this operation as fixing a degenerate polarity in P^n.) The resulting elliptic $(n-1)$-space is the *absolute hyperplane* of extended *Euclidean* n-space $\bar{\mathrm{E}}^n$; its complement is the *ordinary* region E^n.

The coordinates of an ordinary point X in extended affine n-space, with $x_{n+1} \neq 0$, can be normalized as

$$((x),\ 1) = (x_1, x_2, \ldots, x_n,\ 1),$$

so that (x) is the vector of Cartesian coordinates of X, geometrically represented by the directed line segment $\overrightarrow{OX}$ joining X to the origin

O. Homogeneous coordinates of an ordinary hyperplane $\check{U}(r)$ can be expressed as

$$[[u],\ r] = [u_1, u_2, \ldots, u_n, r],$$

with $[u]$ being a nonzero covector of direction numbers and r being a positional parameter. Two hyperplanes $\check{U}(r)$ and $\check{V}(s)$ are parallel, written $\check{U}(r) \parallel \check{V}(s)$, if their direction numbers are proportional; otherwise, they meet in a hyperline, and we write $\check{U}(r) \between \check{V}(s)$. We have $X \Diamond \check{U}(r)$ if and only if $(x)[u] + r = 0$.

Standard bilinear forms on the vector spaces $\mathbb{R}^n$ and $\check{\mathbb{R}}^n$, converting them into dual "inner-product" spaces and giving affine n-space a Euclidean metric, are defined by

$$(x\ y) = (x)(y)\check{}\quad \text{and} \quad [u\ v] = [u]\check{}[v].$$

The distance between two points X and Y in E^n with respective Cartesian coordinates (x) and (y) is then given by

$$|XY| = \sqrt{(y - x\ \ y - x)} = [(y_1 - x_1)^2 + \cdots + (y_n - x_n)^2]^{1/2}, \quad (2.3.4)$$

i.e., the length of the vector $(y{-}x)$ represented by the directed line segment $\overrightarrow{XY}$. The measure (from 0 to π) of the angle $\widehat{XOY}$ with vertex O formed by two rays $\cdot O{\uparrow}X$ ("O through X") and $\cdot O{\uparrow}Y$ ("O through Y"), or by the rays parallel to them emanating from an arbitrary point, is

$$[\![\widehat{XOY}]\!] = \cos^{-1} \frac{(x\ y)}{\sqrt{(x\ x)}\sqrt{(y\ y)}}, \quad (2.3.5)$$

i.e., the angle between the vectors (x) and (y) corresponding to $\overrightarrow{OX}$ and $\overrightarrow{OY}$. Writing $\mathbf{x}$, $\mathbf{y}$, and θ for $\overrightarrow{OX}$, $\overrightarrow{OY}$, and $[\![\widehat{XOY}]\!]$, we see that $(x\ y)$ is the familiar dot product $\mathbf{x} \cdot \mathbf{y} = |\mathbf{x}||\mathbf{y}| \cos\theta$.

The angle (from 0 to $\pi/2$) between two hyperplanes $\check{U}(r)$ and $\check{V}(s)$ with respective directions $[u]$ and $[v]$ is

$$(\check{U}\check{V}) = \cos^{-1} \frac{|[u\ v]|}{\sqrt{[u\ u]}\sqrt{(v\ v]}}, \quad (2.3.6)$$

and the distance between two parallel hyperplanes $\check{U}(r)$ and $\check{U}(s)$ with coordinates $[[u], r]$ and $[[u], s]$ is

$$|\check{U}(r, s)| = \frac{|s - r|}{\sqrt{[u\, u]}}. \tag{2.3.7}$$

The distance from a point X to a hyperplane $\check{U}(r)$ is given by

$$|X\check{U}(r)| = \frac{|(x)[u] + r|}{\sqrt{[u\, u]}}. \tag{2.3.8}$$

If the coordinates $[[u], r]$ of a hyperplane are normalized so that $[u\, u] = 1$, the u's are direction cosines.

While an elliptic polarity on P^n, defined by a bilinear form whose associated quadratic form is positive definite, has no real self-conjugate points or hyperplanes, we may consider P^n as being embedded in complex projective n-space $\mathbb{C}\mathrm{P}^n$. The polarity then has self-conjugate points and hyperplanes that lie on or are tangent to an imaginary $(n-1)$-quadric, the *absolute hypersphere* of the corresponding elliptic n-space eP^n.

In the extended Euclidean plane $\bar{\mathrm{E}}^2$, whose points have homogeneous coordinates (x_1, x_2, x_3), a point on the absolute line eP^1 has coordinates $(x_1, x_2, 0)$ or $[u_1, u_2]$, where $x_1u_1 + x_2u_2 = 0$. The two imaginary self-conjugate points of the absolute involution on eP^1 are the "circular points at infinity" $I = (1, \mathrm{i}, 0)$ and $J = (1, -\mathrm{i}, 0)$, which lie on every circle

$$x_1{}^2 + x_2{}^2 + 2Dx_1x_3 + 2Ex_2x_3 + Fx_3{}^2 = 0.$$

(This is the equation of a real circle provided that $D^2 + E^2 > F$.) Two lines with coordinates $[u_1, u_2, r]$ and $[v_1, v_2, s]$ are orthogonal if and only if their respective absolute points $\dot{U} = [u_1, u_2]$ and $\dot{V} = [v_1, v_2]$ are harmonic conjugates with respect to I and J, i.e., if and only if $\|IJ, \dot{U}\dot{V}\| = -1$. More generally, as shown by Edmond Laguerre in 1853, all angles between lines of E^2 are determined by such cross ratios (cf. Coxeter 1942, pp. 106–108).

EXERCISES 2.3

1. ⊢ If the absolute polarity in eP^n is defined by the identity matrix, the point $P = (p)$ is the absolute pole of the hyperplane $\check{P} = [p]$ and the distance $[P\check{P})$ is $\pi/2$.

2. Taking E as the identity matrix $\backslash 1,\ 1,\ 1\backslash$, find an equation for the circle in eP^2 with center $(0, 0, 1)$ and radius $\pi/3$. Show that this is the same as the circle with axis $[0, 0, 1]$ and coradius $\pi/6$.

3. ⊢ The standard bilinear form (inner product) on $\mathbb{R}^n$ satisfies the *Cauchy-Schwarz inequality* $|(x\ y)| \leq |(x)||(y)|$.

4. ⊢ In E^3, the point R_u in the plane $\check{U}(r) = [u_1,\ u_2,\ u_3,\ r]$ nearest the origin $O = (0, 0, 0)$ has coordinates $\rho(u_1,\ u_2,\ u_3)$, where

$$\rho = -r/({u_1}^2 + {u_2}^2 + {u_3}^2).$$

5. Let $X = (x_1,\ x_2,\ x_3)$ and $Y = (y_1,\ y_2,\ y_3)$ be distinct points in E^3, and let $u_i = y_i - x_i$ $(i = 1, 2, 3)$. Show that the line $\cdot XY\cdot$ is perpendicular to every plane $\check{U}(r) = [u_1,\ u_2,\ u_3,\ r]$.

2.4 HYPERBOLIC GEOMETRY

Eugenio Beltrami (1868a) showed that the hyperbolic plane can be modeled by a circular disk in the Euclidean plane. Ordinary points are represented by the disk's interior points and ordinary lines by its chords, with hyperbolic distances and angles defined in terms of Euclidean measurements (cf. Coxeter 1942, pp. 254–255; Yaglom 1956, supp. 1; Ratcliffe 1994/2006, § 6.1). Hyperbolic n-space is analogously modeled by a ball in E^n (Beltrami 1868b). As subsequently shown by Klein (1871), this *Beltrami-Klein model* for H^n has a natural interpretation in projective n-space.

When a polarity on P^n is defined by a matrix H that is congruent to the *pseudo-identity matrix* $I_{n,1} = \backslash 1, \ldots,\ 1,\ -1\backslash$, the locus of self-conjugate points (or the envelope of self-conjugate hyperplanes) is an *oval* $(n-1)$-quadric ω (a conic if $n = 2$, a pair of points if $n = 1$). Any line of P^n meets ω in at most two points. By fixing such a *hyperbolic* polarity, we convert projective n-space P^n into *complete hyperbolic n-space* hP^n and ω into the *absolute hypersphere*, which separates hP^n into an *ordinary* (interior) region H^n and an *ideal* (exterior) region $\check{\mathrm{H}}^n$. In terms of the dual bilinear forms

$$(x\ddot{\ }y) = (x)H(y)\check{\ } \quad \text{and} \quad [u\ddot{\ }v] = [u]\check{\ }H^{-1}[v],$$

a point X with coordinates (x) or a hyperplane $\check{U}$ with coordinates $[u]$ is

$$\begin{array}{llllll} \text{ordinary} & \text{if} & (x\ddot{\ }x) < 0 & \text{or} & [u\ddot{\ }u] > 0, \\ \text{absolute} & \text{if} & (x\ddot{\ }x) = 0 & \text{or} & [u\ddot{\ }u] = 0, \\ \text{ideal} & \text{if} & (x\ddot{\ }x) > 0 & \text{or} & [u\ddot{\ }u] < 0. \end{array}$$

If instead H is congruent to $-I_{n,1}$, the above inequalities are reversed (cf. Coxeter 1942, pp. 224–225).

When H is taken as the matrix $I_{n,1}$, the ideal hyperplane $x_{n+1} = 0$ can be identified with the absolute hyperplane of extended Euclidean n-space $\bar{\mathrm{E}}^n$. Hyperbolic points with $x_{n+1} \neq 0$ are then represented by ordinary Euclidean points, and their coordinates can be normalized so that $x_{n+1} = 1$. The unit hypersphere $x_1{}^2 + \cdots + x_n{}^2 = 1$ of E^n represents the absolute hypersphere of hP^n, and its interior represents the ordinary region H^n. This Euclidean realization constitutes the Beltrami-Klein model for hyperbolic n-space.

The *discriminant* of two ordinary hyperplanes $\check{U}$ and $\check{V}$ may be defined by

$$|\, u\ddot{\ }v \,| = [u\ddot{\ }u][v\ddot{\ }v] - [u\ddot{\ }v]^2,$$

and the hyperplanes are then

$$\begin{array}{lll} \text{intersecting} & \text{if} & |\, u\ddot{\ }v \,| > 0, \\ \text{parallel} & \text{if} & |\, u\ddot{\ }v \,| = 0, \\ \text{diverging} & \text{if} & |\, u\ddot{\ }v \,| < 0. \end{array}$$

We write the three cases as $\check{U} \between \check{V}$, $\check{U}$)($\check{V}$, and $\check{U}$)($\check{V}$. The distance between two ordinary points X and Y in H^n is given by

$$]XY[\; = \; \cosh^{-1} \frac{|(x\ddot{\ }y)|}{\sqrt{-(x\ddot{\ }x)}\sqrt{-(y\ddot{\ }y)}}, \tag{2.4.1}$$

the angle between two intersecting or parallel hyperplanes $\check{U}$ and $\check{V}$ by

$$(\check{U}\check{V}) \; = \; \cos^{-1} \frac{|[u\ddot{\ }v]|}{\sqrt{[u\ddot{\ }u]}\sqrt{[v\ddot{\ }v]}}, \tag{2.4.2}$$

and the minimum distance between two diverging hyperplanes $\check{U}$ and $\check{V}$ by

$$)\check{U}\check{V}(\; = \; \cosh^{-1} \frac{|[u\ddot{\ }v]|}{\sqrt{[u\ddot{\ }u]}\sqrt{[v\ddot{\ }v]}}. \tag{2.4.3}$$

The distance from a point X to a hyperplane $\check{U}$ is given by

$$]X\check{U}(\; = \; \sinh^{-1} \frac{|(x)[u]|}{\sqrt{-(x\ddot{\ }x)}\sqrt{[u\ddot{\ }u]}}. \tag{2.4.4}$$

Angles range from 0 to $\pi/2$, but distances are unlimited. The imaginary angle $(\check{U}\check{V})$ between two diverging hyperplanes $\check{U}$ and $\check{V}$ is $)\check{U}\check{V}(/\mathrm{i}$.

Klein gave equivalent expressions for distances and angles in terms of logarithms of certain cross ratios (cf. Onishchik & Sulanke 2006, ¶2.6.3); Sossinsky 2012, ¶9.1.1. The formulas given here, like their elliptic and Euclidean counterparts, have the virtue of involving only a single pair of coordinate vectors in each case (as opposed to the two pairs needed for a cross ratio).

Given any hyperplane $\check{U}$ and a point X with $X \not\Diamond \check{U}$, a line through X and one of the absolute points of $\check{U}$ is parallel to $\check{U}$ in that direction. Following Lobachevsky, the angle between the perpendicular from X to $\check{U}$ and the parallel is called the *angle of parallelism* for the distance $x =]X\check{U}($ and is given by

$$\Pi(x) = \cos^{-1} \tanh x = 2 \tan^{-1} \mathrm{e}^{-x}. \tag{2.4.5}$$

As x increases from zero to infinity, $\Pi(x)$ decreases from $\pi/2$ to 0.

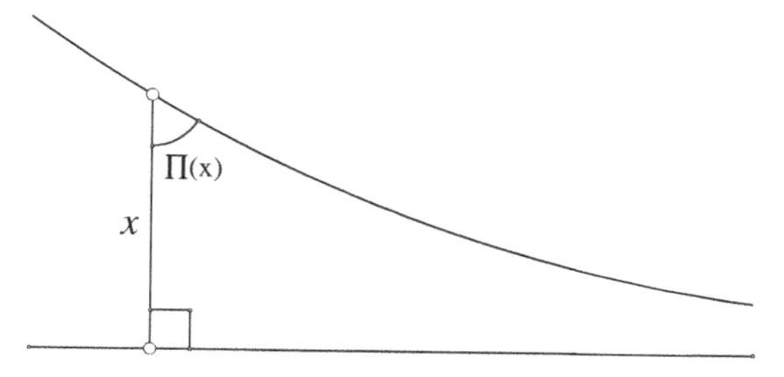

Figure 2.4a Angle of parallelism

A *circle* in the hyperbolic plane is a conic of constant curvature, characterized by its (real or imaginary) points of contact with the absolute circle. Coordinates of the absolute points of a conic defined by a symmetric matrix A satisfy the quadratic equations $(x)A(x)^{\vee} = 0$ and $(x)H(x)^{\vee} = 0$, leading to a homogeneous quartic equation in two of the coordinates. In order for the conic to be a circle, the latter equation must have either two double roots or one quadruple root. In the complete hyperbolic plane, circles (other than the absolute circle) can be *ordinary*, *ideal*, or *exceptional* (Johnson 1981, pp. 460–463).*

For an ordinary circle there are three possibilities. A "proper circle" or *pericycle*, the locus of points at a fixed distance (the *radius*) from a given point (the *center*), is tangent to the absolute circle at two imaginary points. A "limiting curve" or *horocycle*, a circle of infinite radius, is doubly tangent to the absolute circle at one real point. An "equidistant curve" or *pseudocycle*, the locus of points at a fixed distance (the *coradius*) from a given line (the *axis*), is tangent to the absolute circle at two real points (Sommerville 1914, pp. 257–258; Coxeter 1942, pp. 216–217); like a hyperbola, a pseudocycle has two

* Apart from where it touches the absolute circle, all the points and tangent lines of an ordinary circle are ordinary, and those of an ideal circle are all ideal. An exceptional circle is the envelope of ordinary lines making a fixed acute angle with a given line, together with two absolute lines; its nonabsolute points are ideal. In the Beltrami-Klein model, ordinary and ideal circles appear as ellipses (or Euclidean circles), exceptional circles as hyperbolas.

branches.* We denote the respective circles by S^1, $\dot{S}^1$, and $\ddot{S}^1$. A line meets an ordinary circle in at most two ordinary or absolute points.

In the elliptic and Euclidean planes, all circles are bounded, and a circle can be drawn through any three points not on a line (in eP^2, there are four such circles). In the hyperbolic plane, three noncollinear points lie on a unique pericycle, horocycle, or branch of a pseudocycle. In addition, three distinct pseudocycles pass through the vertices of any triangle, with two vertices on one branch and one on the other.

In hyperbolic 3-space, an ordinary sphere is a quadric of constant positive, zero, or negative curvature: a *perisphere* S^2, a *horosphere* $\dot{S}^2$, or a *pseudosphere* $\ddot{S}^2$ (a "proper sphere," a "limiting surface," or an "equidistant surface"). Perispheres and horospheres are simply connected; a pseudosphere has two sheets. Analogously, there are three types of ordinary hypersphere in hyperbolic n-space H^n ($n > 3$). The terms "horocycle" and "horosphere" were coined by Lobachevsky. The names "pseudocycle" and "pseudosphere" are adapted from Beltrami, who called a manifold of constant negative curvature "pseudospherical" space.†

We have referred to the classical Euclidean and non-Euclidean geometries as "metric spaces." A (real) *metric* on a space is a real-valued function d on pairs of points that is nonnegative, nondegenerate, and symmetric and satisfies the Triangle Inequality. That is, $d(X, Y) \geq 0$, $d(X, Y) = 0$ if and only if $X = Y$, $d(X, Y) = d(Y, X)$,

* That a pseudocycle is tangent to the absolute circle is evident in the Beltrami-Klein model for the hyperbolic plane. In the "conformal disk" model, a branch of a pseudocycle (also called a "hypercycle" or an "equidistant") is represented by an arc of a circle that intersects the boundary circle at an angle equal to the angle of parallelism for its coradius (cf. Goldman 1999, p. 24; Onishchik & Sulanke 2006, ¶2.6.6), but this is a property of the model and not of the geometry. If the tangents to two curves at a common absolute point of $\bar{H}^2$ are ordinary lines, they are parallel or coincident and the angle between the curves is zero. But if either of the curves is tangent to the absolute line through that point, as when one of the curves is the absolute circle itself, the angle between them is infinite or indeterminate.

† These seem more appropriate than the alternatives "hypercycle" and "hypersphere," especially as the latter term is ambiguous.

and $d(X,Y) \leq d(X,Z) + d(Z,Y)$ (Ratcliffe 1994/2006, § 1.3). A space with such a metric is a *metric space*. It is evident that distance as it has been defined for each of the linear geometries eP^n, E^n, and H^n has the first three properties, as does distance on S^n (to be defined in the next section). We defer discussion of the Triangle Inequality to § 3.4.

Angles are naturally measured in terms of radians, so that a right angle is $\pi/2$. However, distance in a given metric space is defined only up to multiplication by a positive constant. In a Euclidean space, the unit of distance is arbitrary, but in elliptic or hyperbolic space, there is a natural metric determined by the curvature of the space. The formulas presented here assume an elliptic or hyperbolic unit of distance corresponding to a spatial curvature of ± 1.

The idea of a real metric space being defined by an $(n-1)$-quadric—the *absolute*—in real projective n-space is due to Cayley (1859), who considered the cases where the absolute is an imaginary or degenerate conic.* Klein (1871, 1873) completed what Cayley had begun, obtaining the hyperbolic plane by letting the absolute be real and treating the elliptic plane as a sphere with antipodal points identified. It was also Klein who proposed the names "elliptic" and "hyperbolic" for the non-Euclidean linear geometries. An affine conic is an ellipse, a parabola, or a hyperbola according as it has 0, 1, or 2 real points at infinity, and a line in elliptic, Euclidean, or hyperbolic n-space meets the absolute in 0, 1, or 2 real points.

EXERCISES 2.4

1. Find the angle of parallelism for the distance $x = ½ \log 3$.

2. Taking H as the pseudo-identity matrix $\backslash 1,\ 1,\ -1 \backslash$, find an equation for the pericycle in H^2 with center $(0, 0, 1)$ and radius $\cosh^{-1} 2$.

* The degenerate case had been anticipated by von Staudt (1847), who obtained the Euclidean plane from the real affine plane by fixing an absolute elliptic involution on the line at infinity.

3. Taking H as the pseudo-identity matrix $\backslash 1,\ 1,\ -1\backslash$, find an equation for the pseudocycle with axis $[1, 0, 0]$ and coradius $\sinh^{-1}\ \sqrt{3}$.

4. With the absolute circle of $\bar{\mathrm{H}}^2$ given by the equation $x_1{}^2 + x_2{}^2 - x_3{}^2 = 0$, the three points (2, 2, 3), (−2, 2, 3), and (0, 0, 1) lie on a horocycle. Show that the horocycle has equation $x_1{}^2 + 2x_2{}^2 - 2x_2x_3 = 0$ and determine its center.

5. Let X and Y be two ordinary points of the hyperbolic plane and let U and V be the absolute points of the line $\cdot XY\cdot$, with $XU//YV$. Show that the hyperbolic distance $]XY[$ is given by ½ log $\|XY, UV\|$. (It may be convenient to replace the absolute points U and V by their respective polars $\check{U}$ and $\check{V}$.)

3

CIRCULAR GEOMETRIES

IF PROJECTIVE GEOMETRY is about incidences of points, lines, and planes, inversive geometry deals with point pairs, circles, and spheres. In a manner that closely parallels how a linear metric space can be extracted from projective or affine geometry, a circular metric space can be derived from inversive geometry by fixing a certain *central inversion* (Johnson 1981; cf. Alekseevsky, Vinberg & Solodovnikov 1988, chap. 2, § 2). Depending on the nature of the fixed inversion, the resulting geometry may be either *spherical* or *pseudospherical*.

When represented in Euclidean space, inversive geometry also provides a way to construct conformal (angle-preserving) models of elliptic and hyperbolic spaces. The two non-Euclidean metrics will be seen to have a kind of "pseudo-symmetry" that is evident in formulas relating the angles and sides of a triangle and in measurements of circular arcs and sectors.

3.1 INVERSIVE AND SPHERICAL GEOMETRY

We begin with a hyperbolic polarity in real projective $(n + 1)$-space. Let the points X and hyperplanes $\check{U}$ of P^{n+1} have homogeneous coordinates

$$((x)) = (x_0, x_1, \ldots, x_{n+1}) \quad \text{and} \quad [[u]] = [u_0, u_1, \ldots, u_{n+1}],$$

and let an oval n-quadric Φ have the locus or envelope equation

$$x_1^2 + \cdots + x_{n+1}^2 = x_0^2 \quad \text{or} \quad u_1^2 + \cdots + u_{n+1}^2 = u_0^2,$$

corresponding to the matrix $\backslash -1, 1, \ldots, 1\backslash$. We may regard Φ as the real inversive, or *Möbius*, n-sphere I^n (Veblen & Young 1918, pp. 262–264). The case $n = 2$ is of particular interest; a fuller description of the properties of the real inversive sphere I^2 will be given in Chapter 9.

A. *Inversive geometry.* A projective hyperplane $\check{U}$, with coordinates $[[u]]$, meets Φ in an inversive $(n-1)$-sphere $\mathring{u}$, a *hypersphere* of I^n, which is

$$\begin{array}{lll} \textit{real} & \text{if} & [[u\,u]] > 0, \\ \textit{degenerate} & \text{if} & [[u\,u]] = 0, \\ \textit{imaginary} & \text{if} & [[u\,u]] < 0, \end{array}$$

where $[[u\,u]]$ is the quadratic form associated with the bilinear form

$$[[u\,v]] = -u_0v_0 + u_1v_1 + \cdots + u_{n+1}v_{n+1}.$$

A point X lies on I^n if its coordinates $((x))$ satisfy the locus equation for Φ and on the hypersphere $\mathring{u}$ with hyperplanar coordinates $[[u]]$ if $((x))[[u]] = 0$. The equation $((x))[[u]] = 0$ has at least two, just one, or no real solutions for $((x))$ (on I^n) according as $\mathring{u}$ is real, degenerate, or imaginary; in the last case, the equation has two or more imaginary solutions.

If the discriminant of two real or degenerate inversive hyperspheres $\mathring{u}$ and $\mathring{v}$ is defined by

$$||u\,v|| = [[u\,u]]\,[[v\,v]] - [[u\,v]]^2,$$

then the hyperspheres are

$$\begin{array}{lll} \textit{separating} & \text{if} & ||u\,v|| > 0, \\ \textit{tangent} & \text{if} & ||u\,v|| = 0, \\ \textit{separated} & \text{if} & ||u\,v|| < 0. \end{array}$$

Applied to two point-pairs $\ddot{u} = \{U_1,\ U_2\}$ and $\ddot{v} = \{V_1,\ V_2\}$ on a circle I^1, separation has the usual meaning associated with cyclic order (Coxeter 1966a, p. 218). That is, $\ddot{u}$ and $\ddot{v}$ are separating if $U_1U_2 \,//\, V_1V_2$, tangent if $\ddot{u}$ and $\ddot{v}$ have a point in common, and separated otherwise. For $n \geq 2$, two hyperspheres of I^n are separating, tangent, or separated according as they intersect in a real, degenerate, or imaginary $(n-2)$-sphere.

The angle between two separating inversive hyperspheres $\mathring{u}$ and $\mathring{v}$ is given by

$$(\mathring{u}\ \mathring{v}) = \cos^{-1} \frac{|[\![u\ v]\!]|}{\sqrt{[\![u\ u]\!]}\sqrt{[\![v\ v]\!]}}. \tag{3.1.1}$$

Two separated inversive hyperspheres $\mathring{u}$ and $\mathring{v}$ have an analogous *inversive distance* (Coxeter 1966b), given by

$$)\mathring{u}\ \mathring{v}(= \cosh^{-1} \frac{|[\![u\ v]\!]|}{\sqrt{[\![u\ u]\!]}\sqrt{[\![v\ v]\!]}}. \tag{3.1.2}$$

Another inversive invariant of note is the cross ratio of four points (Wilker 1981, pp. 384–386; Beardon 1983, pp. 32–33). The distance between two points is not defined, and an inversive hypersphere has no center or radius.

B. *Spherical geometry.* A projective collineation of P^{n+1} that commutes with the polarity that defines the n-quadric Φ induces a circle-preserving transformation, or *circularity*, on Φ as the Möbius n-sphere I^n. In particular, a harmonic homology whose center O and median hyperplane $\check{O}$ are pole and polar with respect to Φ induces an *inversion* on I^n. When O and $\check{O}$ have the respective coordinates

$$(1,\ 0, \ldots,\ 0,\ 0) \quad \text{and} \quad [1,\ 0, \ldots,\ 0,\ 0],$$

this is an *elliptic* inversion $Z : \mathrm{I}^n \to \mathrm{I}^n$, i.e., the inversion in the imaginary hypersphere in which $\check{O}$ meets Φ, leaving no real points fixed. If we restrict ourselves to those circularities that commute with Z, we convert the Möbius n-sphere I^n into the *elliptic* n-sphere S^n ("spherical n-space"), with Z becoming the familiar *central inversion* that interchanges antipodal points. (It is a happy accident that two different, seemingly unrelated, meanings of "inversion"—in a hypersphere and in a point—are united in the same transformation.)

For every point X on S^n, $x_0 \neq 0$, and normalized coordinates for X are the last $n+1$ projective coordinates divided by x_0, i.e.,

$$(x) = (x_1, x_2, \ldots, x_{n+1}),$$

where $x_1{}^2 + x_2{}^2 + \cdots + x_{n+1}{}^2 = 1$. The points (x) and $-(x)$ are antipodal. A hyperplane $\breve{U}$ passing through the point O, so that $u_0 = 0$, intersects S^n in a *great hypersphere* $\breve{U}$, with homogeneous coordinates

$$[u] = [u_1, u_2, \ldots, u_{n+1}].$$

We have $X \Diamond \breve{U}$ if and only if $(x)[u] = 0$. Other hyperplanes intersect S^n in real, degenerate, or imaginary "small" hyperspheres.

Two great hyperspheres of S^n are always separating, with any two great circles on an elliptic sphere S^2 meeting in a pair of antipodal points.

Whereas the geometry of I^n is only conformal, the elliptic n-sphere S^n is a full-fledged metric space. In terms of the bilinear forms

$$(x\ y) = x_1y_1 + \cdots + x_{n+1}y_{n+1} \quad \text{and} \quad [u\ v] = u_1v_1 + \cdots + u_{n+1}v_{n+1},$$

with $(x\ x) = (y\ y) = 1$, the distance (from 0 to π) between two points X and Y on S^n is given by

$$[\![X\ Y]\!] = \cos^{-1}(x\ y), \tag{3.1.3}$$

and the angle (from 0 to $\pi/2$) between two great hyperspheres $\breve{U}$ and $\breve{V}$ by

$$((\breve{U}\breve{V})) = \cos^{-1}\frac{|[u\ v]|}{\sqrt{[u\ u]}\sqrt{[v\ v]}}. \tag{3.1.4}$$

The distance from a point X to a great hypersphere $\breve{U}$ is given by

$$[[X\breve{U})) = \sin^{-1}\frac{|(x)[u]|}{\sqrt{[u\,u]}}. \tag{3.1.5}$$

An n-sphere S^n can be embedded in other metric spaces. For instance, the point O and the hyperplane $\breve{O}$ may be taken as the origin and the hyperplane at infinity of Euclidean $(n+1)$-space E^{n+1}, and S^n is then the unit hypersphere $(x\,x) = 1$. If antipodal points are identified, as happens under projection from O to $\breve{O}$, the elliptic n-sphere S^n is transformed into elliptic n-space $eP^n = PS^n = S^n/\langle Z\rangle$. Great hyperspheres of S^n project into hyperplanes of eP^n.

EXERCISES 3.1

1. Show that the circles $[1, 1, 0, 1]$ and $[1, 0, -1, 2]$ on the inversive sphere I^2 are separating and find the angle between them.
2. Show that the circles $[1, 0, -1, 1]$ and $[1, 1, 2, 0]$ on the inversive sphere I^2 are separated and find the distance between them.
3. Find the distance from the point $(\frac{1}{3}, -\frac{2}{3}, \frac{2}{3})$ to the great circle $[4, 0, 3]$ on the elliptic sphere S^2.

3.2 PSEUDOSPHERICAL GEOMETRY

Returning to the Möbius n-sphere as an oval n-quadric Φ in P^{n+1} defined by

$$x_1{}^2 + \cdots + x_{n+1}{}^2 = x_0{}^2 \quad \text{or} \quad u_1{}^2 + \cdots + u_{n+1}{}^2 = u_0{}^2,$$

a harmonic homology whose center O and median hyperplane $\breve{O}$ have the respective coordinates

$$(0, 0, \ldots, 0, 1) \quad \text{and} \quad [0, 0, \ldots, 0, 1]$$

induces a *hyperbolic* inversion $W : I^n \to I^n$—i.e., the inversion in the real inversive hypersphere $\mathring{\omega}$ in which $\breve{O}$ meets Φ, leaving every point of $\mathring{\omega}$ fixed. If W is taken as the central inversion, interchanging antipodal points, then the Möbius n-sphere I^n is converted into the *hyperbolic* n-sphere $\ddot{\mathrm{S}}^n$, separated by $\mathring{\omega}$ into two n-hemispheres. We call $\mathring{\omega}$ the *equatorial hypersphere* of $\ddot{\mathrm{S}}^n$; all other hyperspheres and all points not on $\mathring{\omega}$ are *ordinary*. Up to a scalar factor, the hyperbolic n-sphere has the same metric as an n-pseudosphere (equidistant hypersurface) in H^{n+1}, and $\ddot{\mathrm{S}}^n$ may be called "pseudospherical n-space."

For an ordinary point X on $\ddot{\mathrm{S}}^n$, $x_{n+1} \neq 0$, and normalized coordinates for X are the first $n+1$ projective coordinates (rearranged) divided by x_{n+1}, i.e.,

$$(x) = (x_1, \ldots, x_n \,|\, x_0),$$

with $x_1{}^2 + \cdots + x_n{}^2 - x_0{}^2 = -1$. Two ordinary points X and Y with coordinates (x) and (y) lie in the same n-hemisphere if $x_0 y_0 > 0$. The points (x) and $-(x)$ are antipodal, and points on the equatorial hypersphere are self-antipodal. A hyperplane $\breve{U}$ passing through O, so that $u_{n+1} = 0$, and having a real intersection with $\ddot{\mathrm{S}}^n$, so that $[[u\ u]] > 0$, meets $\ddot{\mathrm{S}}^n$ in a *great hypersphere* $\breve{U}$, having homogeneous coordinates

$$[u] = [u_1, \ldots, u_n \,|\, u_0],$$

with $u_1{}^2 + \cdots + u_n{}^2 - u_0{}^2 > 0$. As before, $X \Diamond \breve{U}$ if and only if $(x)[u] = 0$. Other hyperplanes meet $\ddot{\mathrm{S}}^n$ in various real, degenerate, or imaginary "small" hyperspheres.

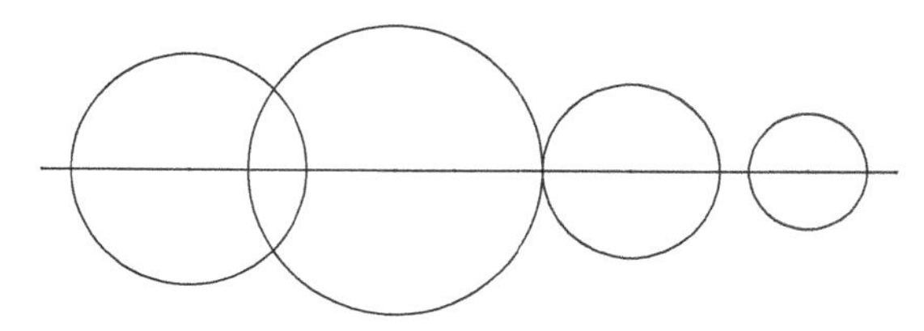

Figure 3.2a Separating, tangent, and separated great circles

Two great hyperspheres of $\ddot{\mathrm{S}}^n$ may be separating, tangent, or separated. (To distinguish the cases, apply the discriminant for inversive hyperspheres, assuming a final hyperplanar coordinate of zero.) Any two separating great circles on a hyperbolic sphere $\ddot{\mathrm{S}}^2$ meet in a pair of antipodal points; tangent great circles have one common equatorial point; separated great circles have no real points in common (see Figure 3.2a). An ordinary small circle on $\ddot{\mathrm{S}}^2$ is a *pseudocycle*, a *horocycle*, or a *pericycle* according as it and the equatorial circle are separating, tangent, or separated (see Figure 3.2b).

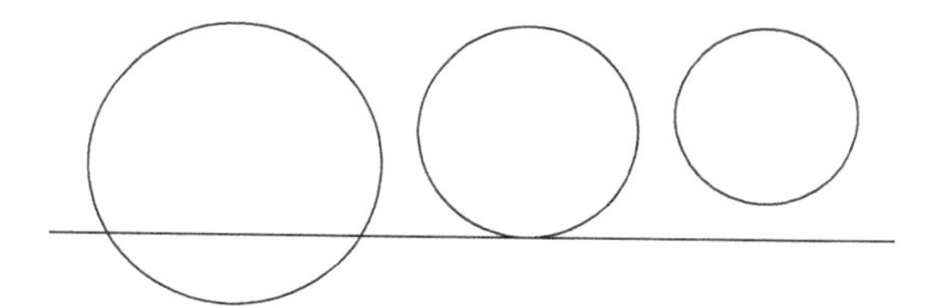

Figure 3.2b Pseudocycle, horocycle, and pericycle

In terms of the bilinear forms

$$(x\ddot{\ }y) = x_1y_1 + \cdots + x_ny_n - x_0y_0 \quad \text{and} \quad [u\ddot{\ }v] = u_1v_1 + \cdots + u_nv_n - u_0v_0,$$

with $(x\ddot{\ }x) = (y\ddot{\ }y) = -1$ (cf. Ratcliffe 1994/2006, § 3.2), the distance between two points X and Y in the same n-hemisphere of $\ddot{\mathrm{S}}^n$ is given by

$$]\!]XY[\![= \cosh^{-1} |(x\ddot{\ }y)|, \tag{3.2.1}$$

the angle (from 0 to $\pi/2$) between two separating or tangent great hyperspheres $\breve{U}$ and $\breve{V}$ by

$$((\breve{U}\breve{V})) = \cos^{-1} \frac{|[u\ddot{\ }v]|}{\sqrt{[u\ddot{\ }u]}\sqrt{[v\ddot{\ }v]}}, \tag{3.2.2}$$

and the minimum distance between two separated great hyperspheres $\breve{U}$ and $\breve{V}$ by

$$]\!]\breve{U}\breve{V}[\![= \cosh^{-1} \frac{|[u\ddot{\ }v]|}{\sqrt{[u\ddot{\ }u]}\sqrt{[v\ddot{\ }v]}}. \tag{3.2.3}$$

The distance from a point X to a great hypersphere $\check{U}$ is given by

$$]]X\check{U}((= \sinh^{-1} \frac{|(x)[u]|}{\sqrt{[u\ddot{\ }u]}}. \tag{3.2.4}$$

The point O and the hyperplane $\check{O}$ can be taken as the origin and the hyperplane at infinity of *Lorentzian* (or "pseudo-Euclidean") $(n + 1)$-space $E^{n,1}$, in which $\ddot{S}^n$ appears as an n-hyperboloid of two sheets (a hyperbola if $n = 1$), asymptotic to the "light cone" $(x\ddot{\ }x) = 0$. Lorentzian space is homogeneous but not isotropic. The (not necessarily real) distance between two points X and Y with respective Cartesian coordinates (x) and (y) is given by

$$|X\ddot{\ }Y| = \sqrt{(y - x\ddot{\ }y - x)} = [(y_1 - x_1)^2 + \cdots + (y_n - x_n)^2 - (y_0 - x_0)^2]^{1/2}, \tag{3.2.5}$$

so that metrically $\ddot{S}^n$ is a hypersphere with center O and imaginary radius i. The properties of such an "imaginary-spherical" geometry were anticipated by Johann Lambert (1728–1777) and extensively explored by Franz Taurinus (1794–1874), but it was Lobachevsky (1835, 1836) who first recognized it as a valid alternative to Euclidean and spherical geometry.

Just as the elliptic n-sphere S^n can be embedded in hyperbolic $(n + 1)$-space as an n-perisphere with center O (and ideal median hyperplane $\check{O}$), so the hyperbolic n-sphere $\ddot{S}^n$ can be embedded in H^{n+1} as an n-pseudosphere with median hyperplane $\check{O}$ (and ideal center O).

Either of the two n-hemispheres into which $\ddot{S}^n$ is separated by the equatorial hypersphere $\mathring{\omega}$ (i.e., either sheet of the n-hyperboloid in $E^{n,1}$ or the n-pseudosphere in H^{n+1}) will serve as an isometric model for hyperbolic n-space. When $n = 2$, the Lorentzian realization is the *hyperboloid model* of the hyperbolic plane (Ratcliffe 1994/2006, § 3.2; cf. Reynolds 1993). If antipodal points are identified, as by projecting from O to $\check{O}$, the hyperbolic n-sphere $\ddot{S}^n$ is transformed into extended hyperbolic n-space $\bar{H}^n = P\ddot{S}^n = \ddot{S}^n/\langle W\rangle$. Great hyperspheres of $\ddot{S}^n$

project into hyperplanes of H^n, and the equatorial hypersphere $\mathring{\omega}$ of $\ddot{\mathrm{S}}^n$ becomes the absolute hypersphere ω of $\bar{\mathrm{H}}^n$.

Still another way to represent $\ddot{\mathrm{S}}^n$ is by Euclidean n-space E^n augmented by a single "exceptional" point that lies on every line, with some hyperplane $\check{O}$ taken to represent the equatorial hypersphere $\mathring{\omega}$. In this conformal "double half-space" model, the points and the k-planes or k-spheres of E^n lying in $\check{O}$ are the equatorial points and k-spheres of $\ddot{\mathrm{S}}^n$. Points that are mirror images with respect to $\check{O}$ are antipodal points of $\ddot{\mathrm{S}}^n$. Euclidean k-planes or k-spheres orthogonal to $\check{O}$ are great k-spheres of $\ddot{\mathrm{S}}^n$.

EXERCISES 3.2

1. Find the angle between the separating great circles $[-1, 1 \mid 1]$ and $[1, 2 \mid 1]$ on the hyperbolic sphere $\ddot{\mathrm{S}}^2$.

2. Find the distance between the separated great circles $[2, 0 \mid 1]$ and $[-2, \sqrt{3} \mid 2]$ on the hyperbolic sphere $\ddot{\mathrm{S}}^2$.

3.3 CONFORMAL MODELS

The elliptic and hyperbolic n-spheres can be projected into Euclidean n-space to provide inversive models for elliptic and hyperbolic n-space. Such mappings preserve circles and angular measure.

A. *Elliptic geometry.* Elliptic n-space eP^n can be conformally represented by a closed ball in E^n, i.e., a hypersphere Ω together with its interior. The k-planes of eP^n are diametral k-disks or caps of k-spheres intersecting Ω in great $(k-1)$-spheres, with diametrically opposite points of Ω being identified. When $n = 2$, this is Klein's "conformal disk" model of the elliptic plane.

To obtain explicit metrics for this and other Euclidean models, let the points of E^n have Cartesian coordinates $(\dot{x}) = (\dot{x}_1, \ldots, \dot{x}_n)$, and let

$(\dot{x}\,\dot{x}) = \dot{x}_1{}^2 + \cdots + \dot{x}_n{}^2$. In the elliptic "conformal ball" model, points of eP^n are represented by Euclidean points $(\dot{x})$ with $(\dot{x}\,\dot{x}) \leq 1$, where $(\dot{x})$ and $-(\dot{x})$ are identified if $(\dot{x}\,\dot{x}) = 1$. Each such point X also has normalized spherical coordinates

$$(x) = (x_1, \ldots, x_n, x_{n+1}),$$

where

$$x_i = \frac{2\dot{x}_i}{1 + (\dot{x}\,\dot{x})} \ (1 \leq i \leq n) \quad \text{and} \quad x_{n+1} = \frac{1 - (\dot{x}\,\dot{x})}{1 + (\dot{x}\,\dot{x})}. \tag{3.3.1}$$

(Optionally, the sign of the last coordinate can be reversed.) Each hyperplane $\check{U}$ has homogeneous spherical coordinates

$$[u] = [u_1, \ldots, u_n, u_{n+1}],$$

with $X \lozenge \check{U}$ if and only if $(x)[u] = 0$.

In terms of the bilinear form

$$(x\,y) = x_1y_1 + \cdots + x_ny_n + x_{n+1}y_{n+1},$$

with $(x\,x) = (y\,y) = 1$, the elliptic distance between two points X and Y is given by

$$[XY] = \cos^{-1} |(x\,y)|. \tag{3.3.2}$$

The angle $(\check{U}\check{V})$ between two hyperplanes $\check{U}$ and $\check{V}$ and the distance $[X\check{U})$ from a point X to a hyperplane $\check{U}$ are respectively equal to the spherical angle $((\check{U}\check{V}))$ and distance $[[X\check{U}))$ defined by (3.1.4) and (3.1.5).

For example (cf. Figure 3.3a), selected points of E^2 with Cartesian coordinates $(\dot{x}_1,\ \dot{x}_2) = (x,\ y)$ correspond to points of eP^2 with coordinates (x_1, x_2, x_3) as follows:

$$(0,\,0) \mapsto (0,\,0,\,1), \quad (1,\,0) \mapsto (1,\,0,\,0), \quad (-1,\,0) \mapsto (-1,\,0,\,0),$$

$$(\tfrac{1}{2}, \tfrac{1}{2}) \mapsto (\tfrac{2}{3}, \tfrac{2}{3}, \tfrac{1}{3}), \quad (-\tfrac{1}{2}, -\tfrac{1}{2}) \mapsto (-\tfrac{2}{3}, -\tfrac{2}{3}, \tfrac{1}{3}),$$

$$(0, \sqrt{2} - 1) \mapsto (0, \tfrac{1}{2}\sqrt{2}, \tfrac{1}{2}\sqrt{2}).$$

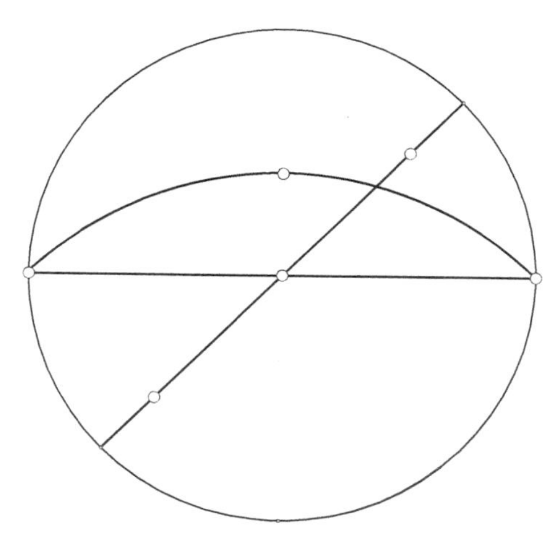

Figure 3.3a Conformal-disk model of eP^2

Coordinates $(1, 0, 0)$ and $(-1, 0, 0)$ represent the same point. Lines through distinct points of eP^2 have coordinates $[u_1, u_2, u_3]$ as follows:

$$(0, 0) \vee (1, 0) = [0, 1, 0], \quad (0, 0) \vee (\tfrac{1}{2}, \tfrac{1}{2}) = [1, -1, 0],$$
$$(1, 0) \vee (0, \sqrt{2} - 1) = [0, 1, -1].$$

Using Formula (3.3.2), we see that the elliptic distance between $(0, 0)$ and either $(\frac{1}{2}, \frac{1}{2})$ or $(-\frac{1}{2}, -\frac{1}{2})$ is $\cos^{-1} \frac{1}{3}$ and that the distance between $(\frac{1}{2}, \frac{1}{2})$ and $(-\frac{1}{2}, -\frac{1}{2})$ is $\cos^{-1} \frac{7}{9}$. The angle between $[0, 1, 0]$ and either $[1, -1, 0]$ or $[0, 1, -1]$ is $\pi/4$, and the angle between $[1, -1, 0]$ and $[0, 1, -1]$ is $\pi/3$.

B. *Hyperbolic geometry.* Hyperbolic n-space H^n can be conformally represented by an open ball in Euclidean n-space, i.e., the interior of a hypersphere Ω. The k-planes of H^n are diametral k-disks or caps of k-spheres orthogonal to Ω. When $n = 2$, this is the "conformal disk" model of the hyperbolic plane. Equivalently, either of the half-spaces into which E^n is separated by a hyperplane $\check{O}$ provides another conformal model. In this case, the k-planes of H^n are represented by Euclidean k-half-planes or k-hemispheres orthogonal to $\check{O}$. When $n = 2$, this is the *Poincaré model*, also called the "upper half-plane" model, of the hyperbolic plane (Poincaré 1882).

Like the projective model developed by Klein, these conformal representations of hyperbolic n-space were first described by Beltrami (1868b; cf. Milnor 1982, pp. 10–11; Stillwell 1996, pp. 35–38). The inversive and projective models are closely related: each arc orthogonal to the boundary circle in E^2 can be transformed into a corresponding chord, converting a disk model of H^2 that preserves angles into one that preserves collinearity.

In the hyperbolic "conformal ball" model (cf. Yaglom 1956, supp. 2; Ratcliffe 1994/2006, § 4.5), points of H^n are represented by Euclidean points $(\dot{x})$ with $(\dot{x}\,\dot{x}) < 1$. Each such point X also has normalized pseudospherical coordinates

$$(x) = (x_1, \ldots, x_n \mid x_0),$$

where

$$x_0 = \frac{1 + (\dot{x}\,\dot{x})}{1 - (\dot{x}\,\dot{x})} \quad \text{and} \quad x_i = \frac{2\dot{x}_i}{1 - (\dot{x}\,\dot{x})}\ (1 \le i \le n). \tag{3.3.3}$$

In the "upper half-space" model (cf. Ratcliffe 1994/2006, § 4.6), points of H^n are represented by Euclidean points $(\dot{x})$ with $\dot{x}_n > 0$. Each such point X also has normalized pseudospherical coordinates (x), where

$$x_0 = \frac{1 + (\dot{x}\,\dot{x})}{2\dot{x}_n}, \quad x_i = \frac{2\dot{x}_i}{2\dot{x}_n}\ (1 \le i \le n-1), \quad x_n = \frac{1 - (\dot{x}\,\dot{x})}{2\dot{x}_n}. \tag{3.3.4}$$

In either case, each hyperplane $\check{U}$ has homogeneous pseudospherical coordinates

$$[u] = [u_1, \ldots, u_n \mid u_0],$$

with $u_1{}^2 + \cdots + u_n{}^2 - u_0{}^2 > 0$. The point X and the hyperplane $\check{U}$ are incident if and only if $(x)[u] = 0$.

In terms of the bilinear form

$$(x\ddot{\ }y) = x_1y_1 + \cdots + x_ny_n - x_0y_0,$$

with $(x\ddot{\ }x) = (y\ddot{\ }y) = -1$, the hyperbolic distance between two points X and Y is given by

$$]XY[= \cosh^{-1} |(x\ddot{\ }y)|. \tag{3.3.5}$$

The angle $(\check{U}\check{V})$ between two intersecting or parallel hyperplanes $\check{U}$ and $\check{V}$, the minimum distance $)\check{U}\check{V}($ between two diverging hyperplanes $\check{U}$ and $\check{V}$, and the distance $]X\check{U}($ from a point X to a hyperplane $\check{U}$ are respectively equal to the pseudospherical angle $((\check{U}\check{V}))$ and distances $))\check{U}\check{V}(($ and $]]X\check{U}(($ defined by (3.2.2), (3.2.3), and (3.2.4).

For example (cf. Figure 3.3b), by (3.3.3) with $n = 2$, in the "conformal disk" model of the hyperbolic plane, points of E^2 with Cartesian coordinates (x, y) correspond to points of H^2 with coordinates $(x_1, x_2 \mid x_0)$ as follows:

$$(0,0) \mapsto (0,0 \mid 1), \quad (\tfrac{1}{2},0) \mapsto (\tfrac{4}{3},0 \mid \tfrac{5}{3}), \quad (-\tfrac{1}{2},0) \mapsto (-\tfrac{4}{3},0 \mid \tfrac{5}{3}),$$
$$(\tfrac{1}{2},\tfrac{1}{2}) \mapsto (2,2 \mid 3), \quad (-\tfrac{1}{2},-\tfrac{1}{2}) \mapsto (-2,-2 \mid 3),$$
$$(\tfrac{1}{5},\tfrac{2}{5}) \mapsto (\tfrac{1}{2},1 \mid \tfrac{3}{2}), \quad (\tfrac{2}{5},\tfrac{1}{5}) \mapsto (1,\tfrac{1}{2} \mid \tfrac{3}{2}),$$
$$(-\tfrac{1}{5},-\tfrac{2}{5}) \mapsto (-\tfrac{1}{2},-1 \mid \tfrac{3}{2}), \quad (-\tfrac{2}{5},-\tfrac{1}{5}) \mapsto (-1,-\tfrac{1}{2} \mid \tfrac{3}{2}).$$

Hyperbolic lines through certain pairs of points have coordinates $[u_1, u_2 \mid u_0]$ as follows:

$$(0,0) \vee (\tfrac{1}{2},0) = [0,1 \mid 0], \quad (0,0) \vee (\tfrac{1}{2},\tfrac{1}{2}) = [1,-1 \mid 0],$$
$$(\tfrac{1}{5},\tfrac{2}{5}) \vee (\tfrac{2}{5},\tfrac{1}{5}) = [1,1 \mid -1], \quad (-\tfrac{1}{5},-\tfrac{2}{5}) \vee (-\tfrac{2}{5},-\tfrac{1}{5}) = [1,1 \mid 1].$$

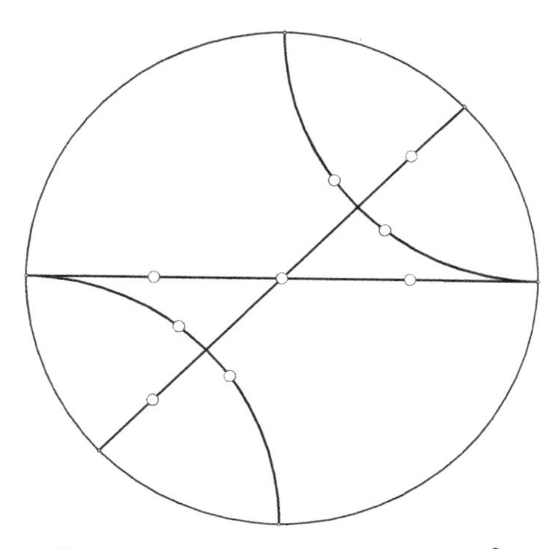

Figure 3.3b Conformal-disk model of H^2

Using Formula (3.3.5), we see that the hyperbolic distance between (0, 0) and either (½, ½) or (−½, −½) is $\cosh^{-1} 3$ and that the distance between (½, ½) and (−½, −½) is $\cosh^{-1} 17$. The angle between $[0, 1 \mid 0]$ and $[1, -1 \mid 0]$ is $\pi/4$, and the angle between $[1, -1 \mid 0]$ and either $[1, 1 \mid -1]$ or $[1, 1 \mid 1]$ is $\pi/2$. The line $[0, 1 \mid 0]$ is parallel (in opposite directions) to both $[1, 1 \mid -1]$ and $[1, 1 \mid 1]$, and the angle between it and either of these lines is therefore 0. The lines $[1, 1 \mid -1]$ and $[1, 1 \mid 1]$ themselves are diverging, and the distance between them is $\cosh^{-1} 3$.

Inversion in the circle $x^2 + (y + 1)^2 = 2$ converts the "conformal disk" representation of H^2 into the "upper half-plane" model. Points (x, y) with $x^2 + y^2 < 1$ are interchanged with points (x', y') with $y' > 0$, where

$$x' = \frac{2x}{x^2 + (y+1)^2}, \quad y' = \frac{2(y+1)}{x^2 + (y+1)^2} - 1. \tag{3.3.6}$$

The correspondence between Cartesian and hyperbolic point coordinates is now given by (3.3.4) with $n = 2$.

For example, the inversion given by (3.3.6) interchanges particular points in the unit disk with points in the upper half-plane as follows:

$$(0,0) \leftrightarrow (0,1), \quad (\tfrac{1}{2},0) \leftrightarrow (\tfrac{4}{5},\tfrac{3}{5}), \quad (-\tfrac{1}{2},0) \leftrightarrow (-\tfrac{4}{5},\tfrac{3}{5}),$$
$$(\tfrac{1}{2},\tfrac{1}{2}) \leftrightarrow (\tfrac{2}{5},\tfrac{1}{5}), \quad (-\tfrac{1}{2},-\tfrac{1}{2}) \leftrightarrow (-2,1),$$
$$(\tfrac{1}{5},\tfrac{2}{5}) \leftrightarrow (\tfrac{1}{5},\tfrac{2}{5}), \quad (-\tfrac{1}{5},\tfrac{2}{5}) \leftrightarrow (-\tfrac{1}{5},\tfrac{2}{5}),$$
$$(-\tfrac{1}{5},-\tfrac{2}{5}) \leftrightarrow (-1,2), \quad (-\tfrac{2}{5},-\tfrac{1}{5}) \leftrightarrow (-1,1).$$

The inversive-disk model of the hyperbolic plane is converted into the projective-disk (Beltrami-Klein) model by the mapping $(x, y) \mapsto (\xi, \eta)$, where

$$\xi = \frac{2x}{x^2 + y^2 + 1}, \quad \eta = \frac{2y}{x^2 + y^2 + 1}. \tag{3.3.7}$$

Each circular arc orthogonal to the unit circle $x^2 + y^2 = 1$ is mapped onto the chord with the same endpoints; each diameter is mapped onto itself. If we write $x = r\cos\theta$ and $y = r\sin\theta$, then the mapping (3.3.7) can be expressed in polar form as $(r,\ \theta) \mapsto (\rho,\ \theta)$, where $\rho = 2r/(r^2+1)$.

C. *Spherical and pseudospherical geometry.* Spherical n-space S^n can be conformally represented by a hypersphere of radius r in Euclidean $(n+1)$-space, and pseudospherical n-space $\ddot{\mathrm{S}}^n$ by a hypersphere of radius ri in Lorentzian $(n+1)$-space. For $n \geq 2$, the (Gaussian) *curvature* of S^n or $\ddot{\mathrm{S}}^n$, or of the derived elliptic or hyperbolic n-space, is the reciprocal of the square of the radius of this n-sphere. Thus S^n and eP^n have constant positive curvature $1/r^2$, while $\ddot{\mathrm{S}}^n$ and H^n have constant negative curvature $-1/r^2$; Euclidean n-space E^n is *flat*, with curvature 0.

For $n \geq 2$, the only simply connected n-manifolds of constant curvature are S^n, E^n, and H^n. Elliptic n-space eP^n is not simply connected, and pseudospherical n-space $\ddot{\mathrm{S}}^n$ is not connected.

For simplicity, all non-Euclidean distance formulas given in this section or elsewhere in the book assume a spatial curvature of ± 1. In spherical or elliptic n-space of curvature 1, a k-sphere of radius r or coradius $\check{r} = \pi/2 - r$ has the same curvature $\csc^2 r = \sec^2 \check{r}$ as a Euclidean k-sphere of radius $\sin r$. In pseudospherical or hyperbolic n-space of curvature -1, a k-perisphere of radius r has the same curvature $\operatorname{csch}^2 r$ as a Euclidean k-sphere of radius $\sinh r$; a k-horosphere has curvature 0; a k-pseudosphere of coradius $\check{r}$ has curvature $-\operatorname{sech}^2 \check{r}$.

Points on the elliptic sphere S^2, as the unit sphere of Euclidean 3-space, have normalized coordinates (x, y, z) with $x^2+y^2+z^2 = 1$. With the terrestrial (or celestial) sphere as a model, the antipodal points $(0, 0, 1)$ and $(0, 0, -1)$ are the “north pole” and “south pole” of S^2. Any great semicircle joining the poles is a “meridian,” the one whose points have $y = 0$ and $x > 0$ being the “prime” meridian.

A point on S^2 can be given polar (or “geographic”) coordinates (ϕ, θ), with $0 \leq \phi \leq \pi$ and $-\pi < \theta \leq \pi$ (see Figure 3.3c). The

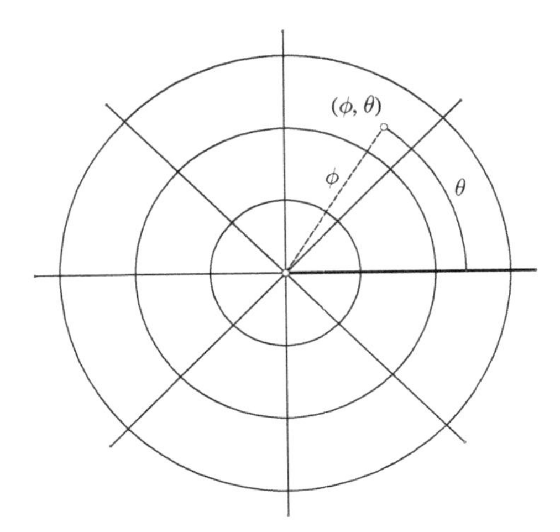

Figure 3.3c Polar coordinates

colatitude ϕ is the distance of the point from the north pole, while the *longitude* θ of a nonpolar point is the directed angle from the prime meridian to the meridian of the point. The longitude of either pole is arbitrary. The *latitude* of any point is the directed distance $\pi/2-\phi$ from the point to the "equator" [0, 0, 1]. A point on the elliptic sphere with polar coordinates (ϕ, θ) has normalized spherical coordinates (x, y, z), where

$$x = \sin\phi\cos\theta, \quad y = \sin\phi\sin\theta, \quad z = \cos\phi. \tag{3.3.8}$$

Similarly, ordinary points on the hyperbolic sphere $\ddot{\mathrm{S}}^2$ have normalized coordinates $(x, y \,|\, t)$ with $x^2 + y^2 - t^2 = -1$. The antipodal points $(0, 0 \,|\, 1)$ and $(0, 0 \,|\, {-1})$ are the "north pole" and "south pole" of $\ddot{\mathrm{S}}^2$. As with the elliptic sphere, a "meridian" is a great semicircle joining the poles, with points on the "prime" meridian having $y = 0$ and $x > 0$.

A point in either hemisphere of $\ddot{\mathrm{S}}^2$ can be given polar coordinates (p, θ), with $p \geq 0$ and $-\pi < \theta \leq \pi$. The *polaritude* p is the distance of the point from the pole in the same hemisphere, while the *longitude* θ of a nonpolar point is the directed angle from the prime meridian to the meridian of the point. The longitude of a pole is arbitrary. Antipodal points have the same polaritude and the same longitude. A point on the hyperbolic sphere with polar coordinates (p, θ) has normalized pseudospherical coordinates $(x, y \mid t)$, where

$$t = \pm \cosh p, \quad x = \sinh p \cos\theta, \quad y = \sinh p \sin\theta, \tag{3.3.9}$$

the sign of t depending on which hemisphere the point lies in. Since either hemisphere of $\ddot{\mathrm{S}}^2$ is an isometric model for H^2, polar coordinates (p, θ) can also be used for points in the hyperbolic plane.

EXERCISES 3.3

1. Show that in the "conformal disk" model of the elliptic plane, a line $[u_1, u_2, u_3]$ corresponds either to a circle with center $(-u_1/u_3, -u_2/u_3)$ and radius $\sqrt{u_1{}^2 + u_2{}^2 + u_3{}^2}/|u_3|$ or (if $u_3 = 0$) to the disk diameter with endpoints $(\pm u_2\sqrt{u_1{}^2 + u_2{}^2}, \mp u_1\sqrt{u_1{}^2 + u_2{}^2})$.

2. Show that in the "conformal disk" model of the hyperbolic plane, a line $[u_1, u_2|u_0]$ corresponds either to a circle with center $(-u_1/u_0, -u_2/u_0)$ and radius $\sqrt{u_1{}^2 + u_2{}^2 - u_0{}^2}/|u_0|$ or (if $u_0 = 0$) to the disk diameter with endpoints , $(\pm u_2\sqrt{u_1{}^2 + u_2{}^2}, \mp u_1\sqrt{u_1{}^2 + u_2{}^2})$.

3. Find the point of H^2 corresponding to $(0, \sqrt{2} - 1)$ in the "conformal disk" model.

4. Verify the correspondences given in the text between Euclidean points (x, y) and (x', y') in the "conformal disk" and "upper half-plane" models of the hyperbolic plane.

5. Show that the mapping $(x, y) \mapsto (\xi, \eta)$ defined by (3.3.7) maps both the interior and the exterior of the unit circle onto the interior.

6. Find the points of E^2 in the projective-disk model for H^2 that correspond to the points $(0, 0)$, $(1/3, 0)$, $(1/2, 0)$, $(1/2, 1/2)$, and $(1/5, 2/5)$ in the inversive-disk model.

7. $\vdash$ The projective-disk model for H^2 is converted into the inversive-disk model by the mapping $(\xi, \eta) \mapsto (x, y)$, where

$$x = \frac{(1 - \sqrt{1 - \xi^2 - \eta^2})\xi}{\xi^2 + \eta^2}, \quad y = \frac{(1 - \sqrt{1 - \xi^2 - \eta^2})\eta}{\xi^2 + \eta^2}$$

if ξ and η are not both zero, with $(0, 0) \mapsto (0, 0)$.

8. $\vdash$ The distance between two points on S^2 with polar coordinates (ϕ_1, θ_1) and (ϕ_2, θ_2) is $\cos^{-1}[\cos\phi_1 \cos\phi_2 + \sin\phi_1 \sin\phi_2 \cos(\theta_2 - \theta_1)]$.

3.4 TRIANGLES AND TRIGONOMETRY

The properties of Euclidean and spherical triangles have been studied since antiquity. Together with their elliptic and hyperbolic counterparts, these elementary geometric figures play a key role in the theory of metric spaces.

A. *Line segments and great arcs.* Any two points A and B in the affine plane A^2 or in the Euclidean or hyperbolic plane E^2 or H^2 lie on a unique line $\cdot AB\cdot$ and, by virtue of the relation of *intermediacy* (or "betweenness") introduced by Moritz Pasch in 1882, separate it into two infinite *rays* $\cdot A/B$ ("A away from B") and $\cdot B/A$ ("B away from A") and a finite *segment* AB (Coxeter 1942, p. 161), so that

$$\cdot AB\cdot = \cdot A/B \cup \{A\} \cup AB \cup \{B\} \cup \cdot B/A.$$

A ray is the same as an *open half-line*; the union $\cdot\overline{A}/B$ of a ray $\cdot A/B$ and its endpoint A is a *closed half-line.* A segment is the same as an *open interval*; the union $\overline{AB}$ of a segment AB and its endpoints A and B is a *closed interval.*

Two nonantipodal points A and B on the elliptic sphere S^2 lie on a unique great circle $\breve{C}$. The points A and B and their antipodes $\bar{A}$ and $\bar{B}$ separate $\breve{C}$ into four *proper* great arcs AB, $B\bar{A}$, $\bar{A}\bar{B}$, $\bar{B}A$—each shorter than a semicircle—with AB congruent to $\bar{A}\bar{B}$, $B\bar{A}$ congruent to $\bar{B}A$, and other pairs of supplementary lengths. The elliptic plane eP^2 is obtained by identifying antipodal points of S^2. Any two points A and B in eP^2 lie on a unique line $\cdot AB\cdot$, separating it into two segments AB and BA of supplementary lengths.

Any two ordinary points A and B in the same hemisphere of the hyperbolic sphere $\ddot{S}^2$ lie on a unique great circle $\breve{C}$. The points A and B, together with their antipodes $\bar{A}$ and $\bar{B}$, separate $\breve{C}$ into two *proper* great arcs AB and $\bar{A}\bar{B}$, each lying entirely in one hemisphere, and two *improper* great arcs $A\bar{A}$ and $B\bar{B}$ crossing the equator. The hyperbolic plane H^2 is obtained by identifying antipodal points of $\ddot{S}^2$, with antipodal pairs of proper great arcs projecting into segments and improper great arcs with antipodal ends projecting into rays.

B. *Metric triangles.* Any three noncollinear points A, B, and C in the affine plane A^2 or in the Euclidean or hyperbolic plane E^2 or H^2 define a unique *triangle* $\langle ABC\rangle$. With more structure than a projective triangle, which is just a configuration of three points and three lines, this is a set

$$\langle ABC\rangle = \{\varnothing; A, B, C; BC, CA, AB; ABC\},$$

partially ordered by dimensionality and incidence, consisting of eight geometric "entities": the *nullity* (the empty set $\varnothing$); three *vertices* (the points A, B, and C); three *sides* (the segments BC, CA, and AB); and the triangle's *body* (the interior region ABC).

Any three points A, B, and C on the elliptic sphere S^2 that do not lie on a great circle are the vertices of a unique triangle $\langle ABC\rangle$, each of whose sides BC, CA, and AB is a proper great arc and whose body ABC is contained within a hemisphere. The three points, together with their antipodes $\bar{A}$, $\bar{B}$, and $\bar{C}$, are the vertices of a tessellation of S^2 by eight spherical triangles, consisting of four antipodal pairs of congruent triangles, namely,

$$\langle ABC\rangle = \{\varnothing; A, B, C; BC, CA, AB; ABC\} \text{ and}$$

$$\langle \bar{A}\bar{B}\bar{C}\rangle = \{\varnothing; \bar{A}, \bar{B}, \bar{C}; \bar{B}\bar{C}, \bar{C}\bar{A}, \bar{A}\bar{B}; \bar{A}\bar{B}\bar{C}\},$$

$$\langle \bar{A}CB \rangle = \{\varnothing; \bar{A}, C, B; BC, B\bar{A}, \bar{A}C; \bar{A}CB\} \text{ and}$$

$$\langle A\bar{C}\bar{B} \rangle = \{\varnothing; A, \bar{C}, \bar{B}; \bar{B}\bar{C}, \bar{B}A, A\bar{C}; A\bar{C}\bar{B}\},$$

$$\langle C\bar{B}A \rangle = \{\varnothing; C, \bar{B}, A; \bar{B}A, CA, C\bar{B}; C\bar{B}A\} \text{ and}$$

$$\langle \bar{C}B\bar{A} \rangle = \{\varnothing; \bar{C}, B, \bar{A}; B\bar{A}, \bar{C}\bar{A}, \bar{C}B; \bar{C}B\bar{A}\},$$

$$\langle BA\bar{C} \rangle = \{\varnothing; B, A, \bar{C}; A\bar{C}, \bar{C}B, AB; BA\bar{C}\} \text{ and}$$

$$\langle \bar{B}\bar{A}C \rangle = \{\varnothing; \bar{B}, \bar{A}, C; \bar{A}C, C\bar{B}, \bar{A}\bar{B}; \bar{B}\bar{A}C\}.$$

Two triangles that have two vertices and one side in common, the other pair of vertices being antipodal and the other pairs of sides being supplementary—e.g., $\langle ABC \rangle$ and each of the three triangles $\langle \bar{A}BC \rangle$, $\langle C\bar{B}A \rangle$, $\langle BA\bar{C} \rangle$—are said to be a *colunar* pair. The region bounded by two great semicircles that is the union of the bodies and the common side of such a pair of triangles is called a *lune* (or "biangle").

In all cases, the body of a triangle is a bounded convex region, and in any of the metric spaces S^2, E^2, or H^2, each interior angle of a triangle has a radian measure between 0 and π. Triangles in the elliptic plane eP^2 also have these properties. Any three noncollinear points A, B, and C determine a tessellation of eP^2 by four triangles, any two of which are colunar (except that antipodal vertices are now coincident). All four triangles have the same vertices A, B, and C. If the sides of one triangle are labeled BC, CA, and AB and its body denoted by ABC—with the supplementary segments being CB, AC, and BA and the colunar regions being ACB, CBA, and BAC—then we have the triangles

$$\langle ABC \rangle = \{\varnothing; A, B, C; BC, CA, AB; ABC\},$$

$$\langle ACB \rangle = \{\varnothing; A, C, B; BC, BA, AC; ACB\},$$

$$\langle CBA \rangle = \{\varnothing; C, B, A; BA, CA, CB; CBA\},$$

$$\langle BAC \rangle = \{\varnothing; B, A, C; AC, CB, AB; BAC\}.$$

Three points A, B, and C in the same hemisphere of the hyperbolic sphere $\ddot{S}^2$ that do not lie on a great circle and the three antipodal points $\bar{A}$, $\bar{B}$, and $\bar{C}$ are the vertices of an antipodal pair of triangles

$$\langle ABC\rangle = \{\varnothing; A, B, C; BC, CA, AB; ABC\} \text{ and}$$

$$\langle \bar{A}\bar{B}\bar{C}\rangle = \{\varnothing; \bar{A}, \bar{B}, \bar{C}; \bar{B}\bar{C}, \bar{C}\bar{A}, \bar{A}\bar{B}; \bar{A}\bar{B}\bar{C}\}.$$

Each side of either triangle is a proper great arc, and each of their bodies is a bounded convex region.

Because S^2, E^2, and H^2 are simply connected, any line or great circle separates the Euclidean or hyperbolic plane or the elliptic sphere into two half-planes or hemispheres. If A, B, and C are the vertices of a triangle, the (open) half-plane or hemisphere bounded by the line or great circle joining A and B that contains C may be denoted by $\cdot AB\cdot \uparrow C$ and the one that does not by $\cdot AB\cdot/C$. The body (interior) of $\langle ABC\rangle$ is then the intersection of $\cdot BC\cdot \uparrow A$, $\cdot CA\cdot \uparrow B$, and $\cdot AB\cdot \uparrow C$, and its exterior is the union of $\cdot BC\cdot/A$, $\cdot CA\cdot/B$, and $AB\cdot/C$.

The *area* of a triangle is the area of its body. If the interior angles at the vertices A, B, and C of a metric triangle $\langle ABC\rangle$ have the respective measures α, β, and γ, then the sum $\alpha+\beta+\gamma$ lies in the range 0 to 3π, with

$$\begin{aligned} 3\pi > \alpha+\beta+\gamma > \pi &\quad \text{for } \langle ABC\rangle \text{ in } S^2 \text{ or } eP^2;\\ \alpha+\beta+\gamma = \pi &\quad \text{for } \langle ABC\rangle \text{ in } E^2; \\ 0 < \alpha+\beta+\gamma < \pi &\quad \text{for } \langle ABC\rangle \text{ in } H^2 \text{ or } \ddot{S}^2. \end{aligned} \tag{3.4.1}$$

The amount by which the angle sum $\alpha+\beta+\gamma$ exceeds or falls short of π for a non-Euclidean triangle $\langle ABC\rangle$ is its *spherical excess* or *hyperbolic defect* and (assuming a spatial curvature of ±1) is equal to its area.

The area of a triangle is also a function of the lengths of its sides, which depend on the chosen unit of distance. Some writers treat non-Euclidean lengths and areas as if they were independent of each other and consequently arrive at mutually inconsistent formulas.

C. *The Triangle Inequality.* If the sides BC, CA, and AB of a metric triangle $\langle ABC \rangle$ have the respective lengths a, b, and c, then because lines in a plane or great circles on a sphere are geodesics and because ABC is a bounded convex region, it follows that for both Euclidean and non-Euclidean triangles we have the strict inequalities

$$b + c > a, \quad c + a > b, \quad a + b > c. \tag{3.4.2}$$

Moreover, most of the familiar Euclidean theorems about congruent triangles—in particular side-angle-side, angle-side-angle, and side-side-side—hold for non-Euclidean triangles as well. (However, angle-angle-side congruence does not generally hold either in the elliptic plane or on the elliptic sphere.) Since the interior angles of a non-Euclidean triangle determine not just its shape but its size, there is also an angle-angle-angle congruence theorem.

In the Euclidean and hyperbolic planes E^2 and H^2 and on the elliptic and hyperbolic spheres S^2 and $\ddot{S}^2$, the length of a side AB of a triangle $\langle ABC \rangle$ is the same as the distance between the points A and B. This makes it easy to extend the Triangle Inequality

$$|AB| \leq |BC| + |CA|, \tag{3.4.3}$$

which holds for any three points A, B, and C in Euclidean n-space, to hyperbolic n-space and to the elliptic and hyperbolic n-spheres (in $\ddot{S}^n$ the points must be restricted to the same n-hemisphere). If A, B, and C do not lie on a line or a great circle, they are the vertices of a unique triangle $\langle ABC \rangle$ and the conclusion follows from (3.4.2). Otherwise, we still have strict inequality except when C lies in the segment or proper great arc AB or, in the case of S^n, when A and B are antipodal, the only situations in which one distance equals the sum of the other two.

Because the total length of an elliptic line is π, the length of a side AB of a triangle in the elliptic plane eP^2 may be greater than the distance $[AB]$ between the points A and B (no distance can exceed $\pi/2$).

Consequently, the validity of the Triangle Inequality for three non-collinear points A, B, and C in eP^n does not immediately follow from the inequalities (3.4.2). Nevertheless, by considering triangles $\langle ABC \rangle$ with sides of various lengths less than, equal to, or greater than $\pi/2$, one can verify that $[AB] < [BC] + [CA]$ in every case. (I leave the details to the reader.) For three collinear points A, B, and C, either the sum of the three distances is π or else one of the distances is the sum of the other two; in either case, we have $[AB] \le [BC] + [CA]$. Combining the two results, we find that the Triangle Inequality holds for any three points in elliptic n-space.* Thus distance in each of the geometries S^n, E^n, H^n, and eP^n has the essential properties that define a metric space.

D. *Trigonometry.* How the sides and the angles of a triangle are related depends on the curvature of the underlying space. For example, in a Euclidean triangle $\langle ABC \rangle$, the side lengths a, b, c and the corresponding angles α, β, γ satisfy the Law of Sines

$$\frac{\sin\alpha}{a} = \frac{\sin\beta}{b} = \frac{\sin\gamma}{c}. \tag{3.4.4}$$

On the elliptic sphere S^2 or in the elliptic plane eP^2, the formula is

$$\frac{\sin\alpha}{\sin a} = \frac{\sin\beta}{\sin b} = \frac{\sin\gamma}{\sin c}, \tag{3.4.5}$$

while on the hyperbolic sphere $\ddot{S}^2$ or in the hyperbolic plane H^2 it is

$$\frac{\sin\alpha}{\sinh a} = \frac{\sin\beta}{\sinh b} = \frac{\sin\gamma}{\sinh c}. \tag{3.4.6}$$

* Two elliptic triangles whose corresponding sides have equal lengths are congruent. However, two triangles having equal distances between corresponding pairs of vertices need not be. Moreover, if A, B, and C are three points on a line such that none of the three distances $[BC]$, $[CA]$, and $[AB]$ is equal to the sum of the other two, then the three segments into which they separate the line can have the same lengths as the sides of a triangle, and the three distances can be equal to the distances between pairs of vertices of a triangle.

Two sides and the included angle of a Euclidean triangle $\langle ABC\rangle$ are related to the third side by the Law of Cosines

$$c^2 = a^2 + b^2 - 2ab\cos\gamma. \tag{3.4.7}$$

When the angle at C is a right angle, this reduces to the Pythagorean Theorem. On the elliptic sphere S^2 or in the elliptic plane eP^2, we have the dual formulas

$$\begin{aligned}\cos c &= \quad \cos a\,\cos b + \sin a\,\sin b\,\cos\gamma,\\ \cos\gamma &= -\cos\alpha\,\cos\beta + \sin\alpha\,\sin\beta\,\cos c.\end{aligned} \tag{3.4.8}$$

On the hyperbolic sphere $\ddot{\mathrm{S}}^2$ or in the hyperbolic plane H^2, the formulas are

$$\begin{aligned}\cosh c &= \cosh a\,\cosh b - \sinh a\,\sinh b\,\cos\gamma,\\ \cos\gamma &= -\cos\alpha\,\cos\beta + \sin\alpha\,\sin\beta\,\cosh c.\end{aligned} \tag{3.4.9}$$

In the hyperbolic plane we can also consider *asymptotic* triangles, with one or more absolute vertices, at which the interior angle is zero. Setting $\beta = \pi/2$ and $\gamma = 0$ in the last formula, we obtain the expression $\sin^{-1}\operatorname{sech} c$ for the angle of parallelism $\Pi(c) = \alpha$. The area of an asymptotic triangle is still finite and equal to the hyperbolic defect, so that a "tricuspal" triangle, with three absolute vertices, has an area of π.

Two Euclidean triangles whose corresponding angles are equal have the lengths of their sides proportional: the triangles are similar but not necessarily congruent. In the elliptic or hyperbolic plane, however, equality of angles implies equality of side lengths. Consequently, it is sometimes said that scale models are impossible in non-Euclidean geometry, but this is not entirely true. A spherical triangle with shorter sides can have the same angles as a plane triangle in elliptic 3-space, and a pseudospherical triangle with longer sides can have the same angles as a plane triangle in hyperbolic 3-space.

The n-dimensional analogue of a triangle, a polytope with $n + 1$ vertices and $n + 1$ sides (or "facets"), is called a *simplex*. A 3-simplex is a *tetrahedron*. The properties of simplexes and other polytopes will be discussed further in later chapters.

EXERCISES 3.4

1. The intermediacy relation "point X lies between points A and B" may be abbreviated as $/AXB/$. Extending the affine (or Euclidean) line $\cdot AB\cdot$ by the point at infinity P_∞, define $/AXB/$ in terms of separating pairs of points. Express the line segment AB and the rays $\cdot A/B$ and $\cdot B/A$ as segments of the circuit $\cdot AB\cdot \cup \{P_\infty\}$.

2. $\vdash$ The area of a spherical lune of angle θ is 2θ.

3. Show that the inequality $[AB] < [BC] + [CA]$ holds for any three non-collinear points A, B, and C in eP^2. (Consider triangles with sides of lengths less than, equal to, or greater than $\pi/2$.)

4. Show that the inequality $[AB] \le [BC]+[CA]$ holds for any three collinear points A, B, and C in eP^2.

5. The area Δ of a Euclidean triangle with sides a and b and included angle γ is given by $\Delta = ½ab \sin \gamma$. Combine this formula with the Law of Cosines (3.4.7) to obtain Heron's Formula $\Delta = \sqrt{s(s-a)(s-b)(s-c)}$, where $s = ½(a+b+c)$ is the triangle's semiperimeter.

6. Find the area of a hyperbolic triangle having two angles of $\pi/4$ adjacent to a side of length $\cosh^{-1} 2$.

7. Prove the identity $\sin^{-1} \operatorname{sech} x = \cos^{-1} \tanh x$.

3.5 NON-EUCLIDEAN CIRCLES

The methods of integral calculus can be used to determine the arc lengths of portions of smooth curves and the areas of regions with nonrectilinear boundaries. In this section we describe some of the metric properties of circles in different geometries.

In the Euclidean plane, a circle with center (h, k) and radius r has equation $(x-h)^2 + (y-k)^2 = r^2$. Given a circle of radius r, an arc subtending a central angle of radian measure θ has length $s = r\theta$, so that the circle's circumference is $2\pi r$. The corresponding circular sector has area $½rs$, making the area of the whole circle equal to πr^2.

In the elliptic plane eP^2, a circle with center $(h) = (h_1, h_2, h_3)$ and radius r or, equivalently, with axis $[h] = [h_1, h_2, h_3]$ and coradius $\check{r} = \pi/2 - r$, has equation

$$|(x\,h)| = \cos r \quad \text{or} \quad |(x)[h]| = \sin \check{r}, \tag{3.5.1}$$

where $(x\,h) = (x)[h] = x_1h_1 + x_2h_2 + x_3h_3$ and where $(x\,x) = (h\,h) = 1$. The length of an arc of a circle of radius r subtending a central angle or axial distance of θ is $s = \theta \sin r = \theta \cos \check{r}$; the circumference is thus seen to be $2\pi \sin r = 2\pi \cos \check{r}$. A circular sector $V(r, s)$ of radius r and arc length s has area

$$|V(r,s)| = s \tan \tfrac{1}{2}r \quad (0 < r < \pi/2). \tag{3.5.2}$$

The area of the whole circle is $4\pi \sin^2 \frac{1}{2}r$. Complementary to such a sector is a *cosector* $U(\check{r}, s)$ of coradius $\check{r}$ and arc length s, whose area is

$$|U(\check{r},s)| = s \tan \check{r} = \theta \sin \check{r} \quad (0 < \check{r} < \pi/2). \tag{3.5.3}$$

The sector and cosector together form a triangle with two right angles, a *great sector* of radius $\pi/2$ (see Figure 3.5a). The combined area is equal to the central angle or axial distance θ. Thus when $\theta = 1$, the

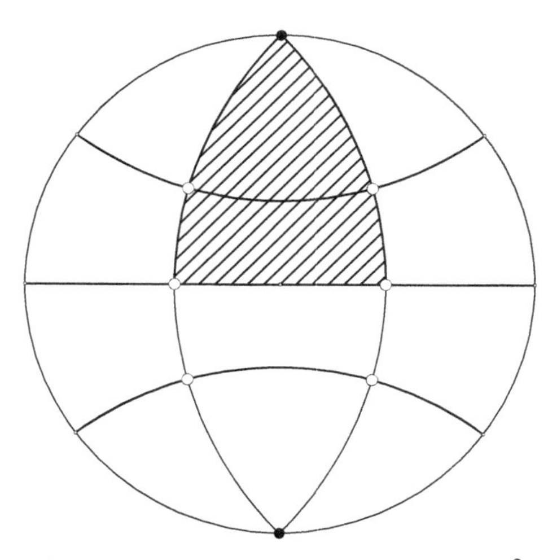

Figure 3.5a Circular sector and cosector in eP^2

area is 1. The above formulas also apply to small circles on the elliptic sphere S^2.

In the hyperbolic plane H^2, a pericycle—i.e., a proper circle—with center $(h) = (h_1,\ h_2 \mid h_0)$ and radius r has equation

$$|(x \ddot{} h)| = \cosh r, \tag{3.5.4}$$

where $(x \ddot{} h) = x_1h_1 + x_2h_2 - x_0h_0$, with coordinates normalized so that $(x \ddot{} x) = (h \ddot{} h) = -1$. An arc of such a pericycle subtending a central angle of θ has length $s = \theta \sinh r$, and the circumference is therefore $2\pi \sinh r$. A pericyclic sector $\ddot{V}(r,s)$ of radius r and arc length s (see Figure 3.5b) has area

$$|\ddot{V}(r,s)| = s \tanh \tfrac{1}{2}r \quad (r > 0). \tag{3.5.5}$$

The area of the whole pericycle is $4\pi \sinh^2 \tfrac{1}{2}r$.

A pseudocycle (equidistant curve) with axis $[m] = [m_1,\ m_2 \mid m_0]$ and coradius $\check{r}$ has equation

$$|(x)[m]| = \sinh \check{r}, \tag{3.5.6}$$

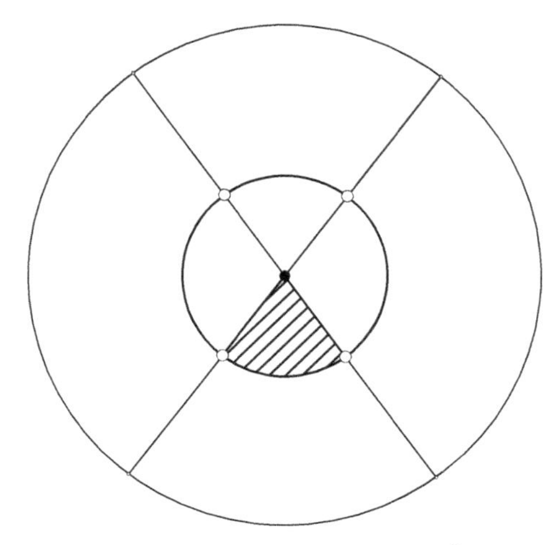

Figure 3.5b Pericyclic sector in H^2

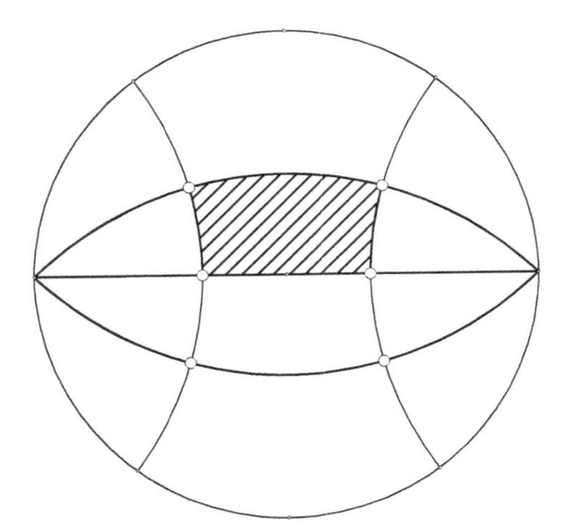

Figure 3.5c Pseudocyclic cosector in H^2

where $(x)[m] = x_0m_0 + x_1m_1 + x_2m_2$, with coordinates normalized so that $(x\ddot{}x) = -1$ and $[m\ddot{}m] = 1$. An arc of such a pseudocycle subtending an axial distance of t has length $s = t \cosh \check{r}$. A pseudocyclic cosector $\ddot{U}(\check{r}, s)$ of coradius $\check{r}$ and arc length s (Figure 3.5c) has area

$$|\ddot{U}(\check{r}, s)| = s \tanh \check{r} = t \sinh \check{r} \quad (\check{r} > 0). \tag{3.5.7}$$

When the arc of a pseudocyclic cosector is replaced by the chord with the same endpoints, the resulting figure is the kind of quadrilateral used by Gerolamo Saccheri (1667–1733) in his attempted proof of Euclid's Fifth Postulate.

Whereas a pericycle has an ordinary point for its center and a pseudocycle has an ordinary line for its axis, the center of a horocycle (limiting curve) is an absolute point, and its axis is the absolute line through this point. Thus a horocycle cannot have an equation analogous to (3.5.4) or (3.5.6). However, all horocycles are congruent, so that, following Coxeter (1942, p. 250), we can deduce their metric properties from a conveniently chosen particular case.

The three points

$$P_\infty = (0,\ 1\,|\,1), \quad P_0 = (0,\ 0\,|\,1), \quad \text{and} \quad P_1 = (1, \tfrac{1}{2}\,|\,\tfrac{3}{2})$$

determine a horocycle whose center is the absolute point P_∞ and whose ordinary points are, for each real number x,

$$P_x = \left(x, \tfrac{1}{2}x^2 \,\middle|\, \tfrac{1}{2}x^2 + 1\right).$$

The radial line joining P_x to the center and the tangent line at P_x are respectively

$$\check{R}_x = \left[1, x \,\middle|\, -x\right] \quad \text{and} \quad \check{T}_x = \left[x, \tfrac{1}{2}x^2 - 1 \,\middle|\, -\tfrac{1}{2}x^2\right].$$

The horocyclic arc with endpoints P_x and P_y has length $s = |y - x|$, and a horocyclic sector $\ddot{W}(s)$ of arc length s (see Figure 3.5d) has area s. Thus the arc from P_0 to P_1 has length 1 and the corresponding sector has area 1. Moreover, the radial line at one end of a horocyclic arc of unit length is parallel to the tangent line at the other end. Thus $\check{R}_0$ is parallel to $\check{T}_1$ and $\check{R}_1$ is parallel to $\check{T}_0$.

The foregoing results for pericycles, pseudocycles, and horocycles also hold for small circles on the hyperbolic sphere $\ddot{\mathrm{S}}^2$.

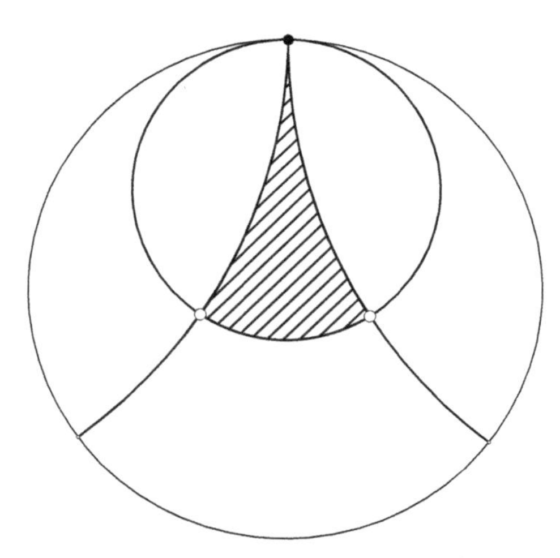

Figure 3.5d Horocyclic sector in H^2

EXERCISES 3.5

1. $\vdash$ The area of the elliptic plane is 2π.
2. Find the length of the chord of a pericycle of radius $\sinh^{-1} 2$ joining the ends of an arc of length π.
3. $\vdash$ The tangent lines at the ends of a horocyclic arc of unit length intersect at an angle of $\pi/3$.
4. Show that the isosceles asymptotic triangle $\langle P_0 P_1 P_\infty \rangle$ has area $\pi - 2 \tan^{-1} 2$.

3.6 SUMMARY OF REAL SPACES

The following diagram (Figure 3.6a) exhibits the relationships among the various real linear and circular geometries. A line joining two geometries indicates that the one below can be derived from the one above by selecting a quadric; by fixing an inversion, a polarity, or a hyperplane; or (for geometries joined by a double arrow) by identifying antipodal points.

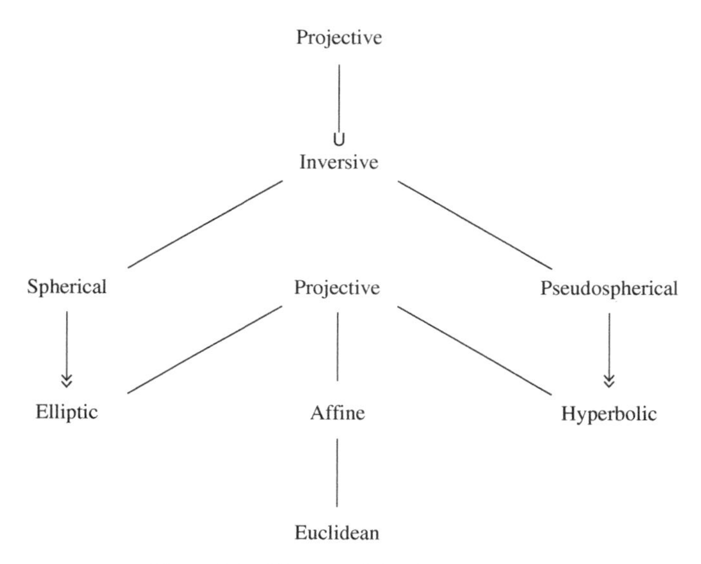

Figure 3.6a Linear and circular geometries

4

REAL COLLINEATION GROUPS

AN INCIDENCE-PRESERVING transformation of an affine or projective n-space over a field F is called a *collineation.* A dualizing transformation of projective n-space mapping points to hyperplanes and hyperplanes to points is a *correlation.* All collineations of $\mathsf{F}A^n$ preserve midpoints; an *affinity* preserves barycenters. Likewise, any collineation or correlation of $\mathsf{F}P^n$ preserves harmonic sets; a *projective* one preserves cross ratios. We shall mainly be concerned in this chapter with operations on real geometries. Given a suitable coordinate system, each collineation of A^n or P^n can be represented by a *linear transformation* of a real vector space. The structure of affine and projective collineation groups and subgroups corresponds to relations among groups of linear transformations.

4.1 LINEAR TRANSFORMATIONS

Geometric operations on projective and affine spaces and the metric spaces derived from them can be expressed in terms of certain transformations of vector spaces, which are not geometries but algebraic systems with quasi-geometric properties. The set $\mathbb{R}^n$ of all lists of n real numbers, regarded as rows $(x) = (x_1, x_2, \ldots, x_n)$, constitutes an n-dimensional left vector space over the real field $\mathbb{R}$, with vector addition and scalar multiplication defined by

$$\begin{aligned}(x)+(y)&=(x_1+y_1,\ldots,x_n+y_n),\\ \lambda(x)&=(\lambda x_1,\ldots,\lambda x_n),\end{aligned} \tag{4.1.1}$$

for all $[(x),\ (y)] \in \mathbb{R}^n \times \mathbb{R}^n$ and all $[\lambda,\ (x)] \in \mathbb{R} \times \mathbb{R}^n$. The lists (x) form an abelian *translation group* T^n with identity $(0) = (0,\ 0,\ldots,\ 0)$ under vector addition, while scalar multiplication satisfies the left distributive, right distributive, associative, and unitive laws

$$\begin{aligned}\lambda[(x)+(y)]&=\lambda(x)+\lambda(y),\\ [\kappa+\lambda](x)&=\kappa(x)+\lambda(x),\\ [\kappa\lambda](x)&=\kappa[\lambda(x)],\\ 1(x)&=(x).\end{aligned} \tag{4.1.2}$$

We do not assume that $\mathbb{R}^n$ comes equipped with an inner product.

Treated as columns $[u] = [u_1,\ u_2,\ldots,\ u_n]$, with similarly defined vector sums $[u]+[v]$ and scalar multiples $[u]\rho$, the same lists make up the dual right vector space $\check{\mathbb{R}}^n$. Vectors of either space correspond to *linear forms* on the other. Each column $[u]$ defines a linear form $\cdot[u] : \mathbb{R}^n \to \mathbb{R}$, with $(x) \mapsto (x)[u]$, and each row (x) defines a linear form $(x)\cdot : \check{\mathbb{R}}^n \to \mathbb{R}$, with $[u] \mapsto (x)[u]$.

Given a set of k rows (x_i), the set $\langle (x_1),\ldots,(x_k)\rangle$ of all linear combinations $\lambda_1(x_1) + \cdots + \lambda_k(x_k)$ is a subspace of $\mathbb{R}^n$, their *span*. For $k = 0$, this is the trivial subspace $0^n = \{(0)\}$. If $\sum \lambda_i(x_i) = (0)$ only when every $\lambda_i = 0$, the rows are *linearly independent*. The span of n linearly independent rows of $\mathbb{R}^n$ is the whole space, and the rows form a *basis* for $\mathbb{R}^n$. Similar definitions apply to columns of the dual space $\check{\mathbb{R}}^n$.

An $m \times n$ matrix A over $\mathbb{R}$ determines a dual pair of *linear transformations*

$$\cdot A : \mathbb{R}^m \to \mathbb{R}^n, \text{ with } (x) \mapsto (x)A \quad \text{and} \quad A\cdot : \check{\mathbb{R}}^n \to \check{\mathbb{R}}^m, \text{ with } [u] \mapsto A[u].$$

Conversely, the matrix A is uniquely determined by the effect of either transformation on a basis for its domain. Linearity means that $\cdot A$ and $A\cdot$ satisfy the distributive and associative laws

$$[(x)+(y)]A = (x)A + (y)A \quad \text{and} \quad A([u]+[v]) = A[u]+A[v],$$

$$[\lambda(x)]A = \lambda[(x)A] \quad \text{and} \quad A([u]\rho) = (A[u])\rho, \quad (4.1.3)$$

thus preserving linear combinations. Because multiplication of real numbers is commutative, linear transformations themselves behave like vectors, with

$$(x)[\cdot A + \cdot B] = (x)A + (x)B \quad \text{and} \quad (A\cdot + B\cdot)[u] = A[u]+B[u],$$

$$(x)[\lambda\cdot A] = \lambda[(x)A] \quad \text{and} \quad (A\cdot\rho)[u] = (A[u])\rho. \quad (4.1.4)$$

The set of all $m \times n$ matrices over $\mathbb{R}$ (or all linear transformations $\mathbb{R}^m \to \mathbb{R}^n$ or $\check{\mathbb{R}}^n \to \check{\mathbb{R}}^m$) forms an mn-dimensional vector space $\mathbb{R}^{m\times n}$, with additive group $\mathrm{T}^{m\times n}$. When $m = n$, it also forms a ring $\mathrm{Mat}_n(\mathbb{R})$. The subset of diagonal matrices

$$\backslash d\backslash = \backslash d_{11}, d_{22}, \ldots, d_{nn}\backslash = \begin{pmatrix} d_{11} & 0 & \cdots & 0 \\ 0 & d_{22} & \cdots & 0 \\ \vdots & \vdots & \ddots & \vdots \\ 0 & 0 & \cdots & d_{nn} \end{pmatrix}$$

forms a commutative subring $\mathrm{Diag}_n(\mathbb{R})$. The center $\mathrm{ZMat}_n(\mathbb{R})$ of the ring $\mathrm{Mat}_n(\mathbb{R})$ is the field of $n \times n$ scalar matrices, isomorphic to $\mathbb{R}$ via the mapping $\lambda I \mapsto \lambda$.

The multiplicative semigroup of $\mathrm{Mat}_n(\mathbb{R})$ is isomorphic to the *linear semigroup* $\mathrm{L}_n = \mathrm{L}(n;\ \mathbb{R})$ of all linear transformations taking $\mathbb{R}^n$ or $\check{\mathbb{R}}^n$ into itself. Each operation on either space is represented by an $n \times n$ matrix over $\mathbb{R}$, with products of operations corresponding to products of matrices. (Operations on $\mathbb{R}^n$ are multiplied from left to right; those on $\check{\mathbb{R}}^n$, from right to left.) The set of all invertible $n \times n$

matrices (with nonzero determinants), or all invertible linear transformations $\mathbb{R}^n \to \mathbb{R}^n$ or $\check{\mathbb{R}}^n \to \check{\mathbb{R}}^n$, forms a group—the (real) *general linear* group $\mathrm{GL}(n;\ \mathbb{R})$, more simply denoted by ${}^{\mathbb{R}}\mathrm{GL}_n$ or just GL_n. Transformations whose matrices have positive determinants constitute a subgroup of index 2, the *direct linear* group $\mathrm{G^+L}_n$. Matrices with determinant ± 1 ("unit" matrices) form a normal subgroup, the *unit linear* group $\bar{\mathrm{S}}\mathrm{L}_n$, and those with determinant 1 ("special" matrices) form the *special linear* group SL_n.

The center of the general linear group GL_n is the *general scalar* group GZ, represented by the nonzero scalar matrices λI. The group GZ is isomorphic to the multiplicative group $\mathrm{GL}(\mathbb{R})$ of nonzero real numbers. It has a subgroup of index 2, the *positive scalar* group GZ^+ of matrices μI with $\mu > 0$, and a subgroup of order 2, the *unit scalar* group $\bar{\mathrm{S}}\mathrm{Z}$ of matrices ιI with $\iota = \pm 1$. In fact, GZ is the direct product of these two subgroups.

EXERCISES 4.1

1. Show that properties (4.1.3) still hold if the field $\mathbb{R}$ is replaced by a skew-field K but that properties (4.1.4) do not.

2. A linear transformation is a vector-space homomorphism. If $A \in \mathbb{R}^{m\times n}$, describe the kernel and image of each of the dual transformations $\cdot A : \mathbb{R}^m \to \mathbb{R}^n$ and $A\cdot : \check{\mathbb{R}}^n \to \check{\mathbb{R}}^m$.

3. A subset of a vector space is a subspace if it is closed under vector addition and scalar multiplication. If $A \in \mathbb{R}^{m\times n}$, show that Ker $\cdot A$ and Img $\cdot A$ are subspaces of $\mathbb{R}^m$ and $\mathbb{R}^n$, respectively, and that Ker $A\cdot$ and Img $A\cdot$ are subspaces of $\check{\mathbb{R}}^n$ and $\check{\mathbb{R}}^m$, respectively.

4. If $A \in \mathbb{R}^{m\times n}$, the *row space* Row A is the subspace of $\mathbb{R}^n$ spanned by the m rows of A, and the *column space* Col A is the subspace of $\check{\mathbb{R}}^m$ spanned by the n columns of A. Show that Row A = Img $\cdot A$ and that Col A = Img $A\cdot$. (The common dimension of the row space and the column space is the *rank* of A.)

5. A system of m linear equations $(a_i)[x] = b_i$ in n unknowns x_j can be written in matrix form as $A[x] = [b]$, where $A \in \mathbb{R}^{m\times n}$, $[x] \in \check{\mathbb{R}}^n$, and $[b] \in \check{\mathbb{R}}^m$. Show that such a system has a solution if and only if $[b] \in \operatorname{Img} A\cdot$.

6. Show that a homogeneous system $A[x] = [0]$ of m equations in n unknowns has a nontrivial solution if and only if $\operatorname{Ker} A\cdot \neq \check{0}^n$.

4.2 AFFINE COLLINEATIONS

In real affine n-space A^n, let $n+1$ points Δ_i $(0 \le i \le n)$ be chosen, not all lying in the same hyperplane. Then (cf. Snapper & Troyer 1971, pp. 16–17) the *simplex of reference* $\Delta = \langle \Delta_0\Delta_1 \cdots \Delta_n\rangle$ defines an affine coordinate system on A^n, associating each point X with a unique vector $(x) \in \mathbb{R}^n$, the *origin* Δ_0 and the ith *basepoint* Δ_i $(1 \le i \le n)$ respectively corresponding to the zero vector (0) and the ith standard basis vector

$$(\delta_i) = (\delta_{i1}, \delta_{i2}, \ldots, \delta_{in}).$$

Here δ_{ij} is the "Kronecker delta" (after Leopold Kronecker, 1823–1891), evaluating to 1 if $i = j$, to 0 if $i \neq j$. Figure 4.2a shows the two-dimensional case. A hyperplane $\check{U}(r)$ with direction $[u] \neq [0]$ and position r is the locus of points X whose coordinates (x) satisfy a linear equation $(x)[u] + r = 0$ and may be associated with the one-dimensional subspace of $\check{\mathbb{R}}^{n+1}$ spanned by the vector $[[u], r]$.

Given a set of m points X_i with coordinates (x_i) and a set of m scalars or "masses" λ_i with $\sum \lambda_i = 1$, the point $\bar{X}$ with coordinates $(\bar{x}) = \sum \lambda_i(x_i)$ is the *barycenter* (or "center of mass") of the m points with the respective masses. When all the scalars are equal to $1/m$, $\bar{X}$ is the *centroid* of the m points; for $m = 2$, $\bar{X}$ is the *midpoint* of the segment X_1X_2.

Let A be an $n \times n$ invertible matrix, and let $(h) \in \mathbb{R}^n$ be any row vector. Then the product of the linear transformation

$$\cdot A : \mathbb{R}^n \to \mathbb{R}^n, \text{ with } (x) \mapsto (x)A,$$

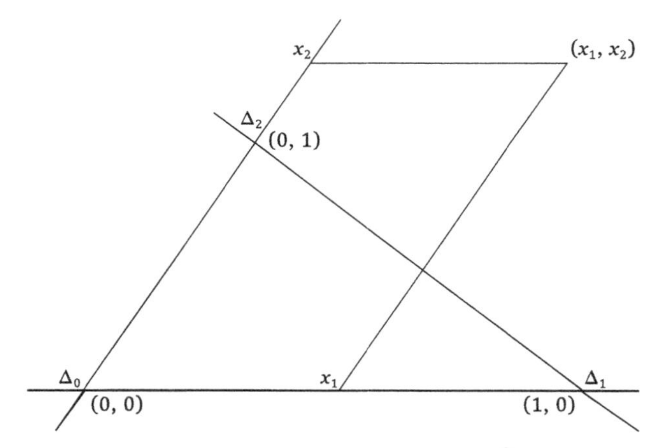

Figure 4.2a Reference triangle for A^2

and the translation

$$+(h) : \mathbb{R}^n \to \mathbb{R}^n, \text{ with } (y) \mapsto (y) + (h),$$

carried out in that order, is a *translinear transformation*

$$[\cdot A,\ +(h)] : \mathbb{R}^n \to \mathbb{R}^n, \text{ with } (x) \mapsto (x)A + (h), \quad \det A \neq 0, \quad (4.2.1)$$

which induces a corresponding operation on A^n—an affine collineation, or *affinity*—preserving barycenters and hence preserving parallelism and ratios of parallel line segments (a line is regarded as parallel to itself). Such a "homobaric" transformation can be represented by an $(n+1) \times (n+1)$ *translinear matrix* (cf. Weyl 1939, p. 47)

$$[A,\ (h)]_1 = \begin{pmatrix} A & [0] \\ (h) & 1 \end{pmatrix}, \quad \det A \neq 0, \qquad (4.2.2)$$

by which the row vector $((x),\ 1) \in \mathbb{R}^{n+1}$ may be postmultiplied to give the row vector $((x)A+(h),\ 1)$. A matrix $[I,\ (h)]_1$ is a *translation matrix*. The translinear matrix representing the product of transformations $[\cdot A, +(h)]$ and $[\cdot B, +(k)]$, in that order, is

$$[A, (h)]_1 \cdot [B, (k)]_1 = \begin{pmatrix} AB & [0] \\ (h)B + (k) & 1 \end{pmatrix}. \qquad (4.2.3)$$

Thus the product is the translinear transformation $[\cdot AB, +(h)B + (k)]$. As applied to points of A^n, affinities are multiplied from left to right.

The transformation $[\cdot A, +(h)]$, which takes the point (x) into the point $(x') = (x)A + (h)$, will be seen to take the hyperplane $[[u], r]$ into the hyperplane $[[u'], r'] = [A^{-1}[u], r - (h)A^{-1}[u]]$. (Note that a translation, with $A = I$, leaves hyperplane directions unchanged.) This can be effected by premultiplying the column vector $[[u], r] \in \check{\mathbb{R}}^{n+1}$ by the inverse translinear matrix

$$[A, (h)]_1{}^{-1} = \begin{pmatrix} A^{-1} & [0] \\ -(h)A^{-1} & 1 \end{pmatrix}, \quad \det A \neq 0. \tag{4.2.4}$$

As applied to hyperplanes, affinities are multiplied from right to left.

When applied to points, the inverse translinear transformation

$$[\cdot A, +(h)]^{-1} : \mathbb{R}^n \to \mathbb{R}^n, \text{ with } (x) \mapsto [(x) - (h)]A^{-1} \tag{4.2.5}$$

effects a *change of coordinates* on A^n, converting the standard coordinates (x) of a point X into new coordinates $(\grave{x}) = [(x) - (h)]A^{-1}$ relative to a new origin $\grave{\Delta}_0$ and new basepoints $\grave{\Delta}_i$ $(1 \leq i \leq n)$, the respective standard coordinates of which are (h) and $(a_i) + (h)$. The matrix $[A, (h)]_1{}^{-1}$ is the *transition matrix* for the change from the standard to the new coordinates.

If $\det A > 0$, then $[A, (h)]_1$ is a *direct* translinear matrix, and the corresponding operation on A^n is a *direct affinity*, preserving sense. Likewise, if $|\det A| = 1$, we have a *unit* translinear matrix, and the operation on A^n is a *unit affinity* or "isobaric" transformation, preserving content (length if $n = 1$, area if $n = 2$, volume if $n = 3$). If $\det A = 1$, then $[A, (h)]_1$ is a *special* translinear matrix, with the operation on A^n being an *equiaffinity*, preserving both content and sense.

The set of all translinear matrices $[A, (h)]_1$ of order $n + 1$ forms a multiplicative group TGL_n isomorphic to the *general affine* group GA_n of all affinities of A^n. The translation matrices $[I, (h)]_1$ constitute an abelian subgroup T_n, which is isomorphic to the additive group T^n of the vector space $\mathbb{R}^n$. The direct and the unit translinear matrices form normal subgroups $\mathrm{TG^+L}_n$ and $\mathrm{T\bar{S}L}_n$, respectively, isomorphic to the *direct affine* and *unit affine* groups $\mathrm{G^+A}_n$ and $\mathrm{\bar{S}A}_n$; their intersection is the subgroup TSL_n of special translinear matrices, isomorphic to the *special affine* group SA_n of equiaffinities. The quotient group $\mathrm{TGL}_n/\mathrm{T}_n \cong \mathrm{GA}_n/\mathrm{T}^n$ is isomorphic to the subgroup of *cislinear* matrices $[A,\ (0)]_1$, i.e., the *general linear* group GL_n. In other words, TGL_n is a semidirect product $\mathrm{T}_n \rtimes \mathrm{GL}_n$.

Let $\lambda \in \mathbb{R}$ be any nonzero scalar, and let $(h) \in \mathbb{R}^n$ be any row vector. Then the product of the scaling operation

$$\lambda\cdot : \mathbb{R}^n \to \mathbb{R}^n, \text{ with } (x) \mapsto \lambda(x),$$

and the translation

$$+(h) : \mathbb{R}^n \to \mathbb{R}^n, \text{ with } (y) \mapsto (y) + (h),$$

is a *transscalar transformation*

$$[\lambda\cdot, +(h)] : \mathbb{R}^n \to \mathbb{R}^n, \text{ with } (x) \mapsto \lambda(x) + (h), \quad \lambda \neq 0, \qquad (4.2.6)$$

which induces a corresponding operation on A^n—a *dilatation*, or "homothetic" transformation—taking each line into a parallel line. Because multiplication in $\mathbb{R}$ is commutative, this is the same as the linear transformation $[\cdot\lambda I, +(h)] : \mathbb{R}^{n+1} \to \mathbb{R}^{n+1}$ associated with the $(n+1) \times (n+1)$ *transscalar matrix*

$$[\lambda I, (h)]_1 = \begin{pmatrix} \lambda I & [0] \\ (h) & 1 \end{pmatrix}, \quad \lambda \neq 0, \qquad (4.2.7)$$

which maps the row vector $((x),\ 1) \in \mathbb{R}^{n+1}$ to the parallel row vector $(\lambda(x)+(h),\ 1)$. If $|\lambda| = 1$, this is a *unit* transscalar matrix, and the operation on A^n is a *unit dilatation*, or "isothetic" transformation. When $\lambda = 1$, the dilatation is the translation with vector (h); when $\lambda = -1$, it is the inversion in the point ½(h). If $\lambda \neq 1$, the dilatation has a unique fixed point $(1 - \lambda)^{-1}(h)$, its *center*. Directions are preserved if $\lambda > 0$, reversed if $\lambda < 0$.

The set of all transscalar matrices $[\lambda I,\ (h)]_1$ of order $n+1$ forms a multiplicative group TGZ_n isomorphic to the *general dilative* group GD_n of all dilatations of A^n; this is a normal subgroup of $\mathrm{TGL}_n \cong \mathrm{GA}_n$. The transscalar matrices with positive scales form a subgroup TGZ_n^+, of index 2, isomorphic to the *positive dilative* group GD_n^+. The unit transscalar matrices form a normal subgroup $\mathrm{T\bar{S}Z}_n$ isomorphic to the *unit dilative* group $\mathrm{\bar{S}D}_n$, and the translation matrices form an abelian subgroup, the *translation* group $\mathrm{T}_n \cong \mathrm{T}^n$. The quotient group $\mathrm{TGZ}_n/\mathrm{T}_n \cong \mathrm{GD}_n/\mathrm{T}^n$ is isomorphic to the general scalar group GZ of $n \times n$ nonzero scalar matrices, so that TGZ_n is a semidirect product $\mathrm{T}_n \rtimes \mathrm{GZ}$.

Any affinity of an affine n-space preserves barycenters. If an affinity takes a figure X into a figure X', the figures are *homobaric*, and we write $\mathsf{X} \frown \mathsf{X}'$. Figures related by a unit affinity are *isobaric*, written $\mathsf{X} \triangleq \mathsf{X}'$. For a direct affinity we write $\mathsf{X} \pm \mathsf{X}'$; for an equiaffinity, $\mathsf{X} \stackrel{+}{\triangleq} \mathsf{X}'$. If the affinity is a dilatation, preserving or reversing directions, the figures are *homothetic*, written $\mathsf{X} \approx \mathsf{X}'$. Figures related by a unit dilatation are *isothetic*, written $\mathsf{X} \approxeq \mathsf{X}'$. For a positive dilatation, the figures are *parallel*, and we write $\mathsf{X} \parallel \mathsf{X}'$; for a translation, they are *equiparallel*, written $\mathsf{X} \# \mathsf{X}'$.

EXERCISES 4.2

1. Relative to the $n+1$ vertices Δ_i of the simplex of reference Δ, the points of A^n have *barycentric coordinates* $(\lambda) = (\lambda_0, \lambda_1, \ldots, \lambda_n)$ with $\sum \lambda_i = 1$. Find the barycentric coordinates of the point X with affine coordinates $(x) = (x_1, \ldots, x_n)$.

2. Find the transition matrix for the change from the standard coordinates (x, y) for A^2 to new coordinates $(\grave{x}, \grave{y})$, relative to the new origin (h, k) and new basepoints $(1+h, k)$ and $(h, 1+k)$.

3. Find the transition matrix for the change from the standard coordinates (x, y) for A^2 to new coordinates $(\grave{x}, \grave{y})$, relative to the same origin $(0, 0)$ and new basepoints $(\cos\theta, \sin\theta)$ and $(-\sin\theta, \cos\theta)$.

4. Is the subgroup of cislinear matrices $[A,\ (0)]_1$ normal in the general translinear group TGL_n?

5. $\vdash$ When n is even, all dilatations of A^n are direct affinities; when n is odd, a positive dilatation is direct and a negative dilatation is opposite.

4.3 HOMOGENEOUS COORDINATES

In real projective $(n-1)$-space P^{n-1}, let $n+1$ *reference points* Δ_i $(1 \le i \le n)$ and ϵ ("epsilon") be chosen, no n of the points lying in the same hyperplane. Dual to the reference points are $n+1$ *reference hyperplanes* $\check{\Delta}_j$ $(1 \le j \le n)$ and $\check{\epsilon}$ ("co-epsilon"), where $\Delta_i \lozenge \check{\Delta}_j$ for $i \ne j$ and where $\check{\epsilon}$ is chosen so that $\|\Delta_i\epsilon,\ \check{\Delta}_j\check{\epsilon}\| = n$ for $i = j$. (The hyperplane $\check{\epsilon}$ is actually determined by any $n-1$ of the n such cross ratios.) This implies that $\Delta_i \not\lozenge \check{\epsilon}$ for any i, that $\epsilon \not\lozenge \check{\Delta}_j$ for any j, and that $\epsilon \not\lozenge \check{\epsilon}$.* The $n+1$ points and $n+1$ hyperplanes together constitute a *frame of reference* $\bar{\Delta}$ that defines a homogeneous coordinate system on P^{n-1} (Veblen & Young 1910, pp. 174–179, 194–198; cf. Coxeter 1949/1993, § 12.2).

A. *Assigning coordinates.* Relative to the chosen frame of reference, each point X of P^{n-1} can be associated with the set of nonzero scalar multiples of a nonzero row vector

$$(x) = (x_1, x_2, \ldots, x_n) \in \mathbb{R}^n,$$

and each hyperplane $\check{U}$ of P^{n-1} with the set of nonzero scalar multiples of a nonzero column vector

$$[u] = [u_1, u_2, \ldots, u_n] \in \check{\mathbb{R}}^n,$$

with X and $\check{U}$ being incident if and only if $(x)[u] = 0$.

* This construction can be carried out not just in real projective space but in any projective $(n-1)$-space over any division ring K; if K is of prime characteristic p, then $\epsilon \lozenge \check{\epsilon}$ when $n \equiv 0 \pmod{p}$.

To begin with, coordinate vectors of the reference points Δ_i $(1 \le i \le n)$ and ϵ (the "unit point") are nonzero scalar multiples of

$$(\delta_i) = (\delta_{i1}, \delta_{i2}, \ldots, \delta_{in}) \quad \text{and} \quad (\epsilon) = (1, 1, \ldots, 1),$$

while coordinate vectors of the reference hyperplanes $\check{\Delta}_j$ $(1 \le j \le n)$ and $\check{\epsilon}$ (the "unit hyperplane") are nonzero scalar multiples of

$$[\delta_j] = [\delta_{1j}, \delta_{2j}, \ldots, \delta_{nj}] \quad \text{and} \quad [\epsilon] = [1,\ 1, \ldots,\ 1].$$

Given these coordinates for the points and hyperplanes of the frame of reference, we proceed to define a *scale* on each of the lines $\Delta_i\Delta_n$ $(1 \le i \le n-1)$, which serve as coordinate axes. The coordinate system is then extended to other lines $\Delta_i\Delta_j$, planes $\Delta_i\Delta_j\Delta_k$, and so on, until all points and hyperplanes have been assigned homogeneous coordinates.

The details of how this is done are as follows. When $n = 2$ (the projective line), we rename the reference points Δ_1 and Δ_2 and the unit point ϵ as

$$P_\infty = (1,\ 0), \quad P_0 = (0,\ 1), \quad \text{and} \quad P_1 = (1,\ 1).$$

Each other point on the line is then associated with some real number x, with the coordinate vector of P_x being a nonzero scalar multiple of $(x, 1)$. Following von Staudt (1857, pp. 166–176, 256–283) and Hessenberg (1905), coordinates are assigned so as to preserve the structure of the real field, allowing points to be added and multiplied like numbers. Sums and products can be defined in terms of quadrangular sets:

$$P_x + P_y = P_{x+y} \quad \text{if and only if} \quad P_\infty P_x P_0 \mathbin{\overline{\wedge}} P_\infty P_y P_{x+y} \tag{4.3.1}$$

and

$$P_x \cdot P_y = P_{xy} \quad \text{if and only if} \quad P_0 P_x P_1 \mathbin{\overline{\wedge}} P_\infty P_y P_{xy} \tag{4.3.2}$$

(Veblen & Young 1910, pp. 141–147; cf. Coxeter 1949/1993, §§ 11.1–11.2). If applied to a line in a projective plane satisfying Axioms I, J, K, and N of Chapter 2, the resulting "geometric calculus" has all the properties of a field, with the possible exception of the commutative law of multiplication, which depends on Axiom P. (With commutativity, the quadrangular relations are involutions.)

The dual reference points $\check{\Delta}_1$ and $\check{\Delta}_2$ and the dual unit point $\check{\epsilon}$ can be renamed as

$$\check{P}_\infty = [1, 0], \quad \check{P}_0 = [0, 1], \quad \text{and} \quad \check{P}_1 = [1, 1],$$

with $\check{P}_\infty = P_0$, $\check{P}_0 = P_\infty$, and $\check{P}_1 = P_{-1}$. Then each other point $\check{P}_u$ is associated with a real number u, with its coordinate vector being a nonzero scalar multiple of $[u, 1]$. The points P_x and $\check{P}_u$ coincide if $xu = -1$.

When $n = 3$ (the projective plane), the lines joining each of the reference points Δ_1, Δ_2, and Δ_3 to the unit point ϵ meet the respective reference lines $\check{\Delta}_1$, $\check{\Delta}_2$, and $\check{\Delta}_3$ in the points $P_1 = (0, 1, 1)$, $Q_1 = (1, 0, 1)$, and $R_1 = (1, 1, 0)$. Using the aliases Q_∞ for Δ_1, P_∞ for Δ_2, and P_0 or Q_0 for Δ_3, we have a one-dimensional frame of reference on two of the three reference lines:

$$P_\infty = (0, 1, 0), \quad P_0 = (0, 0, 1), \quad P_1 = (0, 1, 1);$$

$$Q_\infty = (1, 0, 0), \quad Q_0 = (0, 0, 1), \quad Q_1 = (1, 0, 1).$$

We then construct a scale on $\check{\Delta}_1 = [1, 0, 0]$ so that each point P_y other than P_∞ has a coordinate vector that is a nonzero scalar multiple of $(0, y, 1)$ and a scale on $\check{\Delta}_2 = [0, 1, 0]$ so that the coordinate vector of each point Q other than Q_∞ is a nonzero scalar multiple of $(x, 0, 1)$.

The lines P_yQ_∞ and Q_xP_∞ have the respective coordinates $[0, -1, y]$ and $[-1, 0, x]$ (or scalar multiples thereof). Their intersection is the point $O_{x,y}$ whose coordinate vector is a nonzero scalar multiple of $(x, y, 1)$. The line $P_0O_{1,m}$, with coordinates $[m, -1, 0]$, meets the line $P_\infty Q_\infty$ in the point $R_m = (1, m, 0)$, establishing a scale on the third reference line $\check{\Delta}_3$, with $R_0 = Q_\infty = \Delta_1$ and $R_\infty = P_\infty = \Delta_2$. Any line not through R_∞ meets $\check{\Delta}_1$ and $\check{\Delta}_3$ in distinct points P_b and R_m and is assigned coordinates $[m, -1, b]$.

Every point X or line $\check{U}$ in the plane has now been assigned homogeneous coordinates $(x) = (x_1, x_2, x_3)$ or $[u] = [u_1, u_2, u_3]$, with $X \lozenge \check{U}$ if and only if $(x)[u] = 0$. The points $O_{x,y}$, R_m, and R_∞ can be denoted by the abbreviated coordinates (x, y), (m), and $(\ )$ and the lines P_bR_m, Q_xR_∞, and R_0R_∞ by $[m, b]$, $[x]$, and $[\ ]$. Such "extended affine" coordinates will be discussed further in the next chapter.

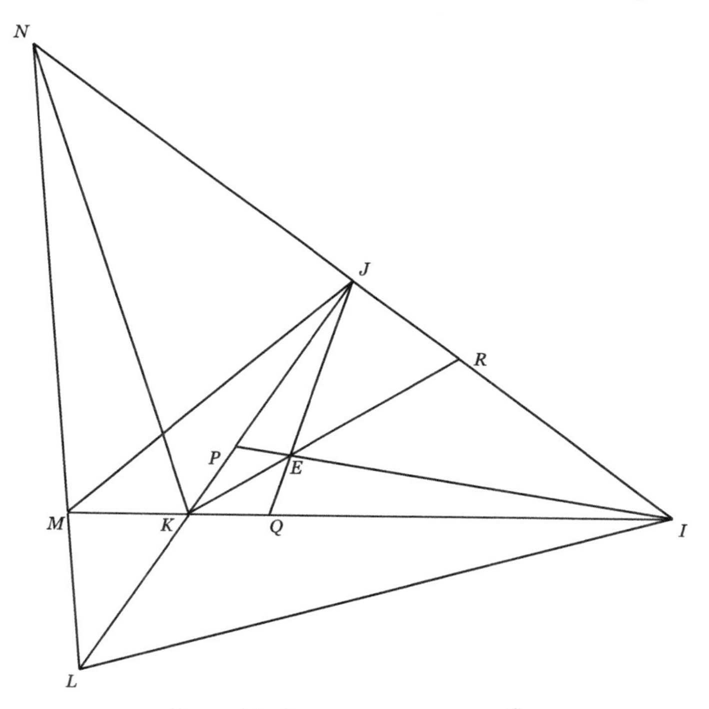

Figure 4.3a Frame of reference for P^2

An analogous procedure can be carried out for any value of n. Thus in projective 3-space we join each of the reference points Δ_1, Δ_2, Δ_3, and Δ_4 to the unit point ϵ to locate unit points in the reference planes $\check{\Delta}_1$, $\check{\Delta}_2$, $\check{\Delta}_3$, and $\check{\Delta}_4$ and use them in turn to locate unit points on the coordinate axes $\Delta_1\Delta_4$, $\Delta_2\Delta_4$, and $\Delta_3\Delta_4$. We then coordinatize each of the planes $\check{\Delta}_1$, $\check{\Delta}_2$, and $\check{\Delta}_3$ and extend the coordinate system to all points not on $\check{\Delta}_4$ and finally to points on $\check{\Delta}_4$ as well. In the process we also construct dual coordinates for planes.

B. *Plane configurations.* In the projective plane P^2 with reference triangle $IJK = \check{I}\check{J}\check{K}$, unit point E, and unit line $\check{E}$ (see Figure 4.3a), the six points

$$I = (1, 0, 0), \quad J = (0, 1, 0), \quad K = (0, 0, 1),$$
$$L = (0, 1, -1), \quad M = (-1, 0, 1), \quad N = (1, -1, 0)$$

are the vertices of the complete quadrilateral $\check{I}\check{J}\check{K}\check{E}$, whose diagonal lines are $\check{P} = [0, 1, 1]$, $\check{Q} = [1, 0, 1]$, and $\check{R} = [1, 1, 0]$. The six lines

$$\check{I} = [1, 0, 0], \quad \check{J} = [0, 1, 0], \quad \check{K} = [0, 0, 1],$$
$$\check{L} = [0, 1, -1], \quad \check{M} = [-1, 0, 1], \quad \check{N} = [1, -1, 0]$$

are likewise the sides of the complete quadrangle $IJKE$, with diagonal points $P = (0, 1, 1)$, $Q = (1, 0, 1)$, and $R = (1, 1, 0)$.

The reference triangle and the *harmonic triangle* $UVW = \check{U}\check{V}\check{W}$, with the three vertices

$$U = (-\tfrac{1}{2}, 1, 1), \quad V = (1, -\tfrac{1}{2}, 1), \quad W = (1, 1, -\tfrac{1}{2})$$

and the three sides

$$\check{U} = [-\tfrac{1}{2}, 1, 1], \quad \check{V} = [1, -\tfrac{1}{2}, 1], \quad \check{W} = [1, 1, -\tfrac{1}{2}],$$

are perspective from the unit point

$$E = (1, 1, 1)$$

and from the unit line

$$\check{E} = [1, 1, 1].$$

The six vertices of the complete quadrilateral $\check{I}\check{J}\check{K}\check{E}$ and the four vertices of the complete quadrangle $UVWE$, together with the six sides of the complete quadrangle $IJKE$ and the four sides of the complete quadrilateral $\check{U}\check{V}\check{W}\check{E}$, form a Desargues configuration 10_3. The perspective triangles IJK and UVW are related by a harmonic homology with center E and axis $\check{E}$.

Several instances of the Pappus configuration 9_3 can be exhibited. The vertices of the triangle $\check{P}\check{Q}\check{R}$ are the points

$$F = (-1,\ 1,\ 1), \quad G = (1,\ -1,\ 1), \quad H = (1,\ 1,\ -1),$$

and the sides of the triangle PQR are the lines

$$\check{F} = [-1,\ 1,\ 1], \quad \check{G} = [1,\ -1,\ 1], \quad \check{H} = [1,\ 1,\ -1].$$

We can take two of the points F, G, H, two of the points I, J, K, and two of the points P, Q, R, lying alternately on two of the lines $\check{L}$, $\check{M}$, $\check{N}$. Pairs of points are joined by one of the lines $\check{F}$, $\check{G}$, $\check{H}$, two of the lines $\check{I}$, $\check{J}$, $\check{K}$, and the three lines $\check{P}$, $\check{Q}$, $\check{R}$, pairs of which meet in the three points L, M, N lying on the line $\check{E}$. (These examples, with all coordinates equal to 0, 1, or -1, are special cases that would also hold in a projective plane coordinatized by a field—or even a skew-field—of characteristic not 2.)

EXERCISES 4.3

1. $\vdash$ On the real projective line P^1, the cross ratio $\|UV,\ WX\|$ of four points $U = (u_1,\ u_2)$, $V = (v_1,\ v_2)$, $W = (w_1,\ w_2)$, and $X = (x_1,\ x_2)$, where either $U \neq W$ and $V \neq X$ or $U \neq X$ and $V \neq W$, is given by

$$\|UV,\ WX\| = \frac{\begin{vmatrix} u_1 & u_2 \\ w_1 & w_2 \end{vmatrix} \begin{vmatrix} v_1 & v_2 \\ x_1 & x_2 \end{vmatrix}}{\begin{vmatrix} u_1 & u_2 \\ x_1 & x_2 \end{vmatrix} \begin{vmatrix} v_1 & v_2 \\ w_1 & w_2 \end{vmatrix}}.$$

2. Show that for any point $P_x = (x,\ 1)$ on P^1, $\|P_\infty P_0,\ P_1 P_x\| = x$.

3. $\vdash$ Addition of points on P^1, as defined by (4.3.1), is commutative.

4. Show that Axiom P of § 2.1 implies that $P_x \cdot P_y = P_y \cdot P_x$.

5. Show that in the real projective plane P^2 the points E, F, G, H, L, M, N, P, Q, R and the lines $\check{E}$, $\check{F}$, $\check{G}$, $\check{H}$, $\check{L}$, $\check{M}$, $\check{N}$, $\check{P}$, $\check{Q}$, $\check{R}$ form a Desargues configuration 10_3.

6. Among the points in P^2 whose coordinates are given in this section, find three collinear triples that satisfy the conditions of Pappus's Theorem.

4.4 PROJECTIVE COLLINEATIONS

A one-to-one transformation of projective $(n-1)$-space P^{n-1} into itself that maps points into points and hyperplanes into hyperplanes, preserving incidence and cross ratios, is a *projective collineation*. By the Fundamental Theorem of Projective Geometry (equivalent to Pappus's Theorem), a projective collineation of P^{n-1} is uniquely determined by its effect on the frame of reference. Specifically, let the desired images of the points Δ_i $(1 \leq i \leq n)$ and ϵ be the $n+1$ points

$$A_1, \quad A_2, \quad \ldots, \quad A_n, \quad \text{and} \quad P,$$

no n lying in the same hyperplane, with respective coordinate vectors

$$(a_1), \quad (a_2), \quad \ldots, \quad (a_n), \quad \text{and} \quad (p).$$

The $n+1$ n-vectors are linearly dependent, while any n of them are linearly independent. It is therefore possible to express (p) as a linear combination of the vectors (a_i) $(1 \leq i \leq n)$. Since the coordinates are homogeneous, we may assume without loss of generality that $(p) = \sum (a_i)$. Thus, if A is the invertible matrix of order n with rows $(a_i) = (\delta_i)A$, so that $\sum (a_i) = (\epsilon)A$, the projective collineation $\Delta_i \mapsto A_i$, $\epsilon \mapsto P$, with $(\delta_i) \mapsto (a_i)$ and $(\epsilon) \mapsto (p)$, is induced by the *projective linear transformation*

$$\langle \cdot A \rangle : \mathrm{P}\mathbb{R}^n \to \mathrm{P}\mathbb{R}^n, \text{ with } \langle (x) \rangle \mapsto \langle (x)A \rangle, \quad \det A \neq 0. \tag{4.4.1}$$

Here $\mathrm{P}\mathbb{R}^n = \mathbb{R}^n/\mathrm{GZ}$ denotes the $(n-1)$-dimensional *projective linear space* whose "points" are one-dimensional subspaces $\langle (x) \rangle$ of $\mathbb{R}^n$. All nonzero scalar multiples of A define the same transformation.

If the columns of A are denoted by $[a_j]$, with $\sum [a_j] = [q]$, then the same projective collineation is also uniquely determined as a mapping $[a_j] \mapsto [\delta_j]$, $[q] \mapsto [\epsilon]$ of hyperplanes. In this guise, the collineation is induced by the "inverse dual" of the transformation (4.4.1), i.e., by the projective linear transformation

$$\langle A^{-1}\cdot\rangle : \mathrm{P}\check{\mathbb{R}}^n \to \mathrm{P}\check{\mathbb{R}}^n, \text{ with } \langle[u]\rangle \mapsto \langle A^{-1}[u]\rangle. \tag{4.4.2}$$

Thus the general projective collineation has two dual aspects: regarded as a transformation $(x) \mapsto (x')$ of points of P^{n-1}, it corresponds to postmultiplication of the row vector (x) by an invertible matrix A; as a transformation $[u] \mapsto [u']$ of hyperplanes of P^{n-1}, it is effected by premultiplying the column vector $[u]$ by the matrix A^{-1}. Any nonzero scalar multiple of either the vector or the matrix will serve equally well. In short, given an invertible matrix A of order n, a projective collineation of P^{n-1} is defined by either of the rules

$$\lambda(x') = (x)A \quad \text{or} \quad A[u'] = [u]\rho, \tag{4.4.3}$$

where λ and ρ are any nonzero scalars. When the dimension $n - 1$ of the space is odd, the collineation is direct or opposite according as $\det A$ is positive or negative. (Projective spaces of even dimension are not orientable.)

In the case of a fixed point or a fixed hyperplane of the projective collineation defined by a real $n \times n$ matrix A, (x') or $[u']$ is a scalar multiple of (x) or $[u]$. Thus $(x)A = \lambda(x)$ or $A[u] = [u]\lambda$ for some λ. Any such scalar λ is an *eigenvalue* of A, with (x) or $[u]$ being a corresponding row or column *eigenvector*. If A is multiplied by some nonzero scalar, then so are the eigenvalues. Projective collineations of P^{n-1} are of various types, with the configuration of fixed points and hyperplanes depending on the solutions of the *characteristic equation* $\det(A - \lambda I) = 0$ (cf. Veblen & Young 1910, pp. 271–276). The only projectivity on P^1 with more than two fixed points is the identity; any other projectivity is said to be *elliptic*, *parabolic*, or *hyperbolic* according as the number of fixed points is 0, 1, or 2.

As previously noted, the set of all invertible real matrices of order n forms a multiplicative group, the *general linear* group $\mathrm{GL}_n \cong \mathrm{GL}(n;\ \mathbb{R})$. The center of GL_n is the *general scalar* group GZ of nonzero scalar matrices. The set of classes $\langle A\rangle$ of nonzero scalar multiples of matrices $A \in \mathrm{GL}_n$ forms the central quotient group $\mathrm{PGL}_n \cong \mathrm{GL}_n/\mathrm{GZ}$, the (real) *projective general linear* group, which is isomorphic to the *general projective* group GP_{n-1}, the group of projective collineations of P^{n-1}.

Matrices of the type $A^{-1}B^{-1}AB$, called "commutators," generate the commutator subgroup of GL_n: the *special linear* group SL_n, comprising all matrices with determinant 1 (cf. Artin 1957, pp. 162–163). When n is odd, the groups PGL_n and SL_n are isomorphic. The center of SL_n is the *special scalar* group $\mathrm{SZ}_n \cong \mathrm{C}_{(n,2)}$, consisting of the scalar matrices of order n with determinant 1, i.e., I alone for n odd, $\pm I$ for n even.* The *projective special linear* group $\mathrm{PSL}_n \cong \mathrm{SL}_n/\mathrm{SZ}_n$, ($n$ even) is isomorphic to the *special projective* group SP_{n-1}, the group of *equiprojective* (sense-preserving) collineations of P^{n-1}.

The *general commutator* quotient group $\mathrm{GCom} \cong \mathrm{GL}_n/\mathrm{SL}_n$ is the multiplicative group of equivalence classes $\underline{\alpha} = \{A \in \mathrm{GL}_n : \det A = \alpha\}$, where $\underline{\alpha} \cdot \underline{\beta} = \underline{\alpha\beta}$. Thus the mapping $\mathrm{GL}_n \to \mathrm{GCom}$, with $A \mapsto \underline{\det A}$, is a homomorphism with kernel $\mathrm{SL}_n = \underline{1}$. (The equivalence classes are just the cosets of SL_n.) The extended mapping $\underline{\det} : \mathrm{L}_n \to \mathrm{Com}$ from the set L_n of all real $n \times n$ matrices (invertible or not) to the set $\mathrm{Com} = \mathrm{GCom} \cup \underline{0}$ of their determinant classes is a semigroup homomorphism. The group $\mathrm{G}^+\mathrm{Com} \cong \mathrm{G}^+\mathrm{L}_n/\mathrm{SL}_n$ of classes of matrices with positive determinants is the *direct commutator* quotient group. The two classes $\underline{1}$ and $\underline{-1}$ form the *unit commutator* quotient group $\bar{\mathrm{S}}\mathrm{Com} \cong \bar{\mathrm{S}}\mathrm{L}_n/\mathrm{SL}_n$. Note that GCom is the direct product of the subgroups $\mathrm{G}^+\mathrm{Com}$ and $\bar{\mathrm{S}}\mathrm{Com}$.

The groups GCom, $\mathrm{G}^+\mathrm{Com}$, and $\bar{\mathrm{S}}\mathrm{Com}$ are respectively isomorphic to the groups GZ, GZ^+, and $\bar{\mathrm{S}}\mathrm{Z}$ of general, positive, and unit

* Here $(n, 2)$ denotes the greatest common divisor of n and 2.

scalar matrices, and the class $\underline{\alpha}$ may be identified with the scalar α. Note, however, that the determinant mappings GZ $\rightarrow$ GCom, $\mathrm{GZ}^+ \rightarrow \mathrm{G}^+\mathrm{Com}$, and $\bar{\mathrm{S}}\mathrm{Z} \rightarrow \bar{\mathrm{S}}\mathrm{Com}$, with $\lambda I \mapsto \underline{\lambda}^n$, are isomorphisms only when n is odd.

When n is even, we have several short exact sequences (the image of one mapping being the kernel of the next), with subgroups and quotient groups of the general linear group GL_n related as in the following commutative diagram.

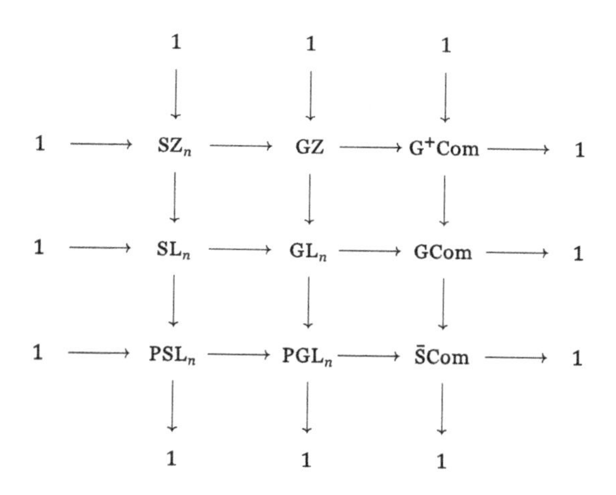

Every collineation of a projective n-space (over a field) preserves harmonic sets, tetrads with cross ratio -1. A *projective* collineation preserves all cross ratios. In the standard notation, invented by von Staudt, if some projective collineation transforms a figure X into a figure X', we write $\mathsf{X} \barwedge \mathsf{X}'$. If related by an elation or a homology, the figures are *perspective*, and we write $\mathsf{X} \doublebarwedge \mathsf{X}'$ (cf. Veblen & Young 1910, p. 57); for a harmonic homology this becomes $\mathsf{X} \overset{\circ}{\wedge} \mathsf{X}'$. Every projective collineation can be expressed as a product of perspective collineations. In some treatments a projective collineation is defined as such a product; since every perspective collineation preserves cross ratios, this is consistent with our definition.

EXERCISES 4.4

1. Find the projective collineation of P^2 that maps the reference points Δ_1, Δ_2, Δ_3, and ϵ into the points A_1, A_2, A_3, and P with respective coordinates $(1, -1, -1)$, $(0, 1, 2)$, $(0, 0, -1)$, and $(2, 1, -2)$.

2. Find the lines into which the reference lines $\check{\Delta}_1$, $\check{\Delta}_2$, $\check{\Delta}_3$, and $\check{\epsilon}$ are mapped by the collineation of Exercise 1.

3. Two real $n \times n$ matrices A and A' are *similar* if there exists an invertible matrix P such that $PAP^{-1} = A'$. Similar matrices have many common properties, including the same trace and the same determinant.

   a. Show that similarity is an equivalence relation on $\mathrm{Mat}_n(\mathbb{R})$, i.e., that it is reflexive, symmetric, and transitive.

   b. Denote by $\langle\!\langle A \rangle\!\rangle$ the equivalence class of matrices similar to A, and let addition and multiplication of similarity classes be given by the rules

   $$\langle\!\langle A \rangle\!\rangle + \langle\!\langle B \rangle\!\rangle = \langle\!\langle A + B \rangle\!\rangle \quad \text{and} \quad \langle\!\langle A \rangle\!\rangle \cdot \langle\!\langle B \rangle\!\rangle = \langle\!\langle AB \rangle\!\rangle.$$

   Show that these operations are well defined, i.e., that they do not depend on the choice of A or B. What are the additive and multiplicative identities?

   c. Define the trace and/or determinant of a similarity class to be the trace and/or determinant of any matrix in the class. Show that

   $$\mathrm{tr}\,\langle\!\langle A + B \rangle\!\rangle = \mathrm{tr}\,\langle\!\langle A \rangle\!\rangle + \mathrm{tr}\,\langle\!\langle B \rangle\!\rangle \quad \text{and} \quad \det\langle\!\langle AB \rangle\!\rangle = \det\langle\!\langle A \rangle\!\rangle \cdot \det\langle\!\langle B \rangle\!\rangle.$$

4.5 PROJECTIVE CORRELATIONS

A one-to-one transformation of projective $(n-1)$-space P^{n-1} into itself that maps points into hyperplanes and hyperplanes into points, dualizing incidences and preserving (transposed) cross ratios, is a *projective correlation*. Like a projective collineation, a projective correlation is uniquely determined by its effect on the frame of reference. Thus, given $n+1$ hyperplanes

$$\check{A}_1, \quad \check{A}_2, \quad \ldots, \quad \check{A}_n, \quad \text{and} \quad \check{P},$$

no n passing through the same point, let

$$(a_1), \quad (a_2), \quad \ldots, \quad (a_n), \quad \text{and} \quad (p).$$

be the transposes of their coordinate vectors, adjusted to make $(p) = \sum(a_i)$, and let A be the $n \times n$ matrix with rows (a_i), so that $\sum(a_i) = (\epsilon)A$. Then the projective correlation $\Delta_i \mapsto \check{A}_i$, $\epsilon \mapsto \check{P}$, with $(\delta_i) \mapsto (a_i)^{\vee}$ and $(\epsilon) \mapsto (p)^{\vee}$, is induced by the *projective linear cotransformation*

$$\langle \cdot A \rangle^{\vee} : \mathrm{P}\mathbb{R}^n \to \mathrm{P}\check{\mathbb{R}}^n, \text{ with } \langle (x) \rangle \mapsto \langle (x)A \rangle^{\vee}, \quad \det A \neq 0. \qquad (4.5.1)$$

Denote the columns of A by $[a_j]$, and let $\sum[a_j] = [q]$. Then the same projective correlation is also uniquely determined as a mapping $[a_j] \mapsto [\delta_j]^{\vee}$, $[q] \mapsto [\epsilon]^{\vee}$, taking hyperplanes into points. This is induced by the projective linear cotransformation

$$\langle A^{-1} \cdot \rangle^{\vee} : \mathrm{P}\check{\mathbb{R}}^n \to \mathrm{P}\mathbb{R}^n, \text{ with } \langle [u] \rangle \mapsto \langle A^{-1}[u] \rangle^{\vee}. \qquad (4.5.2)$$

Thus, given an invertible matrix A of order n, a projective correlation of P^{n-1} is defined by either of the rules

$$\lambda[x']^{\vee} = (x)A \quad \text{or} \quad A(u')^{\vee} = [u]\rho, \qquad (4.5.3)$$

where λ and ρ are any nonzero scalars.

If some projective correlation transforms a figure X into a figure X', we write $\mathsf{X} \veebar \mathsf{X}'$. The product of two projective correlations is a projective collineation. Thus if $\mathsf{X} \veebar \mathsf{X}'$ and $\mathsf{X}' \veebar \mathsf{X}''$, then $\mathsf{X} \barwedge \mathsf{X}''$.

The group PGL_n of dually paired projective linear transformations

$$(\mathrm{P}\mathbb{R}^n, \mathrm{P}\check{\mathbb{R}}^n) \to (\mathrm{P}\mathbb{R}^n, \mathrm{P}\check{\mathbb{R}}^n)$$

is a subgroup of index 2 in the *double projective general linear* group DPGL_n of such transformations and dually paired projective linear cotransformations

$$(\mathrm{P}\mathbb{R}^n, \mathrm{P}\check{\mathbb{R}}^n) \to (\mathrm{P}\check{\mathbb{R}}^n, \mathrm{P}\mathbb{R}^n).$$

For $n > 2$, the group DPGL_n is isomorphic to the *dualized general projective* group $\check{\mathrm{G}}\mathrm{P}_{n-1}$ of projective collineations and correlations, or *projectivities*, of P^{n-1}.

A projective correlation of period 2 is a *polarity*, pairing each point with a unique hyperplane. Two points or two hyperplanes are *conjugate* in a given polarity if each is incident with the hyperplane or point that is polar to the other. A point or a hyperplane that is incident with its own polar or pole is *self-conjugate*. If two figures X and X' are related by a polarity, we write X ♉ X'.

A necessary and sufficient condition for an invertible matrix A of order n to define a polarity of P^{n-1} is that A be either symmetric or skew-symmetric. To see why this is so, suppose that we have a projective correlation $\langle \cdot A\rangle^\vee$ of period 2, taking (x) into $[x']$ and $[x']$ into $(x'') = (x)$, i.e., the polarity with (x) and $[x']$ as pole and polar. Then we must have both

$$\lambda[x']^\vee = (x)A \quad \text{and} \quad A(x)^\vee = [x']\rho.$$

Postmultiplying the first equation by $(x)^\vee$ and premultiplying the second by (x), we get

$$\lambda[x']^\vee(x)^\vee = (x)A(x)^\vee = (x)[x']\rho.$$

If we now take the transpose of each expression, reversing the order, we get

$$\rho[x']^\vee(x)^\vee = (x)A^\vee(x)^\vee = (x)[x']\lambda.$$

This is an identity for all points (x). Unless $[x']^\vee(x)^\vee = (x)[x']$ is always zero, we must have $\lambda = \rho$, and hence $A^\vee = A$; i.e., A is symmetric. But if $(x)A(x)^\vee$ is identically zero, then A must be skew-symmetric: $A^\vee = -A$. In the latter case, which can occur only when n is even, we have a *null polarity*, in which every point (and every hyperplane) is self-conjugate.

On the projective line P^1, hyperplanes are points, so that each point has dual coordinates (x) and $[x']$. The correspondence between the two sets of coordinates is given by the identity regarded as a null polarity

$$\lambda[x']^{\check{}} = (x)J \quad \text{or} \quad J(x)^{\check{}} = [x']\rho, \tag{4.5.4}$$

defined by the skew-symmetric matrix

$$J = \begin{pmatrix} 0 & 1 \\ -1 & 0 \end{pmatrix}.$$

A true polarity on P^1, defined by a symmetric matrix, is an *involution*, a projectivity interchanging conjugate points, at most two being self-conjugate.

In any projective space over a field F, the correlation that interchanges points and hyperplanes with identical coordinates (x) and $[x]$ is a polarity, the *canonical polarity* for the chosen frame of reference. When $\mathsf{F} = \mathbb{R}$, the equation $(x)[x] = 0$ has no nonzero solutions, so that no point or hyperplane is self-conjugate; i.e., the real polarity $(x) \leftrightarrow [x]$ is elliptic.

Because of the fact that the real field $\mathbb{R}$ has no nontrivial automorphisms, *all* collineations and correlations of a real projective space preserve cross ratios; i.e., for $n > 1$, *every collineation or correlation of* P^{n-1} *is projective* (Veblen & Young 1918, pp. 251–252). In projective spaces over some other number systems, such as the complex field $\mathbb{C}$, there are collineations and/or correlations that are not projective.*

EXERCISES 4.5

1. Show that the projective correlation of P^2 that maps the reference points Δ_1, Δ_2, Δ_3, and ϵ into the lines $\check{A}_1$, $\check{A}_2$, $\check{A}_3$, and $\check{P}$ with respective coordinates $[-1, 3, 2]$, $[-1, 1, 1]$, $[2, -3, -4]$, and $[4, -3, -5]$ is a polarity.

* By definition, any collineation of a projective line over a field F preserves harmonic sets. If F has no nontrivial automorphisms, preservation of harmonic sets implies preservation of cross ratios, so that every one-dimensional collineation is a projectivity; in the case of the real field $\mathbb{R}$, this was first shown by von Staudt. In general, if K is a division ring, a one-to-one mapping of a projective line $\mathsf{K}P^1$ onto itself that preserves harmonic sets will preserve cross ratios only up to an automorphism or co-automorphism of K (Artin 1957, p. 85). Onishchik & Sulanke (2006, ¶1.4.5) prove this for the case where K is a field but elsewhere (p. 1) erroneously state that invariance of harmonic sets by itself implies invariance of cross ratios.

2. Find the points into which the reference lines $\check{\Delta}_1$, $\check{\Delta}_2$, $\check{\Delta}_3$, and $\check{\epsilon}$ are mapped by the polarity of Exercise 1.

3. $\vdash$ A skew-symmetric matrix of odd order is singular.

4.6 SUBGROUPS AND QUOTIENT GROUPS

Since every n-dimensional vector space over $\mathbb{R}$ is isomorphic to $\mathbb{R}^n$, and since the choice of a basis or a coordinate system is arbitrary, there is no loss of generality in using $\mathbb{R}^n$ itself and the standard basis $[(\delta_1), \ldots, (\delta_n)]$ to define linear, affine, or projective transformations and groups. The different representations of a given transformation that result from a change of basis vectors or coordinates are equivalent under matrix similarity.

Basic operations on vector spaces or geometries leave certain fundamental properties invariant: linear transformations preserve linear combinations, affine collineations preserve barycenters, and projective collineations preserve cross ratios. These invariant properties characterize the associated linear, affine, and projective transformation groups. Affine collineations or linear transformations preserving sense, size, or slope determine various normal subgroups (direct, unit, or dilative) of the general affine group GA_n or the general linear group $\mathrm{GL}_n \cong \mathrm{GA}_n/\mathrm{T}^n$.

The relationship between the general affine and general linear groups can be expressed as a short exact sequence

$$0 \to \mathrm{T}^n \to \mathrm{GA}_n \to \mathrm{GL}_n \to 1.$$

Likewise, for the general dilative and general scalar groups we have

$$0 \to \mathrm{T}^n \to \mathrm{GD}_n \to \mathrm{GZ} \to 1.$$

Here 0 is the trivial subgroup of the additive group of translations.

The one-dimensional affine and linear groups are identical to the corresponding dilative and scalar groups:

$$\mathrm{GA}_1 \cong \mathrm{GD}_1, \quad \mathrm{G^+A}_1 \cong \mathrm{GD}_1^+, \quad \bar{\mathrm{S}}\mathrm{A}_1 \cong \bar{\mathrm{S}}\mathrm{D}_1, \quad \mathrm{SA}_1 \cong \mathrm{T}^1,$$

$$\mathrm{GL}_1 \cong \mathrm{GZ}, \quad \mathrm{G^+L}_1 \cong \mathrm{GZ}^+, \quad \bar{\mathrm{S}}\mathrm{L}_1 \cong \bar{\mathrm{S}}\mathrm{Z}, \quad \mathrm{SL}_1 \cong 1.$$

Also, the additive group $\mathrm{T}(\mathbb{R}) \cong \mathrm{T}^1$ is isomorphic to the multiplicative group $\mathrm{G^+L}(\mathbb{R}) \cong \mathrm{GZ}^+$ of positive reals, via the mapping $x \mapsto \exp x$. When $n \geq 2$, the various groups of affinities of A^n or linear transformations of $\mathbb{R}^n$ are all distinct. The general, positive, and unit scalar groups GZ, GZ^+, and $\bar{\mathrm{S}}\mathrm{Z}$ do not vary with n.

How the subgroups of the general affine group GA_n or the general linear group GL_n are related to one another depends in part on whether n is even or odd. The following diagrams (Figure 4.6a) show the principal normal subgroups of GA_n. When two groups are joined by a line, the one below is a (normal) subgroup of the one above, and the line is marked with the quotient group. Similar diagrams showing normal subgroups of GL_n (with the same quotient groups) can be constructed by changing group symbols A_n to L_n, D_n to Z, and T^n to 1. For ready reference, the corresponding group names are as follows:

GA_n	General affine	GL_n	General linear
$\mathrm{G^+A}_n$	Direct affine	$\mathrm{G^+L}_n$	Direct linear
$\bar{\mathrm{S}}\mathrm{A}_n$	Unit affine	$\bar{\mathrm{S}}\mathrm{L}_n$	Unit linear
SA_n	Special affine	SL_n	Special linear
GD_n	General dilative	GZ	General scalar
GD_n^+	Positive dilative	GZ^+	Positive scalar
$\bar{\mathrm{S}}\mathrm{D}_n$	Unit dilative	$\bar{\mathrm{S}}\mathrm{Z}$	Unit scalar
T^n	Translation	1	Identity

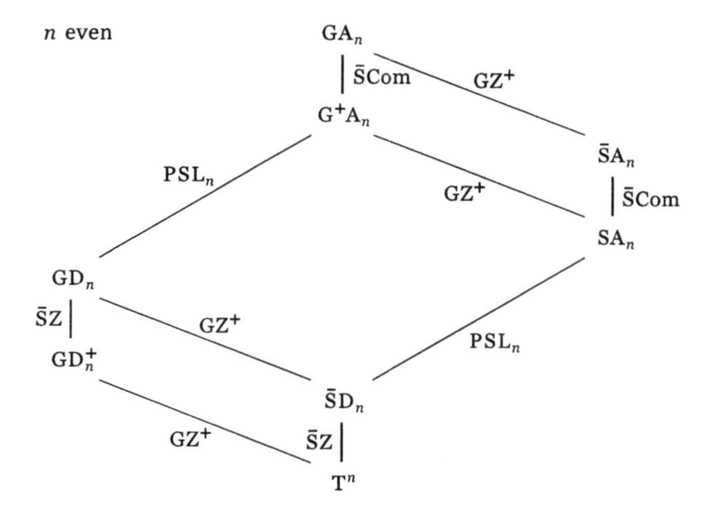

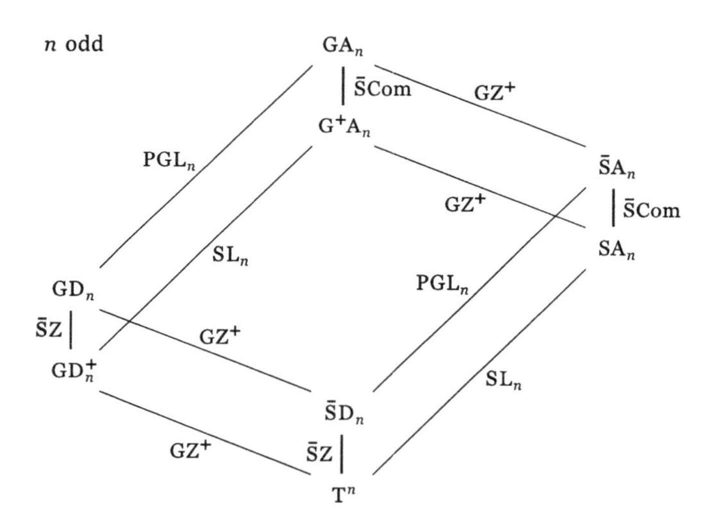

Figure 4.6a Subgroups of GA_n

5

EQUIAREAL COLLINEATIONS

ALTHOUGH DISTANCES AND ANGLES are not defined in the real affine plane, areas can be measured. A collineation of A^2 that preserves area is a *unit affinity*; one that preserves sense as well is an *equiaffinity*. Each unit affinity of A^2 belongs to one of three families, which can be differentiated as *ortholinear*, *paralinear*, and *metalinear* transformations. Triangles and other plane figures in an even-dimensional affine space can be assigned a *twisted area*; the areal metric so defined converts A^{2k} into *areplectic* $2k$-space. A transformation preserving twisted areas is an *equiplexity*, and the group of equiplexities is an extension of the real *symplectic* group Sp_{2k}.

5.1 THE REAL AFFINE PLANE

When the notation of § 4.2 is adapted to an affine plane $\mathsf{F}A^2$ over a field F, each point P has affine coordinates (x, y). It is customary to call x and y the *abscissa* and the *ordinate* of P. The vertices of the triangle of reference are the *origin* $O = (0, 0)$ and the two basepoints

$$I = (1, 0) \quad \text{and} \quad J = (0, 1).$$

An affine line $\check{Q}$ may be given homogeneous coordinates $[u, v, w]$, with P lying on $\check{Q}$ if and only if $xu + yv + w = 0$. If $v \neq 0$, we may normalize the coordinates of $\check{Q}$ as $[m, -1, b]$, which we abbreviate to $[m, b]$; m and b are called the *slope* and the *y-intercept* of $\check{Q}$. If $v = 0$, then $u \neq 0$, and the coordinates are normalized as $[-1, 0, a]$, which we abbreviate to $[a]$; we call a the *x-intercept* of $\check{Q}$. The line $\check{Y} = [0, 0]$ (equation $y = 0$) is the *x-axis*, and the line $\check{X} = [0]$ (equation $x = 0$) is the *y-axis*. Also significant are the *unit point* $E = (1, 1)$ and the *unit line* $\check{E} = [-1, -1]$.

Two lines $[m_1, b_1]$ and $[m_2, b_2]$ are parallel if and only if $m_1 = m_2$, and any two lines $[x_1]$ and $[x_2]$ are parallel. To obtain the extended affine plane $\mathsf{F}\bar{\mathrm{A}}^2$, we adjoin to the *ordinary* points (x, y) a set of *clinary* points (m), one lying on each *usual* line $[m, b]$ of slope m, and a point $(\,)$ lying on every *vertical* line $[x]$. The point $X = (0)$ that lies on every *horizontal* line is the *terminus*, and the point $Y = (\,)$ is the *zenith*. The adjoined points all lie on the *meridian* ("line at infinity") $\check{O} = [\,]$. In the extended plane we see that

$$(x,\ y) \lozenge [m,\ b] \quad \text{if and only if} \quad y = xm + b,$$

$$(x,\ y) \lozenge [x] \text{ for all } y, \qquad (m) \lozenge [m,\ b] \text{ for all } b,$$

$$(x,\ y) \not\lozenge [\,], \quad (m) \not\lozenge [x], \quad (\,) \not\lozenge [m,\ b],$$

$$(\,) \lozenge [x] \text{ for all } x, \qquad (m) \lozenge [\,] \text{ for all } m,$$

$$(\,) \lozenge [\,].$$

The extended affine plane $\mathsf{F}\bar{\mathrm{A}}^2$ can be regarded as a projective plane FP^2 whose frame of reference consists of the four points

$$O = (0,\ 0), \quad X = (0), \quad Y = (\,), \quad \text{and} \quad E = (1,\ 1)$$

and the four lines

$$\check{O} = [\,], \quad \check{X} = [0], \quad \check{Y} = [0,\ 0], \quad \text{and} \quad \check{E} = [-1,\ -1].$$

(see Figure 5.1a). Such extended affine coordinates facilitate the analysis of projective planes over different number systems (cf. Hall 1959, pp. 353–356). Here we shall be primarily concerned with the real affine plane $\mathrm{A}^2 = \mathbb{R}\mathrm{A}^2$.

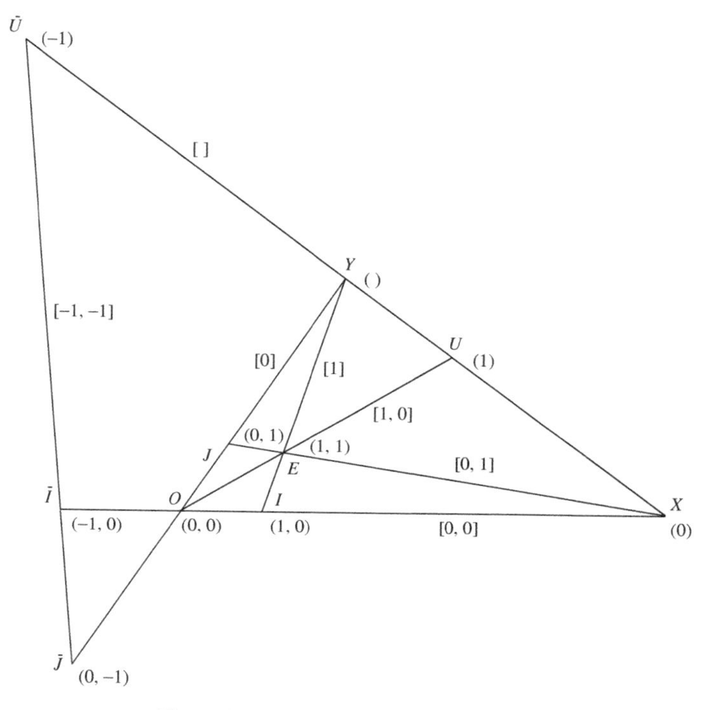

Figure 5.1a Extended affine coordinates

Given two points $P_1 = (x_1, y_1)$ and $P_2 = (x_2, y_2)$ in A^2 and two "weights" λ_1 and λ_2, with $\lambda_1 + \lambda_2 = 1$, the point $\bar{P} = (\bar{x}, \bar{y}) = \lambda_1(x_1, y_1) + \lambda_2(x_2, y_2)$ divides the directed line segment $\overrightarrow{P_1P_2}$ in the ratio $\lambda_2 : \lambda_1$, which we express as $P_1\bar{P}/\bar{P}P_2 = \lambda_2/\lambda_1$. If $\lambda_1 = \lambda_2 = ½$, $\bar{P}$ is the midpoint of the segment. The numbers λ_1 and λ_2 do not necessarily fall between 0 and 1, so that $\bar{P}$ need not lie between P_1 and P_2. As was found by Menelaus of Alexandria at the end of the first century and by Giovanni Ceva in 1678, there are significant consequences when the sides of a triangle are divided in certain ratios.

Theorem 5.1.1 (Ceva). *Given a real affine triangle $\langle ABC \rangle$ and points L, M, and N on the respective lines $\cdot BC\cdot$, $\cdot CA\cdot$, and $\cdot AB\cdot$, with $BL/LC = \lambda$, $CM/MA = \mu$, and $AN/NB = \nu$, the lines $\cdot AL\cdot$, $\cdot BM\cdot$, and $\cdot CN\cdot$ are concurrent or parallel if and only if $\lambda\mu\nu = 1$.*

THEOREM 5.1.2 (Menelaus). *Given a real affine triangle $\langle ABC\rangle$ and points L, M, and N on the respective lines $\cdot BC\cdot$, $\cdot CA\cdot$, and $\cdot AB\cdot$, with $BL/LC = \lambda$, $CM/MA = \mu$, and $AN/NB = \nu$, the points L, M, and N are collinear if and only if $\lambda\mu\nu = -1$.*

Either of these theorems can be proved by assigning suitable coordinates to the points. For example, we can take $\langle ABC\rangle$ to be the reference triangle $\langle OIJ\rangle$ and let

$$L = \left(\frac{1}{1+\lambda}, \frac{\lambda}{1+\lambda}\right), \quad M = \left(0, \frac{1}{1+\mu}\right), \quad N = \left(\frac{\nu}{1+\nu}, 0\right).$$

A *conic* in the real affine plane A^2 is the locus of points whose coordinates (x, y) satisfy a polynomial equation with real coefficients of the form

$$Ax^2 + Bxy + Cy^2 + Dx + Ey + F = 0,$$

at least one of the second-degree terms being nonzero. If the locus is the empty set, the conic is said to be *imaginary*; otherwise it is *real*. If the locus is a single point, a single line, or two lines, the conic is *degenerate*. A real, nondegenerate conic is an *ellipse*, a *parabola*, or a *hyperbola* according as the discriminant $B^2 - 4AC$ is less than, equal to, or greater than zero. A degenerate conic whose locus consists of two parallel lines is a *paratax*.

The translation $+(h, k) : \mathbb{R}^2 \to \mathbb{R}^2$ induces an affinity on A^2, taking each ordinary point (x, y) of A^2 to the point $(x + h, y + k)$. A translation is a direction-preserving collineation, taking each usual line $[m, b]$ to the parallel line $[m, b + k - hm]$ and each vertical line $[x]$ to the vertical line $[x + h]$. Any translation leaves each point (m) or $(\,)$ on the meridian $[\,]$ invariant.

The general affinity of A^2 is induced by the translinear transformation $[\cdot A, +(h, k)] : \mathbb{R}^2 \to \mathbb{R}^2$, where A is a 2×2 invertible matrix. If $\det A = \pm 1$, this is an area-preserving collineation, or *unit affinity*. If $\det A = 1$, it is an *equiaffinity*, either a translation or a type of *affine rotation* that leaves some pencil of lines or conics invariant.

EXERCISES 5.1

1. Taking the triangle in question as the reference triangle $\langle OIJ \rangle$, prove the theorems of Ceva and Menelaus.

2. $\vdash$ The three medians of any triangle are concurrent.

3. $\vdash$ Any line in the affine plane A^2 meets a nondegenerate conic in at most two points.

4. Show that the locus of a conic whose discriminant $B^2 - 4AC$ is greater than zero contains at least two points.

5. A conic whose locus is a single point or two intersecting lines is a *degenerate ellipse* or a *degenerate hyperbola*. If the locus is two parallel or coincident lines, the conic is a *paratax* or a *degenerate paratax* (a "doubly degenerate" conic). A conic whose locus contains no (real) points is an *imaginary ellipse* or an *imaginary paratax* according as its discriminant B^2-4AC is less than or equal to zero. Identify each of the following conics:

a. $x^2 + y^2 = 1.$ **d.** $x^2 = 1.$ **g.** $x^2 - y^2 = 1.$

b. $x^2 + y^2 = 0.$ **e.** $x^2 = 0.$ **h.** $x^2 - y^2 = 0.$

c. $x^2 + y^2 = -1.$ **f.** $x^2 = -1.$ **i.** $x^2 - y = 0.$

5.2 ORTHOLINEAR TRANSFORMATIONS

An $n \times n$ matrix over $\mathbb{R}$ whose eigenvalues are 1 and -1, of respective multiplicities $n - 1$ and 1, so that its square is the identity matrix, is a *linear reflection matrix*. When $n = 2$, such a matrix has the form

$$\begin{pmatrix} a & b \\ c & -a \end{pmatrix}, \text{ with } a^2 + bc = 1. \tag{5.2.1}$$

Depending on the values of a, b, and c, matrices (5.2.1) may belong to one or another of three one-parameter families. Within each family, the product of two linear reflection matrices is a type of *linear rotation matrix*, and the set of all reflection and rotation matrices forms a group. Applied to points of the affine plane, such matrices induce area-preserving collineations.

In the affine plane A^2, a matrix (5.2.1) induces an *affine reflection*, an affinity $(x, y) \mapsto (ax + cy, bx - ay)$ that fixes each point on the mirror

$$[b/(1+a), 0] \quad \text{or} \quad [(1-a)/c, 0]$$

while interchanging pairs of points on each line

$$[-b/(1-a), u] \quad \text{or} \quad [-(1+a)/c, v]$$

of a pencil of parallels, where u or v is the line's y-intercept. Particular choices of the entries a, b, and c produce various special cases.

More generally, an affine collineation $(x, y) \mapsto (ax + cy + h, bx - ay + k)$ with $a^2 + bc = 1$ is an affine reflection if $bh + (1-a)k = 0$ or, equivalently, if $(1+a)h + ck = 0$. Otherwise, it is an *affine transflection*, an opposite unit affinity with one invariant line but no invariant points.

If $b = c$, so that $a^2 + b^2 = 1$, we may take $a = \cos\phi$ and $b = \sin\phi$ or, if we prefer, $a = -\cos\psi$ and $b = \sin\psi$. A matrix (5.2.1) is then an *ortholinear reflection matrix* of the form

$$P = \begin{pmatrix} \cos\phi & \sin\phi \\ \sin\phi & -\cos\phi \end{pmatrix} \quad \text{or} \quad Q = \begin{pmatrix} -\cos\psi & \sin\psi \\ \sin\psi & \cos\psi \end{pmatrix}. \tag{5.2.2}$$

These matrices induce *ortholinear reflections*

$$\cdot\, P : \mathbb{R}^2 \to \mathbb{R}^2, \text{ with } \mu(\cos\theta, \sin\theta) \mapsto \mu(\cos(\phi - \theta), \sin(\phi - \theta)),$$

or (5.2.3)

$$\cdot\, Q : \mathbb{R}^2 \to \mathbb{R}^2, \text{ with } \nu(\sin\theta, \cos\theta) \mapsto \nu(\sin(\psi - \theta), \cos(\psi - \theta)).$$

The product P_1P_2 or Q_1Q_2 of two ortholinear reflection matrices of the same kind is an *ortholinear rotation matrix*

$$R = \begin{pmatrix} \cos\rho & \sin\rho \\ -\sin\rho & \cos\rho \end{pmatrix} \quad \text{or} \quad S = \begin{pmatrix} \cos\sigma & -\sin\sigma \\ \sin\sigma & \cos\sigma \end{pmatrix}, \tag{5.2.4}$$

where $\rho = \phi_2 - \phi_1$ and $\sigma = \psi_2 - \psi_1$. These matrices induce *ortholinear rotations*

$$\cdot R : \mathbb{R}^2 \to \mathbb{R}^2, \text{ with } \mu(\cos\theta,\ \sin\theta) \mapsto \mu(\cos(\theta+\rho),\ \sin(\theta+\rho)),$$

or (5.2.5)

$$\cdot S : \mathbb{R}^2 \to \mathbb{R}^2, \text{ with } \nu(\sin\theta,\ \cos\theta) \mapsto \nu(\sin(\theta+\sigma),\ \cos(\theta+\sigma)).$$

The distinction between the two kinds of reflection matrices or the two kinds of rotation matrices is not hard and fast. If $\phi + \psi = (2k+1)\pi$, then $P = Q$, and likewise if $\rho + \sigma = 2k\pi$, then $R = S$.

In the affine plane A^2, a matrix P or Q induces an affine reflection $\cdot P$ or $\cdot Q$, whose mirror is the line

$$[\tan \tfrac{1}{2}\phi,\ 0] \quad \text{or} \quad [\cot \tfrac{1}{2}\psi,\ 0]$$

and which interchanges points on each line

$$[-\cot \tfrac{1}{2}\phi,\ u] \quad \text{or} \quad [-\tan \tfrac{1}{2}\psi,\ v].$$

A matrix R or S induces an *elliptic rotation* $\cdot R$ or $\cdot S$, whose center is the origin $(0, 0)$ and which leaves invariant any ellipse

$$x^2 + y^2 = w^2.$$

We call such a conic an *ellipse*, rather than a circle, since A^2 has no metric for distance. However, areas can be measured and are invariant under any unit affinity; the reference triangle $\langle OIJ \rangle$ has area ½. Since $\det P = \det Q = -1$ and $\det R = \det S = 1$, we see that as an affine collineation an ortholinear reflection is an opposite unit affinity, preserving area but reversing sense, while an elliptic rotation is an equiaffinity, preserving both area and sense.

Each value of a real parameter θ determines a (possibly multiply covered) sector of the ellipse $x^2 + y^2 = 1$, bounded by radius vectors from the origin O to the points $I = (1,0)$ and $V = (\cos\theta,\ \sin\theta)$ and by the arc from I to V, taken in the counterclockwise direction from I to V if $\theta > 0$, in the clockwise direction if $\theta < 0$ (see Figure 5.2a). Likewise, each value of θ determines a sector bounded by radius vectors

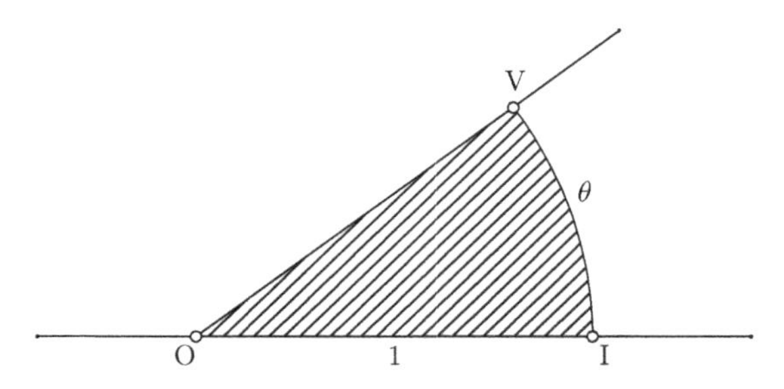

Figure 5.2a Elliptic sector of area $\frac{1}{2}\theta$

from O to the points $J = (0, 1)$ and $U = (\sin\theta, \cos\theta)$ and by the arc from J to U, taken clockwise from J to U if $\theta > 0$, counterclockwise if $\theta < 0$. If $|\theta| < 2\pi$, the area of the sector OIV or OJU is $½|\theta|$, and its directed area, taking orientation into account, is $½\theta$.

The points on the unit ellipse $x^2 + y^2 = 1$ in A^2 can be identified with the points of the elliptic circle S^1, and the group of reflections and rotations $\cdot P$ and $\cdot R$ (or $\cdot Q$ and $\cdot S$) is the *orthogonal* group $O_2 \cong O(2)$ of isometries of S^1. The subgroup of rotations $\cdot R$ (or $\cdot S$) is the (abelian) *special orthogonal* group $O_2^+ \cong SO(2)$ of direct isometries of S^1. The subgroup of order 2 generated by the central inversion $(x, y) \mapsto (-x, -y)$ is the unit scalar group $\bar{S}Z$. The *projective orthogonal* group $PO_2 \cong O_2/\bar{S}Z$ is isomorphic to O_2.

EXERCISES 5.2

1. The *eigenvalues* of an $n \times n$ matrix A are the roots of its characteristic equation $\det(A - \lambda I) = 0$. The Cayley–Hamilton Theorem says that every square matrix satisfies its own characteristic equation. Use this fact to show that if an $n \times n$ matrix A has eigenvalues 1 and -1, of respective multiplicities $n - 1$ and 1, then $A^2 = I$. (Hint: The characteristic equation of A is $(1 - \lambda)^{n-1}(1 + \lambda) = 0$.)

2. Find coordinates for the mirror of the affine reflection taking a point (x, y) into a point $(ax+cy+h,\ bx-ay+k)$, where $a^2+bc = 1$ and $bh+(1-a)k = (1 + a)h + ck = 0$.

3. Show that an elliptic rotation $\cdot R$ or $\cdot S$ defined by an ortholinear rotation matrix (5.2.4) takes each point on the unit ellipse $x^2 + y^2 = 1$ into another such point.

5.3 PARALINEAR TRANSFORMATIONS

A matrix (5.2.1) with $bc = 0$, so that $a^2 = 1$, is a *paralinear reflection matrix*, having one of the forms

$$B = \begin{pmatrix} 1 & b \\ 0 & -1 \end{pmatrix} \quad \text{or} \quad C = \begin{pmatrix} -1 & 0 \\ c & 1 \end{pmatrix},$$
$$D = \begin{pmatrix} -1 & d \\ 0 & 1 \end{pmatrix} \quad \text{or} \quad E = \begin{pmatrix} 1 & 0 \\ e & -1 \end{pmatrix}. \tag{5.3.1}$$

The first two matrices induce *transverse reflections*

$$\cdot B : \mathbb{R}^2 \to \mathbb{R}^2, \text{ with } (x,\ y) \mapsto (x,\ -y + bx),$$
or
$$\cdot C : \mathbb{R}^2 \to \mathbb{R}^2, \text{ with } (x,\ y) \mapsto (-x + cy,\ y). \tag{5.3.2}$$

The second two induce *conjugate reflections*

$$\cdot D : \mathbb{R}^2 \to \mathbb{R}^2, \text{ with } (x,\ y) \mapsto (-x,\ y + dx),$$
or
$$\cdot E : \mathbb{R}^2 \to \mathbb{R}^2, \text{ with } (x,\ y) \mapsto (x + ey,\ -y). \tag{5.3.3}$$

In the affine plane A^2, a matrix B or C induces an affine reflection $\cdot B$ or $\cdot C$, whose mirror is the line

$$y = \tfrac{1}{2}bx \qquad \text{or} \qquad x = \tfrac{1}{2}cy$$

and which leaves invariant each line

$$x = u \qquad \text{or} \qquad y = v.$$

A matrix D or E induces an affine reflection $\cdot D$ or $\cdot E$, whose mirror is the line

$$x = 0 \qquad \text{or} \qquad y = 0$$

and which leaves invariant each line

$$y = -\tfrac{1}{2}dx + v \quad \text{or} \quad x = -\tfrac{1}{2}ey + u.$$

The product B_1B_2 or C_1C_2, D_1D_2 or E_1E_2 of two paralinear reflection matrices of the same kind or the (commutative) product of matrices B and D or of matrices C and E is a *paralinear rotation matrix*

$$F = \begin{pmatrix} 1 & f \\ 0 & 1 \end{pmatrix} \qquad \text{or} \qquad G = \begin{pmatrix} 1 & 0 \\ g & 1 \end{pmatrix},$$
$$H = \begin{pmatrix} -1 & h \\ 0 & -1 \end{pmatrix} \qquad \text{or} \qquad K = \begin{pmatrix} -1 & 0 \\ k & -1 \end{pmatrix}, \tag{5.3.4}$$

where f is $b_2 - b_1$ or $d_1 - d_2$ and g is $c_2 - c_1$ or $e_1 - e_2$ and where $h = b + d$ and $k = c + e$. The first two matrices induce *transvective rotations*

$$\cdot F : \mathbb{R}^2 \to \mathbb{R}^2, \text{ with } (x, y) \mapsto (x, y + fx),$$

or (5.3.5)

$$\cdot G : \mathbb{R}^2 \to \mathbb{R}^2, \text{ with } (x, y) \mapsto (x + gy, y).$$

The second two induce *semitransvective rotations*

$$\cdot H : \mathbb{R}^2 \to \mathbb{R}^2, \text{ with } (x, y) \mapsto (-x, -y + hx),$$

or (5.3.6)

$$\cdot K : \mathbb{R}^2 \to \mathbb{R}^2, \text{ with } (x, y) \mapsto (-x + ky, -y).$$

The (commutative) product F_1F_2 or G_1G_2, H_1H_2 or K_1K_2 of two paralinear rotation matrices of the same kind is a paralinear rotation matrix F or G, where F's entry f is $f_1 + f_2$ or $-(h_1 + h_2)$ and G's entry g is $g_1 + g_2$ or $-(k_1 + k_2)$.

In A^2, a matrix F or G induces a *shear* (or "transvection") $\cdot F$ or $\cdot G$, whose axis of fixed points is the y-axis ($x = 0$) or the x-axis ($y = 0$) and which leaves invariant each parallel line

$$x = u \qquad \text{or} \qquad y = v.$$

A matrix H or K induces a *half-shear* $\cdot H$ or $\cdot K$, whose center is the origin $(0, 0)$ and which interchanges the two lines of any paratax

$$x^2 = u^2 \qquad \text{or} \qquad y^2 = v^2.$$

Lines joining consecutive points in the orbit of a point (x, y) under repetitions of a half-shear pass alternately through two *foci*

$$(0, \pm\tfrac{1}{2}hx) \qquad \text{or} \qquad (\pm\tfrac{1}{2}ky, 0).$$

A shear or a half-shear is a *paratactic rotation*. An affine reflection of one of the types given above is a unit affinity; a paratactic rotation is an equiaffinity.

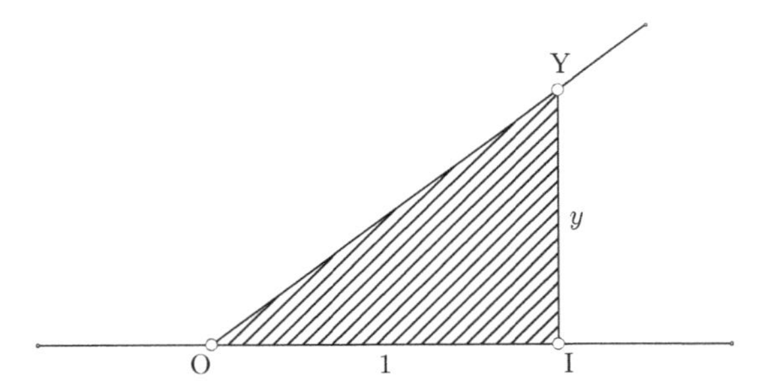

Figure 5.3a Triangular region of area $\frac{1}{2}y$

Each point Y on the line $x = 1$ determines a triangular region bounded by radius vectors from the origin O to the points $I = (1, 0)$ and $Y = (1, y)$ and by the closed interval $\overline{IY}$ (see Figure 5.3a). Similarly, each point X on the line $y = 1$ determines a triangular region bounded by radius vectors from O to $J = (0, 1)$ and $X = (x, 1)$ and by the closed interval $\overline{JX}$. The region OIY has area $\frac{1}{2}|y|$ and directed area ½y. The region OJX has area ½$|x|$ and directed area ½x.

The points on the line $x = 1$ or $y = 1$ in A^2 can be identified with the points of a Euclidean line E^1, and the group of affine reflections and shears $\cdot B$ and $\cdot F$ (or $\cdot C$ and $\cdot G$) is the one-dimensional *Euclidean* group $E_1 \cong \bar{S}A_1 \cong \bar{S}D_1$ of unit dilatations (reflections and translations). The abelian subgroup of shears $\cdot F$ (or $\cdot G$) is the *direct Euclidean* group $E_1^+ \cong SA_1 \cong T^1$ of one-dimensional translations.

The points on the paratax $x^2 = 1$ or $y^2 = 1$ are the points of a "double" *bi-Euclidean line* $2E^1 = S^0 \times E^1$, and the group of affine reflections, shears, and half-shears $\cdot B$, $\cdot D$, $\cdot F$, and $\cdot H$ (or $\cdot C$, $\cdot E$, $\cdot G$, and $\cdot K$) is the *bi-Euclidean* group $2E_1 \cong \bar{S}Z \times E_1$ of unit dilatations (transverse reflections, conjugate reflections, translations, and inversions) of $2E^1$. The subgroup of reflections and shears $\cdot D$ and $\cdot F$ (or $\cdot E$ and $\cdot G$) is the *semidirect bi-Euclidean* group $2E_1^+$ of conjugate reflections and translations, and the subgroup of shears and half-shears $\cdot F$ and $\cdot H$ (or $\cdot G$ and $\cdot K$) is the *direct bi-Euclidean* group $\bar{S}Z \times E_1^+$ of translations and inversions. Here and elsewhere $\bar{S}Z$ is the subgroup of order 2 generated by the central inversion $(x, y) \mapsto (-x, -y)$.

A mapping $\cdot F$ or $\cdot G$ can be combined with a translation $+(h, k)$ to give a "transparalinear rotation" $[\cdot F, +(h, k)]$ or $[\cdot G, +(h, k)]$. To see how this transformation behaves, it is convenient to take $f = 2h$ and $k = h^2$ or to take $g = 2k$ and $h = k^2$, so that the transformation has one of the forms

$$\left[\cdot\begin{pmatrix} 1 & 2h \\ 0 & 1 \end{pmatrix}, +(h, h^2)\right] : \mathbb{R}^2 \to \mathbb{R}^2, \quad \text{with} \quad (x, y) \mapsto (x + h,\ y + 2hx + h^2),$$

or (5.3.7)

$$\left[\cdot\begin{pmatrix} 1 & 0 \\ 2k & 1 \end{pmatrix}, +(k^2, k)\right] : \mathbb{R}^2 \to \mathbb{R}^2, \quad \text{with} \quad (x, y) \mapsto (x + 2ky + k^2,\ y + k).$$

In A^2 this operates as a *parabolic rotation*, the fixed direction of which is that of the y-axis or the x-axis and which leaves invariant any parabola

$$y = x^2 + v \qquad \text{or} \qquad x = y^2 + u,$$

respectively. A parabolic rotation is an equiaffinity.

As transformations of the extended affine plane, certain unit affinities are elations or harmonic homologies. A dilatation (which preserves ratios of areas) is a perspective collineation that fixes each point on the line at infinity (the "meridian" of § 5.1). If an elation, a dilatation is a translation, with center at infinity; if a harmonic homology, it is an inversion (half-turn), with an ordinary center. A shear is an elation and a reflection a harmonic homology, both having an ordinary axis and center at infinity.

EXERCISES 5.3

1. Show that a shear $\cdot F$ or $\cdot G$ defined by a paralinear rotation matrix (5.3.4) takes each point on the line $x = 1$ or $y = 1$ into another such point.
2. Show that points in the orbit of a point $(\pm 1, y)$ or $(x, \pm 1)$ under repetitions of a half-shear $\cdot H$ or $\cdot K$ defined by a paralinear rotation matrix lie alternately on the lines $x = \pm 1$ or $y = \pm 1$, and lines joining consecutive points pass alternately through the points $(0, \pm\frac{1}{2}h)$ or $(\pm\frac{1}{2}k, 0)$.
3. Show that a parabolic rotation $[\cdot F, +(h, h^2)]$ or $[\cdot G, +(k^2, k)]$ defined by a transparalinear rotation matrix (5.3.7) takes each point on the parabola $y = x^2$ or $x = y^2$ into another such point.

5.4 METALINEAR TRANSFORMATIONS

The entries of a matrix (5.2.1) with $b = -c$, i.e., a matrix of the form

$$\begin{pmatrix} a & b \\ -b & -a \end{pmatrix}, \quad \text{with } a^2 - b^2 = 1, \tag{5.4.1}$$

are conveniently expressed in terms of hyperbolic sines and cosines. We then have a *metalinear reflection matrix*

$$P = \begin{pmatrix} \cosh p & \sinh p \\ -\sinh p & -\cosh p \end{pmatrix} \quad \text{or} \quad Q = \begin{pmatrix} -\cosh q & -\sinh q \\ \sinh q & \cosh q \end{pmatrix}. \tag{5.4.2}$$

These matrices induce *metalinear reflections*

$$\cdot P : \mathbb{R}^2 \to \mathbb{R}^2, \text{ with } \begin{cases} \mu(\cosh t, \sinh t) \mapsto \mu(\cosh(p-t), \sinh(p-t)), \\ \nu(\sinh t, \cosh t) \mapsto -\nu(\sinh(p-t), \cosh(p-t)), \end{cases}$$

or (5.4.3)

$$\cdot Q : \mathbb{R}^2 \to \mathbb{R}^2, \text{ with } \begin{cases} \mu(\cosh t, \sinh t) \mapsto -\mu(\cosh(q-t), \sinh(q-t)), \\ \nu(\sinh t, \cosh t) \mapsto \nu(\sinh(q-t), \cosh(q-t)). \end{cases}$$

The product P_1P_2 or Q_1Q_2, P_1Q_2 or Q_1P_2 of two metalinear reflection matrices of the same or different kinds is a *metalinear rotation matrix*

$$R = \begin{pmatrix} \cosh r & \sinh r \\ \sinh r & \cosh r \end{pmatrix} \quad \text{or} \quad S = \begin{pmatrix} -\cosh s & -\sinh s \\ -\sinh s & -\cosh s \end{pmatrix}, \tag{5.4.4}$$

where r is $p_2 - p_1$ or $q_2 - q_1$ and s is $q_2 - p_1$ or $p_2 - q_1$. These matrices induce *metalinear rotations*

$$\cdot R : \mathbb{R}^2 \to \mathbb{R}^2, \text{ with } \begin{cases} \mu(\cosh t, \sinh t) \mapsto \mu(\cosh(t+r), \sinh(t+r)), \\ \nu(\sinh t, \cosh t) \mapsto \nu(\sinh(t+r), \cosh(t+r)), \end{cases}$$

or (5.4.5)

$$\cdot S : \mathbb{R}^2 \to \mathbb{R}^2, \text{ with } \begin{cases} \mu(\cosh t, \sinh t) \mapsto -\mu(\cosh(t+s), \sinh(t+s)), \\ \nu(\sinh t, \cosh t) \mapsto -\nu(\sinh(t+s), \cosh(t+s)). \end{cases}$$

It may be noted that a metalinear reflection or rotation of one kind is one of the other kind combined with the central inversion $(x,\ y) \mapsto (-x,\ -y)$. A reflection or rotation matrix of one kind is the negative of one of the other kind; if $p = q$, then $P = -Q$, and if $r = s$, then $R = -S$. (Metalinear transformations can also be represented by matrices of other forms.)

In the affine plane A^2, a matrix P or Q induces an affine reflection $\cdot P$ or $\cdot Q$, whose mirror is the line

$$[\tanh \tfrac{1}{2}p,\ 0] \quad \text{or} \quad [\coth \tfrac{1}{2}q,\ 0]$$

and which interchanges points on each line

$$[\coth \tfrac{1}{2}p,\ u] \quad \text{or} \quad [\tanh \tfrac{1}{2}q,\ v].$$

A matrix R or S induces a *hyperbolic rotation* $\cdot R$ or $\cdot S$, whose center is the origin (0, 0) and which leaves invariant any hyperbola

$$x^2 - y^2 = u^2 \quad \text{or} \quad y^2 - x^2 = v^2$$

with asymptotes $[\pm 1,\ 0]$. The "straight" rotation $\cdot R$ leaves each branch of the hyperbola invariant, while the "crossed" rotation $\cdot S$ interchanges the two branches. As an affine collineation, a metalinear reflection is a unit affinity; a hyperbolic rotation is an equiaffinity.

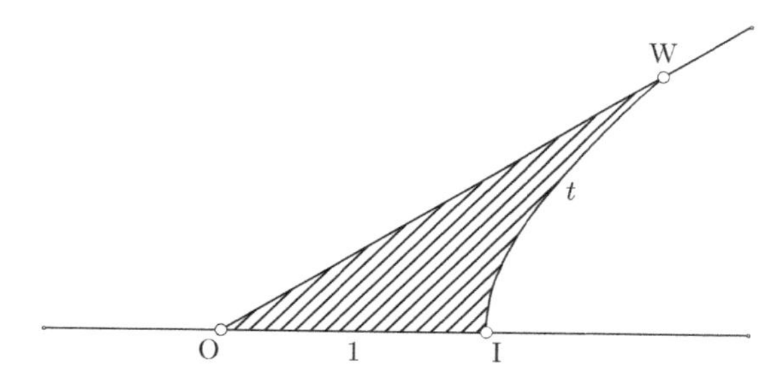

Figure 5.4a Hyperbolic sector of area $\frac{1}{2}t$

Each point W on the branch of the hyperbola $x^2 - y^2 = 1$ with $x > 0$ determines a hyperbolic sector, bounded by radius vectors from the origin O to the points $I = (1,\ 0)$ and $W = (\cosh t,\ \sinh t)$ and by the arc from I to W (see Figure 5.4a). Likewise, each point Z on the branch of the hyperbola $y^2 - x^2 = 1$ with $y > 0$ determines a sector bounded by radius vectors from O to points $J = (0,\ 1)$ and $Z = (\sinh t,\ \cosh t)$ and by the arc from J to Z. The area of the sector OIW or OJZ is $\frac{1}{2}|t|$ and its directed area is $\frac{1}{2}t$.

The points on either of the unit hyperbolas $x^2 - y^2 = 1$ and $y^2 - x^2 = 1$ in A^2 can be identified with the points of the hyperbolic circle $\ddot{S}^1$, and the group of affine reflections and hyperbolic rotations $\cdot P$, $\cdot Q$, $\cdot R$, and $\cdot S$ is the *pseudo-orthogonal* group $O_{1,1} \cong O(1,\ 1)$ of isometries of $\ddot{S}^1$. The subgroup of rotations $\cdot R$ and $\cdot S$ is the (abelian) *special pseudo-orthogonal* group $O_{1,1}^+ \cong SO(1,\ 1)$ of direct isometries of $\ddot{S}^1$. Both of these groups contain the subgroup $\bar{S}Z$ of order 2 generated by the central inversion $(x,\ y) \mapsto (-x,\ -y)$.

A subgroup $O_{1^+,1}$ of reflections $\cdot P$ and rotations $\cdot R$ leaves invariant each branch of the hyperbola $x^2 - y^2 = 1$, and an isomorphic subgroup $O_{1,1^+}$ of reflections $\cdot Q$ and rotations $\cdot R$ leaves invariant each branch of the hyperbola $y^2 - x^2 = 1$. These two subgroups and $O_{1,1}^+$ (each of index 2 in $O_{1,1}$) have a common subgroup $O_{1^+,1^+}$ (of index 4 in $O_{1,1}$) of rotations $\cdot R$, leaving invariant each branch of both hyperbolas.

The points on one branch of either hyperbola can be identified with the points of the hyperbolic line H^1. The group of all isometries (reflections and translations) of H^1 is the *projective pseudo-orthogonal* group $PO_{1,1} \cong O_{1,1}/\bar{S}Z$, which is isomorphic to the groups $O_{1^+,1}$ and $O_{1,1^+}$. The subgroup of direct isometries (translations) is the *special projective pseudo-orthogonal* group $P^+O_{1,1} \cong O_{1,1}^+/\bar{S}Z$, which is isomorphic to $O_{1^+,1^+}$; it is also isomorphic to the additive group T^1.

A comparison of the matrices (5.4.2) and (5.4.4) with their counterparts in § 5.2 makes evident the "pseudo-symmetry" of ortholinear and metalinear transformations, also seen in the parallels between elliptic and hyperbolic geometry. With a change of coordinates, we can obtain even simpler expressions. A matrix (5.2.1) with $a = 0$, i.e., a matrix of the form

$$\begin{pmatrix} 0 & b \\ c & 0 \end{pmatrix}, \text{ with } bc = 1, \tag{5.4.6}$$

provides an alternative representation of a metalinear reflection. In the affine plane such a matrix induces an affine reflection with mirror

$$y = bx \qquad \text{or} \qquad x = cy,$$

interchanging points (x, y) and (cy, bx).

The product of two metalinear reflection matrices (5.4.6) is a metalinear rotation matrix of the form

$$\begin{pmatrix} a & 0 \\ 0 & d \end{pmatrix}, \text{ with } ad = 1. \tag{5.4.7}$$

In the affine plane, a matrix of this kind induces a hyperbolic rotation whose center is the origin $(0, 0)$, taking the point (x, y) to the point (ax, dy) and leaving invariant any hyperbola

$$xy = \pm w^2$$

whose asymptotes are the coordinate axes $[0, 0]$ and $[0]$.

EXERCISES 5.4

1. Show that a hyperbolic rotation $\cdot R$ or $\cdot S$ defined by a metalinear rotation matrix (5.4.4) takes each point on the line $y = x$ or $y = -x$ into another such point.
2. Show that a hyperbolic rotation $\cdot R$ or $\cdot S$ takes each point on the unit hyperbola $x^2 - y^2 = 1$ or $y^2 - x^2 = 1$ into another such point.
3. Show that under the change of coordinates $(\grave{x}, \grave{y}) = \frac{1}{2}\sqrt{2}(x - y,\ x + y)$ the hyperbolic rotation matrix $R = [(\cosh r,\ \sinh r),\ (\sinh r,\ \cosh r)]$ is replaced by the matrix $\grave{R} = \backslash \cosh r - \sinh r,\ \cosh r + \sinh r \backslash$.

5.5 SUMMARY OF EQUIAFFINITIES

A 2×2 linear rotation (or "unimodular") matrix has the general form

$$\begin{pmatrix} a & b \\ c & d \end{pmatrix}, \text{ with } \quad ad - bc = 1. \tag{5.5.1}$$

Such a matrix induces a *unimodular transformation*

$$\cdot[(a,\ b),\ (c,\ d)] : \mathbb{R}^2 \to \mathbb{R}^2,$$

which can be combined with a translation $+(h,\ k)$ to give a "transunimodular" transformation

$$\left[\cdot\begin{pmatrix} a & b \\ c & d \end{pmatrix},\ +(h,\ k)\right] : \mathbb{R}^2 \to \mathbb{R}^2, \qquad ad - bc = 1. \tag{5.5.2}$$

In the affine plane this operates as an equiaffinity, taking the point (x, y) to the point $(ax + cy + h,\ bx + dy + k)$. Every equiaffinity can be expressed as the product of two affine reflections and as the product of at most three shears.

Though we have sorted affine reflections into ortholinear, paralinear, and metalinear families, these are distinctions based on matrix representations; all affine reflections are actually conjugate in the unit affine group $\bar{\mathrm{S}}\mathrm{A}_2$. The only other kind of opposite unit affinity is an affine transflection, the product of an affine reflection and a translation; all affine transflections are likewise conjugate. Equiaffinities, on the other hand, fall into classes with essentially different geometric properties (cf. Artzy 1965, pp. 92–97; Neumann, Stoy & Thompson 1994, pp. 133–135).

The particular kind of equiaffinity represented by (5.5.2) depends mainly on the trace $a + d$ of the matrix. It is an elliptic rotation if $|a + d| < 2$; a translation, a shear, or a parabolic rotation if $a + d = 2$; a half-turn or a half-shear if $a + d = -2$; a hyperbolic rotation if $|a + d| > 2$ ("straight" if $a + d > 2$, "crossed" if $a + d < -2$). With a change of coordinates if need be, any equiaffinity can be put into one of the canonical forms of the preceding sections; the trace of the transformation matrix and the kind of equiaffinity are the same in any case. The set of all unimodular transformations of the vector space $\mathbb{R}^2$ constitutes the special linear group SL_2, and the set of all equiaffinities of A^2 constitutes the special affine group SA_2.

All nontrivial equiaffinities of A^2 are listed in Table 5.5.

Table 5.5 *Plane Equiaffinities*

Equiaffinity	Typical Mapping	Invariant Pencil	Fixed Points
Elliptic rotation	$\mu(\cos\theta,\ \sin\theta) \mapsto$ $\mu(\cos\,(\theta+\rho),\ \sin\,(\theta+\rho))$	Ellipses $x^2+y^2=w^2$	Center $(0, 0)$
Translation	$(x,\ y) \mapsto (x+h,\ y+k)$	Parallel lines $kx-hy=u$	None
Shear	$(x,\ y) \mapsto (x,\ y+fx)$	Parallel lines $x=u$	Axis $x=0$
Parabolic rotation	$(x,\ y) \mapsto (x+h,\ y+2hx+h^2)$	Parabolas $y=x^2+v$	None
Half-turn	$(x,\ y) \mapsto (-x,\ -y)$	Concurrent lines $y=mx$	Center $(0, 0)$
Half-shear	$(x,\ y) \mapsto (-x,\ -y+hx)$	Parataxes $x^2=u^2$	Center $(0, 0)$
Hyperbolic rotation (2 types)	$\mu(\cosh t,\ \sinh t) \mapsto$ $\mu(\cosh\,(t+r),\ \sinh\,(t+r))$, $\nu(\sinh t,\ \cosh t) \mapsto$ $\nu(\sinh\,(t+r),\ \cosh\,(t+r))$ or $\mu(\cosh t,\ \sinh t) \mapsto$ $-\mu(\cosh\,(t+s),\ \sinh\,(t+s))$, $\nu(\sinh t,\ \cosh t) \mapsto$ $-\nu(\sinh\,(t+s),\ \cosh\,(t+s))$	Hyperbolas $x^2-y^2=u^2$ or $y^2-x^2=v^2$	Center $(0, 0)$

If $P_0 = (x_0,\ y_0)$, $P_1 = (x_1,\ y_1)$, and $P_2 = (x_2,\ y_2)$ are any three points of A^2, the directed area $\|P_0P_1P_2\|$ of the ordered triad $(P_0P_1P_2)$ is given by

$$2\,\|P_0P_1P_2\| = \begin{vmatrix} x_1 & y_1 \\ x_2 & y_2 \end{vmatrix} + \begin{vmatrix} x_2 & y_2 \\ x_0 & y_0 \end{vmatrix} + \begin{vmatrix} x_0 & y_0 \\ x_1 & y_1 \end{vmatrix}. \tag{5.5.3}$$

The directed area is zero if the three points are collinear and otherwise is positive or negative according as the sense of the triangle $\langle P_0P_1P_2\rangle$ agrees or disagrees with the sense of the reference triangle $\langle OIJ\rangle$. It is readily seen that an equiaffinity preserves directed areas.

5.6 SYMPLECTIC GEOMETRY

The 2×2 *juxtation matrix*

$$J_2 = \begin{pmatrix} 0 & 1 \\ -1 & 0 \end{pmatrix} \tag{5.6.1}$$

induces a linear rotation $\cdot J_2 : \mathbb{R}^2 \to \mathbb{R}^2$, a quarter-turn with $(x, y) \mapsto (-y, x)$. If n is even, the $n \times n$ skew-symmetric matrix $J = J_n = \backslash J_2, \ldots, J_2 \backslash$, with $\frac{1}{2}n$ diagonal blocks J_2, induces a transformation $\cdot J : \mathbb{R}^n \to \mathbb{R}^n$ of period 4, a *juxtation*, with $(\delta_{2i-1}) \mapsto -(\delta_{2i})$ and $(\delta_{2i}) \mapsto (\delta_{2i-1})$ for $1 \le i \le \frac{1}{2}n$.

A *symplectic matrix* is a square matrix S of even order such that $SJS^{\vee} = J$. The determinant of a symplectic matrix is $+1$. Each $n \times n$ symplectic matrix S induces a *symplectic transformation* $\cdot S : \mathbb{R}^n \to \mathbb{R}^n$. The set of all $n \times n$ symplectic matrices, or symplectic transformations of $\mathbb{R}^n$ (n even), forms a group, the (real) *symplectic* group $\mathrm{Sp}_n \cong \mathrm{Sp}(n; \mathbb{R})$. A 2×2 matrix S is symplectic if and only if $\det S = 1$; thus Sp_2 is the special linear group SL_2. A nonzero scalar multiple of a symplectic matrix is a *gyropetic matrix*, with the set of all $n \times n$ gyropetic matrices forming the (real) *gyropetic* group Gp_n. A 2×2 matrix G is gyropetic if and only if $\det G > 0$; thus Gp_2 is the direct linear group $\mathrm{G^+L}_2$.

A. *Null systems.* In even-dimensional affine space A^n, the origin O and any two points X and Y, with respective coordinates (x) and (y), form an ordered triad (OXY), the *twisted area* $\|OXY\|$ of which is defined by a skew-symmetric bilinear form:

$$2\,\|OXY\| = (x^{\circ}y) = (x)J(y)^{\vee} \tag{5.6.2}$$

(cf. Weyl 1939, p. 165). The twisted area is zero if X and Y are collinear with O, so that (x) and (y) are linearly dependent, but this is not necessary (unless $n = 2$). Any two vectors (x) and (y) with $(x^{\circ}y) \ne 0$ form a *gyral pair* and span a *gyral plane*. If $(x^{\circ}y) = \pm 1$, then (x) and (y) form a *unit pair*. If X, Y, and O are not collinear but $(x^{\circ}y) = 0$, then (x) and (y) form a *null pair* and span a *null plane*. Two gyral pairs are *symplectic* if each vector of one forms a null pair with each vector of the other. The ½n mutually symplectic unit pairs of vectors (δ_{2i-1}) and (δ_{2i}) form a *gyronormal basis* for $\mathbb{R}^n$, with

$$(\delta_{2i-1}{}^\circ\delta_{2j-1}) = (\delta_{2i}{}^\circ\delta_{2j}) = 0 \quad \text{and} \quad (\delta_{2i-1}{}^\circ\delta_{2j}) = \delta_{ij}$$

for all i and j. A symplectic matrix S induces a "symplectic" equiaffinity $\cdot S$, leaving O fixed and taking each other point $X = (x)$ into a point $X' = (x)S$. It is easy to see that $\|OX'Y'\| = \|OXY\|$, so that gyral planes are taken into gyral planes and null planes into null planes.*

These definitions are readily extended. Given any three points X, Y, and Z in A^n (n even), with coordinates (x), (y), and (z), the twisted area $\|XYZ\|$ is defined by

$$2\ \|XYZ\| = ((x\ y\ z)) = (x^\circ y) + (z^\circ x) + (y^\circ z). \tag{5.6.3}$$

Real affine n-space with this areal metric, obtained by fixing a null polarity in the (projective) hyperplane at infinity, is what may be called *areplectic n*-space Ap^n. By combining a symplectic transformation $\cdot S$ with a translation $+(h)$, we obtain an *equiplexity* $[\cdot S, +(h)]$, an equiaffinity that preserves twisted areas. Or a gyropetic transformation $\cdot G$ can be combined with a translation $+(h)$ to give an *areplexity* $[\cdot G, +(h)]$, a direct affinity preserving ratios of twisted areas.

All the areplexities of Ap^n form the *general areplectic* group GAp_n, in which the equiplexities constitute a normal subgroup, the *special areplectic* group SAp_n. Every equiplexity can be expressed as the product of shears (Artin 1957, pp. 138–139). The translation group T^n is a normal subgroup of both GAp_n and SAp_n, with quotient groups $\mathrm{GAp}_n/\mathrm{T}^n \cong \mathrm{Gp}_n$ and $\mathrm{SAp}_n/\mathrm{T}^n \cong \mathrm{Sp}_n$. The groups GAp_2 and SAp_2 are just the direct and special affine groups $\mathrm{G^+A}_2$ and SA_2.

* Artin (1957, pp. 136–142) describes some of the features of symplectic geometry but confusingly uses the term "hyperbolic" for gyral planes (as well as for planes having a pseudo-orthogonal metric, i.e., Lorentzian planes). Artin's term for null planes is "isotropic."

Not every transformation of areplectic n-space that preserves ratios of twisted areas counts as an areplexity, nor is every such transformation a direct affinity. *Antisymplectic* matrices R such that $RJR^{\vee} = -J$ give rise to *antiareplexities*, including *enantioplexities*, which replace twisted areas by their negatives. These affinities, which are direct when n is a multiple of 4 and opposite when not, extend the groups GAp_n and SAp_n to form the *homoplectic* and *isoplectic* groups $\bar{\mathrm{G}}\mathrm{Ap}_n$ and $\bar{\mathrm{S}}\mathrm{Ap}_n$. The quotient groups $\bar{\mathrm{G}}\mathrm{Ap}_n/\mathrm{T}^n \cong \bar{\mathrm{G}}\mathrm{p}_n$ and $\bar{\mathrm{S}}\mathrm{Ap}_n/\mathrm{T}^n \cong \bar{\mathrm{S}}\mathrm{p}_n$ are the *hologyropetic* and *holosymplectic* groups, respectively containing Gp_n and Sp_n as subgroups of index 2. The groups $\bar{\mathrm{G}}\mathrm{Ap}_2$ and $\bar{\mathrm{S}}\mathrm{Ap}_2$ are the general and unit affine groups GA_2 and $\bar{\mathrm{S}}\mathrm{A}_2$.

The geometry of the hyperplane at infinity of Ap^n is *nullary* $(n-1)$-space nP^{n-1}, i.e., real projective $(n-1)$-space with an absolute null polarity. An extensive treatment of geometries of this type is given by Onishchik & Sulanke (2006, § 2.8). The group of collineations that commute with the absolute polarity is the *projective holosymplectic* group $\mathrm{P\bar{S}p}_n \cong \bar{\mathrm{S}}\mathrm{p}_n/\bar{\mathrm{S}}\mathrm{Z}$; the *projective symplectic* group $\mathrm{PSp}_n \cong \mathrm{Sp}_n/\bar{\mathrm{S}}\mathrm{Z}$ is a subgroup of index 2 (cf. Onishchik & Sulanke 2006, ¶ 2.1.3). In nP^3 every plane contains a pencil of self-polar lines (traces of null planes in AP^4), with the center of the pencil being the pole of the plane. The class of self-polar lines in nP^3 is called a *linear complex* (Veblen & Young 1910, pp. 324–325; Coxeter 1942, pp. 69–70; cf. Onishchik & Sulanke 2006, ¶1.8.4); the word "symplectic" (from the Greek *symplektos*, "plaited") was coined by Hermann Weyl in allusion to such complexes.

Areplectic n-space and nullary $(n-1)$-space are related in the same way that Euclidean or Lorentzian n-space is related to elliptic or hyperbolic $(n-1)$-space. In each case, affine n-space is given a metric for distance or area by fixing an absolute polarity—elliptic, hyperbolic, or null—in the projective hyperplane at infinity.

More generally, in place of real affine n-space (n even) equipped with a skew-symmetric bilinear form, one can study differentiable n-manifolds equipped with a symplectic differential form; areas can then be determined by integration. Areplectic n-space as described here is the simplest of such symplectic geometries.

B. *Equivalent forms.* The $n \times n$ block-diagonal matrix $J = \backslash J_2, \ldots, J_2 \backslash$ used in defining a symplectic matrix can be taken as standard, but it is not the only possibility; any invertible skew-symmetric matrix will do. If P is an $n \times n$ orthogonal matrix (i.e., $PP^{\vee} = I$), then the matrix $K = PJP^{\vee}$ is *orthogonally similar* to J. Since J is skew-symmetric, so is K. Suppose that S is a "J-symplectic" matrix, so that $SJS^{\vee} = J$, and let $T = PSP^{\vee}$. Then $TKT^{\vee} = K$, making T a "K-symplectic" matrix. Two cases are of particular importance.

For any real $n \times n$ matrix A with $n = 4k$, let A' be the matrix obtained from A by replacing rows (a_{4i-1}) and (a_{4i-2}) by (a_{4i-2}) and $-(a_{4i-1})$ for $1 \leq i \leq k$ and likewise replacing columns $[a_{4j-1}]$ and $[a_{4j-2}]$ by $[a_{4j-2}]$ and $-[a_{4j-1}]$ for $1 \leq j \leq k$. Then $J' = J'_{4k}$ is a block-diagonal matrix $\backslash J'_4, \ldots, J'_4 \backslash$, where

$$J'_4 = \begin{pmatrix} O & \dot{I} \\ -\dot{I} & O \end{pmatrix} = \begin{pmatrix} 0 & 0 & 1 & 0 \\ 0 & 0 & 0 & -1 \\ -1 & 0 & 0 & 0 \\ 0 & 1 & 0 & 0 \end{pmatrix}. \tag{5.6.4}$$

If S is J-symplectic, then S' is J'-symplectic. For n a multiple of 4, the set of $n \times n$ J'-symplectic matrices forms the *bisymplectic* group Sp'_n, and the set of their nonzero scalar multiples forms the *bigyropetic* group Gp'_n.

Alternatively, given a real $n \times n$ matrix A with $n = 4k$, let A'' be the matrix obtained from A by replacing rows (a_{4i}) and (a_{4i-2}) by (a_{4i-2}) and $-(a_{4i})$ and replacing columns $[a_{4j}]$ and $[a_{4j-2}]$ by $[a_{4j-2}]$ and $-[a_{4j}]$. Then $J'' = J''_{4k}$ is a block-diagonal matrix $\backslash J''_4, \ldots, J''_4 \backslash$, where

$$J_4'' = \begin{pmatrix} O & \dot{J} \\ -\dot{J} & O \end{pmatrix} = \begin{pmatrix} 0 & 0 & 0 & 1 \\ 0 & 0 & 1 & 0 \\ 0 & -1 & 0 & 0 \\ -1 & 0 & 0 & 0 \end{pmatrix}. \tag{5.6.5}$$

If S is J-symplectic, then S'' is J''-symplectic. For n a multiple of 4, the set of $n \times n$ J''-symplectic matrices forms the *cobisymplectic* group Sp_n'', and the set of their nonzero scalar multiples forms the *cobigyropetic* group Gp_n''. As subgroups of SL_n, the bisymplectic group Sp_n' and the cobisymplectic group Sp_n'' are both conjugate to the symplectic group Sp_n, so that all three groups are actually isomorphic, as are the groups Gp_n, Gp_n', and Gp_n''.

Symplectic matrices over the complex field $\mathbb{C}$ can be defined in the same manner as those over the real field; again, the determinant of a symplectic matrix is $+1$. Each $n \times n$ complex symplectic matrix S induces a *symplectic transformation* $\cdot S : \mathbb{C}^n \to \mathbb{C}^n$. The set of all $n \times n$ complex symplectic matrices, or symplectic transformations of $\mathbb{C}^n$ (n even), forms the *complex symplectic* group $\mathrm{Sp}(n;\ \mathbb{C})$, which we abbreviate as ${}^{\mathbb{C}}\mathrm{Sp}_n$ or $\breve{\mathrm{Sp}}_n$. The set of nonzero real scalar multiples of such matrices forms the *complex gyropetic* group ${}^{\mathbb{C}}\mathrm{Gp}_n \cong \breve{\mathrm{Gp}}_n$. The group ${}^{\mathbb{C}}\mathrm{Sp}_{2k}$ is isomorphic to the group $\mathrm{Sp}_{4k}' \cap \mathrm{Sp}_{4k}''$ of real $4k \times 4k$ matrices that are both bisymplectic and cobisymplectic, and likewise ${}^{\mathbb{C}}\mathrm{Gp}_{2k}$ is isomorphic to $\mathrm{Gp}_{4k}' \cap \mathrm{Gp}_{4k}''$.

The term "symplectic" is also used for certain matrices over the division ring $\mathbb{H}$ of quaternions. The set of all $n \times n$ quaternionic symplectic matrices ($n \geq 1$) forms the *quaternionic symplectic* group $\mathrm{Sp}(n;\ \mathbb{H})$, which can be abbreviated as ${}^{\mathbb{H}}\mathrm{Sp}_n$ or $\tilde{\mathrm{Sp}}_n$, with the nonzero real scalar multiples of such matrices forming the *quaternionic gyropetic* group ${}^{\mathbb{H}}\mathrm{Gp}_n \cong \tilde{\mathrm{Gp}}_n$. These groups will be defined and discussed in § 10.4.

EXERCISES 5.6

1. If S is an $n \times n$ symplectic matrix (n even), show that the equiaffinity $\cdot S$ preserves twisted areas $\|OXY\|$.

2. Show that an equiplexity $[\cdot S, +(h)]$ preserves twisted areas $\|XYZ\|$.

3. Show that an areplexity $[\cdot G, +(h)]$ preserves ratios of twisted areas.

4. In areplectic 4-space AP^4, the origin (0) and pairs of distinct base-points (δ_i) and (δ_j) determine six coordinate planes. With the areal metric defined by the bilinear form $(x^\circ y)$, which of these are gyral planes and which are null planes?

5. $\vdash$ A real 4×4 matrix that is both symplectic and bisymplectic is cobisymplectic as well.

6

REAL ISOMETRY GROUPS

EACH REAL METRIC SPACE has a group of distance-preserving transformations, or *isometries*, corresponding to some multiplicative group of matrices. For the elliptic n-sphere S^n, this is the *orthogonal* group O_{n+1}. Extending the orthogonal group O_n by the translation group T^n, we obtain the isometry group of Euclidean n-space E^n—the *Euclidean* group E_n, a normal subgroup of the larger group $\tilde{E}_n$ of *similarities*. The group of isometries of the hyperbolic n-sphere $\ddot{S}^n$ is the *pseudo-orthogonal* group $O_{n,1}$. Identifying antipodal points of S^n or $\ddot{S}^n$ gives elliptic n-space eP^n or hyperbolic n-space H^n, the respective isometry groups of which are the *projective orthogonal* group PO_{n+1} and the *projective pseudo-orthogonal* group $PO_{n,1}$. Every isometry can be expressed as the product of reflections, and every reflection is represented by a typical *reflection matrix*.

6.1 SPHERICAL AND ELLIPTIC ISOMETRIES

The standard *inner product* of two rows (x) and (y) of $\mathbb{R}^n$ or two columns $[u]$ and $[v]$ of $\check{\mathbb{R}}^n$, commonly expressed as the "dot product" $\mathbf{x} \cdot \mathbf{y}$ or $\check{\mathbf{u}} \cdot \check{\mathbf{v}}$ of two vectors or covectors, is given by the symmetric bilinear form

$$(x\ y) = (x)(y)^\vee \quad \text{or} \quad [u\ v] = [u]^\wedge[v], \tag{6.1.1}$$

whose associated quadratic form $(x\ x)$ or $[u\ u]$ is positive definite. A *unit* row (x) or column $[u]$ has $(x\ x) = 1$ or $[u\ u] = 1$. Two rows (x) and

(y) or two columns $[u]$ and $[v]$ are *orthogonal* if $(x\,y) = 0$ or $[u\,v] = 0$. An *orthogonal transformation* of $\mathbb{R}^n$ or $\check{\mathbb{R}}^n$ is a linear transformation that takes unit rows into unit rows or unit columns into unit columns and consequently takes orthogonal rows or columns into orthogonal rows or columns. A necessary and sufficient condition for a real $n \times n$ matrix P to determine an orthogonal transformation $\cdot P : \mathbb{R}^n \to \mathbb{R}^n$ or $P\cdot : \check{\mathbb{R}}^n \to \check{\mathbb{R}}^n$ is that $PP^{\vee} = I$, in which case P is called an *orthogonal matrix*. The determinant of an orthogonal matrix is ± 1.

The set of all orthogonal transformations of $\mathbb{R}^n$ or $\check{\mathbb{R}}^n$ forms a group, the *orthogonal* group $\mathrm{O}_n \cong \mathrm{O}(n;\ \mathbb{R})$, with each group operation being represented by an $n \times n$ orthogonal matrix. Those orthogonal transformations for which the corresponding matrix has determinant 1 form a subgroup of index 2, the *special orthogonal* group $\mathrm{O}_n^+ \cong \mathrm{SO}(n;\ \mathbb{R})$. The center of the orthogonal group is the unit scalar group $\bar{\mathrm{S}}\mathrm{Z}$, of order 2, represented by the two matrices $\pm I$.

A. *Spherical isometries.* The points of the elliptic n-sphere S^n can be coordinatized by the unit rows $(x) = (x_1,\ x_2, \ldots,\ x_{n+1})$ of $\mathbb{R}^{n+1}$, and homogeneous coordinates for great hyperspheres are given by the nonzero columns $[u] = [u_1,\ u_2, \ldots,\ u_{n+1}]$ of $\check{\mathbb{R}}^{n+1}$. If P is an $(n+1) \times (n+1)$ orthogonal matrix, then the orthogonal transformation $\cdot P : \mathbb{R}^{n+1} \to \mathbb{R}^{n+1}$ defines an *isometry* of S^n, taking a point X with coordinates (x) into a point X' with coordinates $(x)P$ and preserving spherical distances (3.1.3) between points. The same isometry takes a great hypersphere $\check{U}$ with coordinates $[u]$ into a great hypersphere $\check{U}'$ with coordinates $P^{\vee}[u]$. A spherical isometry is direct or opposite according as the determinant of its matrix is $+1$ or -1.

The classification of isometries can be facilitated by considering the eigenvalues and eigenvectors of the corresponding matrices. A (real or complex) number λ is an *eigenvalue* of a square matrix A if there is a nonzero row (x) or column $[u]$, an *eigenvector* of A, such that $(x)A = \lambda(x)$ or $A[u] = [u]\lambda$. Each eigenvalue λ is a root of the *characteristic equation* $\det(A - \lambda I) = 0$. The sum of the eigenvalues is equal to the trace of A, and their product is the determinant of A. Although the eigenvectors associated with an isometry or other transformation depend on its matrix representation, the eigenvalues are invariants of the transformation itself.

The spherical reflection whose mirror is a given great hypersphere $\breve{P}$ with coordinates $[p] = [p_1, p_2, \ldots, p_{n+1}]$ is represented by an *orthogonal reflection matrix*, a symmetric orthogonal matrix P of order $n + 1$ with entries

$$p_{ij} = \delta_{ij} - \frac{2p_i p_j}{[p\,p]}. \qquad (6.1.2)$$

It will be seen that (x) is a row eigenvector of the matrix P with eigenvalue 1 if $(x)[p] = 0$, that $[u]$ is a column eigenvector with eigenvalue 1 if $[p\,u] = 0$, that $[p]$ is a column eigenvector with eigenvalue -1, and that $(p) = [p]^{\vee}$ is a row eigenvector with eigenvalue -1. A reflection is an opposite isometry.

The rotation that is the product of reflections in given great hyperspheres $\breve{P}$ and $\breve{Q}$ with coordinates $[p]$ and $[q]$ is represented by an *orthogonal rotation matrix*, which is an orthogonal matrix R with entries

$$r_{ij} = (p_i)[q_j], \qquad (6.1.3)$$

where (p_i) and $[q_j]$ are the ith row and the jth column of the corresponding reflection matrices P and Q; i.e., $R = PQ$. It will be found that (x) is a row eigenvector of the matrix R with eigenvalue 1 if $(x)[p] = (x)[q] = 0$ and that R also has eigenvalues $\cos\theta \pm \mathrm{i}\sin\theta$, where θ is the angle of the rotation. When the mirrors are orthogonal, the rotation is a half-turn ($\theta = \pi$), and -1 is an eigenvalue of multiplicity 2. A rotation is a direct isometry.

For example, on the sphere S^2, reflections $\cdot P_1$, $\cdot P_2$, and $\cdot P_3$ in the great circles $\breve{P}_1 = [0, 1, 0]$, $\breve{P}_2 = [1, -1, 0]$, and $\breve{P}_3 = [0, 1, 1]$ are represented by the matrices

$$P_1 = \begin{pmatrix} 1 & 0 & 0 \\ 0 & -1 & 0 \\ 0 & 0 & 1 \end{pmatrix}, \quad P_2 = \begin{pmatrix} 0 & 1 & 0 \\ 1 & 0 & 0 \\ 0 & 0 & 1 \end{pmatrix}, \quad P_3 = \begin{pmatrix} -1 & 0 & 0 \\ 0 & 0 & -1 \\ 0 & -1 & 0 \end{pmatrix}.$$

As products of pairs of these reflections, we have the rotations $\cdot R_{12} = \cdot P_1 P_2$, $\cdot R_{13} = \cdot P_1 P_3$, and $\cdot R_{23} = \cdot P_2 P_3$, represented by the matrices

$$R_{12} = \begin{pmatrix} 0 & 1 & 0 \\ -1 & 0 & 0 \\ 0 & 0 & 1 \end{pmatrix}, \quad R_{13} = \begin{pmatrix} 1 & 0 & 0 \\ 0 & 0 & 1 \\ 0 & -1 & 0 \end{pmatrix}, \quad R_{23} = \begin{pmatrix} 0 & 0 & -1 \\ 1 & 0 & 0 \\ 0 & -1 & 0 \end{pmatrix}.$$

Since $(\check{P}_1\check{P}_2) = (\check{P}_1\check{P}_3) = \pi/4$ and $(\check{P}_2\check{P}_3) = \pi/3$, these rotations are of periods 4, 4, and 3, corresponding to rotation angles of $\pi/2$, $\pi/2$, and $2\pi/3$.

The *central inversion* (or "negation"), which interchanges antipodal points and takes each great hypersphere into itself, is represented by the *negation matrix* $-I$; the central inversion commutes with every isometry. It is a direct isometry of S^n if n is odd, an opposite isometry if n is even.

The group of all isometries of S^n is the orthogonal group O_{n+1}, and the subgroup of direct isometries is the special orthogonal group O^+_{n+1}, which is abelian when $n < 2$. The group Ω_{n+1} generated by squares of rotations is the commutator subgroup of O_{n+1}; as special cases, $\Omega_1 \cong 1$, $\Omega_2 \cong \mathrm{O}^+_2$, and $\Omega_3 \cong \mathrm{O}^+_3$ (Snapper & Troyer 1971, pp. 337–340). The group of order 2 generated by the central inversion is the unit scalar group $\bar{\mathrm{S}}\mathrm{Z}$.

B. *Elliptic isometries.* Real projective n-space P^n with an orthogonal metric—i.e., the elliptic n-sphere S^n with antipodal points identified—is elliptic n-space eP^n. As in P^n, homogeneous coordinates

$$(x) = (x_1, x_2, \ldots, x_{n+1}) \quad \text{and} \quad [u] = [u_1, u_2, \ldots, u_{n+1}]$$

can be assigned to points X and hyperplanes $\check{U}$ of eP^n, but now any two reference points E_i and E_j $(i \neq j)$ are separated by a distance of $\pi/2$ and any two reference hyperplanes $\check{E}_i$ and $\check{E}_j$ ($i \neq j$) are orthogonal.

A projective collineation that preserves elliptic distances (2.3.1) is an *isometry* of eP^n. If P is an $(n+1) \times (n+1)$ orthogonal matrix, then the *projective orthogonal transformation*

$$\langle \cdot P \rangle : \mathrm{P}\mathbb{R}^{n+1} \to \mathrm{P}\mathbb{R}^{n+1}, \text{ with } \langle (x) \rangle \mapsto \langle (x)P \rangle, \quad PP^{\check{}} = I, \tag{6.1.4}$$

determines an isometry of eP^n, taking a point X with coordinates (x) into a point X' with coordinates $(x)P$ and preserving distances between points. The same isometry takes a hyperplane $\check{U}$ with coordinates $[u]$ into a hyperplane $\check{U}'$ with coordinates $P^{\check{}}[u]$. Since all coordinates $\lambda(x)$ represent the same point, the matrices $\pm P$ determine the same isometry. Since $\det(-P) = (-1)^{n+1} \det P$, the two determinants have the same sign when n is odd, opposite signs when n is even. An isometry of eP^n (n odd) is direct or opposite according as the determinant of either of its matrices is $+1$ or -1.

The reflection whose mirror is a given hyperplane $\check{P}$ with coordinates $[p]$ is represented by an orthogonal reflection matrix (6.1.2). The rotation that is the product of reflections in given hyperplanes with coordinates $[p]$ and $[q]$ is represented by an orthogonal rotation matrix (6.1.3). When n is odd, a reflection is an opposite isometry and a rotation is a direct isometry.

The group of all isometries of eP^n is the group O_{n+1} of isometries of S^n with the group generated by the central inversion factored out; this is the *projective orthogonal* group $PO_{n+1} \cong O_{n+1}/\bar{S}Z$. For odd n, the subgroup of direct isometries is the *special projective orthogonal* group $P^+O_{n+1} \cong PO^+_{n+1} \cong O^+_{n+1}/\bar{S}Z$.

EXERCISES 6.1

1. ⊢ Two row or column eigenvectors corresponding to distinct eigenvalues of a square matrix are linearly independent.

2. $\vdash$ If (x) is a row eigenvector and $[u]$ a column eigenvector corresponding to distinct eigenvalues of a square matrix A, then $(x)[u] = 0$.

3. Use Formula (6.1.2) to obtain the matrices given in this section for reflections in the great circles $[0, 1, 0]$, $[1, -1, 0]$, and $[0, 1, 1]$.

4. Show directly that the rotations $\cdot R_{12}$, $\cdot R_{13}$, and $\cdot R_{23}$ given in this section have the respective periods 4, 4, and 3.

5. Show that in the elliptic plane eP^2 a reflection $\langle \cdot P \rangle$ can also be expressed as a half-turn.

6.2 EUCLIDEAN TRANSFORMATIONS

In Euclidean n-space E^n let an arbitrary point O be selected, and let n points E_i $(1 \le i \le n)$ on the unit hypersphere with center O be chosen so that the n lines $\cdot OE_i \cdot$ are mutually orthogonal. Then O is the origin and the points E_i are the basepoints of a rectangular (Cartesian) coordinate system on E^n whereby each point X is associated with a unique row vector $(x) \in \mathbb{R}^n$, with the points O and E_i having coordinate vectors (0) and (δ_i). The n mutually orthogonal unit vectors (δ_i) form an *orthonormal basis* for $\mathbb{R}^n$, with $(\delta_i\ \delta_j) = \delta_{ij}$ for all i and j. Likewise, a hyperplane $\check{U}(r)$ is represented by a nonzero scalar multiple of the column vector $[[u], r] \in \check{\mathbb{R}}^{n+1}$, where the components of the nonzero covector $[u] \in \check{\mathbb{R}}^n$ are direction numbers of $\check{U}(r)$ and the point on $\check{U}(r)$ nearest the origin has coordinate vector $\rho(u)$, where $\rho = -r/[u\ u]$.

A. *Euclidean isometries.* Let P be an $n \times n$ orthogonal matrix and let $(h) \in \mathbb{R}^n$ be any row vector. Then the product of the orthogonal transformation

$$\cdot P : \mathbb{R}^n \to \mathbb{R}^n, \text{ with } (x) \mapsto (x)P,$$

and the translation

$$+(h) : \mathbb{R}^n \to \mathbb{R}^n, \text{ with } (y) \mapsto (y) + (h),$$

is a *transorthogonal transformation*

$$[\cdot P,\ +(h)] : \mathbb{R}^n \to \mathbb{R}^n, \text{ with } (x) \mapsto (x)P + (h), \quad PP^\vee = I, \quad (6.2.1)$$

which induces a corresponding operation on E^n—an *isometry*—preserving the Euclidean distance (2.3.4) between two points and taking any figure to a *congruent* figure. Such a transformation can be represented by an $(n+1) \times (n+1)$ *transorthogonal matrix*

$$[P,\ (h)]_1 = \begin{pmatrix} P & [0] \\ (h) & 1 \end{pmatrix}, \quad PP^\vee = I, \quad (6.2.2)$$

by which the row vector $((x),\ 1) \in \mathbb{R}^{n+1}$ may be postmultiplied to produce the row vector $((x)P + (h),\ 1)$. A transorthogonal matrix (6.2.2) is a special case of a translinear matrix (4.2.2).

The set of all transorthogonal matrices $[P,\ (h)]_1$ forms a multiplicative group TO_n isomorphic to the (full) *Euclidean* group E_n of all isometries of E^n, with the translation matrices $[I,\ (h)]_1$ constituting a normal subgroup T_n. The quotient group $TO_n/T_n \cong E_n/T^n$ is isomorphic to the orthogonal group O_n of $n \times n$ orthogonal matrices P. The Euclidean group E_n is a subgroup (but not a normal subgroup) of the unit affine group $\bar{S}A_n$.

The isometry $[\cdot P,\ +(h)]$, which takes a point X with coordinates (x) into a point X' with coordinates $(x') = (x)P + (h)$, takes a hyperplane $\check{U}(r)$ with coordinates $[[u],\ r]$ into a hyperplane $\check{U}'(r')$ with coordinates $[[u'],\ r'] = [P^\vee[u],\ r - (h)P^\vee[u]]$. This can be effected by premultiplying the column vector $[[u],\ r] \in \check{\mathbb{R}}^{n+1}$ by the inverse transorthogonal matrix

$$[P,\ (h)]_1^{-1} = \begin{pmatrix} P^\vee & [0] \\ -(h)P^\vee & 1 \end{pmatrix}, \quad PP^\vee = I. \quad (6.2.3)$$

The reflection whose mirror is a given hyperplane $\check{P}(h)$ with coordinates $[[p],\ h]$ is represented by a *transorthogonal reflection matrix*, this being a transorthogonal matrix $[P,\ (h)]_1$, where $P \in \mathrm{O}_n$ and $(h) \in \mathbb{R}^n$, with entries

$$p_{ij} = \delta_{ij} - \frac{2p_ip_j}{[p\ p]} \quad \text{and} \quad h_j = \frac{2hp_j}{[p\ p]}. \tag{6.2.4}$$

Thus P is an orthogonal reflection matrix, and the reflection $\cdot P$ reverses the displacement vector (h); i.e., $(h)P = -(h)$. It may also be seen that the matrix $[P,\ (h)]_1$ is its own inverse, that $((x),\ 1)$ is a row eigenvector of this matrix with eigenvalue 1 if $(x)[p]+h = 0$, that $[[u],r]$ is a column eigenvector with eigenvalue 1 if $[p\ u] = 0$, and that $[[p],h]$ is a column eigenvector with eigenvalue -1.

Every isometry can be expressed as the product of reflections. Since a reflection reverses sense, an isometry is direct or opposite—i.e., sense preserving or sense reversing—according as it is the product of an even or an odd number of reflections. Since a reflection matrix has determinant -1, a transorthogonal matrix has determinant $+1$ or -1 according as the corresponding isometry is direct or opposite. The set of *direct* transorthogonal matrices, with determinant 1, forms a group TO_n^+, a subgroup of index 2 in TO_n, isomorphic to the *direct Euclidean* group E_n^+ of direct isometries of E^n.

In particular, the product of two reflections is a direct isometry of E^n. Consider the reflections represented by matrices $[P,\ (f)]_1$ and $[Q,\ (g)]_1$. If $P = Q$ and $(f) = (g)$, then the mirrors coincide, and the isometry is the identity. If $P = Q$ but $(f) \neq (g)$, then the mirrors are parallel, and the isometry is a translation orthogonal to the mirrors through twice the distance between them. Otherwise, the two mirrors are intersecting, and the isometry is a rotation about their common hyperline through twice the angle between them.

The rotation that is the product of reflections in given intersecting hyperplanes $\check{P}(f)$ and $\check{Q}(g)$ with coordinates $[[p],\ f]$ and $[[q],g]$ is represented by a *transorthogonal rotation matrix*, which is a transorthogonal matrix $[R,\ (h)]_1 = [P,\ (f)]_1 \cdot [Q,\ (g)]_1$ with

$$r_{ij} = (p_i)[q_j] \quad \text{and} \quad h_j = (f)[q_j] + g_j, \tag{6.2.5}$$

where (p_i) and $[q_j]$ are the ith row and the jth column of the corresponding orthogonal reflection matrices P and Q and where (f) and (g) are the corresponding displacement vectors. Thus $R = PQ$ is an orthogonal rotation matrix. We find that $((x), 1)$ is a row eigenvector of the matrix $[R, (h)]_1$ with eigenvalue 1 if $(x)[p] + f = (x)[q] + g = 0$ and that the matrix also has eigenvalues $\cos\theta \pm \mathrm{i}\sin\theta$, where θ is the angle of the rotation. When the hyperplanes are orthogonal, the rotation is a half-turn ($\theta = \pi$), and -1 is an eigenvalue of multiplicity 2.

For example, in the Euclidean plane E^2, reflections $\cdot U$, $\cdot V$, and $\cdot W$ in the x-axis $\check{U} = [0,\ 1,\ 0]$ ($y = 0$), the y-axis $\check{V} = [1,\ 0,\ 0]$ ($x = 0$), and the line $\check{W} = [\sqrt{3},\ -1,\ 2]$ ($\sqrt{3}x - y + 2 = 0$), are represented by the matrices

$$U = \begin{pmatrix} 1 & 0 & 0 \\ 0 & -1 & 0 \\ 0 & 0 & 1 \end{pmatrix}, \quad V = \begin{pmatrix} -1 & 0 & 0 \\ 0 & 1 & 0 \\ 0 & 0 & 1 \end{pmatrix}, \quad W = \begin{pmatrix} -\frac{1}{2} & \frac{1}{2}\sqrt{3} & 0 \\ \frac{1}{2}\sqrt{3} & \frac{1}{2} & 0 \\ -\sqrt{3} & 1 & 1 \end{pmatrix}.$$

In E^2 the product of reflections in two perpendicular lines is a half-turn, and the product of a reflection in a horizontal or vertical line and a reflection in a line of slope m is a rotation through an angle of $2\tan^{-1}|m|$ or $2\cot^{-1}|m|$. For the x-axis $\check{U}$, the y-axis $\check{V}$, and the line $\check{W}$ (slope $\sqrt{3}$), the corresponding matrices are

$$UV = \begin{pmatrix} -1 & 0 & 0 \\ 0 & -1 & 0 \\ 0 & 0 & 1 \end{pmatrix}, \quad UW = \begin{pmatrix} -\frac{1}{2} & \frac{1}{2}\sqrt{3} & 0 \\ -\frac{1}{2}\sqrt{3} & -\frac{1}{2} & 0 \\ -\sqrt{3} & 1 & 1 \end{pmatrix},$$

$$VW = \begin{pmatrix} \frac{1}{2} & -\frac{1}{2}\sqrt{3} & 0 \\ \frac{1}{2}\sqrt{3} & \frac{1}{2} & 0 \\ -\sqrt{3} & 1 & 1 \end{pmatrix}.$$

The respective rotations, through angles of π, $2\pi/3$, and $\pi/3$, are of periods 2, 3, and 6.

The translation that is the product of reflections in given parallel hyperplanes $\check{P}(f)$ and $\check{P}(g)$ with coordinates $[[p], f]$ and $[[p], g]$ is represented by a *translation matrix*, which is a transorthogonal matrix $[I,\ (h)]_1$, with $(h) = (f)P + (g)$, where P is the corresponding orthogonal reflection matrix and where (f) and (g) are the respective displacement vectors. The matrix $[I,\ (h)]_1$ has 1 as its only eigenvalue, with $((x), 0)$ being a row eigenvector for any nonzero (x) and $[[u], r]$ being a column eigenvector if $(h)[u] = 0$.

The product of reflections in n mutually orthogonal hyperplanes of E^n is an inversion in their common point; such an isometry is direct or opposite according as n is even or odd. The inversion in the point $\frac{1}{2}(h)$ is represented by a *transversion matrix*, a transorthogonal matrix $[-I,\ (h)]_1$. The set of all translation and transversion matrices forms a group $\mathrm{T\bar{S}Z}_n$, isomorphic to the unit dilative group $\mathrm{\bar{S}D}_n$ of translations and inversions and containing the translation group T^n as a subgroup of index 2.

B. *Similarities.* Given a nonzero scalar $\lambda \in \mathbb{R}$ and an $n \times n$ orthogonal matrix P, the product (in either order) of the scaling operation $\lambda\cdot : \mathbb{R}^n \to \mathbb{R}^n$ and the orthogonal transformation $\cdot P : \mathbb{R}^n \to \mathbb{R}^n$ is an *orthopetic transformation*

$$\lambda{\cdot}P : \mathbb{R}^n \to \mathbb{R}^n, \text{ with } (x) \mapsto \lambda(x)P, \quad \lambda \neq 0,\ PP^{\check{}} = I,$$

taking orthogonal rows into orthogonal rows. This is the same as the linear transformation $\cdot\lambda P : \mathbb{R}^n \to \mathbb{R}^n$ associated with the *orthopetic matrix* λP. The product of this transformation and the translation

$$+(h) : \mathbb{R}^n \to \mathbb{R}^n, \text{ with } (y) \mapsto (y) + (h),$$

is a *transorthopetic transformation*

$$[\lambda{\cdot}P,\ +(h)] : \mathbb{R}^n \to \mathbb{R}^n, \text{ with } (x) \mapsto \lambda(x)P + (h), \quad \lambda \neq 0,\ PP^{\check{}} = I, \tag{6.2.6}$$

which induces a corresponding operation on E^n—a *similarity*—preserving ratios of distances and taking any figure to a *similar* figure.

Such a transformation can be represented by an $(n+1) \times (n+1)$ *transorthopetic matrix*

$$[\lambda P,\ (h)]_1 = \begin{pmatrix} \lambda P & [0] \\ (h) & 1 \end{pmatrix}, \qquad \lambda \neq 0,\ PP^{\vee} = I, \tag{6.2.7}$$

by which one may postmultiply the row vector $((x),\ 1) \in \mathbb{R}^{n+1}$ to obtain the row vector $(\lambda(x)P + (h),\ 1)$. If figure X is similar to figure X', we write $\mathsf{X} \sim \mathsf{X}'$. If $|\lambda| = 1$, the similarity is an isometry ($\mathsf{X} \simeq \mathsf{X}'$); if $P = I$, it is a dilatation ($\mathsf{X} \approx \mathsf{X}'$); if $\lambda P = \pm I$, it is a unit dilatation ($\mathsf{X} \approxeq \mathsf{X}'$). The similarity $\lambda \cdot P = [\lambda \cdot P,\ +(0)]$, which leaves the origin invariant, is transformed by the translation $+(h)$ into the conjugate similarity

$$[-(h),\ \lambda \cdot P,\ +(h)] = [\lambda \cdot P,\ -\lambda(h)P + (h)],$$

which leaves the point (h) invariant.

The set of all transorthopetic matrices of order $n+1$ forms a multiplicative group $\mathrm{T\tilde{O}}_n$ isomorphic to the *euclopetic* (Euclidean similarity) group $\tilde{\mathrm{E}}_n$. The transorthopetic matrices with positive determinants form a subgroup $\mathrm{T\tilde{O}}_n^+$ of index 2 isomorphic to the *direct euclopetic* group $\tilde{\mathrm{E}}_n^+$. The groups $\mathrm{TO}_n \cong \mathrm{E}_n$ and $\mathrm{TO}_n^+ \cong \mathrm{E}_n^+$ of transorthogonal and direct transorthogonal matrices are of course normal subgroups of $\mathrm{T\tilde{O}}_n \cong \tilde{\mathrm{E}}_n$ and $\mathrm{T\tilde{O}}_n^+ \cong \tilde{\mathrm{E}}_n^+$. The full euclopetic group $\tilde{\mathrm{E}}_n$ is a (nonnormal) subgroup of the general affine group GA_n.

Likewise, the transscalar matrices form a normal subgroup TGZ_n isomorphic to the general dilative group GD_n, and the transscalar matrices with positive scales form a subgroup TGZ_n^+ isomorphic to the positive dilative group GD_n^+. The unit transscalar matrices form a normal subgroup $\mathrm{T\bar{S}Z}_n$, isomorphic to the unit dilative group $\bar{\mathrm{S}}\mathrm{D}_n$, and the translation matrices form an abelian subgroup T_n. If n is even, all dilatations are direct similarities; if n is odd, positive dilatations are direct and negative dilatations are opposite.

Dilatations, which preserve or reverse directions, are affine transformations. The general similarity preserves angular measure, which is not an affine concept but instead depends on the definition of perpendicularity. An isometry also preserves distances. Some writers contend that "similarity geometry" is to be distinguished from Euclidean metric geometry, on the grounds that isometries require that some unit of length be chosen (e.g., see the introductory essay in Yaglom 1956, part III). This is not so: one does not need a yardstick to tell whether two children are the same height.

The quotient groups $\mathrm{T\tilde{O}}_n/\mathrm{T}_n \cong \tilde{\mathrm{E}}_n/\mathrm{T}^n$ and $\mathrm{T\tilde{O}}_n^+/\mathrm{T}_n \cong \tilde{\mathrm{E}}_n^+/\mathrm{T}^n$ are isomorphic to the *orthopetic* group $\tilde{\mathrm{O}}_n$ of nonzero scalar multiples of orthogonal matrices and the *direct orthopetic* group $\tilde{\mathrm{O}}_n^+$ of such matrices with positive determinants, and $\mathrm{TGZ}_n/\mathrm{T}_n \cong \mathrm{GD}_n/\mathrm{T}^n$ is isomorphic to the general scalar group GZ of nonzero scalar matrices λI. The quotient group $\mathrm{T\tilde{O}}_n/\mathrm{TGZ}_n \cong \tilde{\mathrm{E}}_n/\mathrm{GD}_n$ is the *projective orthogonal* group $\mathrm{PO}_n \cong \tilde{\mathrm{O}}_n/\mathrm{GZ} \cong \mathrm{O}_n/\bar{\mathrm{S}}\mathrm{Z}$, where $\bar{\mathrm{S}}\mathrm{Z}$ is the group of unit scalar matrices $\pm I$.

There are several short exact sequences involving the foregoing groups. When n is odd, we have the following commutative diagram.

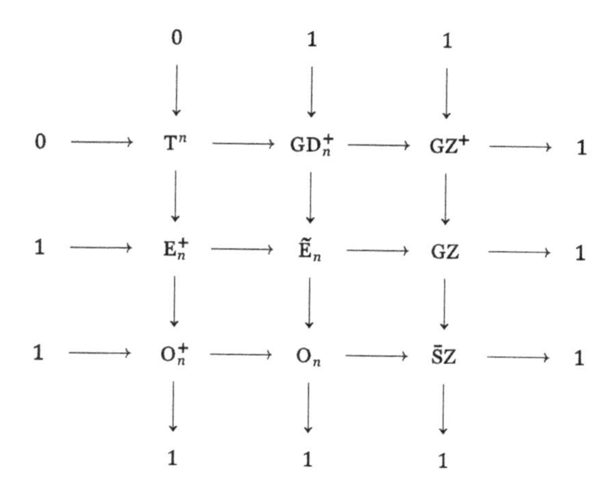

C. *Markovian coordinates.* It is sometimes convenient to consider the points of Euclidean n-space as lying in a hyperplane $\check{\epsilon}$ ("co-epsilon") of E^{n+1}. Each hyperplane of E^{n+1} that is not parallel to $\check{\epsilon}$ intersects it in a unique hyperline, i.e., in a hyperplane of E^n. Specifically, denoting by (ϵ) or $[\epsilon]$ the row or column whose entries are $n+1$ ones, let $\check{\epsilon}$ be the hyperplane $[[\epsilon], 0]$ of E^{n+1} with equation

$$u_0 + u_1 + \cdots + u_n = 0.$$

Then the points U of E^n have unique coordinate vectors

$$(u) = (u_0, u_1, \ldots, u_n) \in \mathbb{R}^{n+1}, \quad (u)[\epsilon] = 0, \tag{6.2.8}$$

and the hyperplanes $\check{P}(h)$ of E^n have homogeneous coordinate vectors

$$[[p], h] = [p_0, p_1, \ldots, p_n, h] \in \check{\mathbb{R}}^{n+2}, \quad (\epsilon)[p] = 0. \tag{6.2.9}$$

The $(n+2) \times (n+2)$ matrix $[P, (h)]_1$ of a reflection is now not only transorthogonal but also *Markovian*, i.e., invertible and having the property that the sum of the entries in any row is equal to 1. (In probability theory, such matrices, with nonnegative entries, are transition matrices for Markov chains.) Since any isometry can be expressed as the product of reflections, and since the set of matrices with the row-sum property is closed under multiplication, the matrix of every isometry is Markovian.

EXERCISES 6.2

1. Use Formula (6.2.4) to obtain the matrices given in this section for reflections in the lines $[0, 1, 0]$, $[1, 0, 0]$, and $[\sqrt{3}, -1, 2]$.
2. Show directly that the rotations whose matrices are given in this section have the respective periods 2, 3, and 6.
3. $\vdash$ When n is even, a dilatation $[\lambda \cdot I, +(h)]$ of E^n is always a direct similarity; when n is odd, it is direct or opposite according as λ is positive or negative.
4. $\vdash$ Every similarity is the product of an isometry and a dilatation.
5. $\vdash$ The product of two Markovian matrices is Markovian.

6.3 HYPERBOLIC ISOMETRIES

The $(n+1) \times (n+1)$ pseudo-identity matrix $\dot{I} = I_{n,1} = \backslash 1, \ldots, 1, -1 \backslash$ can be used to define a *pseudo-inner product*

$$(x \ddot{} y) = (x)\dot{I}(y)^{\check{}} \quad \text{or} \quad [u \ddot{} v] = [u]^{\check{}}\dot{I}[v] \tag{6.3.1}$$

of rows (x) and (y) of $\mathbb{R}^{n+1}$ or columns $[u]$ and $[v]$ of $\check{\mathbb{R}}^{n+1}$. Two rows (x) and (y) or two columns $[u]$ and $[v]$ are *pseudo-orthogonal* if $(x \ddot{} y) = 0$ or $[u \ddot{} v] = 0$. A *pseudo-orthogonal transformation* of $\mathbb{R}^{n+1}$ or $\check{\mathbb{R}}^{n+1}$ is a linear transformation that preserves the relevant bilinear form and therefore takes pseudo-orthogonal rows or columns into pseudo-orthogonal rows or columns (cf. Alekseevsky, Vinberg & Solodovnikov 1988, chap. 4, § 1; Onishchik & Sulanke 2006, ¶2.1.4). A necessary and sufficient condition for a real $(n+1) \times (n+1)$ matrix P to determine a pseudo-orthogonal transformation $\cdot P : \mathbb{R}^{n+1} \to \mathbb{R}^{n+1}$ is that $P\ddot{P}^{\check{}} = I$, where $\ddot{P}^{\check{}} = \dot{I}P^{\check{}}\dot{I}$ is the *pseudotranspose* of P, and P is then called a *pseudo-orthogonal matrix*. The determinant of a pseudo-orthogonal matrix is ± 1.*

The set of all pseudo-orthogonal transformations of $\mathbb{R}^{n+1}$ or $\check{\mathbb{R}}^{n+1}$ forms a group, the *pseudo-orthogonal* group $\mathrm{O}_{n,1} \cong \mathrm{O}(n,\ 1;\ \mathbb{R})$, with each group operation being represented by an $(n+1) \times (n+1)$ pseudo-orthogonal matrix. The pseudo-orthogonal transformations for which the matrix has determinant 1 form a subgroup of index 2, the *special pseudo-orthogonal* group $\mathrm{O}^{+}_{n,1} \cong \mathrm{SO}(n,\ 1;\ \mathbb{R})$.

Also of significance in a pseudo-orthogonal matrix P are the signs of the entry $p_{n+1,n+1}$ and its minor det $P_{(n+1,n+1)}$. The pseudo-orthogonal group $\mathrm{O}_{n,1}$ has two other subgroups of index 2: the group $\mathrm{O}_{n^+,1}$ of operations represented by matrices P with det $P_{(n+1,n+1)} > 0$

* In the literature, the term "pseudo-orthogonal" is used more generally for bilinear forms, transformations, and matrices corresponding to diagonal matrices involving any mixture of ± 1's. Those of the type considered here, with just one -1 (or just one $+1$) are also called "Lorentzian" (cf. Ratcliffe 1994/2006, § 3.1) after H. A. Lorentz (1853–1928) and his celebrated formulas for transforming space-time coordinates.

and the group $O_{n,1^+}$ of operations whose matrices have $p_{n+1,n+1} > 0$. When $n = 1$, these two subgroups are isomorphic. The three subgroups of index 2 have a common subgroup $O_{n^+,1^+}$, of index 4 in $O_{n,1}$. (The group $O_{3^+,1^+} \cong \Omega_{3,1}$, the commutator subgroup of $O_{3,1}$, is the "Lorentz group" of relativity theory.) The center of the pseudo-orthogonal group is the unit scalar group $\bar{S}Z$, of order 2, represented by the two matrices $\pm I$.

Points of the hyperbolic n-sphere $\ddot{S}^n$ can be coordinatized by rows (x) with $(x\ddot{\ }x) = -1$. A pseudo-orthogonal transformation $(x) \mapsto (x)P$, preserving the distance (3.2.1) between points in the same n-hemisphere, is then an *isometry* of $\ddot{S}^n$. The pseudo-orthogonal group $O_{n,1}$ is the group of all isometries of $\ddot{S}^n$, and the special pseudo-orthogonal group $O^+_{n,1}$ is the subgroup of direct isometries. The subgroup of "positive" isometries that do not interchange the two n-hemispheres of $\ddot{S}^n$ is $O_{n,1^+}$, and the subgroup of direct isometries of this kind is $O_{n^+,1^+}$. Since either n-hemisphere is an isometric model for hyperbolic n-space, we see that $O_{n,1^+}$ and $O_{n^+,1^+}$ are, respectively, the "hyperbolic group" of isometries of H^n and the subgroup of direct isometries.

Real projective n-space P^n with a pseudo-orthogonal metric is complete hyperbolic n-space hP^n. Homogeneous coordinates

$$(x) = (x_1, \ldots, x_n | x_{n+1}) \quad \text{and} \quad [u] = [u_1, \ldots, u_n | u_{n+1}]$$

can be assigned to points X and hyperplanes $\check{U}$, with $(x\ddot{\ }x) < 0$ and $[u\ddot{\ }u] > 0$ for points and hyperplanes in the ordinary region H^n.

An *isometry* of H^n is a projective collineation that preserves hyperbolic distances (2.4.1). If P is a pseudo-orthogonal matrix of order $n + 1$, then the *projective pseudo-orthogonal transformation*

$$\langle \cdot P \rangle : P\mathbb{R}^{n+1} \to P\mathbb{R}^{n+1}, \text{ with } \langle (x) \rangle \mapsto \langle (x)P \rangle, \quad P\ddot{P}^{\vee} = I, \tag{6.3.2}$$

determines an isometry of H^n, taking a point X with coordinates (x) into a point X' with coordinates $(x)P$. The same isometry takes a hyperplane $\check{U}$ with coordinates $[u]$ into a hyperplane $\check{U}'$ with coordinates $P\check{}[u]$. Since point coordinates are homogeneous, the matrices $\pm P$ determine the same isometry. Without loss of generality, we may assume that $p_{n+1,n+1} > 0$; an isometry of H^n is then direct or opposite according as $\det P$ is $+1$ or -1.

The hyperbolic reflection whose mirror is a given hyperplane $\check{P}$ with coordinates $[p] = [p_1, \ldots, p_n | p_{n+1}]$ is represented by a *pseudo-orthogonal reflection matrix*. This is a pseudo-orthogonal matrix P of order $n+1$ with entries

$$p_{ij} = \delta_{ij} - \frac{2p_i p_j}{[p\ddot{}p]} \quad (1 \le j \le n), \quad p_{i,n+1} = \delta_{i,n+1} + \frac{2p_i p_{n+1}}{[p\ddot{}p]}. \qquad (6.3.3)$$

We find that (x) is a row eigenvector of the matrix P with eigenvalue 1 if $(x)[p] = 0$, that $[u]$ is a column eigenvector with eigenvalue 1 if $[p\ddot{}u] = 0$, and that $[p]$ is a column eigenvector with eigenvalue -1.

A reflection in H^n is an involutory opposite isometry fixing a hyperplane. The product of two different reflections is a direct isometry—a rotation, a striation, or a translation according as the mirrors are intersecting, parallel, or diverging. (In the hyperbolic plane, every nontrivial direct isometry is of one of these types.) Criteria for distinguishing the three cases were given in § 2.4. Each of the three types of isometry is characterized by its eigenvalues, whose product is unity and whose sum is the trace of the associated matrix.

The rotation that is the product of reflections in given intersecting hyperplanes $\check{P}$ and $\check{Q}$ with coordinates $[p]$ and $[q]$ is represented by a *pseudo-orthogonal rotation matrix*, which is a pseudo-orthogonal matrix R of order $n+1$ with entries

$$r_{ij} = (p_i)[q_j], \text{ with } \left|\sum r_{ii}\right| < n+1, \qquad (6.3.4)$$

where (p_i) and $[q_j]$ are the ith row and the jth column of the corresponding "intersecting" reflection matrices P and Q. It will be found that (x) is a row eigenvector of the matrix R with eigenvalue 1 if

$(x)[p] = (x)[q] = 0$ and that R also has eigenvalues $\cos\theta \pm i \sin\theta$, where θ is the angle of the rotation (twice the angle between $\check{P}$ and $\check{Q}$). When the mirrors are orthogonal, the rotation is a half-turn ($\theta = \pi$), and -1 is an eigenvalue of multiplicity 2.

The striation that is the product of reflections in given parallel hyperplanes $\check{P}$ and $\check{Q}$ with coordinates $[p]$ and $[q]$ is represented by a *pseudo-orthogonal striation matrix*, which is a pseudo-orthogonal matrix S with entries

$$s_{ij} = (p_i)[q_j], \text{ with } \sum s_{ii} = n + 1, \tag{6.3.5}$$

i.e., the product of the corresponding "parallel" reflection matrices P and Q. The matrix S has 1 as its only eigenvalue, with $[r]$ as a column eigenvector and $(\dot{r}) = [r]^{\check{}}\dot{I}$ as a row eigenvector if $[p \ddot{} r] = [q \ddot{} r] = [r \ddot{} r] = 0$. That is, column eigenvectors are coordinates of the absolute hyperplane whose pole is the absolute point common to the invariant horocycles of the striation, and row eigenvectors are coordinates of this point.

The translation that is the product of reflections in given diverging hyperplanes $\check{P}$ and $\check{Q}$ with coordinates $[p]$ and $[q]$ is represented by a *pseudo-orthogonal translation matrix*, which is a pseudo-orthogonal matrix T with entries

$$t_{ij} = (p_i)[q_j], \text{ with } \sum t_{ii} > n + 1, \tag{6.3.6}$$

i.e., the product of the corresponding "diverging" reflection matrices P and Q. Here $[u]$ is a column eigenvector of the matrix T with eigenvalue 1 if $[p \ddot{} u] = [q \ddot{} u] = 0$. The matrix T also has eigenvalues e^t and e^{-t}, where t is the axial distance of the translation (twice the minimum distance between $\check{P}$ and $\check{Q}$). The corresponding column eigenvectors $[r]$ and $[s]$ are the coordinates of the two absolute hyperplanes whose poles are the absolute points of the axis and the invariant pseudocycles of the translation, and the corresponding row eigenvectors $(\dot{r}) = [r]^{\check{}}\dot{I}$ and $(\dot{s}) = [s]^{\check{}}\dot{I}$ are the coordinates of these points.

For example, in the hyperbolic plane H^2, reflections $\cdot P_1$, $\cdot P_2$, $\cdot P_3$, and $\cdot P_4$ in the lines $\check{P}_1 = [0,\ 1\,|\,0]$, $\check{P}_2 = [1,\ -1\,|\,0]$, $\check{P}_3 = [1,\ 1\,|\,1]$, and $\check{P}_4 = [1,\ 1\,|\,-1]$ are represented by the matrices

$$P_1 = \begin{pmatrix} 1 & 0 & 0 \\ 0 & -1 & 0 \\ 0 & 0 & 1 \end{pmatrix}, \quad P_2 = \begin{pmatrix} 0 & 1 & 0 \\ 1 & 0 & 0 \\ 0 & 0 & 1 \end{pmatrix},$$

$$P_3 = \begin{pmatrix} -1 & -2 & -2 \\ -2 & -1 & -2 \\ 2 & 2 & 3 \end{pmatrix}, \quad P_4 = \begin{pmatrix} -1 & -2 & 2 \\ -2 & -1 & 2 \\ -2 & -2 & 3 \end{pmatrix}.$$

Among the products of pairs of these reflections are the rotations $\cdot R_{12} = \cdot P_1 P_2$, $\cdot R_{23} = \cdot P_2 P_3$, and $\cdot R_{24} = \cdot P_2 P_4$, represented by the matrices

$$R_{12} = \begin{pmatrix} 0 & 1 & 0 \\ -1 & 0 & 0 \\ 0 & 0 & 1 \end{pmatrix}, \quad R_{23} = \begin{pmatrix} -2 & -1 & -2 \\ -1 & -2 & -2 \\ 2 & 2 & 3 \end{pmatrix}, \quad R_{24} = \begin{pmatrix} -2 & -1 & 2 \\ -1 & -2 & 2 \\ -2 & -2 & 3 \end{pmatrix}.$$

Since $(\check{P}_1\check{P}_2) = \pi/4$ and $(\check{P}_2\check{P}_3) = (\check{P}_2\check{P}_4) = \pi/2$, the first rotation is of period 4, and the other two are half-turns. Since the line $\check{P}_1$ is parallel to both $\check{P}_3$ and $\check{P}_4$, the products $\cdot S_{13} = \cdot P_1 P_3$ and $\cdot S_{14} = \cdot P_1 P_4$ are striations, represented by the matrices

$$S_{13} = \begin{pmatrix} -1 & -2 & -2 \\ 2 & 1 & 2 \\ 2 & 2 & 3 \end{pmatrix}, \quad S_{14} = \begin{pmatrix} -1 & -2 & 2 \\ 2 & 1 & -2 \\ -2 & -2 & 3 \end{pmatrix}.$$

The lines $\check{P}_3$ and $\check{P}_4$ are diverging, with $)\check{P}_3\check{P}_4(= \cosh^{-1} 3$, so that the product $\cdot T_{34} = \cdot P_3 P_4$ is a translation, represented by the matrix

$$T_{34} = \begin{pmatrix} 9 & 8 & -12 \\ 8 & 9 & -12 \\ -12 & -12 & 17 \end{pmatrix}.$$

As noted in § 6.1, the group of isometries $\cdot P$ of the elliptic n-sphere S^n is the orthogonal group O_{n+1}, represented by orthogonal matrices P of order $n + 1$. The group of isometries $\langle \cdot P \rangle$ of elliptic n-space eP^n

is the *projective orthogonal* group $\mathrm{PO}_{n+1} \cong \mathrm{O}_{n+1}/\bar{\mathrm{S}}\mathrm{Z}$, represented by pairs $\{\pm P\}$ of orthogonal matrices. The latter group has a hyperbolic analogue.

Per (6.3.2), the matrix group $\mathrm{O}_{n,1^+}$ of all isometries of hyperbolic n-space can be represented equally well by a quotient group, the *projective pseudo-orthogonal* group $\mathrm{PO}_{n,1} \cong \mathrm{O}_{n,1}/\bar{\mathrm{S}}\mathrm{Z}$, with each isometry $\langle \cdot P \rangle$ corresponding to a pair $\{\pm P\}$ of pseudo-orthogonal matrices. The subgroup $\mathrm{O}_{n^+,1^+}$ of direct isometries then takes the form of the *special projective pseudo-orthogonal* group $\mathrm{P}^+\mathrm{O}_{n,1}$, which for odd n is isomorphic to $\mathrm{O}^+_{n,1}/\bar{\mathrm{S}}\mathrm{Z}$ and for even n to $\mathrm{O}_{n^+,1}/\bar{\mathrm{S}}\mathrm{Z}$. As we shall see in § 9.3, the planar groups $\mathrm{PO}_{2,1}$ and $\mathrm{P}^+\mathrm{O}_{2,1}$ are respectively isomorphic to the projective general and projective special linear groups PGL_2 and PSL_2.

The absolute hypersphere of hyperbolic n-space H^n is essentially a real inversive (Möbius) $(n-1)$-sphere I^{n-1}, on which the trace of an ordinary k-plane H^k $(1 \le k \le n-1)$ is a real inversive $(k-1)$-sphere I^{k-1}. Each isometry of H^n induces a circle-preserving transformation, or circularity, of I^{n-1}. For $n \ge 2$, the group of circularities of I^{n-1} is the group $\mathrm{PO}_{n,1}$ of isometries of H^n, and the subgroup of direct circularities ("homographies") is the group $\mathrm{P}^+\mathrm{O}_{n,1}$ of direct isometries of H^n.

EXERCISES 6.3

1. Use Formula (6.3.3) to obtain the matrices given in this section for reflections in the lines $[0, 1 \mid 0]$, $[1, -1 \mid 0]$, $[1, 1 \mid 1]$, and $[1, 1 \mid -1]$.

2. Show directly that the rotations $\cdot R_{12}$, $\cdot R_{23}$, and $\cdot R_{24}$ given in this section have the respective periods 4, 2, and 2.

3. Show directly that the striations $\cdot S_{13}$ and $\cdot S_{14}$ are aperiodic.

4. Find the axial distance of the translation $\cdot T_{34}$ and show directly that it is aperiodic.

7

COMPLEX SPACES

EACH OF THE LINEAR AND CIRCULAR real geometries that were described in Chapters 2 and 3 has a complex analogue. Just as a metric can be defined on real projective n-space by choosing an absolute polarity, so complex projective n-space $\mathbb{C}P^n$ can be given a real-valued *unitary* metric by fixing an absolute *antipolarity*. In this way we obtain *elliptic*, *Euclidean*, or *hyperbolic* unitary n-space. Analogous to the real inversive geometry of the Möbius n-sphere is the inversive unitary geometry of the Möbius *n-antisphere* $\mathrm{I}\textsc{u}^n$. Specializing a family of *unitary inversions* on $\mathrm{I}\textsc{u}^n$ produces the *elliptic* or *hyperbolic* n-antisphere.

7.1 ANTILINEAR GEOMETRIES

Spaces coordinatized over the complex field $\mathbb{C}$ can be derived from complex projective n-space $\mathbb{C}P^n$. A one-to-one incidence-preserving transformation of $\mathbb{C}P^n$ taking k-planes into k-planes is a *collineation*; one that takes k-planes into $(n-k-1)$-planes is a *correlation*. If $n = 1$, we assume that harmonic sets go to harmonic sets. If cross ratios are preserved, the transformation is *projective*; if cross ratios are replaced by their complex conjugates, the transformation is *antiprojective*.

An antiprojective correlation of $\mathbb{C}P^n$ of period 2 is an *antipolarity* (an *anti-involution* if $n = 1$). The self-conjugate points of an antipolarity, if any, lie on an $(n-1)$-*antiquadric* (an *anticonic* if $n = 2$, a *chain* if $n = 1$). If the points in which any complex projective line meets the $(n-1)$-antiquadric all lie on a chain, the antiquadric is said to be *oval.*

A. *Complex projective geometry.* Given three distinct collinear points U, V, W in $\mathbb{C}\mathrm{P}^n$, we may characterize the chain $\mathbb{R}(UVW)$ as the set of all points X on the line UV such that the cross ratio $\| UV, WX \|$ is real (or infinite). A chain is essentially a real projective line embedded in a complex projective space (cf. Veblen & Young 1918, pp. 16–23; Coxeter 1949/1993, app. 1). Some writers, e.g., Schwerdtfeger (1962, p. 1) and Goldman (1999, p. 9), prefer to speak of "circles," a term reserved here for 1-spheres in a real inversive or real metric space.

We observe that the complex projective plane $\mathbb{C}\mathrm{P}^2$ has all the postulated properties of a real plane except for not satisfying Axiom S of § 2.1. Given four distinct collinear points U, V, W, and Z not all belonging to the same chain, let us define the *shell* $\mathbb{C}(UVWZ)$ to be the set of points on the line UV that contains those four points together with all the points of the chain through any three points of the set. (A shell has the topology of a real sphere S^2.) Then we obtain the complex projective plane by replacing Axiom S with

Axiom S^2. *There is more than one chain on a line but not more than one shell.*

Points and hyperplanes of $\mathbb{C}\mathrm{P}^n$ can be given homogeneous coordinates in $\mathbb{C}^{n+1}$ and $\check{\mathbb{C}}^{n+1}$, and collineations and correlations are then associated with square matrices. The *antitranspose* A^* of a complex matrix A is its transposed conjugate $\bar{A}^{\vee}$; a matrix A is *Hermitian* ("antisymmetric") if $A^* = A$. Any invertible Hermitian matrix A of order $n+1$ defines an antipolarity on $\mathbb{C}\mathrm{P}^n$ (an anti-involution on $\mathbb{C}\mathrm{P}^1$). The *antipolar* of a point Z with coordinates (z) is the hyperplane $\check{Z}'$ with coordinates $[z'] = A(z)^*$, and the *antipole* of a hyperplane $\check{W}$ with coordinates $[w]$ is the point W' with coordinates $(w') = [w]^* A^{-1}$. Two points Z and W or two hyperplanes $\check{W}$ and $\check{Z}$ are conjugate in the antipolarity, i.e., each is incident with the antipolar or antipole of the other, if and only if

$$(z)A(w)^* = 0 \quad \text{or} \quad [w]^*A^{-1}[z] = 0.$$

Thus conjugacy in the antipolarity is defined by means of a Hermitian *sesquilinear form* $\cdot A\odot : \mathbb{C}^{n+1} \to \mathbb{C}^{n+1}$, with $[(z), (w)] \mapsto \sum\sum z_i a_{ij} \bar{w}_j$. In a *unitary* space, distances and/or angles can be measured by means of a fixed antipolarity.

B. *Unitary metric spaces.* When an antipolarity is defined by a matrix E that is anticongruent to the identity matrix $I = I_{n+1}$ (i.e., $RER^* = I$ for some invertible matrix R), the antiquadratic forms $(z\ \bar{z})$ and $[\bar{w}\ w]$ corresponding to the sesquilinear forms

$$(z\ \bar{w}) = (z)E(w)^* \quad \text{and} \quad [\bar{w}\ z] = [w]^*E^{-1}[z]$$

are positive definite: $(z\ \bar{z})$ and $[\bar{w}\ w]$ are real and positive for all nonzero (z) and $[w]$, so that there are no self-conjugate points or hyperplanes. The fixing of such an *elliptic* antipolarity converts complex projective n-space $\mathbb{C}\mathrm{P}^n$ into *elliptic unitary* n-space Pu^n. The distance between two points Z and W in Pu^n is given by

$$[ZW] = \cos^{-1} \frac{|(z\ \bar{w})|}{\sqrt{(z\ \bar{z})}\sqrt{(w\ \bar{w})}}, \tag{7.1.1}$$

and the angle between two hyperplanes $\check{W}$ and $\check{Z}$ by

$$(\check{W}\check{Z}) = \cos^{-1} \frac{|[\bar{w}\ z]|}{\sqrt{(\bar{w}\ w)}\sqrt{(\bar{z}\ z)}}. \tag{7.1.2}$$

Elliptic unitary n-space is isometric to the real elliptic $2n$-sphere S^{2n}. It is the only complex linear geometry with a simple angular measure.

We may instead first fix the projective hyperplane $z_{n+1} = 0$, so that $\mathbb{C}\mathrm{P}^n$ becomes extended *complex affine n*-space $\mathbb{C}\bar{\mathrm{A}}^n$, and then fix in it the absolute antipolarity defined by the identity matrix. The resulting elliptic unitary $(n-1)$-space is the *absolute hyperplane* of extended

Euclidean unitary n-space $\bar{\mathrm{E}}\mathrm{U}^n$, the complement of the absolute hyperplane being the *ordinary* region EU^n. The coordinates of an ordinary point Z, with $z_{n+1} \neq 0$, can be normalized as

$$((z),\ 1) = (z_1, z_2, \ldots, z_n,\ 1),$$

so that (z) is the vector of complex Cartesian coordinates of Z. Homogeneous coordinates of an ordinary hyperplane $\check{W}(r)$ can be expressed as

$$[[w],\ r] = [w_1, w_2, \ldots, w_n, r],$$

with $[w] \neq [0]$. We have $Z \lozenge \check{W}(r)$ if and only if $(z)[w] + r = 0$.

On the vector space $\mathbb{C}^n$ and its dual $\check{\mathbb{C}}^n$ we have the standard sesquilinear forms

$$(z\ \bar{w}) = (z)(w)^* \quad \text{and} \quad [\bar{w}\ z] = [w]^*[z].$$

The (real) distance between two points Z and W in EU^n with respective complex Cartesian coordinates (z) and (w) is given by

$$|ZW| = \sqrt{(w - z\ \ \bar{w} - \bar{z})} = [|w_1 - z_1|^2 + \cdots + |w_n - z_n|^2]^{1/2}, \quad (7.1.3)$$

and the distance between two parallel hyperplanes $\check{W}(r)$ and $\check{W}(s)$ by

$$|\check{W}(r,\ s)| = \frac{|s - r|}{\sqrt{[\bar{w}\ w]}}. \quad (7.1.4)$$

Euclidean unitary n-space is isometric to real Euclidean $2n$-space E^{2n}.

When an antipolarity on $\mathbb{C}\mathrm{P}^n$ is defined by a matrix H that is anticongruent to the pseudo-identity matrix $I_{n,1}$, the locus of self-conjugate points (or the envelope of self-conjugate hyperplanes) is an oval $(n-1)$-antiquadric ω. Any line of $\mathbb{C}\mathrm{P}^n$ meets ω in at most the points of a chain. Fixing such a *hyperbolic* antipolarity converts complex projective n-space $\mathbb{C}\mathrm{P}^n$ into *complete hyperbolic unitary n*-space $\mathrm{P\ddot{U}}^n$, separated by ω as the *absolute hyperchain* (the *absolute chain* if

$n = 1$) into *ordinary* and *ideal* regions HU^n and $\check{\mathrm{H}}\mathrm{U}^n$. In terms of the dual sesquilinear forms

$$(z\ddot{\ }\bar{w}) = (z)H(w)^* \quad \text{and} \quad [\bar{w}\ddot{\ }z] = [w]^*H^{-1}[z],$$

a point Z with coordinates (z) or a hyperplane $\check{W}$ with coordinates $[w]$ is

$$\begin{array}{lllll} \text{ordinary} & \text{if} & (z\ddot{\ }\bar{z}) < 0 & \text{or} & [\bar{w}\ddot{\ }w] > 0, \\ \text{absolute} & \text{if} & (z\ddot{\ }\bar{z}) = 0 & \text{or} & [\bar{w}\ddot{\ }w] = 0, \\ \text{ideal} & \text{if} & (z\ddot{\ }\bar{z}) > 0 & \text{or} & [\bar{w}\ddot{\ }w] < 0. \end{array}$$

Because the sesquilinear forms are Hermitian, $(z\ddot{\ }\bar{w})$ and $(w\ddot{\ }\bar{z})$ are complex conjugates, as are $[\bar{w}\ddot{\ }z]$ and $[\bar{z}\ddot{\ }w]$, so that

$$(z\ddot{\ }\bar{w})(w\ddot{\ }\bar{z}) = |(z\ddot{\ }\bar{w})|^2 \quad \text{and} \quad [\bar{w}\ddot{\ }z][\bar{z}\ddot{\ }w] = |[\bar{w}\ddot{\ }z]|^2.$$

The *discriminant* of two ordinary hyperplanes $\check{W}$ and $\check{Z}$ is the real number

$$|\,\bar{w}\ddot{\ }z\,| = [\bar{w}\ddot{\ }w][\bar{z}\ddot{\ }z] - [\bar{w}\ddot{\ }z][\bar{z}\ddot{\ }w],$$

and the hyperplanes are intersecting, parallel, or diverging according as $|\,\bar{w}\ddot{\ }z\,|$ is greater than, equal to, or less than zero. The distance between two ordinary points Z and W in HU^n is given by

$$[ZW] = \cosh^{-1} \frac{|(z\ddot{\ }\bar{w})|}{\sqrt{-(z\ddot{\ }\bar{z})}\sqrt{-(w\ddot{\ }\bar{w})}} \tag{7.1.5}$$

(cf. Goldman 1999, pp. 77–78), and the minimum distance between two diverging hyperplanes $\check{W}$ and $\check{Z}$ by

$$)\check{W}\check{Z}(= \cosh^{-1} \frac{|[\bar{w}\ddot{\ }z]|}{\sqrt{[\bar{w}\ddot{\ }w]}\sqrt{[\bar{z}\ddot{\ }z]}}. \tag{7.1.6}$$

Some of the metric properties of elliptic and hyperbolic unitary n-space correspond to properties of the real elliptic and hyperbolic $2n$-spheres S^{2n} and $\ddot{\mathrm{S}}^{2n}$; the case $n = 1$ is treated in detail by Goldman (1999, chap. 1). But the trigonometry of HU^n ($n > 1$) is considerably more complicated than that of a real non-Euclidean space (Goldman 1999, pp. 84–99), and no angle formulas will be given here.

EXERCISES 7.1

1. ⊢ The elliptic unitary line Pu^1, with an absolute anti-involution defined by the sesquilinear form $(z\,\bar{w}) = z_1\bar{w}_1 + z_2\bar{w}_2$, is isometric to the real elliptic sphere S^2.

2. ⊢ The Euclidean unitary line Eu^1 (the Argand diagram) is isometric to the real Euclidean plane E^2.

3. ⊢ The complete hyperbolic unitary line $\mathrm{P\ddot{u}}^1$, with an absolute anti-involution defined by the sesquilinear form $(z\ddot{}\bar{w}) = z_1\bar{w}_1 - z_2\bar{w}_2$, is isometric to the real hyperbolic sphere $\ddot{\mathrm{S}}^2$.

7.2 ANTICIRCULAR GEOMETRIES

Let the points Z and hyperplanes $\check{W}$ of complex projective $(n+1)$-space $\mathbb{C}\mathrm{P}^{n+1}$ have homogeneous coordinates

$$((z)) = (z_0, z_1, \ldots, z_{n+1}) \quad \text{and} \quad [[w]] = [w_0, w_1, \ldots, w_{n+1}],$$

and let an oval n-antiquadric Ψ have the locus or envelope equation

$$|z_1|^2 + \cdots + |z_{n+1}|^2 = |z_0|^2 \quad \text{or} \quad |w_1|^2 + \cdots + |w_{n+1}|^2 = |w_0|^2.$$

Proceeding by way of *inversive unitary* geometry, we may identify Ψ with the *Möbius n-antisphere* Iu^n, the one-point compactification of *Heisenberg n*-space (cf. Goldman 1999, pp. 59–66, 118–125).* A projective hyperplane $\check{W}$ with coordinates $[[w]]$ meets Iu^n in a (possibly degenerate) inversive $(n-1)$-antisphere $\mathring{\omega}$, provided that the antiquadratic form $[[\bar{w}\,w]]$ associated with the sesquilinear form

$$[[\bar{w}\,z]] = -\bar{w}_0 z_0 + \bar{w}_1 z_1 + \cdots + \bar{w}_{n+1} z_{n+1}$$

has a nonnegative value. A 1-antisphere is an *anticircle*; a 0-antisphere is a *chain*.

* The latter geometry, with a real symplectic structure, is named after Werner Heisenberg (1901–1976), since its group of automorphisms is related to the Heisenberg group of quantum field theory. Inversive unitary geometry itself will be discussed further in Chapter 9.

A. *The elliptic n-antisphere.* A projective collineation of $\mathbb{C}P^{n+1}$ ($n \geq 0$) that commutes with the antipolarity defined by the oval n-antiquadric Ψ induces a *holomorphism* of Ψ regarded as the Möbius n-antisphere $\mathrm{I}\mathrm{U}^n$. In particular, any homology of unit cross ratio ι whose center O is antipolar to its median hyperplane $\check{O}$ induces a *unitary inversion* on $\mathrm{I}\mathrm{U}^n$. For a harmonic homology, with cross ratio -1, this is an *orthogonal inversion*, of period 2. When O and $\check{O}$ are, respectively, interior and exterior to Ψ, as when they have the respective coordinates

$$(1, 0, \ldots, 0, 0) \quad \text{and} \quad [1, 0, \ldots, 0, 0],$$

the inversions so induced are *elliptic*, leaving no points of $\mathrm{I}\mathrm{U}^n$ fixed. By restricting ourselves to those holomorphisms that commute with these *central inversions*, we convert the Möbius n-antisphere $\mathrm{I}\mathrm{U}^n$ into the *elliptic* n-antisphere $\mathrm{S}\mathrm{U}^n$ ($n \geq 0$).

As applied to the complex projective line $\mathbb{C}P^1$, this procedure is carried out on an inversive chain $\mathrm{I}\mathrm{U}^0$, which is essentially a real inversive circle I^1. The central inversions on $\mathrm{I}\mathrm{U}^0$ are induced by projectivities that leave a certain pair of points of $\mathbb{C}P^1$ fixed, yielding the elliptic chain $\mathrm{S}\mathrm{U}^0$, with the same metric as the real elliptic circle S^1.

A point Z on $\mathrm{S}\mathrm{U}^n$ has normalized coordinates

$$(z) = (z_1, z_2, \ldots, z_{n+1}),$$

where $|z_1|^2 + |z_2|^2 + \cdots + |z_{n+1}|^2 = 1$. If $|\iota| = 1$, the points (z) and $\iota(z)$ are *confibral*; all points that are confibral to a given point Z lie on a *great chain* $\langle Z \rangle$. This is what Coolidge (1924, pp. 140–141) calls a "normal" chain or a "geodesic thread." The points (z) and $-(z)$ are *antipodal*. For $n > 0$, a hyperplane $\check{W}$ passing through O, so that $w_0 = 0$, intersects $\mathrm{S}\mathrm{U}^n$ in a *great hyperchain* $\check{W}$, with homogeneous coordinates

$$[w] = [w_1, w_2, \ldots, w_{n+1}].$$

We have $Z \lozenge \check{W}$ if and only if $(z)[w] = 0$. In terms of the sesquilinear forms

$$(z\,\bar{w}) = z_1\bar{w}_1 + \cdots + z_{n+1}\bar{w}_{n+1} \quad \text{and} \quad [\bar{w}\,z] = \bar{w}_1 z_1 + \cdots + \bar{w}_{n+1} z_{n+1},$$

with $(z\,\bar{z}) = (w\,\bar{w}) = 1$, the distance between two points Z and W on $\mathrm{S}\textsc{u}^n$ is given by

$$[\![ZW]\!] = \cos^{-1}\operatorname{Re}(z\,\bar{w}), \tag{7.2.1}$$

and the angle between two great hyperchains $\breve{W}$ and $\breve{Z}$ by

$$((\breve{W}\breve{Z})) = \cos^{-1}\frac{|[\bar{w}\,z]|}{\sqrt{[\bar{w}\,w]}\sqrt{[\bar{z}\,z]}}. \tag{7.2.2}$$

The point O and the hyperplane $\breve{O}$ may be taken as the origin and the hyperplane at infinity of Euclidean unitary $(n+1)$-space $\mathrm{E}\textsc{u}^{n+1}$, and $\mathrm{S}\textsc{u}^n$ is then the unit hyperchain $(z\,\bar{z}) = 1$, isometrically modeled by the unit hypersphere in real Euclidean $(2n+2)$-space. Thus the elliptic n-antisphere is isometric to the real elliptic $(2n+1)$-sphere, with great chains of $\mathrm{S}\textsc{u}^n$ corresponding to great circles of S^{2n+1}, and great hyperchains to great $(2n-1)$-spheres. If confibral points are identified, $\mathrm{S}\textsc{u}^n$ ("spherical unitary" n-space) is transformed into elliptic unitary n-space $\mathrm{P}\textsc{u}^n$, which is isometric to S^{2n}. Great chains of $\mathrm{S}\textsc{u}^n$ project into points of $\mathrm{P}\textsc{u}^n$, great hyperchains of $\mathrm{S}\textsc{u}^n$ into hyperplanes of $\mathrm{P}\textsc{u}^n$. All the great chains of $\mathrm{S}\textsc{u}^1$ form a *Hopf fibration* of S^3 (Hopf 1931, p. 655), and the corresponding projection $\mathrm{S}^3 \to \mathrm{S}^2$ is the *Hopf mapping* (Banchoff 1988, p. 227; cf. Onishchik & Sulanke 2006, ¶1.10.1).

B. *The hyperbolic n-antisphere.* Again considering the Möbius n-antisphere as an oval n-antiquadric Ψ in $\mathbb{C}\mathrm{P}^{n+1}$ $(n > 0)$ defined by

$$|z_1|^2 + \cdots + |z_{n+1}|^2 = |z_0|^2 \quad \text{or} \quad |w_1|^2 + \cdots + |w_{n+1}|^2 = |w_0|^2,$$

a homology of unit cross ratio whose center O and median hyperplane $\breve{O}$ have the respective coordinates

$$(0,\,0,\ldots,\,0,\,1) \quad \text{and} \quad [0,\,0,\ldots,\,0,\,1],$$

induces a unitary inversion on Ψ that fixes every point on the inversive hyperchain $\mathring{\omega}$ in which $\check{O}$ meets Ψ. If such *hyperbolic* inversions are taken as central inversions, then the Möbius n-antisphere $\mathrm{I}\textsc{u}^n$ is converted into the *hyperbolic* n-antisphere $\ddot{\mathrm{S}}\textsc{u}^n$ $(n > 0)$. We call $\mathring{\omega}$ the *equatorial hyperchain* of $\ddot{\mathrm{S}}\textsc{u}^n$; all points not on $\mathring{\omega}$ are *ordinary*. In contrast to the two n-hemispheres of the real hyperbolic n-sphere $\ddot{\mathrm{S}}^n$, the ordinary part of $\ddot{\mathrm{S}}\textsc{u}^n$ is connected. Another name for $\ddot{\mathrm{S}}\textsc{u}^n$ is *pseudospherical unitary* n-space.

Normalized coordinates for an ordinary point Z on $\ddot{\mathrm{S}}\textsc{u}^n$ are

$$(z) = (z_1, \ldots, z_n \mid z_0),$$

with $|z_1|^2 + \cdots + |z_{n+1}|^2 - |z_0|^2 = -1$. If $|\iota| = 1$, the points (z) and $\iota(z)$ are *confibral*; all points confibral to a given point Z lie on a *great chain* $\langle Z \rangle$. The points (z) and $-(z)$ are *antipodal*, and equatorial points are self-antipodal. A hyperplane $\check{W}$ that contains a great chain, implying that $w_{n+1} = 0$, meets $\ddot{\mathrm{S}}\textsc{u}^n$ in a hyperbolic $(n-1)$-antisphere $\check{W}$, a *great hyperchain*, having homogeneous coordinates

$$[w] = [w_1, \ldots, w_n \mid w_0],$$

with $|w_1|^2 + \cdots + |w_{n+1}|^2 - |w_0|^2 > 0$. We have $Z \lozenge \check{W}$ if and only if $(z)[w] = 0$. Unfortunately, there are no simple formulas for distances and angles on $\ddot{\mathrm{S}}\textsc{u}^n$.

If confibral points are identified, the hyperbolic n-antisphere $\ddot{\mathrm{S}}\textsc{u}^n$ is transformed into extended hyperbolic unitary n-space $\bar{\mathrm{H}}\textsc{u}^n = \mathrm{P}\ddot{\mathrm{S}}\textsc{u}^n$. Great chains of $\ddot{\mathrm{S}}\textsc{u}^n$ project into points of $\mathrm{H}\textsc{u}^n$, great hyperchains of $\ddot{\mathrm{S}}\textsc{u}^n$ into hyperplanes of $\mathrm{H}\textsc{u}^n$. The equatorial hyperchain $\mathring{\omega}$ of $\ddot{\mathrm{S}}\textsc{u}^n$ becomes the absolute hyperchain ω of $\bar{\mathrm{H}}\textsc{u}^n$.

It is important to note that, whereas the real n-spheres S^n and $\ddot{\mathrm{S}}^n$ are just "double" versions of the real n-spaces eP^n and $\bar{\mathrm{H}}^n$, there is a fundamental difference between the n-antispheres $\mathrm{S}\textsc{u}^n$ and $\ddot{\mathrm{S}}\textsc{u}^n$ and the unitary n-spaces $\mathrm{P}\textsc{u}^n$ and $\bar{\mathrm{H}}\textsc{u}^n$. Spherical or pseudospherical unitary n-space has real dimension $2n + 1$, while elliptic or hyperbolic unitary n-space has real dimension $2n$.

EXERCISES 7.2

1. ⊢ The elliptic anticircle Su^1 is isometric to the real elliptic 3-sphere S^3.

2. ⊢ When confibral points are identified, the elliptic anticircle Su^1 is transformed into the elliptic unitary line Pu^1.

3. ⊢ When confibral points are identified, the hyperbolic anticircle $\ddot{\mathrm{S}}\mathrm{u}^1$ is transformed into the extended hyperbolic unitary line $\bar{\mathrm{H}}\mathrm{u}^1$.

7.3 SUMMARY OF COMPLEX SPACES

The following diagram (Figure 7.3a) shows how the various complex geometries discussed here are related. A line joining two geometries

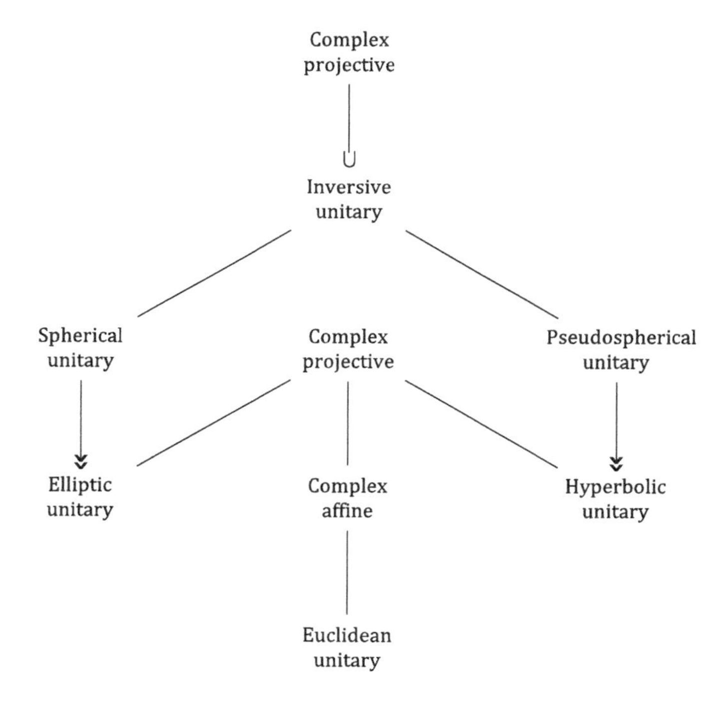

Figure 7.3a Antilinear and anticircular geometries

indicates that the one below can be derived from the one above by selecting an antiquadric; by fixing a family of unitary inversions, an antipolarity, or a hyperplane; or (for geometries joined by a double arrow) by identifying confibral points.

8

COMPLEX COLLINEATION GROUPS

LINEAR TRANSFORMATIONS of complex vector spaces have many of the same properties as their real counterparts, but the existence of complex conjugates creates further possibilities. By combining a linear transformation of $\mathbb{C}^n$ with the conjugation automorphism, we obtain an *antilinear* transformation. Collineations of complex affine spaces include not only affinities but also *antiaffinities*, and complex projective spaces have both projective and *antiprojective* collineations and correlations. Mappings of complex spaces can be represented either by complex matrices or, by resolving each complex number into its real and imaginary parts, by real *duplex* matrices.

8.1 LINEAR AND AFFINE TRANSFORMATIONS

The set $\mathbb{C}^n$ of all lists $(z) = (z_1, z_2, \ldots, z_n)$ of n complex numbers, taken as rows, constitutes an n-dimensional left vector space over $\mathbb{C}$; taken as columns $[w] = [w_1, w_2, \ldots, w_n]$, such lists are the elements of the dual right vector space $\check{\mathbb{C}}^n$. Each column $[w]$ defines a linear form $\cdot[w] : \mathbb{C}^n \to \mathbb{C}$, with $(z) \mapsto (z)[w]$, and each row (z) defines a dual linear form $(z)\cdot : \check{\mathbb{C}}^n \to \mathbb{C}$, with $[w] \mapsto (z)[w]$. The vector space $\mathbb{C}^n$ or $\check{\mathbb{C}}^n$ has an involutory automorphism

$$\bigcirc : \mathbb{C}^n \to \mathbb{C}^n, \text{ with } (z) \mapsto (\bar{z}), \quad \text{or} \quad \square : \check{\mathbb{C}}^n \to \check{\mathbb{C}}^n, \text{ with } [w] \mapsto [\bar{w}],$$

replacing the entries of a row or column by their complex conjugates, and a linear form $\cdot[w]$ or $(z)\cdot$ can be combined with it to give an *antilinear form*

$$\odot[w] : \mathbb{C}^n \to \mathbb{C}, \quad \text{with } (z) \mapsto (\bar{z})[w], \quad \text{or}$$
$$(z)\boxdot : \check{\mathbb{C}}^n \to \mathbb{C}, \quad \text{with } [w] \mapsto (z)[\bar{w}].$$

A. *Linear transformations.* An $n \times n$ invertible matrix A over the complex field $\mathbb{C}$ determines a pair of invertible linear transformations

$$\cdot A : \mathbb{C}^n \to \mathbb{C}^n, \quad \text{with } (z) \mapsto (z)A, \quad \text{and}$$
$$A\cdot : \check{\mathbb{C}}^n \to \check{\mathbb{C}}^n, \quad \text{with } [w] \mapsto A[w],$$

of the dual vector spaces $\mathbb{C}^n$ and $\check{\mathbb{C}}^n$. The set of all such transformations forms a group, the *complex general linear* group $\mathrm{GL}(n;\ \mathbb{C})$, abbreviated as $\mathrm{GL}_n(\mathbb{C})$ or ${}^{\mathbb{C}}\mathrm{GL}_n$, each group operation corresponding to an $n \times n$ invertible matrix over $\mathbb{C}$. Matrices with positive real determinants determine the *complex direct linear* group $\mathrm{G}^+\mathrm{L}_n(\mathbb{C})$. Matrices A with $|\det A| = 1$ determine the *complex unit linear* group $\bar{\mathrm{S}}\mathrm{L}_n(\mathbb{C})$, and those with $\det A = 1$, the *complex special linear* group $\mathrm{SL}_n(\mathbb{C})$.

The center of the complex general linear group is the *complex general scalar* group $\mathrm{GZ}(\mathbb{C})$, represented by the nonzero scalar matrices λI. Matrices ιI with $|\iota| = 1$ form the *complex unit scalar* group $\bar{\mathrm{S}}\mathrm{Z}(\mathbb{C})$. These groups are isomorphic, respectively, to the multiplicative group $\mathrm{GL}(\mathbb{C})$ of nonzero complex numbers and its subgroup $\bar{\mathrm{S}}\mathrm{L}(\mathbb{C})$ of complex numbers of absolute value 1; the latter is isomorphic to the additive group $\mathrm{S}^1 \cong \mathrm{T}^1/2\pi\mathbb{Z}$ (the "circle group"). The center of the complex special linear group $\mathrm{SL}_n(\mathbb{C})$ is the *complex special scalar* group $\mathrm{SZ}_n(\mathbb{C}) \cong \mathrm{C}_n$, represented by the $n \times n$ matrices ωI with $\omega^n = 1$.

Any linear transformation of a complex vector space can be combined with the conjugation automorphism $\bigcirc$ or $\square$. Thus an $n \times n$ invertible matrix A over $\mathbb{C}$ determines a pair of invertible *antilinear transformations*

$$\odot A : \mathbb{C}^n \to \mathbb{C}^n, \quad \text{with } (z) \mapsto (\bar{z})A, \quad \text{and}$$
$$A\boxdot : \check{\mathbb{C}}^n \to \check{\mathbb{C}}^n, \quad \text{with } [w] \mapsto A[\bar{w}],$$

of the dual vector spaces $\mathbb{C}^n$ and $\check{\mathbb{C}}^n$. The product of antilinear transformations $\odot A$ and $\odot B$ is the linear transformation $\cdot\bar{A}B$; the product of antilinear transformations $A\boxdot$ and $B\boxdot$ is the linear transformation $B\bar{A}\cdot$.

The set of all linear and antilinear transformations of $\mathbb{C}^n$ or $\check{\mathbb{C}}^n$ forms a group, the *general semilinear* group $\mathrm{G\bar{L}}_n(\mathbb{C})$, containing the general linear group $\mathrm{GL}_n(\mathbb{C})$ as a subgroup of index 2. The general scalar group $\mathrm{GZ}(\mathbb{C})$ can be augmented by *antiscalar* transformations $(z) \mapsto \lambda(\bar{z})$ or $[w] \mapsto [\bar{w}]\rho$ to give the *general semiscalar* group $\mathrm{G\bar{Z}}(\mathbb{C})$. Each of the subgroups of $\mathrm{GL}_n(\mathbb{C})$ or $\mathrm{GZ}(\mathbb{C})$ likewise has a semilinear or semiscalar extension.

Each complex number $z = x + y\mathrm{i}$ can be represented by a 2×2 real *bivalent matrix*

$$Z = \{z\} = (x,\ y)_{(2)} = \begin{pmatrix} x & y \\ -y & x \end{pmatrix}. \tag{8.1.1}$$

The transpose $Z^{\check{}} = (x,\ -y)_{(2)}$ represents the complex conjugate $\bar{z} = x - y\mathrm{i}$, and $\det Z = |z|^2$. The set of all such matrices is closed under both addition and multiplication, defining a field $\mathbb{R}^{(2)}$ isomorphic to the complex field $\mathbb{C}$ via the correspondence $Z \mapsto z$. More generally, any $m \times n$ matrix A over $\mathbb{C}$ can be represented by a $2m \times 2n$ *duplex matrix* $\{A\}$ over $\mathbb{R}$ comprising mn 2×2 bivalent blocks $\{a_{ij}\} = (\mathrm{Re}\, a_{ij},\ \mathrm{Im}\, a_{ij})_{(2)}$ (cf. Neumann, Stoy & Thompson 1994, p. 208). If $m = n$, then $\det\{A\} = |\det A|^2$.

The group $\mathrm{GL}_n(\mathbb{C})$ of invertible matrices of order n over $\mathbb{C}$ is isomorphic to the *general duplex* group $\mathrm{GL}_{(2)n}$ of invertible duplex matrices of order $2n$ over $\mathbb{R}$. The unit linear subgroup $\mathrm{\bar{S}L}_n(\mathbb{C})$ is isomorphic to the *special duplex* group $\mathrm{SL}_{(2)n}$ of such matrices with determinant 1. The group $\mathrm{GZ}(\mathbb{C})$ of nonzero scalar matrices over $\mathbb{C}$ is isomorphic to the *general bivalent* group $\mathrm{GZ}_{(2)}$ of nonzero diagonal block bivalent matrices $\backslash(\mu,\ \nu)_{(2)}\backslash$. The group $\bar{\mathrm{S}}\mathrm{Z}(\mathbb{C})$ is isomorphic to the *special bivalent* group $\mathrm{SZ}_{(2)}$ of such matrices with $\mu^2 + \nu^2 = 1$, so that $(\mu,\ \nu)_{(2)}$ is a 2×2 orthogonal rotation matrix; thus $\bar{\mathrm{S}}\mathrm{Z}(\mathbb{C}) \cong \mathrm{O}_2^+$.

B. *Affine transformations.* Complex affine n-space $\mathbb{C}\mathrm{A}^n$ can be coordinatized so that each point Z is associated with a unique row vector $(z) \in \mathbb{C}^n$ and each hyperplane $\check{W}(r)$ with the nonzero scalar multiples of a column vector $[[w],\ r] \in \check{\mathbb{C}}^{n+1}$. The point Z lies on the hyperplane $\check{W}(r)$ if and only if $(z)[w] + r = 0$. By resolving each vector $(z) = (x) + (y)\mathrm{i}$ into its real and imaginary parts, we can represent the points of $\mathbb{C}\mathrm{A}^n$ by the points of real affine $2n$-space A^{2n} with *semiduplex coordinates*

$$((x,\ y)) = (x_1,\ y_1,\ \ldots,\ x_n,\ y_n). \tag{8.1.2}$$

A complex hyperplane $[[w],\ r] = [[u],\ p] + \mathrm{i}[[v],\ q]$ is then represented by a real $(2n-2)$-plane, the intersection of two real hyperplanes with coordinates

$$[[[u, -v]],\ p] = [u_1, -v_1,\ \ldots,\ u_n, -v_n,\ p]$$

and

$$[[[v,\ u]],\ q] = [v_1,\ u_1,\ \ldots,\ v_n,\ u_n,\ q].$$

We assign this $(2n-2)$-plane of A^{2n}, or the corresponding hyperplane of $\mathbb{C}\mathrm{A}^n$, the *duplex coordinates*

$$[[(u,\ v)_{(2)}],\ (p,\ q)] = \begin{pmatrix} u_1 & v_1 \\ -v_1 & u_1 \\ \vdots & \vdots \\ u_n & v_n \\ -v_n & u_n \\ p & q \end{pmatrix}. \tag{8.1.3}$$

The point Z with semiduplex coordinates $\mathbf{(}(x,\ y)\mathbf{)}$ lies on the hyperplane $\check{W}(r)$ with these duplex coordinates if and only if

$$\mathbf{(}(x,\ y)\mathbf{)}\mathbf{[}(u,\ v)_{(2)}\mathbf{]} + (p,\ q) = (0,\ 0).$$

If A is an $n \times n$ invertible matrix over $\mathbb{C}$ and $\mathbf{(}h\mathbf{)} \in \mathbb{C}^n$ is any row vector, then the product of the linear transformation $\cdot A : \mathbb{C}^n \to \mathbb{C}^n$ and the translation $+\mathbf{(}h\mathbf{)} : \mathbb{C}^n \to \mathbb{C}^n$ is a *complex translinear transformation*

$$[\cdot A,\ +\mathbf{(}h\mathbf{)}] : \mathbb{C}^n \to \mathbb{C}^n, \text{ with } \mathbf{(}z\mathbf{)} \mapsto \mathbf{(}z\mathbf{)}A + \mathbf{(}h\mathbf{)}, \quad \det A \neq 0, \quad (8.1.4)$$

which induces an affinity on $\mathbb{C}\mathrm{A}^n$. Such a transformation can be represented either by an $(n+1) \times (n+1)$ translinear matrix (2.3.2) with complex entries or by a $(2n+1) \times (2n+1)$ real *transduplex matrix*

$$[\{A\},\ \mathbf{(}(f,\ g)\mathbf{)}]_1 = \begin{pmatrix} \{A\} & \mathbf{[}[0,\ 0]\mathbf{]} \\ \mathbf{(}(f,\ g)\mathbf{)} & 1 \end{pmatrix}, \quad (8.1.5)$$

where $\{A\} = (\mathrm{Re}\,A,\ \mathrm{Im}\,A)_{(2)}$ is the duplex equivalent to A, $\mathbf{[}[0,\ 0]\mathbf{]}$ is a column of $2n$ zeros, and $\mathbf{(}f\mathbf{)}$ and $\mathbf{(}g\mathbf{)}$ are the real and imaginary parts of $\mathbf{(}h\mathbf{)}$. If we postmultiply the row vector $(\mathbf{(}(x,\ y)\mathbf{)},\ 1) \in \mathbb{R}^{2n+1}$ by this matrix, we obtain the row vector $(\mathbf{(}(x,\ y)\mathbf{)}\{A\} + \mathbf{(}(f,\ g)\mathbf{)},\ 1)$. If $\det\{A\} = 1$, then $[\{A\},\ \mathbf{(}(f, g)\mathbf{)}]_1$ is a *special* transduplex matrix, and the operation on $\mathbb{C}\mathrm{A}^n$ is a unit affinity. When $\{A\}$ is the identity matrix, (8.1.5) is a translation matrix.

The transformation $[\cdot A,\ +\mathbf{(}h\mathbf{)}]$ takes the hyperplane $[\mathbf{[}w\mathbf{]},\ r]$ of $\mathbb{C}\mathrm{A}^n$ into the hyperplane $[A^{-1}\mathbf{[}w\mathbf{]},\ r - \mathbf{(}h\mathbf{)}A^{-1}\mathbf{[}w\mathbf{]}]$. The corresponding mapping of $(2n-2)$-planes in A^{2n} is effected by premultiplying the $(2n+1) \times 2$ duplex coordinate matrix $[\mathbf{[}(u,\ v)_{(2)}\mathbf{]},\ (p,\ q)]$ by the inverse transduplex matrix

$$[\{A\},\ \mathbf{(}(f,\ g)\mathbf{)}]_1^{-1} = \begin{pmatrix} \{A\}^{-1} & \mathbf{[}[0,\ 0]\mathbf{]} \\ -\mathbf{(}(f,\ g)\mathbf{)}\{A\}^{-1} & 1 \end{pmatrix}. \quad (8.1.6)$$

The set of all transduplex matrices of order $2n+1$ forms a multiplicative group $\mathrm{TGL}_{(2)n}$ isomorphic to the *complex general affine* group $\mathrm{GA}_n(\mathbb{C})$ of all affinities of $\mathbb{C}\mathrm{A}^n$, in which the translation matrices constitute an abelian subgroup $\mathrm{T}_{(2)n}$. The latter group is isomorphic to the additive group $\mathrm{T}^n(\mathbb{C})$ of the vector space $\mathbb{C}^n = \mathbb{R}^{(2)n}$, which is the direct product of n copies of the additive group $\mathrm{T}(\mathbb{C})$ of the complex field $\mathbb{C}$. The special transduplex matrices form a normal subgroup $\mathrm{TSL}_{(2)n}$, isomorphic to the *complex unit affine* group $\bar{\mathrm{S}}\mathrm{A}_n(\mathbb{C})$. The quotient group $\mathrm{TGL}_{(2)n}/\mathrm{T}_{(2)n} \cong \mathrm{GA}_n(\mathbb{C})/\mathrm{T}^n(\mathbb{C})$ is isomorphic to the complex general linear group $\mathrm{GL}_n(\mathbb{C})$ of $n \times n$ invertible matrices over $\mathbb{C}$.

Those transduplex matrices $[\{A\},\ \mathbf{((}f,\ g\mathbf{))}]_1$ for which the corresponding complex translinear matrices $[A,\ \mathbf{(}h\mathbf{)}]_1$ have positive real determinants form another normal subgroup of $\mathrm{TGL}_{(2)n}$, isomorphic to the *complex direct affine* group $\mathrm{G}^+\mathrm{A}_n(\mathbb{C})$. The ones corresponding to complex translinear matrices having determinant 1 form a normal subgroup isomorphic to the *complex special affine* group $\mathrm{SA}_n(\mathbb{C})$.

Let $\lambda \in \mathbb{C}$ be any nonzero scalar, and let $\mathbf{(}h\mathbf{)} \in \mathbb{C}^n$ be any row vector. Then the product of the scaling operation $\lambda\cdot : \mathbb{C}^n \to \mathbb{C}^n$ and the translation $+\mathbf{(}h\mathbf{)} : \mathbb{C}^n \to \mathbb{C}^n$ is a *complex transscalar transformation*

$$[\lambda\cdot,\ +\mathbf{(}h\mathbf{)}] : \mathbb{C}^n \to \mathbb{C}^n, \text{ with } \mathbf{(}z\mathbf{)} \mapsto \lambda\mathbf{(}z\mathbf{)} + \mathbf{(}h\mathbf{)}, \quad \lambda \neq 0, \qquad (8.1.7)$$

which induces a dilatation on $\mathbb{C}\mathrm{A}^n$. This is the same as the linear transformation $[\cdot\lambda I,\ +\mathbf{(}h\mathbf{)}] : \mathbb{C}^n \to \mathbb{C}^n$ associated with an $(n+1) \times (n+1)$ complex transscalar matrix or with the $(2n+1) \times (2n+1)$ real *transbivalent matrix*

$$[\backslash(\mu,\ \nu)_{(2)}\backslash,\ \mathbf{((}f,\ g\mathbf{))}]_1 = \begin{pmatrix} \backslash(\mu,\ \nu)_{(2)}\backslash & \mathbf{[[}0,\ 0\mathbf{]]} \\ \mathbf{((}f,\ g\mathbf{))} & 1 \end{pmatrix}, \qquad (8.1.8)$$

where $(\mu,\ \nu) = (\operatorname{Re}\lambda,\ \operatorname{Im}\lambda)$, and $\mathbf{(}f\mathbf{)}$ and $\mathbf{(}g\mathbf{)}$ are the real and imaginary parts of $\mathbf{(}h\mathbf{)}$. Postmultiplying the row vector $(\mathbf{((}x,\ y\mathbf{))},\ 1) \in \mathbb{R}^{2n+1}$ by this matrix yields the row vector $((\mu,\ \nu)\mathbf{((}x,\ y)_{(2)}\mathbf{)} + \mathbf{((}f,\ g\mathbf{))},\ 1)$. If $\mu^2 + \nu^2 = 1$, this is a *special* transbivalent matrix, and the operation on $\mathbb{C}\mathrm{A}^n$ is a unit dilatation. When $\mu = 1$ and $\nu = 0$, it is the translation with vector $\mathbf{(}h\mathbf{)}$.

The set of all transbivalent matrices of order $2n+1$ forms a multiplicative group $\mathrm{TGZ}_{(2)n}$ isomorphic to the *complex general dilative* group $\mathrm{GD}_n(\mathbb{C})$ of all dilatations of $\mathbb{C}\mathrm{A}^n$; this is a normal subgroup of $\mathrm{TGL}_{(2)n} \cong \mathrm{GA}_n(\mathbb{C})$. The matrices with $\mu > 0$ and $\nu = 0$ form a normal subgroup isomorphic to the *complex positive dilative* group $\mathrm{GD}_n^+(\mathbb{C})$. The special transbivalent matrices form another normal subgroup $\mathrm{TSZ}_{(2)n}$, isomorphic to the *complex unit dilative* group $\bar{\mathrm{S}}\mathrm{D}_n(\mathbb{C})$, and the translation matrices form an abelian subgroup $\mathrm{T}_{(2)n}$. The quotient group $\mathrm{TGZ}_{(2)n}/\mathrm{T}_{(2)n} \cong \mathrm{GD}_n(\mathbb{C})/\mathrm{T}^n(\mathbb{C})$ is isomorphic to the complex general scalar group $\mathrm{GZ}(\mathbb{C})$ of $n \times n$ nonzero scalar matrices over $\mathbb{C}$.

If A is an $n \times n$ invertible matrix over $\mathbb{C}$ and $(h) \in \mathbb{C}^n$ is any row vector, then the product of the antilinear transformation $\odot A : \mathbb{C}^n \to \mathbb{C}^n$ and the translation $+(h) : \mathbb{C}^n \to \mathbb{C}^n$ is a *transantilinear transformation*

$$[\odot A,\ +(h)] : \mathbb{C}^n \to \mathbb{C}^n, \text{ with } (z) \mapsto (\bar{z})A + (h), \quad \det A \neq 0, \quad (8.1.9)$$

which induces an *antiaffinity* on $\mathbb{C}\mathrm{A}^n$. The hyperplane $[[w],\ r]$ is taken into the hyperplane $[A^{-1}[\bar{w}],\ \bar{r} - (h)A^{-1}[\bar{w}]]$. An antiaffinity preserves parallelism and centroids but does not preserve barycenters in general.

The set of all affinities and antiaffinities of $\mathbb{C}\mathrm{A}^n$ forms a group, the *general semiaffine* group $\mathrm{G\bar{A}}_n(\mathbb{C})$, containing the general affine group $\mathrm{GA}_n(\mathbb{C})$ as a subgroup of index 2. Likewise, adjoining *antidilatations* $(z) \mapsto \lambda(\bar{z}) + (h)$ to the general dilative group $\mathrm{GD}_n(\mathbb{C})$ gives the *general semidilative* group $\mathrm{G\bar{D}}_n(\mathbb{C})$. Each of the various complex affine and dilative subgroups has an analogous semiaffine or semidilative extension.

EXERCISES 8.1

1. ⊢ The antilinear transformation $(z) \mapsto (\bar{z})A$ maps $(z) + (w)$ into $(\bar{z})A + (\bar{w})A$ and $\lambda(z)$ into $\bar{\lambda}[(\bar{z})A]$, so that linear combinations are taken to conjugate linear combinations.

2. $\vdash$ The set of bivalent matrices $\{z\} = (x,\ y)_{(2)}$ is closed under addition and multiplication.

3. Show that the translinear transformation $(z) \mapsto (z)A + (h)$ takes the hyperplane $[[w],\ r]$ into the hyperplane $[A^{-1}[w],\ r - (h)A^{-1}[w]]$.

4. Show that the transantilinear transformation $(z) \mapsto (\bar{z})A + (h)$ does not generally preserve barycenters.

8.2 PROJECTIVE TRANSFORMATIONS

Complex projective $(n-1)$-space $\mathbb{C}\mathrm{P}^{n-1}$ can be coordinatized so that each point Z is represented by the nonzero scalar multiples of a nonzero row vector $(z) \in \mathbb{C}^n$ and each hyperplane $\check{W}$ by the nonzero scalar multiples of a nonzero column vector $[w] \in \check{\mathbb{C}}^n$, with $Z \lozenge \check{W}$ if and only if $(z)[w] = 0$. If $(z) = (x) + (y)\mathrm{i}$ and $[w] = [u] + \mathrm{i}[v]$, then the points and hyperplanes of $\mathbb{C}\mathrm{P}^{n-1}$ can be given "bihomogeneous" semiduplex coordinates

$$((x,\ y)) = (x_1,\ y_1,\ \ldots,\ x_n,\ y_n) \quad \text{and} \quad [[u,\ v]] = [u_1,\ v_1,\ \ldots,\ u_n,\ v_n].$$

All row vectors $(\mu,\ \nu)((x,\ y)_{(2)})$ with $(\mu,\ \nu) \neq (0,\ 0)$ represent the same point, and all column vectors $[[u,\ v]_{(2)}][\sigma,\ \tau]$ with $[\sigma,\ \tau] \neq [0,\ 0]$ represent the same hyperplane. The point Z with semiduplex coordinates $((x,\ y))$ is incident with the hyperplane $\check{W}$ with semiduplex coordinates $[[u,\ v]]$ if and only if

$$((x,\ y)_{(2)})[(u,\ v)_{(2)}] = (0,\ 0)_{(2)}$$

or, more simply, if and only if $((x,\ y))[(u,\ v)_{(2)}] = (0,\ 0)$. It should be noted that $[(u,\ v)_{(2)}]$ is not the same matrix as $[[u,\ v]_{(2)}] = [(u,\ -v)_{(2)}]$.

A one-to-one transformation of complex projective $(n-1)$-space $\mathbb{C}\mathrm{P}^{n-1}$ into itself taking points into points and hyperplanes into hyperplanes, preserving incidence and cross ratios, is a *projective collineation*. If A is an invertible matrix of order n, a projective collineation $\mathbb{C}P^{n-1} \to \mathbb{C}P^{n-1}$ is induced by the *complex projective linear transformation*

$$\langle \cdot A \rangle : P\mathbb{C}^n \to P\mathbb{C}^n, \text{ with } \langle (z) \rangle \mapsto \langle (z)A \rangle, \quad \det A \neq 0, \tag{8.2.1}$$

of the projective linear space $P\mathbb{C}^n = \mathbb{C}^n/GZ(\mathbb{C})$ or by the "inverse dual" transformation

$$\langle A^{-1} \cdot \rangle : P\check{\mathbb{C}}^n \to P\check{\mathbb{C}}^n, \text{ with } \langle [w] \rangle \mapsto \langle A^{-1}[w] \rangle, \tag{8.2.2}$$

of the projective linear space $P\check{\mathbb{C}}^n = \check{\mathbb{C}}^n/GZ(\mathbb{C})$. A projective collineation of $\mathbb{C}P^{n-1}$ is thus defined by either of the rules

$$\lambda(z') = (z)A \quad \text{or} \quad A[w'] = [w]\rho, \tag{8.2.3}$$

λ and ρ being any nonzero scalars. If $(z) = (x) + (y)i$, $[w] = [u] + i[v]$, and $\{A\}$ is the duplex equivalent to A, then the collineation can also be defined by the rule

$$(\mu, \nu)((x', y')_{(2)}) = ((x, y)_{(2)})\{A\}$$

or

$$\{A\}[[u', v']_{(2)}] = [[u, v]_{(2)}][\sigma, \tau]. \tag{8.2.4}$$

By definition, the set of all invertible matrices of order n over $\mathbb{C}$ forms the complex general linear group $GL_n(\mathbb{C})$. The set of classes $\langle A \rangle$ of nonzero scalar multiples of matrices $A \in GL_n(\mathbb{C})$ forms its central quotient group, the *complex projective general linear* group $PGL_n(\mathbb{C}) \cong GL_n(\mathbb{C})/GZ(\mathbb{C})$, which is isomorphic to the *complex general projective* group $GP_{n-1}(\mathbb{C})$, the group of projective collineations of $\mathbb{C}P^{n-1}$. Matrices with determinant 1 form the complex special linear group $SL_n(\mathbb{C})$, the commutator subgroup of $GL_n(\mathbb{C})$, whose central quotient group $PSL_n(\mathbb{C}) \cong SL_n(\mathbb{C})/SZ_n(\mathbb{C})$ is isomorphic to $PGL_n(\mathbb{C})$ for all n. The general and unit commutator quotient groups $GCom(\mathbb{C}) \cong GL_n(\mathbb{C})/SL_n(\mathbb{C})$ and $\bar{S}Com(\mathbb{C}) \cong \bar{S}L_n(\mathbb{C})/SL_n(\mathbb{C})$ are isomorphic, respectively, to the general and unit scalar groups $GZ(\mathbb{C})$ and $\bar{S}Z(\mathbb{C})$.

A one-to-one transformation of complex projective $(n-1)$-space $\mathbb{C}\mathrm{P}^{n-1}$ into itself taking points into hyperplanes and hyperplanes into points, dualizing incidences and preserving cross ratios, is a *projective correlation.* If A is an invertible matrix of order n, a projective correlation $\mathbb{C}\mathrm{P}^{n-1} \to \mathbb{C}\mathrm{P}^{n-1}$ is induced by the *complex projective linear cotransformation*

$$\langle \cdot A \rangle^{\vee} : \mathrm{P}\mathbb{C}^n \to \mathrm{P}\check{\mathbb{C}}^n, \text{ with } \langle (z) \rangle \mapsto \langle (z)A \rangle^{\vee}, \quad \det A \neq 0, \tag{8.2.5}$$

or, equivalently, by the cotransformation

$$\langle A^{-1} \cdot \rangle^{\vee} : \mathrm{P}\check{\mathbb{C}}^n \to \mathrm{P}\mathbb{C}^n, \text{ with } \langle [w] \rangle \mapsto \langle A^{-1}[w] \rangle^{\vee}. \tag{8.2.6}$$

Thus a projective correlation of $\mathbb{C}\mathrm{P}^{n-1}$ is defined by either of the rules

$$\lambda [z']^{\vee} = (z)A \quad \text{or} \quad A(w')^{\vee} = [w]\rho, \tag{8.2.7}$$

where λ and ρ are any nonzero scalars. If $\{A\}$ is the duplex equivalent to A, then the correlation can also be defined by the rule

$$(\mu, \nu)[[x', y']_{(2)}]^{\vee} = ((x, y)_{(2)})\{A\}$$

or

$$\{A\}((u', v')_{(2)})^{\vee} = [[u, v]_{(2)}][\sigma, \tau]. \tag{8.2.8}$$

The product of two projective correlations is a projective collineation.

The group $\mathrm{PGL}_n(\mathbb{C})$ of dually paired projective linear transformations

$$(\mathrm{P}\mathbb{C}^n, \mathrm{P}\check{\mathbb{C}}^n) \to (\mathrm{P}\mathbb{C}^n, \mathrm{P}\check{\mathbb{C}}^n)$$

is a subgroup of index 2 in the *complex double projective general linear* group $\mathrm{DPGL}_n(\mathbb{C})$ of such transformations and dually paired projective linear cotransformations

$$(\mathrm{P}\mathbb{C}^n, \mathrm{P}\check{\mathbb{C}}^n) \to (\mathrm{P}\check{\mathbb{C}}^n, \mathrm{P}\mathbb{C}^n).$$

For $n > 2$, the group $\mathrm{DPGL}_n(\mathbb{C})$ is isomorphic to the *complex dualized general projective* group $\check{\mathrm{G}}\mathrm{P}_{n-1}(\mathbb{C})$ of projective collineations and correlations of $\mathbb{C}\mathrm{P}^{n-1}$.

A projective correlation of period 2 is a polarity. As with real polarities, the matrix of a polarity of $\mathbb{C}\mathrm{P}^{n-1}$ is either symmetric or (if n is even) skew-symmetric. In the latter case we have a *null polarity*, in which every point or hyperplane is incident with its own polar or pole.

EXERCISES 8.2

1. If $(z) = (x) + (y)\mathrm{i}$ and $\{A\}$ is the duplex equivalent of the $n \times n$ complex matrix A, show that the projective collineation $\langle(z)\rangle \mapsto \langle(z)A\rangle$ of $\mathbb{C}\mathrm{P}^{n-1}$ corresponds to the real projective collineation

$$\langle((x,\ y)_{(2)})\rangle \mapsto \langle((x,\ y)_{(2)})\{A\}\rangle.$$

2. Show that the projective correlation $\langle(z)\rangle \mapsto \langle(z)A\rangle\check{}$ of $\mathbb{C}\mathrm{P}^{n-1}$ corresponds to the real projective correlation

$$\langle((x,\ y)_{(2)})\rangle \mapsto \langle((x,\ y)_{(2)})\{A\}\rangle\check{}.$$

8.3 ANTIPROJECTIVE TRANSFORMATIONS

A collineation or correlation of complex projective $(n-1)$-space $\mathbb{C}\mathrm{P}^{n-1}$ that replaces cross ratios by their complex conjugates is said to be *antiprojective* (cf. Veblen & Young 1918, pp. 250–253). If A is an invertible matrix of order n, an antiprojective collineation $\mathbb{C}\mathrm{P}^{n-1} \to \mathbb{C}\mathrm{P}^{n-1}$ is induced by the *projective antilinear transformation*

$$\langle\odot A\rangle : \mathrm{P}\mathbb{C}^n \to \mathrm{P}\mathbb{C}^n, \text{ with } \langle(z)\rangle \mapsto \langle(\bar{z})A\rangle, \quad \det A \neq 0, \qquad (8.3.1)$$

or by the "inverse antidual" transformation

$$\langle A^{-1}\boxdot\rangle : \mathrm{P}\check{\mathbb{C}}^n \to \mathrm{P}\check{\mathbb{C}}^n, \text{ with } \langle[w]\rangle \mapsto \langle A^{-1}[\bar{w}]\rangle. \qquad (8.3.2)$$

That is, an antiprojective collineation of $\mathbb{C}\mathrm{P}^{n-1}$ is defined by either of the rules

$$\lambda(z') = (\bar{z})A \quad \text{or} \quad A[w'] = [\bar{w}]\rho, \qquad (8.3.3)$$

where λ and ρ are any nonzero scalars. If $\{A\}$ is the duplex equivalent to A, then the collineation can also be defined by the rules

$$(\mu,\ \nu)((x',\ y')_{(2)}) = ((x,\ -y)_{(2)})\{A\}$$

or (8.3.4)

$$\{A\}[[u',\ v']_{(2)}] = [[u,\ -v]_{(2)}][\sigma,\ \tau].$$

The product of two antiprojective collineations is a projective collineation.

If A is an invertible matrix of order n, an antiprojective correlation $\mathbb{C}P^{n-1} \to \mathbb{C}P^{n-1}$ is induced by the *projective antilinear cotransformation*

$$\langle\cdot A\rangle^* : P\mathbb{C}^n \to P\check{\mathbb{C}}^n, \text{ with } \langle(z)\rangle \mapsto \langle(z)A\rangle^*, \quad \det A \neq 0, \tag{8.3.5}$$

or, equivalently, by the cotransformation

$$\langle A^{-1}\cdot\rangle^* : P\check{\mathbb{C}}^n \to P\mathbb{C}^n, \text{ with } \langle[w]\rangle \mapsto \langle A^{-1}[w]\rangle^*, \tag{8.3.6}$$

where the star denotes the antitranspose. Thus an antiprojective correlation of $\mathbb{C}P^{n-1}$ is defined by either of the rules

$$\lambda[z']^* = (z)A \quad \text{or} \quad A(w')^* = [w]\rho, \tag{8.3.7}$$

where λ and ρ are any nonzero scalars. If $\{A\}$ is the duplex equivalent to A, then the correlation can also be defined by the rules

$$(\mu,\ \nu)[[x',\ -y']_{(2)}]^{\check{}} = ((x,\ y)_{(2)})\{A\}$$

or (8.3.8)

$$\{A\}((u',\ -v')_{(2)})^{\check{}} = [[u,\ v]_{(2)}][\sigma,\ \tau].$$

The product of two antiprojective correlations is a projective collineation.

An antiprojective correlation of period 2 is an antipolarity. A sufficient condition for the correlation (8.3.7) to be an antipolarity of $\mathbb{C}P^{n-1}$ is that the matrix A be Hermitian.

If some antiprojective collineation transforms a figure X into a figure X', we write $\mathsf{X} \overset{*}{\barwedge} \mathsf{X}'$. If one figure is transformed into the other by an antiprojective correlation, we write $\mathsf{X} \overset{*}{\veebar} \mathsf{X}'$ or, if the figures are related by an antipolarity, $\mathsf{X} \overset{*}{\circledcirc} \mathsf{X}'$.

The set of all collineations of $\mathbb{C}\mathrm{P}^{n-1}$, projective or antiprojective, forms the *general semiprojective* group $\mathrm{G\bar{P}}_{n-1}(\mathbb{C})$, isomorphic to the *projective general semilinear* group $\mathrm{PG\bar{L}}_n(\mathbb{C})$, containing the projective general linear group $\mathrm{PGL}_n(\mathbb{C})$ as a subgroup of index 2. The group $\mathrm{PG\bar{L}}_n(\mathbb{C})$ in turn is a subgroup of index 2 in the *double projective general semilinear* group $\mathrm{DPG\bar{L}}_n(\mathbb{C})$, which for $n > 2$ is isomorphic to the *dualized general semiprojective* group $\mathrm{\check{G}\bar{P}}_{n-1}(\mathbb{C})$ of all collineations and correlations of $\mathbb{C}\mathrm{P}^{n-1}$.

EXERCISES 8.3

1. If $(z) = (x) + (y)\mathrm{i}$ and $\{A\}$ is the duplex equivalent of the $n \times n$ complex matrix A, show that the antiprojective collineation $\langle(z)\rangle \mapsto \langle(\bar{z})A\rangle$ of $\mathbb{C}\mathrm{P}^{n-1}$ corresponds to the real projective collineation

$$\langle((x,\ y)_{(2)})\rangle \mapsto \langle((x,\ -y)_{(2)})\{A\}\rangle.$$

2. If $\{\bar{A}\}$ is the duplex equivalent of the conjugate of the $n \times n$ complex matrix A, show that the antiprojective correlation $\langle(z)\rangle \mapsto \langle(z)A\rangle^*$ of $\mathbb{C}\mathrm{P}^{n-1}$ corresponds to the real projective correlation

$$\langle((x,\ y)_{(2)})\rangle \mapsto \langle((x,\ -y)_{(2)})\{\bar{A}\}\rangle\check{}.$$

8.4 SUBGROUPS AND QUOTIENT GROUPS

The one-dimensional complex affine and linear groups are the same as the corresponding complex dilative and scalar groups, and the unit dilative and unit scalar groups can be identified with the two-dimensional direct Euclidean and special orthogonal groups:

$$\mathrm{GA}_1(\mathbb{C}) \cong \mathrm{GD}_1(\mathbb{C}) \cong \tilde{\mathrm{E}}_2^+, \quad \bar{\mathrm{S}}\mathrm{A}_1(\mathbb{C}) \cong \bar{\mathrm{S}}\mathrm{D}_1(\mathbb{C}) \cong \mathrm{E}_2^+,$$
$$\mathrm{GL}_1(\mathbb{C}) \cong \mathrm{GZ}(\mathbb{C}) \cong \tilde{\mathrm{O}}_2^+, \quad \bar{\mathrm{S}}\mathrm{L}_1(\mathbb{C}) \cong \bar{\mathrm{S}}\mathrm{Z}(\mathbb{C}) \cong \mathrm{O}_2^+.$$

In fact, the direct orthopetic group $\tilde{\mathrm{O}}_2^+$ is isomorphic to its subgroup O_2^+ (cf. Beardon 2005, p. 227). When $n \geq 2$, the various groups of affinities of $\mathbb{C}\mathrm{A}^n$ or linear transformations of $\mathbb{C}^n$ are distinct. The scalar groups $\mathrm{GZ}(\mathbb{C})$ and $\bar{\mathrm{S}}\mathrm{Z}(\mathbb{C})$ do not depend on n. Each affine or linear group can be extended by the *semi-identity* group $\bar{1}$ of order 2 generated by the conjugation automorphism, giving a semiaffine or semilinear group.

Figure 8.4a shows the principal normal subgroups of the complex general affine group $\mathrm{GA}_n(\mathbb{C})$ or, briefly, ${}^{\mathbb{C}}\mathrm{GA}_n$. When one group is a (normal) subgroup of another, the two are connected by a line labeled with the quotient group. A similar diagram for the complex general linear group $\mathrm{GL}_n(\mathbb{C}) \cong \mathrm{GA}_n(\mathbb{C})/\mathrm{T}^n(\mathbb{C})$ is obtained when group symbols A_n, D_n, and T^n are respectively replaced by L_n, Z_n or Z, and 1.

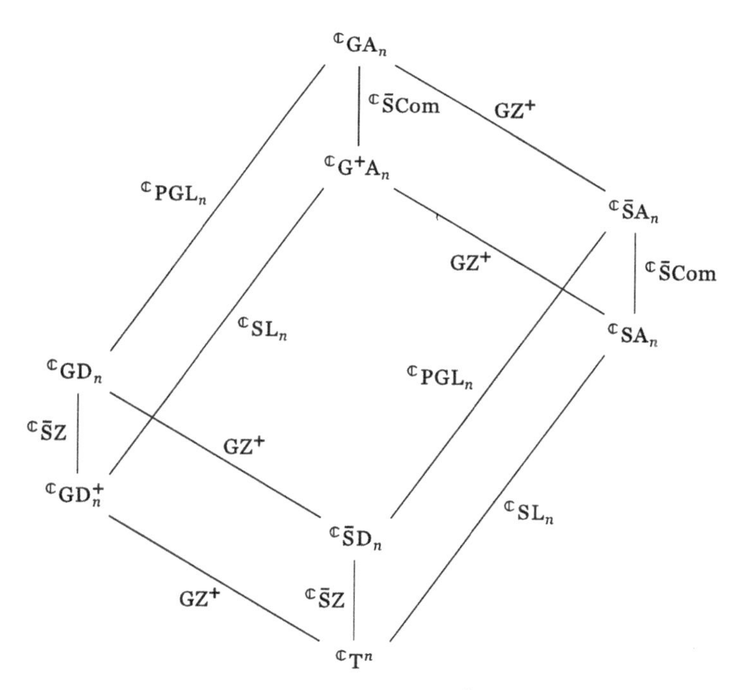

Figure 8.4a Subgroups of ${}^{\mathbb{C}}\mathrm{GA}_n$

Subgroup diagrams for the general semiaffine group $G\bar{A}_n(\mathbb{C})$ and the general semilinear group $G\bar{L}_n(\mathbb{C})$ follow the same pattern. In the first case the group symbols A_n, D_n, and T^n are replaced by $\bar{A}_n$, $\bar{D}_n$, and $\bar{T}^n$; in the second case they are replaced by $\bar{L}_n$, $\bar{Z}_n$ or $\bar{Z}$, and $\bar{1}$. (Some writers replace G by Γ to denote general semilinear and semiaffine groups, but this notation does not readily extend to subgroups.)

9

CIRCULARITIES AND CONCATENATIONS

THE REAL INVERSIVE n-SPHERE can be given a Euclidean metric as the *parabolic* n-sphere $\dot{S}^n$. For $n = 2$, this is the Euclidean plane augmented by a single *exceptional point* $\dot{O}$ lying on every line, with such extended lines regarded as *great circles* of $\dot{S}^2$. Circle-preserving transformations that leave $\dot{O}$ fixed are *similarities* of the parabolic sphere, while those that treat all points alike are *circularities* (homographies and antihomographies) of the real inversive sphere I^2. Inversive circles correspond to chains of the complex projective line $\mathbb{C}P^1$, and the group of all circularities of I^2 is isomorphic to the group of *concatenations* (projectivities and antiprojectivities) of $\mathbb{C}P^1$. Inversive unitary n-space is conformal to the real inversive $(2n+1)$-sphere, and subspaces of Iu^n correspond to odd-dimensional subspaces of I^{2n+1}.

9.1 THE PARABOLIC n-SPHERE

In § 3.1, we saw that an oval n-quadric Φ in real projective $(n+1)$-space P^{n+1} can be regarded as the Möbius (real inversive) n-sphere I^n. The k-planes of P^{n+1} meet Φ in inversive $(k-1)$-spheres—*circles* if $k = 2$. A projective collineation that commutes with the polarity for which Φ is the locus of self-conjugate points induces a circle-preserving transformation, or *circularity*, on I^n. Any projective hyperplane $\check{O}$ determines an inversive $(n-1)$-sphere $\mathring{\omega}$, which is real, degenerate, or imaginary according as $\check{O}$ intersects, is tangent to, or does not meet

Φ. If $\mathring{\omega}$ is nondegenerate, we can fix the inversion in $\mathring{\omega}$ as the *central inversion*, and those circularities that commute with it are isometries of a "central" n-sphere. When $\mathring{\omega}$ is imaginary, this is the usual *elliptic* n-sphere S^n; when $\mathring{\omega}$ is real, it is the *hyperbolic* n-sphere $\ddot{\mathrm{S}}^n$. Although an inversion must be of one of these two types, there is yet a third kind of metric n-sphere to be derived from I^n.

When a hyperplane $\check{O}$ is *tangent* to the n-quadric Φ, the *point of contact* $\dot{O}$ is the only real point of a degenerate $(n-1)$-sphere $\dot{\omega}$, and the circularities of I^n that leave $\dot{\omega}$ fixed are transformations of the *parabolic* n-sphere $\dot{\mathrm{S}}^n$. Let the points X and hyperplanes $\check{U}$ of P^{n+1} have homogeneous coordinates

$$((x)) = (x_0, x_1, \ldots, x_{n+1}) \quad \text{and} \quad [[u]] = [u_0, u_1, \ldots, u_{n+1}],$$

and, as before, take the locus or envelope equation of Φ to be

$$x_1{}^2 + \cdots + x_{n+1}{}^2 = x_0{}^2 \quad \text{or} \quad u_1{}^2 + \cdots + u_{n+1}{}^2 = u_0{}^2.$$

Then it will be convenient to let $\dot{O}$ and $\check{O}$ have the respective coordinates

$$(1,\ 0, \ldots,\ 0,\ 1) \quad \text{and} \quad [1,\ 0, \ldots,\ 0,\ -1].$$

We call $\dot{O}$ the *exceptional point* and $\dot{\omega}$ the *exceptional hypersphere* of $\dot{\mathrm{S}}^n$; all other real points and real or degenerate $(n-1)$-spheres are *ordinary*.

For every point X on $\dot{\mathrm{S}}^n$, $x_0 \neq 0$, and normalized coordinates for X are its projective coordinates divided by x_0, i.e.,

$$((x)) = (1,\ x_1, \ldots,\ x_n,\ x_{n+1}), \tag{9.1.1}$$

where $x_1{}^2 + \cdots + x_{n+1}{}^2 = 1$. Then X is ordinary if and only if $x_{n+1} \neq 1$. A hyperplane $\check{U}$ passing through $\dot{O}$, so that $u_0 + u_{n+1} = 0$, intersects $\dot{\mathrm{S}}^n$ in a *great hypersphere* $\check{U}$, with homogeneous coordinates

$$[[u]] = [u_0,\ u_1, \ldots,\ u_n,\ -u_0], \tag{9.1.2}$$

$\check{U}$ being ordinary if and only if the n-vector $[u] = [u_1, \ldots, u_n]$ is nonzero. As usual, we have $X \diamond \check{U}$ if and only if $((x))[[u]] = 0$.

The parabolic n-sphere $\dot{\mathrm{S}}^n$ can be given a Euclidean metric. In terms of normalized coordinates and the bilinear forms

$$((x\,\dot{}\,y)) = x_1y_1 + \cdots + x_{n+1}y_{n+1} \quad \text{and} \quad [u\ v] = u_1v_1 + \cdots + u_nv_n,$$

with $((x\,\dot{}\,x)) = ((y\,\dot{}\,y)) = 1$, the distance between two ordinary points X and Y on $\dot{\mathrm{S}}^n$ is given by

$$|XY| = \sqrt{\frac{2 - 2((x\,\dot{}\,y))}{(1 - x_{n+1})(1 - y_{n+1})}}, \tag{9.1.3}$$

and the angle between intersecting (separating) great hyperspheres $\check{U}$ and $\check{V}$ by

$$(\check{U}\check{V}) = \cos^{-1} \frac{|[u\ v]|}{\sqrt{[u\ u]}\sqrt{[v\ v]}}. \tag{9.1.4}$$

Two great hyperspheres $\check{U}$ and $\check{V}$ are parallel (tangent at $\dot{O}$) if one of the n-vectors $[u]$ and $[v]$ is a scalar multiple of the other.

The exceptional point $\dot{O}$ may be taken as the single point at infinity of an n-horosphere (a horocycle if $n = 1$) in hyperbolic $(n+1)$-space H^{n+1}, the ordinary part of which (by Wachter's Theorem) provides an isometric model for Euclidean n-space E^n. The great hyperspheres of $\dot{\mathrm{S}}^n$, $(n-1)$-horospheres in H^{n+1}, are the hyperplanes of E^n (Coxeter 1966a, p. 224).

That the geometry of the parabolic n-sphere is Euclidean may be seen more clearly if we divide the normalized coordinates $((x))$ of an ordinary point X by $1 - x_{n+1}$ to obtain the *parabolic coordinates*

$$((\dot{x})) = \left(\tfrac{1}{2}((\dot{x}\,\dot{x}) + 1),\ \dot{x}_1, \ldots,\ \dot{x}_n,\ \tfrac{1}{2}((\dot{x}\,\dot{x}) - 1)\right), \tag{9.1.5}$$

where

$$\dot{x}_i = \frac{x_i}{1 - x_{n+1}} \ (1 \le i \le n) \quad \text{and} \quad (\dot{x}\,\dot{x}) = \dot{x}_1{}^2 + \cdots + \dot{x}_n{}^2 = \frac{1 + x_{n+1}}{1 - x_{n+1}}.$$

Then $(\dot{x}) = (\dot{x}_1, \ldots, \dot{x}_n)$ is the vector of Cartesian coordinates for X. Likewise, if we divide the homogeneous coordinates $[\![u]\!]$ of an ordinary great hypersphere $\breve{U}$ by $\sqrt{[u\,u]} = [u_1{}^2 + \cdots + u_n{}^2]^{1/2}$, we obtain the parabolic coordinates

$$[\![\dot{u}]\!] = [\dot{u}_0,\ \dot{u}_1, \ldots,\ \dot{u}_n,\ -\dot{u}_0], \tag{9.1.6}$$

of a Euclidean hyperplane $\breve{U}$, with $[\dot{u}] = [\dot{u}_1, \ldots, \dot{u}_n]$ being a unit covector of direction cosines (Johnson 1981, pp. 455–456).

EXERCISES 9.1

1. Formula (9.1.3) gives the distance between two ordinary points on $\dot{\mathrm{S}}^n$ with normalized coordinates $((x))$ and $((y))$. Find an expression for the distance in terms of parabolic coordinates $((\dot{x}))$ and $((\dot{y}))$.

2. Formula (9.1.4) gives the angle between two intersecting great hyperspheres of $\dot{\mathrm{S}}^n$ with homogeneous coordinates $[\![u]\!]$ and $[\![v]\!]$. How is this formula affected by the change to parabolic coordinates $[\![\dot{u}]\!]$ and $[\![\dot{v}]\!]$?

9.2 THE REAL INVERSIVE SPHERE

The parabolic sphere $\dot{\mathrm{S}}^2$ can be represented by the ordinary Euclidean plane $\hat{\mathrm{E}}^2$ augmented by a single point $\dot{O}$ at infinity—as a topological space this is the *one-point compactification* $\hat{\mathrm{E}}^2$.

A. *Points and circles.* An ordinary point of $\dot{\mathrm{S}}^2$ with Cartesian coordinates (x, y) has parabolic coordinates

$$\left(\tfrac{1}{2}(x^2 + y^2 + 1),\ x,\ y,\ \tfrac{1}{2}(x^2 + y^2 - 1)\right),$$

while the exceptional point $\dot{O}$ is

$$(1,\ 0,\ 0,\ 1).$$

Every point, ordinary or exceptional, has normalized coordinates

$$(1,\ \xi,\ \eta,\ \zeta),$$

with $\xi^2 + \eta^2 + \zeta^2 = 1$. For an ordinary point

$$\xi = \frac{2x}{x^2 + y^2 + 1}, \quad \eta = \frac{2y}{x^2 + y^2 + 1}, \quad \zeta = \frac{x^2 + y^2 - 1}{x^2 + y^2 + 1}.$$

Circles on $\dot{\mathrm{S}}^2$ may be either *great* or *small*. A small circle with center (h, k) and radius r has parabolic coordinates

$$\left[\tfrac{1}{2}(r^2 - h^2 - k^2 - 1),\ h,\ k,\ -\tfrac{1}{2}(r^2 - h^2 - k^2 + 1)\right].$$

The equation of a small circle is $(x - h)^2 + (y - k)^2 = r^2$. If $r = 0$, the circle is degenerate. An ordinary great circle, i.e., a line, with inclination θ and displacement c has coordinates

$$[c,\ \sin\theta,\ -\cos\theta,\ -c],$$

with $0 \leq \theta < \pi$. The equation of a line is $y \cos\theta = x \sin\theta + c$; if $\theta \neq \pi/2$, the line has slope $m = \tan\theta$ and y-intercept $b = c \sec\theta$. The exceptional point $(1, 0, 0, 1)$, which lies on every ordinary great circle, is the only point of the exceptional great circle

$$[1,\ 0,\ 0,\ -1].$$

Any three distinct points of $\dot{\mathrm{S}}^2$ lie on a unique circle, great or small. A necessary and sufficient condition for four points A, B, C, and D with parabolic or normalized coordinates $((a))$, $((b))$, $((c))$, and $((d))$ to be concyclic is that

$$\begin{vmatrix} a_0 & a_1 & a_2 & a_3 \\ b_0 & b_1 & b_2 & b_3 \\ c_0 & c_1 & c_2 & c_3 \\ d_0 & d_1 & d_2 & d_3 \end{vmatrix} = 0. \tag{9.2.1}$$

That is, four points are concyclic if and only if their coordinates are linearly dependent. If D is the exceptional point $(1, 0, 0, 1)$, this is the condition for three points A, B, C to lie on a great circle, i.e., to be collinear.

B. *Parabolic similarities.* Let λ be a nonzero scalar, P a 2×2 orthogonal matrix

$$\begin{pmatrix} \cos\phi & \sin\phi \\ -\sin\phi & \cos\phi \end{pmatrix} \quad \text{or} \quad \begin{pmatrix} \cos\phi & \sin\phi \\ \sin\phi & -\cos\phi \end{pmatrix},$$

and (h, k) an arbitrary vector. Then the transformation $\cdot[\lambda]P(h, k) : \dot{S}^2 \to \dot{S}^2$, with $(x, y) \mapsto \lambda(x, y)P + (h, k)$ and $\dot{O} \mapsto \dot{O}$, is a *parabolic similarity*, which can be induced by the *parabolic similarity matrix*

$$[\lambda]P(h, k) = \begin{pmatrix} \dfrac{\lambda^2 + h^2 + k^2 + 1}{2\lambda} & \dfrac{h}{\lambda} & \dfrac{k}{\lambda} & \dfrac{\lambda^2 + h^2 + k^2 - 1}{2\lambda} \\ h\cos\phi + k\sin\phi & \cos\phi & \sin\phi & h\cos\phi + k\sin\phi \\ \mp h\sin\phi \pm k\cos\phi & \mp\sin\phi & \pm\cos\phi & \mp h\sin\phi + k\cos\phi \\ \dfrac{\lambda^2 - h^2 - k^2 - 1}{2\lambda} & -\dfrac{h}{\lambda} & -\dfrac{k}{\lambda} & \dfrac{\lambda^2 - h^2 - k^2 + 1}{2\lambda} \end{pmatrix}. \tag{9.2.2}$$

That is, a coordinate vector for any point of $\dot{S}^2$ can be postmultiplied by the matrix (9.2.2) to produce a coordinate vector of the image point. The exceptional point $\dot{O} = (1, 0, 0, 1)$ is invariant.

The parabolic similarity $\cdot[\lambda]P(h, k)$ is the product of the scaling operation $\cdot[\lambda]I$, the orthogonal transformation $\cdot[1]P$, and the translation $\cdot I(h, k)$, respectively, induced by the *parabolic scalar matrix*

$$[\lambda]I = \begin{pmatrix} \dfrac{\lambda^2 + 1}{2\lambda} & 0 & 0 & \dfrac{\lambda^2 - 1}{2\lambda} \\ 0 & 1 & 0 & 0 \\ 0 & 0 & 1 & 0 \\ \dfrac{\lambda^2 - 1}{2\lambda} & 0 & 0 & \dfrac{\lambda^2 + 1}{2\lambda} \end{pmatrix}, \tag{9.2.3}$$

the *parabolic orthogonal matrix*

$$[1]P = \begin{pmatrix} 1 & 0 & 0 & 0 \\ 0 & \cos\phi & \sin\phi & 0 \\ 0 & \mp\sin\phi & \pm\cos\phi & 0 \\ 0 & 0 & 0 & 1 \end{pmatrix}, \qquad (9.2.4)$$

and the *parabolic translation matrix*

$$I(h,\ k) = \begin{pmatrix} 1+\frac{1}{2}(h^2+k^2) & h & k & \frac{1}{2}(h^2+k^2) \\ h & 1 & 0 & h \\ k & 0 & 1 & k \\ -\frac{1}{2}(h^2+k^2) & -h & -k & 1-\frac{1}{2}(h^2+k^2) \end{pmatrix}. \qquad (9.2.5)$$

Note that $\det\,[\lambda]I = 1$, $\det\,[1]P = \det P = \pm 1$, and $\det I(h,\ k) = 1$.

Matrices $[\lambda]I$ and $[1]P$ commute, the product in either order being a *parabolic orthopetic matrix* $[\lambda]P$. The same similarity is induced by a parabolic similarity matrix (9.2.2) and its negative, with the scalar λ replaced by $-\lambda$ and the angle ϕ by $\phi+\pi$. Thus the group of all matrices $[\lambda]P(h,\ k)$ is two-to-one homomorphic to the euclopetic group $\tilde{\mathrm{E}}_2$ of Euclidean similarities.

A similarity $\cdot[\lambda]P(h,\ k)$ takes great circles into great circles, i.e., lines into lines, and is direct or opposite according as $\det P$ is $+1$ or -1. When $|\lambda| = 1$, it is a *parabolic isometry*; when $P = \pm I$, it is a *dilatation*. If $(h,\ k) = (0,\ 0)$, the origin $O = (1,\ 0,\ 0,\ -1)$ is an invariant point.

C. *Reflections and inversions.* The Euclidean group E_2 of all parabolic isometries is generated by reflections in lines. The reflection $\cdot P_c : \dot{\mathrm{S}}^2 \to \dot{\mathrm{S}}^2$, whose mirror is the line $\check{P}(c)$ with coordinates $[c,\ \sin \tfrac{1}{2}\phi,\ -\cos \tfrac{1}{2}\phi,\ -c]$, is induced by the *parabolic reflection matrix*

$$P_c = \begin{pmatrix} 1+2c^2 & -2c\sin\frac{1}{2}\phi & 2c\cos\frac{1}{2}\phi & 2c^2 \\ 2c\sin\frac{1}{2}\phi & \cos\phi & \sin\phi & 2c\sin\frac{1}{2}\phi \\ -2c\cos\frac{1}{2}\phi & \sin\phi & -\cos\phi & -2c\cos\frac{1}{2}\phi \\ -2c^2 & 2c\sin\frac{1}{2}\phi & -2c\cos\frac{1}{2}\phi & 1-2c^2 \end{pmatrix}. \tag{9.2.6}$$

The reflection $\cdot P_c$ leaves all points on the line $\breve{P}(c)$ invariant and takes each small circle whose center lies on $\breve{P}(c)$ and each line perpendicular to $\breve{P}(c)$ into itself. The product of an even number of reflections is a direct isometry—a rotation or a translation. The product of an odd number of reflections is an opposite isometry—a reflection or a transflection.

A similarity $\cdot[\lambda]P(h,\ k)$ of the parabolic sphere $\dot{\mathrm{S}}^2$ is a circularity of the inversive sphere I^2. It is not the most general circularity, however, because great circles always go into great circles and small circles into small circles. A wider class of transformations, utilized by Ludwig Magnus as early as 1831 and later described in detail by Möbius (1855), is obtained by considering a reflection in a line to be a particular case of an *inversion in a circle*, which may be any real circle, great or small, or indeed any nondegenerate circle, real or imaginary.

A small circle $\mathring{q}\ (h,\ k)$ with center $(h,\ k)$ and radius r has coordinates $[\tfrac{1}{2}(q-1),\ h,\ k,\ -\tfrac{1}{2}(q+1)]$, where $q = r^2 - h^2 - k^2$. The circular inversion $\cdot Q_{(h,k)} : \dot{\mathrm{S}}^2 \to \dot{\mathrm{S}}^2$ is induced by the *small inversion matrix*

$$Q_{(h,k)} = \begin{pmatrix} 1+\dfrac{(q-1)^2}{2r^2} & -\dfrac{h(q-1)}{r^2} & -\dfrac{k(q-1)}{r^2} & \dfrac{q^2-1}{2r^2} \\ \dfrac{h(q-1)}{r^2} & 1-\dfrac{2h^2}{r^2} & -\dfrac{2hk}{r^2} & \dfrac{h(q+1)}{r^2} \\ \dfrac{k(q-1)}{r^2} & -\dfrac{2hk}{r^2} & 1-\dfrac{2k^2}{r^2} & \dfrac{k(q+1)}{r^2} \\ -\dfrac{q^2-1}{2r^2} & \dfrac{h(q+1)}{r^2} & \dfrac{k(q+1)}{r^2} & 1-\dfrac{(q+1)^2}{2r^2} \end{pmatrix}. \tag{9.2.7}$$

The inversion $\cdot Q_{(h,k)}$ leaves all points on the circle $\mathring{q}\ (h,\ k)$ invariant and takes each (great or small) circle orthogonal to $\mathring{q}\ (h,\ k)$ into itself; it interchanges the ordinary point $\left(\frac{1}{2}(h^2+k^2+1),\ h,\ k,\ -\frac{1}{2}(h^2+k^2-1)\right)$, alias (h, k), and the exceptional point $(1, 0, 0, 1)$. In general, inversion in the circle $\mathring{q}\ (h,\ k)$ interchanges points (x, y) and (x', y'), where

$$x' = \frac{r^2(x-h)}{(x-h)^2+(y-k)^2} + h, \quad y' = \frac{r^2(y-k)}{(x-h)^2+(y-k)^2} + k. \quad (9.2.8)$$

If r^2 is replaced by $-r^2$, then $\mathring{q}\ (h,\ k)$ is an imaginary circle with center (h, k) and radius ri. The number r^2 or $-r^2$ is the *power* of the inversion.

Since a reflection in a line can be regarded as an inversion in a great circle of $\dot{\mathrm{S}}^2$, we may call (9.2.6) a *great inversion matrix*. The four matrices

$$\begin{pmatrix} 1 & 0 & 0 & 0 \\ 0 & 1 & 0 & 0 \\ 0 & 0 & 1 & 0 \\ 0 & 0 & 0 & -1 \end{pmatrix}, \quad \begin{pmatrix} 1 & 0 & 0 & 0 \\ 0 & 1 & 0 & 0 \\ 0 & 0 & -1 & 0 \\ 0 & 0 & 0 & 1 \end{pmatrix}, \quad \begin{pmatrix} 1 & 0 & 0 & 0 \\ 0 & -1 & 0 & 0 \\ 0 & 0 & 1 & 0 \\ 0 & 0 & 0 & 1 \end{pmatrix}, \quad \begin{pmatrix} -1 & 0 & 0 & 0 \\ 0 & 1 & 0 & 0 \\ 0 & 0 & 1 & 0 \\ 0 & 0 & 0 & 1 \end{pmatrix}$$

respectively correspond to inversion in the real unit circle $[0, 0, 0, 1]$, reflection in the x-axis $[0, 0, 1, 0]$, reflection in the y-axis $[0, 1, 0, 0]$, and inversion in the imaginary unit circle $[1, 0, 0, 0]$.

While an inversion in a great circle is an isometry of the parabolic sphere $\dot{\mathrm{S}}^2$, an inversion in a small circle is not a proper transformation of $\dot{\mathrm{S}}^2$ since it does not leave the exceptional point fixed. An inversion of either type is a circularity of the inversive sphere I^2, but inversive geometry recognizes no distinction between great circles and small circles. It is thus desirable to give a single definition of inversion in terms of generic *circles*.

Let $\mathring{c}$ be a circle of I^2, with coordinates $[c_0,\ c_1,\ c_2,\ c_3]$, and let $[[u\ u]]$ be the quadratic form associated with the bilinear form

$$[\![u\ v]\!] = -u_0v_0 + u_1v_1 + u_2v_2 + u_3v_3.$$

Then $\mathring{c}$ is real, degenerate, or imaginary according as $[\![c\ c]\!]$ is positive, zero, or negative. In the nondegenerate cases, define an *inversion* in $\mathring{c}$ to be the involutory circularity $\langle \cdot C \rangle : \mathrm{I}^2 \to \mathrm{I}^2$ induced by the (generic) *inversion matrix*

$$C = \frac{1}{[\![c\ c]\!]}\begin{pmatrix} c_0{}^2 + c_1{}^2 + c_2{}^2 + c_3{}^2 & -2c_0c_1 & -2c_0c_2 & -2c_0c_3 \\ 2c_1c_0 & -c_0{}^2 - c_1{}^2 + c_2{}^2 + c_3{}^2 & -2c_1c_2 & -2c_1c_3 \\ 2c_2c_0 & -2c_2c_1 & -c_0{}^2 + c_1{}^2 - c_2{}^2 + c_3{}^2 & -2c_2c_3 \\ 2c_3c_0 & -2c_3c_1 & -2c_3c_2 & -c_0{}^2 + c_1{}^2 + c_2{}^2 - c_3{}^2 \end{pmatrix} \tag{9.2.9}$$

(cf. Schwerdtfeger 1962, pp. 117–118). If $[\![c\ c]\!] > 0$, $\langle \cdot C \rangle$ is a *hyperbolic inversion*, leaving all points on the real circle $\mathring{c}$ invariant and taking each circle orthogonal to $\mathring{c}$ into itself. If $[\![c\ c]\!] < 0$, $\langle \cdot C \rangle$ is an *elliptic inversion* in the imaginary circle $\mathring{c}$, leaving no real points invariant.

D. *Homographies and antihomographies.* The product of an even number of hyperbolic inversions is a direct circularity of I^2, generally known as a *homography* or "Möbius transformation"; the product of an odd number is an opposite circularity, or *antihomography*. An elliptic inversion can be expressed as the product of inversions in three mutually orthogonal real circles and is thus an antihomography. Every circularity is the product of at most four hyperbolic inversions.

The formulas and criteria of § 3.1 are applicable. Two real circles $\mathring{u}$ and $\mathring{v}$, with respective coordinates $[\![u]\!]$ and $[\![v]\!]$, are separating, tangent, or separated according as the discriminant

$$[\![u\ u]\!]\,[\![v\ v]\!] - [\![u\ v]\!]^2$$

is positive, zero, or negative. The cosine of the angle between two separating circles is

$$\frac{|[\![u\ v]\!]|}{\sqrt{[\![u\ u]\!]}\sqrt{[\![v\ v]\!]}},$$

so that the circles are orthogonal if $[\![u\ v]\!] = 0$.

The product of inversions in two real circles is a *rotary* (or "elliptic") homography if the circles are separating, an *elative* (or "parabolic") homography if they are tangent, a *dilative* (or "hyperbolic") homography if they are separated. The product of inversions in three real circles, two of which may be taken to be orthogonal to the third, is an *elliptic*, *parabolic*, or *hyperbolic* antihomography according as the first two are separating, tangent, or separated. The product of inversions in four real circles—two separating circles each orthogonal to both of two separated circles—is a *spiral* (or "loxodromic") homography, the commutative product of a rotary and a dilative homography (Veblen & Young 1918, pp. 246–248; Coxeter 1966a, pp. 230, 235–237).

The matrix of any circularity of the real inversive sphere can be expressed as the product of inversion matrices (9.2.9), taking one of two forms depending on whether the circularity is direct or opposite. In terms of the diagonal matrix $H = -I_{1,3} = \backslash -1,\ 1,\ 1,\ 1\backslash$, a matrix M of determinant 1 such that $MHM^{\breve{}} = H$ induces a homography on I^2, and a matrix N of determinant -1 such that $NHN^{\breve{}} = H$ an antihomography. We may call M and N *homographic* and *antihomographic* matrices. Since H (being congruent to $I_{3,1}$) is the matrix of a hyperbolic polarity, and since the same transformation is induced by a matrix M or N and any of its nonzero scalar multiples, the *general inversive* group GI_2 of all circularities of I^2 is the projective pseudo-orthogonal group $\mathrm{PO}_{3,1}$. The *special inversive* group SI_2 of homographies is the special projective pseudo-orthogonal group $\mathrm{P^+O}_{3,1}$.

EXERCISES 9.2

1. Show that when coordinates of points are taken as rows and coordinates of circles as columns, a point $((x))$ of the parabolic sphere $\dot{\mathrm{S}}^2$ lies on a (small or great) circle $[[u]]$ if and only if $((x))[[u]] = 0$.

2. Show that the parabolic similarity $\cdot[\lambda]P(h,\ k)$ defined by (9.2.2) is the product of the orthopetic transformation $\cdot[\lambda]P$ and the translation $\cdot I(h,\ k)$.

3. Show that the reflection $\cdot P_c$ defined by (9.2.6) leaves the line $\check{P}(c)$ with coordinates $[c, \sin \frac{1}{2}\phi, -\cos \frac{1}{2}\phi, -c]$ pointwise invariant.

4. Show that the inversion $\cdot Q_{(h,k)}$ defined by (9.2.7) leaves the circle $\mathring{q}\ (h, k)$ with coordinates $[\frac{1}{2}(q-1), h, k, -\frac{1}{2}(q+1)]$ pointwise invariant.

5. Show that the inversion $\langle \cdot C \rangle$ defined by (9.2.9) with $[\![c\ c]\!] > 0$ leaves the circle $\mathring{c}$ with coordinates $[c_0, c_1, c_2, c_3]$ pointwise invariant.

9.3 THE COMPLEX PROJECTIVE LINE

As what is commonly called the *Riemann sphere* (after Bernhard Riemann, 1826–1866), the parabolic sphere $\dot{\mathrm{S}}^2$ provides a conformal model for the complex projective line $\mathbb{C}\mathrm{P}^1$. The points of $\mathbb{C}\mathrm{P}^1$ may be given homogeneous coordinates $(z) = (z_1, z_2)$; a point Z with $z_2 \neq 0$ may also be associated with the single complex number $z = z_1/z_2$, and the remaining point may be associated with the extended value ∞. Then an ordinary point of $\dot{\mathrm{S}}^2$ with Cartesian coordinates (x, y) represents the point of $\mathbb{C}\mathrm{P}^1$ associated with $x+y\mathrm{i}$, and the exceptional point of $\dot{\mathrm{S}}^2$ corresponds to the "point at infinity" of $\mathbb{C}\mathrm{P}^1$.

In terms of normalized coordinates $(1, \xi, \eta, \zeta)$ for points of $\dot{\mathrm{S}}^2$, with $\xi^2 + \eta^2 + \zeta^2 = 1$, we have the correspondence

$$z \leftrightarrow (1, \xi, \eta, \zeta),$$

where, if $\zeta \neq 1$,

$$\operatorname{Re} z = \frac{\xi}{1-\zeta}, \quad \operatorname{Im} z = \frac{\eta}{1-\zeta}, \quad |z|^2 = \frac{1+\zeta}{1-\zeta}.$$

If $\zeta = 1$, then $z = \infty$. In particular, we have the identifications

$$\begin{aligned} 1 &\leftrightarrow (1, 1, 0, 0), & -1 &\leftrightarrow (1, -1, 0, 0), \\ \mathrm{i} &\leftrightarrow (1, 0, 1, 0), & -\mathrm{i} &\leftrightarrow (1, 0, -1, 0), \\ \infty &\leftrightarrow (1, 0, 0, 1), & 0 &\leftrightarrow (1, 0, 0, -1). \end{aligned}$$

The connection between the complex projective line $\mathbb{C}\mathrm{P}^1$ and the parabolic sphere $\dot{\mathrm{S}}^2$ has a simple geometric interpretation in

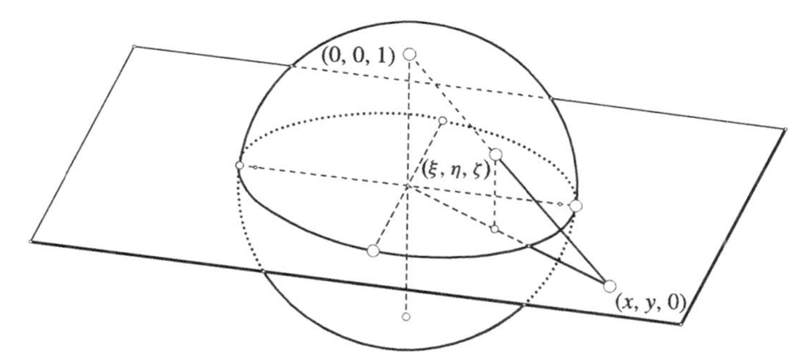

Figure 9.3a Stereographic projection of $\mathbb{C}\mathrm{P}^1$ onto $\dot{\mathrm{S}}^2$

Euclidean 3-space. Let $\dot{\mathrm{S}}^2$ be represented by the unit sphere $\xi^2 + \eta^2 + \zeta^2 = 1$, with the "north pole" (0, 0, 1) being the exceptional point, and let the equatorial plane $\zeta = 0$ serve as an Argand diagram for $\mathbb{C}$ (see Figure 9.3a). Then the mapping $\mathbb{C}\mathrm{P}^1 \to \dot{\mathrm{S}}^2$ can be achieved by stereographic projection, in which a line through $(x, y, 0)$ and (0, 0, 1) meets the sphere again in the point (ξ, η, ζ), with (0, 0, 1) itself corresponding to the point at infinity (Klein 1884, part I, chap. 2, § 2; Neumann, Stoy & Thompson 1994, pp. 210–211; Carter, Segal & Macdonald 1995, pp. 55–56; cf. Schwerdtfeger 1962, pp. 22–29; Burn 1985, pp. 77–84). Going the other way, the mapping from the sphere to the plane gives a conformal Euclidean model for the *elliptic* sphere S^2; the unit disk $x^2 + y^2 \leq 1$ with diametrically opposite points identified is then a conformal model for the elliptic plane eP^2 (Coxeter 1942, pp. 258–260).

The cross ratio $\| UV,\ WZ \|$ of four points on $\mathbb{C}\mathrm{P}^1$ with respective homogeneous coordinates (u), (v), (w), and (z) is given by

$$\|UV,\ WZ\| = \frac{\begin{vmatrix} u_1 & u_2 \\ w_1 & w_2 \end{vmatrix} \begin{vmatrix} v_1 & v_2 \\ z_1 & z_2 \end{vmatrix}}{\begin{vmatrix} u_1 & u_2 \\ z_1 & z_2 \end{vmatrix} \begin{vmatrix} v_1 & v_2 \\ w_1 & w_2 \end{vmatrix}}, \tag{9.3.1}$$

provided that either $U \neq W$ and $V \neq Z$ or $U \neq Z$ and $V \neq W$. If $U = W$ or $V = Z$, then $\| UV,\ WZ \| = 0$; if $U = V$ or $W = Z$, then

$\| UV, WZ \| = 1$; if $U = Z$ or $V = W$, then $\| UV, WZ \| = \infty$. If each point is denoted by its unique associated value in $\mathbb{C} \cup \{\infty\}$, it is easily shown that $\| \infty 0, 1 z\| = z$ for all z.

Real circles of $\dot{\mathrm{S}}^2$ correspond to *chains* of $\mathbb{C}\mathrm{P}^1$. Any three distinct points U, V, W lie on a unique chain $\overset{\circ}{c} = \mathbb{R}(UVW)$, which may be assigned real homogeneous coordinates $[c_0, c_1, c_2, c_3]$, with $c_0{}^2 < c_1{}^2 + c_2{}^2 + c_3{}^2$. A point Z associated with a finite complex number z lies on $\overset{\circ}{c}$ if and only if

$$(c_0 + c_3)|z|^2 + 2c_1(\operatorname{Re} z) + 2c_2(\operatorname{Im} z) + (c_0 - c_3) = 0,$$

and the point at infinity lies on $\overset{\circ}{c}$ if and only if $c_0 + c_3 = 0$. Points that lie on the same chain are said to be *concatenal.* A necessary and sufficient condition for four points U, V, W, and Z to be concatenal is that the cross ratio $\| UV, WZ\|$ be real or infinite.

If $n = \cos\theta + \mathrm{i}\sin\theta$ is a complex number of absolute value 1 and δ is any real number, a great circle (line) $\breve{N}(\delta)$ with inclination $\theta + \frac{1}{2}\pi$ and displacement δ is the chain $\{z : (\operatorname{Re} n\bar{z}) + \delta = 0\} \cup \{\infty\}$. If C is a point associated with a finite complex number c and ρ is a positive real number, a small circle $\overset{\circ}{\mu}(\mathrm{C})$ with center C and radius ρ is the chain $\{z : |z - c| = \rho\}$.

If $A = [(a_{11}, a_{12}), (a_{21}, a_{22})]$ is an invertible matrix over $\mathbb{C}$, a projectivity $\mathbb{C}\mathrm{P}^1 \to \mathbb{C}\mathrm{P}^1$ is induced by the projective linear transformation

$$\langle\cdot A\rangle : \mathrm{P}\mathbb{C}^2 \to \mathrm{P}\mathbb{C}^2, \text{ with } \langle(z)\rangle \mapsto \langle(z)A\rangle, \qquad \det A \neq 0. \tag{9.3.2}$$

The same projectivity is induced by the *linear fractional transformation*

$$\cdot\langle A\rangle : \mathbb{C} \cup \{\infty\} \to \mathbb{C} \cup \{\infty\}, \text{ with } z \mapsto \frac{a_{11}z + a_{21}}{a_{12}z + a_{22}}, \quad a_{11}a_{22} \neq a_{12}a_{21}, \tag{9.3.3}$$

where $z \mapsto \infty$ if $a_{12}z + a_{22} = 0$, $\infty \mapsto a_{11}/a_{12}$ if $a_{12} \neq 0$, and $\infty \mapsto \infty$ if $a_{12} = 0$ (Klein 1884, part I, chap. 2, § 1; cf. Burn 1985, pp. 50–52; Neumann, Stoy & Thompson 1994, pp. 19–20, 210–215). An antiprojectivity $\mathbb{C}\mathrm{P}^1 \to \mathbb{C}\mathrm{P}^1$ is likewise induced by the projective antilinear transformation

$$\langle \odot A\rangle : \mathrm{P}\mathbb{C}^2 \to \mathrm{P}\mathbb{C}^2, \text{ with } \langle (z)\rangle \mapsto \langle (\bar{z})A\rangle, \quad \det A \neq 0, \qquad (9.3.4)$$

as well as by the *antilinear fractional transformation*

$$\odot\langle A\rangle : \mathbb{C}\cup\{\infty\} \to \mathbb{C}\cup\{\infty\}, \text{ with } z \mapsto \frac{a_{11}\bar{z}+a_{21}}{a_{12}\bar{z}+a_{22}}, \quad a_{11}a_{22} \neq a_{12}a_{21}. \qquad (9.3.5)$$

Since a projectivity preserves all cross ratios and an antiprojectivity preserves real cross ratios, both of these transformations take chains into chains, and indeed they are the only transformations of $\mathbb{C}\mathrm{P}^1$ that do so (Veblen & Young 1918, p. 252). We call such a chain-preserving transformation a *concatenation*. A concatenation of the complex projective line $\mathbb{C}\mathrm{P}^1$ is a circularity of the real inversive (Möbius) sphere I^2, a projectivity corresponding to a homography and an antiprojectivity to an antihomography.

The antiprojectivities $z \mapsto -n(n\bar{z}+2\delta)$, where δ is real and $|n| = 1$, and $z \mapsto (c\bar{z}+\mu)/(\bar{z}-\bar{c})$, where μ is real and $|c|^2+\mu \neq 0$, defined by the matrices

$$N(\delta) = \begin{pmatrix} n & 0 \\ 2\delta & -\bar{n} \end{pmatrix} \quad \text{and} \quad \mu(C) = \begin{pmatrix} c & 1 \\ \mu & -\bar{c} \end{pmatrix}, \qquad (9.3.6)$$

are respectively the reflection in the line $\check{N}(\delta) = [\delta,\ \mathrm{Re}\,n,\ \mathrm{Im}\,n,\ -\delta]$ and the inversion in the small circle $\mathring{\mu}(C) = [½(\mu-1),\ \mathrm{Re}\,c,\ \mathrm{Im}\,c,\ -½(\mu+1)]$.

Given a concatenation of $\mathbb{C}\mathrm{P}^1$ defined by a 2×2 complex matrix A, we may without loss of generality assume that $\det A = 1$. For a projectivity the corresponding homography $\langle \cdot M\rangle : \mathrm{I}^2 \to \mathrm{I}^2$ is then defined by any nonzero scalar multiple of the real homographic matrix

$$\begin{pmatrix} (++++) & \mathrm{Re}(a_{11}\bar{a}_{12}+a_{21}\bar{a}_{22}) & -\mathrm{Im}(a_{11}\bar{a}_{12}+a_{21}\bar{a}_{22}) & (+-+-) \\ \mathrm{Re}(a_{11}\bar{a}_{21}+a_{12}\bar{a}_{22}) & \mathrm{Re}(a_{11}\bar{a}_{22}+a_{12}\bar{a}_{21}) & -\mathrm{Im}(a_{11}\bar{a}_{22}-a_{12}\bar{a}_{21}) & \mathrm{Re}(a_{11}\bar{a}_{21}-a_{12}\bar{a}_{22}) \\ \mathrm{Im}(a_{11}\bar{a}_{21}+a_{12}\bar{a}_{22}) & \mathrm{Im}(a_{11}\bar{a}_{22}+a_{12}\bar{a}_{21}) & \mathrm{Re}(a_{11}\bar{a}_{22}-a_{12}\bar{a}_{21}) & \mathrm{Im}(a_{11}\bar{a}_{21}-a_{12}\bar{a}_{22}) \\ (++--) & \mathrm{Re}(a_{11}\bar{a}_{12}-a_{21}\bar{a}_{22}) & -\mathrm{Im}(a_{11}\bar{a}_{12}-a_{21}\bar{a}_{22}) & (+--+) \end{pmatrix} \qquad (9.3.7)$$

(Coxeter 1966a, pp. 237–238; Wilker 1993, p. 136). The four corner symbols $(+\pm\pm\pm)$ stand for

$$\tfrac{1}{2}(|a_{11}|^2 \pm |a_{12}|^2 \pm |a_{21}|^2 \pm |a_{22}|^2),$$

with the indicated choice of signs. The antihomography $\langle\cdot N\rangle : \mathrm{I}^2 \to \mathrm{I}^2$ corresponding to an antiprojectivity with matrix A is defined by the real antihomographic matrix obtained from (9.3.7) by changing the sign of each entry in the third column.

The group of all projectivities of $\mathbb{C}\mathrm{P}^1$ is the *complex projective general linear* group $\mathrm{PGL}_2(\mathbb{C})$—alias the *complex linear fractional* group $\mathrm{LF}(\mathbb{C})$. The group of all concatenations of $\mathbb{C}\mathrm{P}^1$—i.e., the group of all projectivities and antiprojectivities—is the *complex projective general semilinear* group $\mathrm{P\Gamma L}_2(\mathbb{C})$, containing $\mathrm{PGL}_2(\mathbb{C})$ as a subgroup of index 2. From the association between 2×2 invertible complex matrices and 4×4 homographic matrices, we see that $\mathrm{PGL}_2(\mathbb{C})$ is isomorphic to the special inversive group $\mathrm{SI}_2 \cong \mathrm{P}^+\mathrm{O}_{3,1}$ of all homographies of I^2 (the "Möbius group"), while $\mathrm{P\Gamma L}_2(\mathbb{C})$ is isomorphic to the general inversive group $\mathrm{GI}_2 \cong \mathrm{PO}_{3,1}$ of all circularities of I^2, i.e., the group of all homographies and antihomographies.

In the Poincaré model, the hyperbolic plane H^2 is represented conformally by the "upper half-plane" $\{z : \operatorname{Im} z > 0\}$, with the real axis representing the absolute circle. A linear fractional transformation $\cdot\langle A\rangle : \mathbb{C} \cup \{\infty\} \to \mathbb{C} \cup \{\infty\}$ with A real and $\det A > 0$ maps the upper half-plane onto itself; there is no loss of generality in assuming that $\det A = 1$. The group of such transformations is the (real) projective special linear group PSL_2, isomorphic to the special projective pseudo-orthogonal group $\mathrm{P}^+\mathrm{O}_{2,1}$ of direct isometries of H^2. Restricted to the real axis $\mathbb{R} \cup \{\infty\}$, this is also the group of homographies of the real inversive circle I^1 or the group of equiprojectivities of the real projective line P^1.

The linear fractional transformations $\cdot\langle A\rangle : \mathbb{C}\cup\{\infty\} \to \mathbb{C}\cup\{\infty\}$ with A real and $\det A \neq 0$ constitute the projective general linear group PGL_2. When $\det A < 0$, $\cdot\langle A\rangle$ interchanges the upper and lower half-planes; if the "antipodal" points z and $\bar{z}$ are identified, such a transformation can be interpreted as an opposite isometry of H^2. The projective pseudo-orthogonal group $\mathrm{PO}_{2,1}$ of all isometries of H^2 or all circularities of I^1 is thus isomorphic to PGL_2.

EXERCISES 9.3

1. Given the points $\Delta_1 = (1, 0)$, $\Delta_2 = (0, 1)$, $\epsilon = (1, 1)$, and $Z = (z, 1)$ on $\mathbb{C}\mathrm{P}^1$, show that $\|\Delta_1\Delta_2, \epsilon Z\| = z$.

2. $\vdash$ The product of linear fractional transformations $\cdot\langle A\rangle$ and $\cdot\langle B\rangle$ is the linear fractional transformation $\cdot\langle AB\rangle$.

3. $\vdash$ The product of antilinear fractional transformations $\odot\langle A\rangle$ and $\odot\langle B\rangle$ is the linear fractional transformation $\cdot\langle \bar{A}B\rangle$.

4. Given a 2×2 complex matrix A with $\det A = 1$, show that the projectivity $\cdot A : \mathbb{C}\mathrm{P}^1 \to \mathbb{C}\mathrm{P}^1$ corresponds to the homography $\langle\cdot M\rangle : \mathrm{I}^2 \to \mathrm{I}^2$, where M is given by (9.3.7).

9.4 INVERSIVE UNITARY GEOMETRY

Its role in complex analysis gives a special significance to the Möbius sphere I^2 but tends to obscure the larger picture. The theory of circularities readily extends to higher dimensions; Wilker (1981) and Beardon (1983, chap. 3) have provided comprehensive accounts of n-dimensional real inversive geometry. The identification of real inversive circles with complex projective chains generalizes as a correspondence between odd-dimensional real inversive spaces and the inversive unitary spaces that were introduced in § 7.2.*

* One must be careful not to confuse inversive unitary spaces with the unrelated complex inversive spaces of Veblen & Young (1918, pp. 264–267) and Scherk (1960).

The points X and the real hyperspheres $\mathring{u}$ of the Möbius (real inversive) n-sphere I^n ($n \geq 1$) can be given homogeneous real coordinates

$$((x)) = (x_0, x_1, \ldots, x_{n+1}) \quad \text{and} \quad [[u]] = [u_0, u_1, \ldots, u_{n+1}],$$

where $x_1{}^2 + \cdots + x_{n+1}{}^2 - x_0{}^2 = 0$ and $u_1{}^2 + \cdots + u_{n+1}{}^2 - u_0{}^2 > 0$. The point X lies on the hypersphere $\mathring{u}$ if and only if $((x))[[u]] = 0$.

As with I^2, a transformation of I^n taking circles (inversive 1-spheres) into circles is a *circularity*. A circularity that preserves sense is a *homography*, and one that reverses sense is an *antihomography*. A homography

$$\langle \cdot M \rangle : \mathrm{I}^n \to \mathrm{I}^n, \text{ with } \langle ((x)) \rangle \mapsto \langle ((x))M \rangle, \tag{9.4.1}$$

is induced by an $(n+2) \times (n+2)$ *homographic matrix* M, such that $MHM\check{} = H$ with $\det M = 1$, where $H = -I_{1,n+1} = \backslash -1, 1, \ldots, 1 \backslash$, and an antihomography

$$\langle \cdot N \rangle : \mathrm{I}^n \to \mathrm{I}^n, \text{ with } \langle ((x)) \rangle \mapsto \langle ((x))N \rangle, \tag{9.4.2}$$

is induced by an *antihomographic matrix* N, where $NHN\check{} = H$ with $\det N = -1$.

The group GI_n of all circularities of I^n ($n \geq 1$) is the projective pseudo-orthogonal group $\mathrm{PO}_{n+1,1}$. The subgroup SI_n of all homographies is the special projective pseudo-orthogonal group $\mathrm{P}^+\mathrm{O}_{n+1,1}$.

What Goldman (1999) describes as the absolute (or "boundary") of hyperbolic unitary $(n+1)$-space HU^{n+1} is inversive unitary n-space. The points Z and the hyperchains $\mathring{\omega}$ of the Möbius n-antisphere IU^n ($n \geq 0$) can be given homogeneous complex coordinates

$$((z)) = (z_0, z_1, \ldots, z_{n+1}) \quad \text{and} \quad [[w]] = [w_0, w_1, \ldots, w_{n+1}],$$

with $|z_1|^2 + \cdots + |z_{n+1}|^2 - |z_0|^2 = 0$ and $|w_1|^2 + \cdots + |w_{n+1}|^2 - |w_0|^2 > 0$. The point Z lies on the hyperchain $\mathring{\omega}$ if and only if $((z))[[w]] = 0$.

The Möbius n-antisphere $\mathrm{I}\mathrm{U}^n$ is conformally modeled by the real inversive $(2n+1)$-sphere I^{2n+1}, and its k-subspaces $\mathrm{I}\mathrm{U}^k$ by real inversive $(2k+1)$-spheres I^{2k+1} $(0 \le k \le n-1)$. A 0-subspace is a chain, a 1-subspace an anticircle, and an $(n-1)$-subspace a hyperchain. In addition, $\mathrm{I}\mathrm{U}^n$ has "semisubspaces"—*spinal* $2k$*-spheres* $\mathrm{I}\mathrm{U}^{(2k-1)/2}$—that are conformal to real inversive $2k$-spheres I^{2k} $(0 \le k \le n)$. A spinal 0-sphere $\mathrm{I}\mathrm{U}^{-1/2}$ is a pair of points; a spinal 2-sphere $\mathrm{I}\mathrm{U}^{1/2}$ is a shell; a spinal $2n$-sphere $\mathrm{I}\mathrm{U}^{(2n-1)/2}$ is a *hypershell.* When $\mathrm{I}\mathrm{U}^n$ is taken as the absolute n-antisphere of $\mathrm{H}\mathrm{U}^{n+1}$, a hypershell is the trace on $\mathrm{I}\mathrm{U}^n$ of a *bisector* $\mathrm{H}\mathrm{U}^{(2n+1)/2}$, the locus of points in $\mathrm{H}\mathrm{U}^{n+1}$ equidistant from two given points (Mostow 1980, pp. 183–190; Goldman 1999, chap. 5).

A transformation of $\mathrm{I}\mathrm{U}^n$ taking chains into chains is a *concatenation.* If $\mathrm{I}\mathrm{U}^n$ is represented by an oval n-antiquadric Ψ in complex projective $(n+1)$-space $\mathbb{C}\mathrm{P}^{n+1}$, a concatenation of $\mathrm{I}\mathrm{U}^n$ is a collineation of $\mathbb{C}\mathrm{P}^{n+1}$ that leaves Ψ invariant. An $(n+2) \times (n+2)$ *zygographic matrix* P such that $PHP^* = H$, where $H = \backslash -1, 1, \ldots, 1\backslash$, induces a *holomorphism*

$$\langle \cdot P \rangle : \mathrm{I}\mathrm{U}^n \to \mathrm{I}\mathrm{U}^n, \text{ with } \langle ((z)) \rangle \mapsto \langle ((z))P \rangle, \tag{9.4.3}$$

or an *antiholomorphism*

$$\langle \odot P \rangle : \mathrm{I}\mathrm{U}^n \to \mathrm{I}\mathrm{U}^n, \text{ with } \langle ((z)) \rangle \mapsto \langle ((\bar{z}))P \rangle. \tag{9.4.4}$$

The group of all holomorphisms of $\mathrm{I}\mathrm{U}^n$ $(n \ge 0)$ is the *projective pseudo-unitary* group $\mathrm{PU}_{n+1,1}$, to be described further in § 10.3. The group of all concatenations—holomorphisms and antiholomorphisms—is the *semiprojective pseudo-unitary* group $\bar{\mathrm{P}}\mathrm{U}_{n+1,1}$. The groups $\bar{\mathrm{P}}\mathrm{U}_{1,1}$ and $\mathrm{PU}_{1,1}$ of concatenations and holomorphisms of the chain $\mathrm{I}\mathrm{U}^0$ are, respectively, isomorphic to the groups $\mathrm{PO}_{2,1} \cong \mathrm{PGL}_2$ and $\mathrm{P}^+\mathrm{O}_{2,1} \cong \mathrm{PSL}_2$ of circularities and homographies of the real inversive circle I^1.

EXERCISES 9.4

1. Find a necessary and sufficient condition for $n+2$ points of a Möbius n-sphere to lie on a hypersphere.

2. $\vdash$ Any three points of a Möbius n-antisphere lie on a chain.

10

UNITARY ISOMETRY GROUPS

WE SAW IN CHAPTER 6 how isometries of a real metric space are determined by orthogonal, Euclidean, or pseudo-orthogonal transformations of a real vector space. Analogously, each of the unitary metric spaces described in Chapter 7 has a group of isometries that can be defined in terms of *unitary*, *transunitary*, or *pseudo-unitary* transformations of a complex vector space. The complex field $\mathbb{C}$ being a two-dimensional vector space over $\mathbb{R}$, such transformations can be represented either by complex matrices or by real *duplex* matrices. The group U_n of $n \times n$ unitary matrices is isomorphic to the *orthogonal duplex* group $\mathrm{O}_{(2)n}$ of real $2n \times 2n$ matrices that are both orthogonal and symplectic.

The division ring $\mathbb{H}$ of *quaternions* is a four-dimensional vector space over $\mathbb{R}$ endowed with an associative (but noncommutative) multiplication, as well as a two-dimensional vector space over $\mathbb{C}$. The group $\mathrm{S\tilde{p}}_n$ of $n \times n$ quaternionic symplectic matrices is isomorphic to the *unitary duplex* group $\mathrm{U}_{(2)n}$ of complex $2n \times 2n$ matrices that are both unitary and symplectic or the *orthogonal quadruplex* group $\mathrm{O}_{(4)n}$ of real $4n \times 4n$ matrices that are both orthogonal and bisymplectic (or cobisymplectic).

10.1 UNITARY TRANSFORMATIONS

Orthogonal transformations of complex vector spaces, orthogonal matrices with complex entries, and the complex orthogonal group

$O(n;\ \mathbb{C})$ can be defined as in the real case. Complex symplectic transformations and matrices and the complex symplectic group $Sp(n;\ \mathbb{C}) \cong \breve{Sp}_n$ were mentioned at the end of § 5.6. More relevant to our purposes, however, are transformations, matrices, and groups that involve complex conjugation.

The standard *inner product* of two rows (z) and (w) of $\mathbb{C}^n$ or two columns $[w]$ and $[z]$ of $\check{\mathbb{C}}^n$ is given by the Hermitian sesquilinear form

$$(z\ \bar{w}) = (z)(w)^* \quad \text{or} \quad [\bar{w}\ z] = [w]^*[z], \tag{10.1.1}$$

where the star denotes the antitranspose. The associated real-valued antiquadratic forms $(z\ \bar{z})$ and $[\bar{w}\ w]$ are positive definite. A *unit* row (z) or column $[w]$ has $(z\ \bar{z}) = 1$ or $[\bar{w}\ w] = 1$. A *unitary transformation* of $\mathbb{C}^n$ or $\check{\mathbb{C}}^n$ is a linear transformation that takes unit rows into unit rows or unit columns into unit columns. A necessary and sufficient condition for a complex $n \times n$ matrix R to determine a unitary transformation $\cdot R : \mathbb{C}^n \to \mathbb{C}^n$ or $R\cdot : \check{\mathbb{C}}^n \to \check{\mathbb{C}}^n$ is that $RR^* = I$, and a matrix with this property is called a *unitary matrix*. If R is a unitary matrix, then $|\det R| = 1$.

The set of all unitary transformations of $\mathbb{C}^n$ or $\check{\mathbb{C}}^n$, or all $n \times n$ unitary matrices, forms a group, the *unitary* group $U_n \cong U(n)$. Unitary transformations whose matrices have determinant 1 form a normal subgroup, the *special unitary* group $SU_n \cong SU(n)$. The center of the unitary group is the complex unit scalar group $\bar{S}Z(\mathbb{C})$, represented by the unit scalar matrices ιI where $|\iota| = 1$. The *cyclotomic* group $C_m(\mathbb{C})$, represented by matrices ωI with $\omega^m = 1$, is a finite cyclic subgroup of $\bar{S}Z(\mathbb{C})$ for each positive integer m.

Corresponding to each $n \times n$ unitary matrix R is a $2n \times 2n$ real duplex matrix $\{R\} = (\operatorname{Re} R,\ \operatorname{Im} R)_{(2)}$ that is both orthogonal and symplectic. The group of all such matrices is the *orthogonal duplex* group $O_{(2)n} \cong U_n$. The center of $O_{(2)n}$ is the special bivalent group $SZ_{(2)n} \cong O_2^+$.

The unit rows $(z) = (z_1, z_2, \ldots, z_{n+1})$ of $\mathbb{C}^{n+1}$ and the nonzero columns $[w] = [w_1, w_2, \ldots, w_{n+1}]$ of $\check{\mathbb{C}}^{n+1}$ can be taken as normalized coordinates for the points and homogeneous coordinates for the great hyperchains of the elliptic n-antisphere $\mathrm{S}\mathrm{U}^n$. Given an $(n+1) \times (n+1)$ unitary matrix R, the unitary transformation $\cdot R : \mathbb{C}^{n+1} \to \mathbb{C}^{n+1}$, with $(z) \mapsto (z)R$, defines an *isometry* of $\mathrm{S}\mathrm{U}^n$. A point Z with coordinates (z) is taken into a point Z' with coordinates $(z)R$, and a great hyperchain $\breve{W}$ with coordinates $[w]$ is taken into a great hyperchain $\breve{W}'$ with coordinates $R^*[w]$. An isometry of $\mathrm{S}\mathrm{U}^n$ preserves distances (7.2.1) between points, but not every distance-preserving transformation is an isometry. The *anti-isometry* of $\mathrm{S}\mathrm{U}^n$ defined by the *anti-unitary transformation* $\odot R : \mathbb{C}^{n+1} \to \mathbb{C}^{n+1}$, with $(z) \mapsto (\bar{z})R$, also preserves distances.

An isometry other than the identity that leaves invariant every point of a great hyperchain is a *unitary reflection*. A unitary reflection $\cdot R(2\pi/\theta)$ is determined by its *mirror* $\breve{R}$ and its *angle* $\tfrac{1}{2}\theta$, where $0 < \theta < 2\pi$. If $\breve{R}$ has coordinates $[r]$, the reflection is represented by a *unitary reflection matrix*, a unitary matrix $R(2\pi/\theta)$ of order $n+1$ with entries

$$r_{ij} = \delta_{ij} + (\iota - 1)\frac{\bar{r}_i r_j}{[\bar{r}\, r]}, \tag{10.1.2}$$

where $\iota = \exp i\theta$. When the angle $\tfrac{1}{2}\theta$ is a submultiple of π, so that $2\pi/\theta$ is an integer p, the reflection has period p. If $p = 2$, corres ponding to an angle of $\pi/2$, (10.1.2) defines an *orthogonal reflection*.

A *central inversion*, which permutes the confibral points of each great chain and takes each great hyperchain into itself, is represented by a unit scalar matrix ιI. The *negation*, with $\iota = -1$, interchanges anti podal points. Central inversions commute with every isometry.

The group of all isometries of $\mathrm{S}\mathrm{U}^n$ is the unitary group U_{n+1}. The abelian subgroup of central inversions is the complex unit scalar group $\bar{\mathrm{S}}\mathrm{Z}(\mathbb{C})$. The subgroup generated by the negation is the real unit scalar group $\bar{\mathrm{S}}\mathrm{Z}$.

Complex projective n-space $\mathbb{C}\mathrm{P}^n$ with a unitary metric—i.e., the elliptic n-antisphere $\mathrm{S}\mathrm{U}^n$ with confibral points identified—is elliptic unitary n-space $\mathrm{P}\mathrm{U}^n$. A projective collineation that preserves elliptic unitary distances (7.1.1) is an *isometry* of $\mathrm{P}\mathrm{U}^n$. If R is an $(n+1)\times(n+1)$ unitary matrix, the *projective unitary transformation*

$$\langle\cdot R\rangle : \mathrm{P}\mathbb{C}^{n+1} \to \mathrm{P}\mathbb{C}^{n+1}, \text{ with } \langle(z)\rangle \mapsto \langle(z)R\rangle, \quad RR^* = I, \quad (10.1.3)$$

determines an isometry of $\mathrm{P}\mathrm{U}^n$, taking a point Z with homogeneous coordinates (z) into a point Z' with coordinates $(z)R$. The same isometry takes a hyperplane $\check{W}$ with homogeneous coordinates $[w]$ into a hyperplane $\check{W}'$ with coordinates $R^*[w]$. A unitary reflection whose mirror is a given hyperplane with coordinates $[r]$ is represented by a unitary reflection matrix with entries given by (10.1.2) for some unit scalar $\iota \neq 1$.

The group of all isometries of $\mathrm{P}\mathrm{U}^n$ is the group of isometries of $\mathrm{S}\mathrm{U}^n$ with the group $\bar{\mathrm{S}}\mathrm{Z}(\mathbb{C})$ of central inversions factored out. This is the *projective unitary* group $\mathrm{PU}_{n+1} \cong \mathrm{U}_{n+1}/\bar{\mathrm{S}}\mathrm{Z}(\mathbb{C})$.

EXERCISES 10.1

1. $\vdash$ Every 2×2 special unitary matrix is of the form $[(u,\ v),\ (-\bar{v},\ \bar{u})]$, with $|u|^2 + |v|^2 = 1$.
2. $\vdash$ A central inversion $\cdot\iota I : \mathrm{S}\mathrm{U}^n \to \mathrm{S}\mathrm{U}^n$, with $(z) \mapsto \iota(z)$, takes each great hyperchain $[w]$ into itself.
3. In the elliptic unitary plane $\mathrm{P}\mathrm{U}^2$, let $\check{R}$ be the line with coordinates $[r] = [1,\ 0,\ \mathrm{i}]$. Use Formula (10.1.1) to find the matrix $R(3)$ of the unitary reflection with mirror $\check{R}$ and angle $½\theta = \pi/3$.

10.2 TRANSUNITARY TRANSFORMATIONS

Each point Z of Euclidean unitary n-space $\mathrm{E}\mathrm{U}^n$ may be associated with a unique row vector $(z) \in \mathbb{C}^n$, and each hyperplane $\check{W}(r)$ with the set of nonzero scalar multiples of a column vector $[[w],\ r] \in \check{\mathbb{C}}^{n+1}$. The point Z lies on the hyperplane $\check{W}(r)$ if and only if $(z)[w]+r = 0$. Points and hyperplanes can also be given semiduplex and duplex coordinates

$$((x,\ y)) \in \mathbb{R}^{2n} \quad \text{and} \quad [[(u,\ v)_{(2)}],\ (p,\ q)] \in \mathbb{R}^{(2n+1)\times 2},$$

with Z lying on $\check{W}(r)$ if and only if $((x,\ y))[(u,\ v)_{(2)}] + (p, q) = (0, 0)$.

If R is an $n \times n$ unitary matrix and $(h) \in \mathbb{C}^n$ is any row vector, the mapping

$$[\cdot R,\ +(h)] : \mathbb{C}^n \to \mathbb{C}^n,\ \text{with}\ (z) \mapsto (z)R + (h), \quad RR^* = I, \quad (10.2.1)$$

is a *transunitary transformation*, and the corresponding operation on $\mathrm{E}\mathrm{U}^n$ preserving Euclidean unitary distances (7.1.3) is an *isometry*. The matrix representation of such a transformation is given either by an $(n+1) \times (n+1)$ *transunitary matrix*

$$[R,\ (h)]_1 = \begin{pmatrix} R & [0] \\ (h) & 1 \end{pmatrix}, \quad RR^* = I, \quad (10.2.2)$$

or by a $(2n+1) \times (2n+1)$ real *transorthogonal duplex matrix*

$$[\{R\},\ ((f,\ g))]_1 = \begin{pmatrix} \{R\} & [[0,\ 0]] \\ ((f,\ g)) & 1 \end{pmatrix}, \quad (10.2.3)$$

where $\{R\}$ is the orthogonal duplex equivalent to the unitary matrix R, $[[0,\ 0]]$ is a column of $2n$ zeros, and (f) and (g) are the real and imaginary parts of the row vector (h).

A reflection of angle ½θ in a hyperplane with coordinates $[[r],\ h]$ is represented by a *transunitary reflection matrix* $[R(2\pi/\theta),\ (h)]_1$ with entries

$$r_{ij} = \delta_{ij} + (\iota - 1)\frac{\bar{r}_i r_j}{[\bar{r}\ r]} \quad \text{and} \quad h_j = (\iota - 1)\frac{h r_j}{[\bar{r}\ r]}, \quad (10.2.4)$$

where $\iota = \exp \mathrm{i}\theta$. Relative to duplex coordinates $[(p,\ q)_{(2)}],\ (f,\ g)]$ for the mirror, the reflection can also be represented by a *transorthogonal duplex reflection matrix* $[\{R(2\pi/\theta)\},\ ((f,\ g))]_1$.

The set of all transorthogonal duplex matrices of order $2n+1$ or of all transunitary matrices of order $n+1$ forms a multiplicative group $\mathrm{TO}_{(2)n} \cong \mathrm{TU}_n$ isomorphic to the *Euclidean unitary* group $\mathrm{E}\mathrm{U}_n$ of all isometries of $\mathrm{E}\mathrm{U}^n$, and the translation matrices $[\{I\},\ ((f,\ g))]_1$

or $[I,\ (h)]_1$ form a normal subgroup $\mathrm{T}_{(2)n} \cong \mathrm{T}_n(\mathbb{C})$. The quotient group $\mathrm{TO}_{(2)n}/\mathrm{T}_{(2)n} \cong \mathrm{TU}_n/\mathrm{T}_n(\mathbb{C}) \cong \mathrm{E}\mathrm{U}_n/\mathrm{T}^n(\mathbb{C})$ is isomorphic to the unitary group U_n of $n \times n$ unitary matrices.

If λ is a nonzero scalar and R is an $n \times n$ unitary matrix, the product of the scaling operation $\lambda\cdot : \mathbb{C}^n \to \mathbb{C}^n$ and the unitary transformation $\cdot R : \mathbb{C}^n \to \mathbb{C}^n$ is a *zygopetic transformation*

$$\lambda \cdot R : \mathbb{C}^n \to \mathbb{C}^n, \text{ with } (z) \mapsto \lambda(z)R, \quad \lambda \neq 0,\ RR^* = I,$$

preserving angles between vectors. (Without loss of generality, we may assume that λ is real and even that $\lambda > 0$.) This is equivalent to the linear transformation $\cdot\lambda R : \mathbb{C}^n \to \mathbb{C}^n$ induced by the zygopetic matrix λR. The product of a zygopetic transformation and a translation $+(h) : \mathbb{C}^n \to \mathbb{C}^n$ is a *transzygopetic transformation*

$$[\lambda \cdot R,\ +(h)] : \mathbb{C}^n \to \mathbb{C}^n, \text{ with } (z) \mapsto \lambda(z)R + (h), \quad \lambda \neq 0,\ RR^* = I, \tag{10.2.5}$$

which induces a *unitary similarity* on $\mathrm{E}\mathrm{U}^n$. A unitary similarity can be represented by an $(n+1) \times (n+1)$ *transzygopetic matrix*

$$[\lambda R,\ (h)]_1 = \begin{pmatrix} \lambda R & [0] \\ (h) & 1 \end{pmatrix}, \quad \lambda \neq 0,\ RR^* = I, \tag{10.2.6}$$

or by a $(2n+1) \times (2n+1)$ real *transorthopetic duplex matrix*

$$[\{\lambda R\},\ ((f,\ g))]_1 = \begin{pmatrix} \{\lambda R\} & [[0,\ 0]] \\ ((f,\ g)) & 1 \end{pmatrix}, \tag{10.2.7}$$

where the orthopetic duplex matrix $\{\lambda R\}$ is a bivalent multiple of the orthogonal duplex matrix $\{R\}$. When $|\lambda| = 1$, the similarity is an isometry; when R is a unit scalar matrix λI, it is a dilatation.

The set of all transorthopetic duplex matrices of order $2n+1$ or of all transzygopetic matrices of order $n+1$ forms a multiplicative group $\mathrm{T}\tilde{\mathrm{O}}_{(2)n} \cong \mathrm{T}\tilde{\mathrm{U}}_n$, which is isomorphic to the *euclopetic unitary* group $\tilde{\mathrm{E}}\mathrm{U}_n$. The quotient group $\mathrm{T}\tilde{\mathrm{O}}_{(2)n}/\mathrm{T}_{(2)n} \cong \mathrm{T}\tilde{\mathrm{U}}_n/\mathrm{T}_n(\mathbb{C}) \cong \tilde{\mathrm{E}}\mathrm{U}_n/\mathrm{T}^n(\mathbb{C})$

is isomorphic to the *zygopetic* group $\tilde{\mathrm{U}}_n \cong \mathrm{U}_n \times \mathrm{GZ}^+$. The *proto-zygopetic* group $\widetilde{\mathrm{SU}}_n \cong \mathrm{SU}_n \times \mathrm{GZ}^+$ is the group of positive real scalar multiples of $n \times n$ special unitary matrices.

EXERCISES 10.2

1. In the Euclidean unitary plane Eu^2, let $\check{R}$ be the line with coordinates $[[r], h] = [1, \mathrm{i}, 2+3\mathrm{i}]$. Use Formula (10.2.4) to find the matrix $[R(4), (h)]_1$ of the transunitary reflection with mirror $\check{R}$ and angle $\tfrac{1}{2}\theta = \pi/4$.
2. Given a nonzero complex number κ and an $n \times n$ unitary matrix S, show that the transformation $\kappa\cdot S : \mathbb{C}^n \to \mathbb{C}^n$, with $(z) \mapsto \kappa(z)S$, can be expressed equivalently as $(z) \mapsto \lambda(z)R$, where R is unitary and λ is real and positive.

10.3 PSEUDO-UNITARY TRANSFORMATIONS

By means of the pseudo-identity matrix $\dot{I} = I_{n,1} = \backslash 1, 1, \ldots, -1\backslash$, we may define a *pseudo-inner product*

$$(z\ddot{\ }\bar{w}) = (z)\dot{I}(w)^* \quad \text{or} \quad [\bar{w}\ddot{\ }z] = [w]^*\dot{I}[z] \tag{10.3.1}$$

of rows (z) and (w) of $\mathbb{C}^{n+1}$ or columns $[w]$ and $[z]$ of $\check{\mathbb{C}}^{n+1}$. A *pseudo-unit* row (z) or column $[w]$ has $(z\ddot{\ }\bar{z}) = \pm 1$ or $[\bar{w}\ddot{\ }w] = \pm 1$. A *pseudo-unitary transformation* of $\mathbb{C}^{n+1}$ or $\check{\mathbb{C}}^{n+1}$ is a linear transformation that preserves the relevant sesquilinear form and so takes pseudo-unit rows or columns into pseudo-unit rows or columns. A complex matrix R of order $n+1$ determines a pseudo-unitary transformation $\cdot R : \mathbb{C}^{n+1} \to \mathbb{C}^{n+1}$ or $R\cdot : \check{\mathbb{C}}^{n+1} \to \check{\mathbb{C}}^{n+1}$ if and only if $R\ddot{R}^* = I$, where $\ddot{R}^* = \dot{I}R^*\dot{I}$ is the *pseudo-antitranspose* of R. Such a matrix R is called a *pseudo-unitary matrix*, and $|\det R| = 1$.

The set of all pseudo-unitary transformations of $\mathbb{C}^{n+1}$ or $\check{\mathbb{C}}^{n+1}$ forms the *pseudo-unitary* group $\mathrm{U}_{n,1} \cong \mathrm{U}(n, 1)$, with group operations represented by pseudo-unitary matrices. The pseudo-unitary transformations whose matrices have determinant 1 form a normal subgroup, the *special pseudo-unitary* group $\mathrm{SU}_{n,1} \cong \mathrm{SU}(n, 1)$. The group $\mathrm{SU}_{1,1}$ is isomorphic to the real special linear group SL_2, these being conjugate subgroups of the complex special linear group $\mathrm{SL}_2(\mathbb{C})$. The center of the pseudo-unitary group is the complex unit scalar group $\bar{\mathrm{S}}\mathrm{Z}(\mathbb{C})$.

The points Z and great hyperchains $\breve{W}$ of the hyperbolic n-antisphere $\ddot{\mathrm{S}}\mathrm{U}^n$ can be given normalized coordinates (z) with $(z\ddot{\ }\bar{z}) = -1$ and homogeneous coordinates $[w]$ with $[\bar{w}\ddot{\ }w] > 0$. Even though we lack an explicit distance formula, a pseudo-unitary transformation $(z) \mapsto (z)R$, with $[w] \mapsto \ddot{R}^*[w]$, will be called an *isometry* of $\ddot{\mathrm{S}}\mathrm{U}^n$, with the group of all such isometries being the pseudo-unitary group $\mathrm{U}_{n,1}$.

Complex projective n-space $\mathbb{C}\mathrm{P}^n$ with a pseudo-unitary metric is complete hyperbolic unitary n-space $\mathrm{P}\ddot{\mathrm{U}}^n$. Points Z and hyperplanes $\breve{W}$ can be assigned homogeneous coordinates

$$(z) = (z_1, \ldots, z_n \mid z_{n+1}) \quad \text{and} \quad [w] = [w_1, \ldots, w_n \mid w_{n+1}],$$

with $(z\ddot{\ }\bar{z}) < 0$ and $[\bar{w}\ddot{\ }w] > 0$ for points lying in and hyperplanes intersecting the ordinary region $\mathrm{H}\mathrm{U}^n$ interior to the absolute hyperchain.

A projective collineation that preserves hyperbolic unitary distances (7.1.5) is an *isometry* of $\mathrm{H}\mathrm{U}^n$. Given a pseudo-unitary matrix R of order $n+1$, the *projective pseudo-unitary transformation*

$$\langle\cdot R\rangle : \mathrm{P}\mathbb{C}^{n+1} \to \mathrm{P}\mathbb{C}^{n+1}, \text{ with } \langle(z)\rangle \mapsto \langle(z)R\rangle, \quad R\ddot{R}^* = I, \tag{10.3.2}$$

determines an isometry of $\mathrm{H}\mathrm{U}^n$, taking a point Z with coordinates (z) into a point Z' with coordinates $(z)R$. A hyperplane $\breve{W}$ with coordinates $[w]$ is taken into a hyperplane $\breve{W}'$ with coordinates $\ddot{R}^*[w]$.

A reflection of angle ½θ whose mirror is a given hyperplane with coordinates $[r]$ is represented by a *pseudo-unitary reflection matrix*, a pseudo-unitary matrix $R(2\pi/\theta)$ with entries

$$r_{ij} = \delta_{ij} + (\iota - 1)\frac{\bar{r}_i r_j}{[\bar{r}\ddot{\ }r]} \quad (1 \le j \le n), \quad r_{i,n+1} = \delta_{i,n+1} - (\iota - 1)\frac{\bar{r}_i r_{n+1}}{[\bar{r}\ddot{\ }r]}, \tag{10.3.3}$$

where $\iota = \exp \mathrm{i}\theta$.

The group of all isometries of $\mathrm{H\,U}^n$ is the group $\mathrm{U}_{n,1}$ with the complex unit scalar group $\bar{\mathrm{S}}\mathrm{Z}(\mathbb{C})$ factored out, i.e., the *projective pseudo-unitary* group $\mathrm{PU}_{n,1} \cong \mathrm{U}_{n,1}/\bar{\mathrm{S}}\mathrm{Z}(\mathbb{C})$.

The absolute hyperchain, or boundary, of HU^n is an inversive $(n-1)$-antisphere $\mathrm{I\,U}^{n-1}$ on which the trace of an ordinary k-plane $\mathrm{H}U^k$ $(1 \le k \le n-1)$ is a $(k-1)$-antisphere IU^{k-1}. Each isometry of HU^n induces a holomorphism of IU^{n-1}. The group of holomorphisms of IU^{n-1} is the same as the group $\mathrm{PU}_{n,1}$ of isometries of HU^n.

EXERCISES 10.3

1. The condition for an $(n+1) \times (n+1)$ complex matrix R to be pseudo-unitary is that $R\ddot{R}^* = I$. Show that this implies that $(r_i \ddot{\ } \bar{r}_j) = \delta_{ij}$ for any two rows (r_i) and (r_j) and that $[\bar{r}_i \ddot{\ } r_j] = \delta_{ij}$ for any two columns $[r_i]$ and $[r_j]$, except that $(r_{n+1} \ddot{\ } \bar{r}_{n+1}) = [\bar{r}_{n+1} \ddot{\ } r_{n+1}] = -1$.
2. In the hyperbolic unitary plane HU^2, let $\check{R}$ be the line with coordinates $[r] = [1, \mathrm{i}, 0]$. Use Formula (10.3.3) to find the matrix $R(6)$ of the pseudo-unitary reflection with mirror $\check{R}$ and angle $½\theta = \pi/6$.

10.4 QUATERNIONS AND RELATED SYSTEMS

The division ring $\mathbb{H}$ of quaternions is a four-dimensional vector space over $\mathbb{R}$, with basis $[1, \mathrm{i}, \mathrm{j}, \mathrm{k}]$ having an associative multiplication of vectors satisfying Hamilton's famous equations

$$\mathrm{i}^2 = \mathrm{j}^2 = \mathrm{k}^2 = \mathrm{ijk} = -1. \qquad (10.4.1)$$

The multiplication so defined is noncommutative; e.g., $\mathrm{ij} = \mathrm{k} = -\mathrm{ji}$.

A. *Quaternionic groups.* Each quaternion $\mathrm{Q} = t + x\mathrm{i} + y\mathrm{j} + z\mathrm{k}$ can be identified with a point (t, x, y, z) of Euclidean 4-space and can be represented by a 4×4 real *quadrivalent matrix*

$$Q = \{\{\mathrm{Q}\}\} = (t,\ x,\ y,\ z)_{(4)} = \begin{pmatrix} t & x & y & z \\ -x & t & -z & y \\ -y & z & t & -x \\ -z & -y & x & t \end{pmatrix}. \qquad (10.4.2)$$

Note that Q is a real duplex matrix, so that setting $u = t + x\mathrm{i}$ and $v = y + z\mathrm{i}$, we can also represent $\mathrm{Q} = u + v\mathrm{j}$ by a 2×2 complex *bivalent matrix*

$$\mathbf{Q} = \{\mathrm{Q}\} = (u,\ v)_{(2)} = \begin{pmatrix} u & v \\ -\bar{v} & \bar{u} \end{pmatrix}. \qquad (10.4.3)$$

Every automorphism of $\mathbb{H}$ is an inner automorphism. The ring $\mathbb{H}$ also has an involutory co-automorphism (sum-preserving and product-reversing) $✡ : \mathbb{H} \to \mathbb{H}$, with $\mathrm{Q} \mapsto \tilde{\mathrm{Q}}$, the *quaternionic conjugate* $\tilde{\mathrm{Q}} = t - x\mathrm{i} - y\mathrm{j} - z\mathrm{k}$ being represented by the real matrix $Q^{\vee}$ or the complex matrix $\mathbf{Q}^*$. The *norm* of a quaternion Q is the nonnegative real number $\mathrm{Q}\tilde{\mathrm{Q}} = t^2 + x^2 + y^2 + z^2 = \det \mathbf{Q}$, whose square root is its *absolute value* $|\mathrm{Q}|$. A *unit* quaternion has $|\mathrm{Q}| = 1$. We call $t = \mathrm{re}\ \mathrm{Q}$ and $(x,\ y,\ z) = \mathrm{pu}\ \mathrm{Q}$ the *real part* and the *pure part* of Q. A *pure* quaternion $\mathrm{P} = x\mathrm{i} + y\mathrm{j} + z\mathrm{k}$ has $\mathrm{P}^2 = -\mathrm{P}\tilde{\mathrm{P}} = -(x^2 + y^2 + z^2)$. Pure quaternions behave like vectors in $\mathbb{R}^3$, with $\mathrm{i} = (1,\ 0,\ 0)$, $\mathrm{j} = (0,\ 1,\ 0)$, and $\mathrm{k} = (0,\ 0,\ 1)$ and with $\mathrm{re}\ \mathrm{PQ} = -\mathrm{P} \cdot \mathrm{Q}$ and $\mathrm{pu}\ \mathrm{PQ} = \mathrm{P} \times \mathrm{Q}$.

The set of all matrices (10.4.2) or (10.4.3) is closed under both addition and multiplication, defining a skew-field $\mathbb{R}^{(4)}$ or $\mathbb{C}^{(2)}$ isomorphic to $\mathbb{H}$. More generally, an $m \times n$ matrix A over $\mathbb{H}$ can be represented by a $2m \times 2n$ *duplex matrix* $\{A\}$ over $\mathbb{C}$ comprising mn bivalent blocks or by a $4m \times 4n$ *quadruplex matrix* $\{\{A\}\}$ over $\mathbb{R}$ comprising mn quadrivalent blocks. The set of all invertible $4n \times 4n$ real quadruplex matrices forms a multiplicative group $\mathrm{GL}_{(4)n} \cong \mathrm{GL}_{(4)n}(\mathbb{R})$, and the set of all invertible $2n \times 2n$ complex duplex matrices forms a multiplicative group $\mathrm{GL}_{(2)n}(\mathbb{C})$, both of which are isomorphic to the *quaternionic general linear* group $\mathrm{GL}_n(\mathbb{H})$ of invertible $n \times n$ quaternionic matrices.*

In accordance with the theory of Dieudonné (1943; cf. Artin 1957, pp. 151–158; Aslaksen 1996), the determinant of a square quaternionic matrix is a nonnegative real number. If $\{A\}$ is the $2n \times 2n$ complex duplex matrix corresponding to an $n \times n$ quaternionic matrix A, then $\det\{A\} = (\det A)^2$. A square matrix A is invertible if and only if $\det A \neq 0$. The commutator subgroup of $\mathrm{GL}_n(\mathbb{H})$ is the *quaternionic special linear* group $\mathrm{SL}_n(\mathbb{H})$ of $n \times n$ quaternionic matrices with determinant 1. The general commutator quotient group $\mathrm{GCom}(\mathbb{H}) \cong \mathrm{GL}_n(\mathbb{H})/\mathrm{SL}_n(\mathbb{H})$, isomorphic to the positive scalar group GZ^+, is the multiplicative group of equivalence classes $\underline{\alpha}$ of invertible matrices with the same determinant α.

The *contrapose* $A^{\star}$ of an $m \times n$ quaternionic matrix A is the $n \times m$ matrix obtained from A by transposing and taking quaternionic conjugates. If $m = n$, then $\det A^{\star} = \det A$. As separate operations, transposition and conjugation generally do not preserve quaternionic determinants or even matrix invertibility. For example,

$$\begin{vmatrix} 1 & \mathrm{i} \\ \mathrm{j} & \mathrm{k} \end{vmatrix} = \begin{vmatrix} 1 & -\mathrm{j} \\ -\mathrm{i} & -\mathrm{k} \end{vmatrix} = 2 \quad \text{but} \quad \begin{vmatrix} 1 & \mathrm{j} \\ \mathrm{i} & \mathrm{k} \end{vmatrix} = \begin{vmatrix} 1 & -\mathrm{i} \\ -\mathrm{j} & -\mathrm{k} \end{vmatrix} = 0.$$

A square matrix S with quaternionic entries is *symplectic* if $SS^{\star} = I$. A symplectic matrix has determinant 1. Each symplectic matrix S induces a *symplectic transformation*

$$\cdot S : \mathbb{H}^n \to \mathbb{H}^n, \text{ with } (\mathrm{Q}) \mapsto (\mathrm{Q})S, \quad \text{or}$$
$$S^{\star}\cdot : \check{\mathbb{H}}^n \to \check{\mathbb{H}}^n, \text{ with } [\mathrm{Q}] \mapsto S^{\star}[\mathrm{Q}].$$

* The convenience of these matrix representations is largely the result of our choosing to identify quaternions with *rows* (t, x, y, z) or (u, v), belonging to a *left* vector space over $\mathbb{R}$ or $\mathbb{C}$. Because quaternionic multiplication is noncommutative, adopting the opposite convention—identifying quaternions with *columns* and treating $\mathbb{H}$ as a *right* vector space—is not an inconsequential change, and the corresponding matrix forms are much less intuitive (cf. Du Val 1964, pp. 39–42).

The (quaternionic) *symplectic* group $\tilde{\mathrm{Sp}}_n \cong \mathrm{Sp}(n;\ \mathbb{H})$, the multiplicative group of $n \times n$ symplectic matrices over $\mathbb{H}$, is the group of symplectic transformations of the left linear space $\mathbb{H}^n$ of rows $(\mathrm{Q}) = (\mathrm{Q}_1, \ldots, \mathrm{Q}_n)$ or the right linear space $\check{\mathbb{H}}^n$ of columns $[\mathrm{Q}] = [\mathrm{Q}_1, \ldots, \mathrm{Q}_n]$.

Real and complex symplectic matrices and the real and complex groups Sp_n and $\breve{\mathrm{Sp}}_n$ were defined in § 5.6. If a real duplex matrix is orthogonal, it is also symplectic, and conversely, the *orthogonal duplex* group $\mathrm{O}_{(2)n} \cong \mathrm{O}_{2n} \cap \mathrm{Sp}_{2n}$ is isomorphic to the unitary group U_n. Likewise, a complex duplex matrix is unitary if and only if it is symplectic. The *unitary duplex* group $U_{(2)n} \cong U_{2n} \cap \breve{\mathrm{Sp}}_{2n}$ is isomorphic to the symplectic group $\tilde{\mathrm{Sp}}_n$ (cf. Weyl 1939, pp. 171–173), as is the *orthogonal quadruplex* group $\mathrm{O}_{(4)n} \cong \mathrm{O}_{4n} \cap \mathrm{Sp}'_{4n} \cong \mathrm{O}_{4n} \cap \mathrm{Sp}''_{4n}$. Relationships among some of these groups are shown in Figure 10.4a.

The group $\tilde{\mathrm{Sp}}_1 \cong \mathrm{SU}_2$ is the multiplicative group $\mathrm{SL}(\mathbb{H})$ of unit quaternions (cf. Beardon 1983, pp. 17–18; Neumann, Stoy & Thompson 1994, pp. 196–197). For given unit quaternions L and R, the mappings $\mathrm{Q} \mapsto \tilde{\mathrm{L}}\mathrm{Q}$ and $\mathrm{Q} \mapsto \mathrm{QR}$ correspond respectively to left and right double rotations of S^3, represented by the unit 3-sphere $t^2 + x^2 + y^2 + z^2 = 1$ in E^4. Every direct isometry of S^3 is the product of a left rotation and a right rotation, corresponding to a mapping $\mathrm{Q} \mapsto \tilde{\mathrm{L}}\mathrm{QR}$. Since $(-\tilde{\mathrm{L}})\mathrm{Q}(-\mathrm{R}) = \tilde{\mathrm{L}}\mathrm{QR}$, this is a two-to-one homomorphism $\tilde{\mathrm{Sp}}_1 \times \tilde{\mathrm{Sp}}_1 \to \mathrm{O}_4^+$. For each unit quaternion s, the mapping $\mathrm{Q} \mapsto \tilde{\mathrm{S}}\mathrm{QS}$ corresponds to a simple rotation that takes the great 2-sphere $t = 0$, $x^2 + y^2 + z^2 = 1$ into itself, giving a two-to-one homomorphism $\tilde{\mathrm{Sp}}_1 \to \mathrm{O}_3^+$ (cf. Conway & Smith 2003, pp. 23–24). Factoring out the central inversion $\mathrm{Q} \mapsto -\mathrm{Q}$, we then have $\mathrm{P}^+\mathrm{O}_4 \cong \mathrm{O}_3^+ \times \mathrm{O}_3^+$.

For any given nonzero pure quaternion P, the set of unit quaternions $\mathrm{T} = t + \tau\mathrm{P}$, with t and τ real, forms a multiplicative group, a subgroup of the group $\tilde{\mathrm{Sp}}_1$ of all unit quaternions. The transformations $\mathbb{H} \to \mathbb{H}$ defined by the mappings $\mathrm{Q} \mapsto \tilde{\mathrm{T}}\mathrm{Q}$ and $\mathrm{Q} \mapsto \mathrm{QT}$ correspond to left and right Clifford translations of the 3-sphere S^3 or elliptic 3-space eP^3 (Artzy 1965, pp. 191–193).

A nonzero real scalar multiple λS of a quaternionic symplectic matrix S is a *gyropetic matrix*, and the multiplicative group of $n \times n$ gyropetic matrices over $\mathbb{H}$ is the (quaternionic) *gyropetic* group $\mathrm{G\tilde{p}}_n$. The group $\mathrm{G\tilde{p}}_1 \cong \widetilde{\mathrm{SU}}_2$ is the multiplicative group $\mathrm{GL}(\mathbb{H})$ of nonzero quaternions.

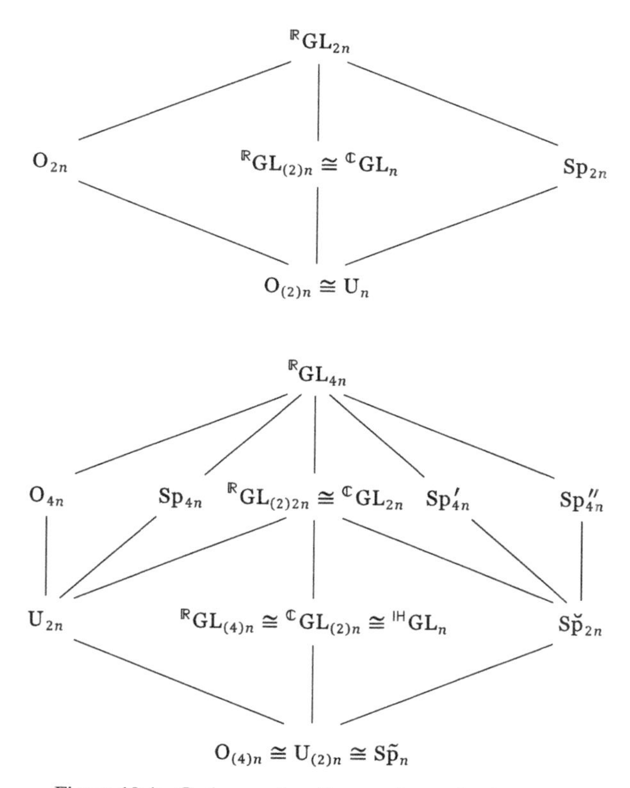

Figure 10.4a Orthogonal, unitary, and symplectic groups

B. *The quaternionic projective line.* John Wilker (1993) showed that the real inversive 4-sphere I^4, represented by the parabolic 4-sphere $\dot{\mathrm{S}}^4$, provides a conformal model for the quaternionic projective line $\mathbb{H}\mathrm{P}^1$. Real circles of I^4 are *chains* of $\mathbb{H}\mathrm{P}^1$. A chain-preserving transformation of $\mathbb{H}\mathrm{P}^1$ is a *concatenation*, which is either a projectivity or an antiprojectivity, according as cross-ratio conjugacy classes are preserved or replaced by their quaternionic conjugates. (Quaternionic cross ratios are defined only up to an inner automorphism.) A concatenation of $\mathbb{H}P^1$ is a circularity of I^4.

If $M = [(A,\ B),\ (C,\ D)]$ is a 2×2 invertible matrix over $\mathbb{H}$, a projectivity $\mathbb{H}P^1 \to \mathbb{H}P^1$ is induced by the *linear fractional transformation*

$$\cdot\langle M\rangle : \mathbb{H}\cup\{\infty\} \to \mathbb{H}\cup\{\infty\}, \text{ with } Q \mapsto (QB+D)\backslash(QA+C), \quad \det M \neq 0, \tag{10.4.4}$$

where $Q \mapsto \infty$ if $QB + D = 0$, $\infty \mapsto B\backslash A$ if $B \neq 0$, and $\infty \mapsto \infty$ if $B = 0$. (The left quotient $B\backslash A$ is the product of B^{-1} and A, in that order.) An antiprojectivity $\mathbb{H}P^1 \to \mathbb{H}P^1$ is likewise induced by the *antilinear fractional transformation*

$$\odot\langle M\rangle : \mathbb{H}\cup\{\infty\} \to \mathbb{H}\cup\{\infty\}, \text{ with } Q \mapsto (\tilde{Q}B+D)\backslash(\tilde{Q}A+C), \quad \det M \neq 0. \tag{10.4.5}$$

Note that the linear fractional transformation defined by a nonreal scalar matrix $\backslash S,\ S\backslash$ is an inner automorphism. Since a projectivity preserves all cross-ratio classes and an antiprojectivity preserves real cross-ratio classes, each of these transformations takes chains into chains.

If N is a unit quaternion and δ is a real number, the inversive 3-sphere $\breve{N}(\delta) = \{Q : (\text{re } \tilde{Q}N) + \delta = 0\} \cup \{\infty\}$ is a great hypersphere (hyperplane) of $\dot{S}^4$. If C is a point associated with a finite quaternion C and ρ is a positive real number, the inversive 3-sphere $\mathring{\mu}(C) = \{Q : |Q - C| = \rho\}$ is a small hypersphere with center C and radius ρ. A reflection in a hyperplane $\breve{N}(\delta)$ and an inversion in a hypersphere $\mathring{\mu}(C)$ are respectively given by

$$Q \mapsto -N(\tilde{Q}N + 2\delta) \quad \text{and} \quad Q \mapsto (\tilde{Q} - \tilde{C})\backslash(\tilde{Q}C + \mu),$$

where δ is real and $|N| = 1$ and where μ is real and $|C|^2 + \mu = \rho^2 \neq 0$. The respective antiprojectivities are defined by the matrices

$$N(\delta) = \begin{pmatrix} \mathrm{N} & 0 \\ 2\delta & -\tilde{\mathrm{N}} \end{pmatrix} \quad \text{and} \quad \mu(C) = \begin{pmatrix} \mathrm{C} & 1 \\ \mu & -\tilde{\mathrm{C}} \end{pmatrix} \tag{10.4.6}$$

(cf. Wilker 1993, pp. 115–116). Either of these matrices may be multiplied by a nonzero *real* scalar, but they cannot otherwise be modified.

Wilker showed how a homography or antihomography of the Möbius 4-sphere I^4 (or a direct or opposite isometry of hyperbolic 5-space) can be represented by a linear fractional or antilinear fractional transformation $\mathbb{H} \cup \{\infty\} \to \mathbb{H} \cup \{\infty\}$ determined by a 2×2 invertible matrix over $\mathbb{H}$ (or a nonzero real scalar multiple thereof). The special inversive group $\mathrm{SI}_4 \cong \mathrm{P}^+\mathrm{O}_{5,1}$ is thus isomorphic to the *quaternionic projective general linear* group $\mathrm{PGL}_2(\mathbb{H})$. The general inversive group $\mathrm{GI}_4 \cong \mathrm{PO}_{5,1}$ of all circularities of I^4 (or all isometries of H^5) is isomorphic to the *quaternionic projective general semilinear* group $\mathrm{P\Gamma L}_2(\mathbb{H})$, containing $\mathrm{PGL}_2(\mathbb{H})$ as a subgroup of index 2.

Following Ahlfors (1985, pp. 66–69; cf. Maclachlan, Waterman & Wielenberg 1989, p. 740), we define the *reverse* of a quaternion $\mathrm{Q} = u+v\mathrm{j} = t+x\mathrm{i}+y\mathrm{j}+z\mathrm{k}$ to be the quaternion $\mathrm{Q}^* = u+\mathrm{j}v = t+x\mathrm{i}+y\mathrm{j}-z\mathrm{k}$. When Q is represented by a complex bivalent matrix, the matrix for Q^* is the matrix for $\tilde{\mathrm{Q}}$ turned upside down. Any 2×2 quaternionic matrix

$$M = [(\mathrm{A}, \mathrm{B}), (\mathrm{C}, \mathrm{D})] = \begin{pmatrix} \mathrm{A} & \mathrm{B} \\ \mathrm{C} & \mathrm{D} \end{pmatrix}$$

has a *pseudo-determinant* $\Delta(M) = \mathrm{AD}^* - \mathrm{BC}^*$. A matrix M is *reversible* if $\mathrm{AB}^* = \mathrm{BA}^*$ and $\mathrm{CD}^* = \mathrm{DC}^*$, which implies that $\mathrm{A}^*\mathrm{C} = \mathrm{C}^*\mathrm{A}$ and $\mathrm{B}^*\mathrm{D} = \mathrm{D}^*\mathrm{B}$.

In the semigroup of reversible 2×2 matrices with real pseudo-determinants, the matrices M with $\Delta(M) \neq 0$ form the *generalized linear* group $\mathrm{G}^*\mathrm{L}_2(\mathbb{H})$, those with $\Delta(M) = \pm 1$ form the *unitized linear* group $\bar{\mathrm{S}}^*\mathrm{L}_2(\mathbb{H})$, and those with $\Delta(M) = 1$ form the *specialized linear* group $\mathrm{S}^*\mathrm{L}_2(\mathbb{H})$. The central quotient groups are the *projective generalized linear* and *projective specialized linear* groups

$$\mathrm{PG^*L_2}(\mathbb{H}) \cong \mathrm{G^*L_2}(\mathbb{H})/\mathrm{GZ}(\mathbb{H}) \cong \bar{\mathrm{S}}^*\mathrm{L_2}(\mathbb{H})/\mathrm{SZ_2}(\mathbb{H})$$

and

$$\mathrm{PS{*}L_2}(\mathbb{H}) \cong \mathrm{S^*L_2}(\mathbb{H})/\mathrm{SZ_2}(\mathbb{H}),$$

where $\mathrm{GZ}(\mathbb{H})$ is the subgroup of nonzero real scalar matrices and $\mathrm{SZ_2}(\mathbb{H}) \cong \{\pm I\}$. The two groups $\mathrm{PG^*L_2}(\mathbb{H})$ and $\mathrm{PS^*L_2}(\mathbb{H})$ are respectively isomorphic to the group $\mathrm{GI_3} \cong \mathrm{PO_{4,1}}$ of all circularities of the Möbius 3-sphere I^3 or isometries of hyperbolic 4-space H^4 and the subgroup $\mathrm{SI_3} \cong \mathrm{P^+O_{4,1}}$ of homographies of I^3 or direct isometries of H^4 (Ahlfors 1985, p. 72; Waterman 1993, pp. 94–96).

C. *Clifford algebras.* Each of the projective pseudo-orthogonal or special projective pseudo-orthogonal groups $\mathrm{PO}_{n,1}$ or $\mathrm{P^+O}_{n,1}$ $(2 \le n \le 5)$ is isomorphic to what can be construed as a group of linear fractional or antilinear fractional transformations:

$$\begin{aligned} \mathrm{PO_{2,1}} &\cong \mathrm{PGL_2}(\mathbb{R}), & \mathrm{P^+O_{2,1}} &\cong \mathrm{PSL_2}(\mathbb{R}), \\ \mathrm{PO_{3,1}} &\cong \mathrm{PG\bar{L}_2}(\mathbb{C}), & \mathrm{P^+O_{3,1}} &\cong \mathrm{PGL_2}(\mathbb{C}), \\ \mathrm{PO_{4,1}} &\cong \mathrm{PG^*L_2}(\mathbb{H}), & \mathrm{P^+O_{4,1}} &\cong \mathrm{PS^*L_2}(\mathbb{H}), \\ \mathrm{PO_{5,1}} &\cong \mathrm{PG\bar{L}_2}(\mathbb{H}), & \mathrm{P^+O_{5,1}} &\cong \mathrm{PGL_2}(\mathbb{H}). \end{aligned} \tag{10.4.7}$$

That is, for n equal to 2, 3, 4, or 5, each isometry of hyperbolic n-space H^n or circularity of the Möbius $(n-1)$-sphere I^{n-1} corresponds to the set of nonzero real scalar multiples of some 2×2 invertible matrix with real, complex, or quaternionic entries.

The correspondences (10.4.7) can also be described in terms of *Clifford algebras*. W. K. Clifford (1882) defined a family of linear associative algebras $\mathbb{C}_n$ of dimension 2^n over the real field $\mathbb{R}$ generated, for $n > 0$, by elements i_s $(1 \le s \le n)$, with $\mathrm{i}_s{}^2 = -1$ for each s and $\mathrm{i}_s\mathrm{i}_r = -\mathrm{i}_r\mathrm{i}_s$ for $r \ne s$ (cf. Neumann, Stoy & Thompson 1994, pp. 200–204; Baez 2002, pp. 156–159). The Clifford algebra $\mathbb{C}_0$ is $\mathbb{R}$ itself; $\mathbb{C}_1 = \mathbb{C}$ is the complex field; $\mathbb{C}_2$ is the division ring $\mathbb{H}$ of quaternions.

The algebra $\mathbb{C}_3$ of *octals* is an eight-dimensional vector space over $\mathbb{R}$ with basis

$$[1,\ \mathrm{i}_1,\ \mathrm{i}_2,\ \mathrm{i}_{12},\ \mathrm{i}_3,\ \mathrm{i}_{31},\ \mathrm{i}_{23},\ \mathrm{i}_{123}],$$

where $\mathrm{i}_{12} = \mathrm{i}_1\mathrm{i}_2$, etc. Any two of the elements i_1, i_2, and i_3 or any two of the elements i_{12}, i_{31}, and i_{23} generate a four-dimensional subalgebra isomorphic to $\mathbb{H}$, so that $\mathbb{C}_3$ can be expressed as the direct sum $\mathbb{H} \oplus \mathbb{H}$ of two quaternionic algebras. Each octal

$$\mathbf{c} = c_0 + c_1\mathrm{i}_1 + c_2\mathrm{i}_2 + c_{12}\mathrm{i}_{12} + c_3\mathrm{i}_3 + c_{31}\mathrm{i}_{31} + c_{23}\mathrm{i}_{23} + c_{123}\mathrm{i}_{123}$$

can be identified with a point $(c) = (c_0,\ c_1,\ c_2,\ c_{12},\ c_3,\ c_{31},\ c_{23},\ c_{123})$ of Euclidean 8-space and can be represented by an 8×8 real *octivalent matrix*

$$C = \{\{\{\mathbf{c}\}\}\}$$
$$= \begin{pmatrix} c_0 & c_1 & c_2 & c_{12} & c_3 & c_{31} & c_{23} & c_{123} \\ -c_1 & c_0 & -c_{12} & c_2 & c_{31} & -c_3 & -c_{123} & c_{23} \\ -c_2 & c_{12} & c_0 & -c_1 & -c_{23} & -c_{123} & c_3 & c_{31} \\ -c_{12} & -c_2 & c_1 & c_0 & -c_{123} & c_{23} & -c_{31} & c_3 \\ -c_3 & -c_{31} & c_{23} & -c_{123} & c_0 & c_1 & -c_2 & c_{12} \\ -c_{31} & c_3 & -c_{123} & -c_{23} & -c_1 & c_0 & c_{12} & c_2 \\ -c_{23} & -c_{123} & -c_3 & c_{31} & c_2 & -c_{12} & c_0 & c_1 \\ c_{123} & -c_{23} & -c_{31} & -c_3 & -c_{12} & -c_2 & -c_1 & c_0 \end{pmatrix}. \tag{10.4.8}$$

The Clifford algebra $\mathbb{C}_3$ is isomorphic to the ring of such matrices. The set of invertible matrices of this type forms a multiplicative group $\mathrm{GL}_{(8)}$ isomorphic to the group $\mathrm{GL}(\mathbb{C}_3)$ of invertible octals. Each element $\mathbf{c}$ has an *octal norm*

$$(c_0{}^2 + c_1{}^2 + c_2{}^2 + c_{12}{}^2 + c_3{}^2 + c_{31}{}^2 + c_{23}{}^2 + c_{123}{}^2)^2$$
$$- 4(c_0c_{123} - c_1c_{23} - c_2c_{31} - c_{12}c_3)^2,$$

whose square is the determinant of C.

Theodor Vahlen (1902) showed that homographies of the Möbius $(n-1)$-sphere I^{n-1} can be represented as linear fractional transformations over $\mathbb{C}_{n-2}$, and Lars Ahlfors (1985) identified homographies of I^{n-1} (or direct isometries of H^n) with certain 2×2 "Clifford matrices" over $\mathbb{C}_{n-2}$.

D. *Octonions.* All Clifford algebras are associative, but only $\mathbb{C}_0$, $\mathbb{C}_1$, and $\mathbb{C}_2$ (i.e., $\mathbb{R}$, $\mathbb{C}$, and $\mathbb{H}$) are division rings. The only other normed division algebra (with unity) over the reals, discovered independently by John Graves in 1843 and Arthur Cayley in 1845, is the alternative division ring $\mathbb{O}$ of *octonions*. This is an eight-dimensional vector space over $\mathbb{R}$ with basis

$$[1,\ \mathrm{e}_1,\ \mathrm{e}_2,\ \mathrm{e}_3,\ \mathrm{e}_4,\ \mathrm{e}_5,\ \mathrm{e}_6,\ \mathrm{e}_7],$$

having a noncommutative and nonassociative multiplication of vectors.

Each pair $[1, \mathrm{e}_i]$ is a basis for a two-dimensional subalgebra isomorphic to $\mathbb{C}$. A quadruple $[1, \mathrm{e}_i, \mathrm{e}_j, \mathrm{e}_k]$ is a basis for a four-dimensional subalgebra isomorphic to $\mathbb{H}$ if $(i\ j\ k)$ is one of the seven associative triples

$$(1\,2\,4),\ (2\,3\,5),\ (3\,4\,6),\ (4\,5\,7),\ (5\,6\,1),\ (6\,7\,2),\ (7\,1\,3).$$

In other cases with distinct subscripts we have $(\mathrm{e}_i\mathrm{e}_j)\mathrm{e}_k = -\mathrm{e}_i(\mathrm{e}_j\mathrm{e}_k)$, so that $\mathbb{O}$ is not generally associative.

EXERCISES 10.4

1. $\vdash$ The set of quadrivalent matrices $(t,\ x,\ y,\ z)_{(4)}$ is closed under addition and multiplication.

2. $\vdash$ Let $M = [(\mathrm{A},\ \mathrm{B}),\ (\mathrm{C},\ \mathrm{D})] = ([\mathrm{A},\ \mathrm{C}], [\mathrm{B},\ \mathrm{D}])$ be any 2×2 matrix over $\mathbb{H}$, and let Q be any quaternion. Then the following properties hold.

a. If $R = [(\mathrm{C},\ \mathrm{D}),\ (\mathrm{A},\ \mathrm{B})]$ or $R = ([\mathrm{B},\ \mathrm{D}],\ [\mathrm{A},\ \mathrm{C}])$ (two rows or two columns interchanged), then $\det R = \det M$.

b. If $S = [\mathrm{Q}(\mathrm{A},\ \mathrm{B}),\ (\mathrm{C},\ \mathrm{D})]$ or $S = ([\mathrm{A},\ \mathrm{C}]\mathrm{Q},\ [\mathrm{B},\ \mathrm{D}])$ (row or column multiplied by scalar), then $\det S = |\mathrm{Q}|\ \det M$.

c. If $T = [(\mathrm{A},\ \mathrm{B}),\ (\mathrm{QA}+\mathrm{C},\ \mathrm{QB}+\mathrm{D})]$ or $T = ([\mathrm{A},\ \mathrm{C}],\ [\mathrm{AQ}+\mathrm{B},\ \mathrm{CQ}+\mathrm{D}])$ (scalar multiple of one row or column added to another), then $\det T = \det M$.

3. ⊢ The *bisymplectic duplex* group $\mathrm{Sp}'_{(2)2n} \cong \mathrm{Sp}'_{4n} \cap \mathrm{Sp}''_{4n}$ is isomorphic to the complex symplectic group $\check{\mathrm{Sp}}_{2n}$.

4. ⊢ The quaternionic symplectic group $\tilde{\mathrm{Sp}}_1$ is isomorphic to the special unitary group SU_2.

5. ⊢ Corresponding to the mapping $\mathrm{Q} \mapsto \tilde{\mathrm{s}}\mathrm{Q}\mathrm{s}$, where s is a unit quaternion $\alpha+\beta\mathrm{i}+\gamma\mathrm{j}+\delta\mathrm{k}$, a simple rotation $(x,\ y,\ z) \mapsto (x,\ y,\ z)S$ of the unit sphere is effected by the matrix

$$S = \begin{pmatrix} \alpha^2+\beta^2-\gamma^2-\delta^2 & -2\alpha\delta+2\beta\gamma & 2\alpha\gamma+2\beta\delta \\ 2\alpha\delta+2\beta\gamma & \alpha^2-\beta^2+\gamma^2-\delta^2 & -2\alpha\beta+2\gamma\delta \\ -2\alpha\gamma+2\beta\delta & 2\alpha\beta+2\gamma\delta & \alpha^2-\beta^2-\gamma^2+\delta^2 \end{pmatrix}.$$

(As shown by Olinde Rodrigues in 1840, every 3×3 special orthogonal matrix can be expressed in this form for some choice of parameters α, β, γ, δ.)

6. ⊢ When the quaternionic projective line $\mathbb{HP}^1$ is modeled by the parabolic 4-sphere $\dot{\mathrm{S}}^4$, each quaternion Q can be represented by a point with normalized coordinates $(1,\ \tau,\ \xi,\ \eta,\ \zeta,\ \omega)$, where

$$\mathrm{re}\ \mathrm{Q} = \frac{\tau}{1-\omega}, \quad \mathrm{pu}\ \mathrm{Q} = \left(\frac{\xi}{1-\omega},\ \frac{\eta}{1-\omega},\ \frac{\zeta}{1-\omega}\right), \quad |\mathrm{Q}|^2 = \frac{1+\omega}{1-\omega}$$

(with $\mathrm{Q} = \infty$ if $\omega = 1$), giving the particular identifications

$$\begin{aligned} 1 &\leftrightarrow (1,\ 1,\ 0,\ 0,\ 0,\ 0), & -1 &\leftrightarrow (1,\ -1,\ 0,\ 0,\ 0,\ 0), \\ \mathrm{i} &\leftrightarrow (1,\ 0,\ 1,\ 0,\ 0,\ 0), & -\mathrm{i} &\leftrightarrow (1,\ 0,\ -1,\ 0,\ 0,\ 0), \\ \mathrm{j} &\leftrightarrow (1,\ 0,\ 0,\ 1,\ 0,\ 0), & -\mathrm{j} &\leftrightarrow (1,\ 0,\ 0,\ -1,\ 0,\ 0), \\ \mathrm{k} &\leftrightarrow (1,\ 0,\ 0,\ 0,\ 1,\ 0), & -\mathrm{k} &\leftrightarrow (1,\ 0,\ 0,\ 0,\ -1,\ 0), \\ \infty &\leftrightarrow (1,\ 0,\ 0,\ 0,\ 0,\ 1), & 0 &\leftrightarrow (1,\ 0,\ 0,\ 0,\ 0,\ -1). \end{aligned}$$

7. ⊢ The set of reversible matrices $M = [(\mathrm{A},\ \mathrm{B}),\ (\mathrm{C},\ \mathrm{D})]$ with real pseudo-determinants $\Delta(M) = \mathrm{AD}^* - \mathrm{BC}^*$ is closed under multiplication.

8. $\vdash$ The set of octivalent matrices $\{\{\{\mathbf{c}\}\}\}$ is closed under addition and multiplication.

9. $\vdash$ The center of the Clifford algebra $\mathbb{C}_3$ is the two-dimensional subalgebra spanned by 1 and i_{123}.

11

FINITE SYMMETRY GROUPS

DISCRETE GROUPS OF ISOMETRIES can be associated with the symmetries of certain geometric figures, such as polygons and polyhedra. Typically, these are *Coxeter groups*, generated by reflections in the facets of a simplex (e.g., a triangle or a tetrahedron) whose dihedral angles are submultiples of π, or else subgroups or extensions of such groups. As a discrete subgroup of the continuous group of isometries of some real metric space, a Coxeter group may be *spherical*, *Euclidean*, or *hyperbolic*. Spherical Coxeter groups, described in this chapter, are finite. Their infinite Euclidean and hyperbolic counterparts will be discussed in Chapters 12 and 13.

11.1 POLYTOPES AND HONEYCOMBS

A polygon or a polyhedron is a two- or three-dimensional instance of a *polytope*, a geometric figure consisting of points, line segments, planar regions, etc., having a particular hierarchical structure. When realized in some Euclidean or non-Euclidean space, a polytope also has certain metric properties, such as edge lengths and (dihedral) angles. A partition of a line or a circle into segments or arcs or a tessellation of a plane or a sphere by polygons joined edge to edge is a one- or two-dimensional version of a *honeycomb*, a kind of degenerate polytope.

Abstractly, an *n-polytope* can be defined as a partially ordered set of j-dimensional elements $(-1 \leq j \leq n)$, its *j-faces*, including just one of rank -1 and just one of rank n. Each j-face J has a unique *span* $\langle J \rangle$, a partially ordered subset of i-faces $(-1 \leq i \leq j)$ constituting a j-polytope, and a unique *cospan* $\rangle J \langle$, a partially ordered subset of k-faces $(j \leq k \leq n)$ having the structure of an $(n-j-1)$-polytope. An i-face I is *subsidiary* to a j-face J, written $I \leq J$, whenever $I \in \langle J \rangle$. If $I \leq J$ but $I \neq J$, we write $I < J$. Each j-face of a *real* n-polytope is part of a j-subspace of a real metric n-space.

All real n-polytopes are *dyadic*; i.e., for each j from 0 to $n-1$, given a $(j-1)$-face I and a $(j+1)$-face K with $I < K$, there are exactly two *adjacent* j-faces J_1 and J_2 such that $I < J_1 < K$ and $I < J_2 < K$. (This condition holds vacuously when $n < 1$.) An n-polytope must be *properly connected*; i.e., no proper subset of the j-faces has all the foregoing properties.

As a partially ordered set, every n-polytope has a *Hasse diagram* (after Helmut Hasse, 1898–1979), with $n+2$ levels of nodes representing the j-faces of each rank, nodes on successive levels being connected by a branch if the corresponding j-faces are incident. For example, Figure 11.1a shows the Hasse diagram for a tetrahedron $\langle ABCD \rangle$, with four 0-faces, six 1-faces, and four 2-faces. Contained within the Hasse diagram of any polytope are subdiagrams for the span and cospan of each of its elements.

The unique (-1)-face of every real n-polytope is the *nullity*, alias the empty set $\varnothing$. The unique (-1)-polytope is the *nullitope* $\langle\,\rangle$, the set $\{\varnothing\}$ whose only member is the empty set. A 0-face, or *vertex*, is a point, which may be either ordinary or absolute. A 0-polytope (or *monon*) $\langle A \rangle$ is the set $\{\varnothing;\ A\}$ spanned by a single point A. A 1-face, or *edge*, is a line segment or a minor arc of a great circle. A 1-polytope (or *dion*) $\langle AB \rangle = \{\varnothing;\ A,\ B;\ AB\}$ comprises the empty set, two points A and B, and the segment or arc AB joining them. A 2-polytope is a *polygon*, a 3-polytope is a *polyhedron*, and a 4-polytope is a *polychoron*.

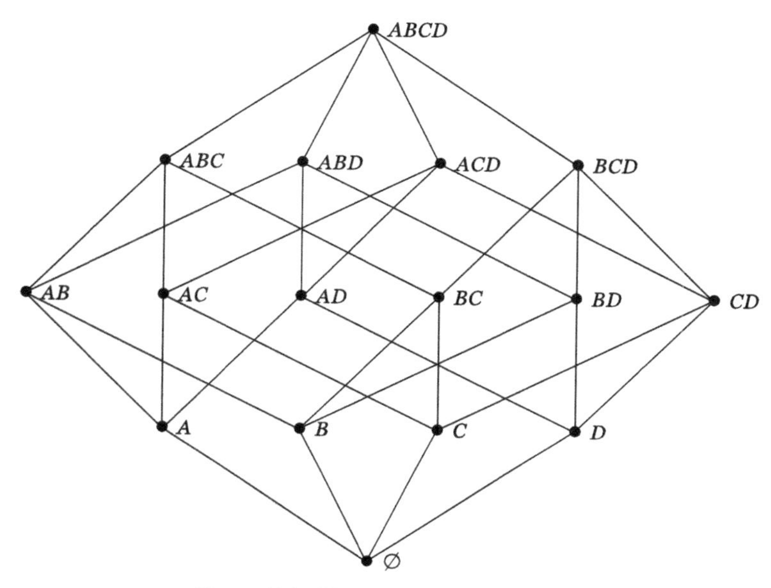

Figure 11.1a Hasse diagram for $\langle ABCD \rangle$

The vertices of a dion are its *endpoints*, and the edges of a polygon are its *sides*. The 2-faces of a polyhedron are just called *faces*, and the 3-faces of a polychoron are *cells*. The single n-face of an n-polytope is its *body*, essentially its "interior"; an $(n - 1)$-face ("hyperface") is a *facet*, and the $(n - 2)$-face between two adjacent facets is a *ridge*.

An *n-honeycomb*, or polytopal space-filling, is abstractly the same as an $(n+1)$-polytope but has a *scope* (all or part of n-space) instead of a body. The only 0-honeycomb is an *antipodion*, consisting of the two points of S^0. A 1-honeycomb is a *partition* of a line or a circle, and its 1-faces are called *parts*. A 2-honeycomb is a *tessellation* of a plane or a sphere, and a 3-honeycomb is a *cellulation*. The n-faces ("hypercells") of an n-honeycomb are its *cellules*, and the $(n - 1)$-faces separating adjacent cellules are *walls*.

A totally ordered subset of j-faces of an n-polytope or $(n - 1)$-honeycomb, one of each rank from -1 to n, is called a *flag*. A polytope or honeycomb is *regular* if there are isometries permuting the vertices and taking any flag into any other flag. The nullitope)(is trivially regular, as are all monons () and any dion { } with two ordinary or two absolute endpoints (but not one of each).

A polygon is regular if it is both equiangular and equilateral, and a partition is regular if it is equilateral. A regular *p-gon*, either a polygon with p vertices and p sides or a partition of the circle S^1 into arcs of length $2\pi/p$, is denoted $\{p\}$. A regular *apeirogon* $\{\infty\}$ is either a partition of the Euclidean line E^1 into infinitely many equal-length segments or an infinite polygon inscribed in a horocycle or in the absolute circle of the hyperbolic plane. A regular *pseudogon* $\{\pi i/\lambda\}$ is a partition of the hyperbolic line H^1 into segments of length 2λ.

Each regular figure has a *Schläfli symbol* (after Ludwig Schläfli, 1814–1895). A regular polyhedron or tessellation whose face polygons are $\{p\}$'s, q meeting at each vertex, so that its "vertex figures" are $\{q\}$'s, is denoted $\{p, q\}$. In general, a regular polytope or honeycomb whose facet or cellule polytopes are $\{p, q, \ldots, u\}$'s and whose vertex figures are $\{q, \ldots, u, v\}$'s is denoted $\{p, q, \ldots, u, v\}$.* Each regular n-polytope P has a *dual* P̌, with the j-faces of P̌ corresponding to the $(n-j-1)$-faces of P. The Schläfli symbol for P̌ is the reverse of the symbol for P.

EXERCISES 11.1

1. Identify the regular polyhedra (or spherical tessellations) $\{3, 3\}$, $\{3, 4\}$, $\{4, 3\}$, $\{3, 5\}$, and $\{5, 3\}$.

2. Describe the Euclidean tessellations $\{4, 4\}$, $\{3, 6\}$, and $\{6, 3\}$.

11.2 POLYGONAL GROUPS

The *symmetry group* of a figure in a metric space (exclusive of "fractal" figures with self-similarities) is the group of isometries that

* The curly brackets conventionally used in Schläfli symbols should not be confused with the standard notation for set membership.

take the figure into itself. We shall be particularly interested in figures whose symmetry groups are discrete. Since every isometry of a Euclidean or non-Euclidean space is the product of reflections, any discrete symmetry group is a subgroup of a discrete group generated by reflections or an extension of such a group by a group of automorphisms (other isometries). These groups are known as *Coxeter groups*, after H. S. M. Coxeter (1907–2003), who developed the general theory of reflection groups in the classical real metric spaces.

It will be sufficient to consider discrete groups operating in spherical, Euclidean, or hyperbolic space. Any group operating in elliptic n-space eP^n is either isomorphic to a group G that operates on the n-sphere S^n or (if G contains the central inversion) is a quotient group $G/2$ of half the order. Any nontrivial Coxeter group W operating in S^n, E^n, or H^n has a *direct subgroup* W^+ of index 2, generated by rotations or other direct isometries.

The trivial group of order 1 generated by no reflections is the *identity* group][, the symmetry group of the nullitope)(or a monon (). The group generated by a single reflection ρ in any space is the *bilateral* group [ ], of order 2, satisfying the relation $\rho^2 = 1$. The group [ ] is the symmetry group of a dion { } or of any figure that has only bilateral symmetry. Abstractly, [ ] is the dihedral group D_1, isomorphic to the cyclic group C_2, and the direct subgroup $[\,]^+ \cong\,][$ is the cyclic group $\mathrm{C}_1 \cong 1$.

The group generated by reflections ρ_1 and ρ_2 in orthogonal great circles of S^2 or orthogonal lines of E^2 or H^2, satisfying the relations

$$\rho_1{}^2 = \rho_2{}^2 = (\rho_1\rho_2)^2 = 1, \tag{11.2.1}$$

is the *rectangular* group (or "four-group") $[2] \cong \mathrm{D}_2$, of order 4, the direct product [ ] × [ ] of two bilateral groups and the symmetry group of a rectangle { } × { }. The direct subgroup $[2]^+ \cong \mathrm{C}_2$ generated by the half-turn $\sigma = \rho_1\rho_2$, satisfying the relation $\sigma^2 = 1$, is the *rhombic* group, of order 2.

For $p \geq 3$, the nonabelian dihedral group $[p] \cong \mathrm{D}_p$, of order $2p$, generated by reflections ρ_1 and ρ_2 in great circles or lines intersecting at an angle of π/p, satisfies the relations

$$\rho_1{}^2 = \rho_2{}^2 = (\rho_1\rho_2)^p = 1. \tag{11.2.2}$$

The cyclic subgroup $[p]^+ \cong \mathrm{C}_p$, of order p, generated by the rotation $\sigma = \rho_1\rho_2$, satisfying the relation $\sigma^p = 1$, is the *p-gonal* group, and (11.2.2) is the *full p-gonal* group. The groups $[p]$ and $[p]^+$ are respectively the full symmetry group and the rotation group of a regular p-gon $\{p\}$. Any of these groups can also be taken as operating on the circle S^1, where the mirror of a reflection is an antipodal pair of points.

For $p \geq 2$, the reflections ρ_1 and ρ_2 are conveniently represented by the orthogonal reflection matrices

$$R_1 = \begin{pmatrix} 1 & 0 \\ 0 & -1 \end{pmatrix} \quad \text{and} \quad R_2 = \begin{pmatrix} \cos 2\pi/p & \sin 2\pi/p \\ \sin 2\pi/p & -\cos 2\pi/p \end{pmatrix},$$

and the circular rotation σ by the orthogonal rotation matrix

$$S = R_1R_2 = \begin{pmatrix} \cos 2\pi/p & \sin 2\pi/p \\ -\sin 2\pi/p & \cos 2\pi/p \end{pmatrix}.$$

For particular values of p, sines and cosines are as follows:

p	2	3	4	5	6	8	10	12	∞
$2\sin 2\pi/p$	0	$\sqrt{3}$	2	$\sqrt{3+\bar{\tau}}$	$\sqrt{3}$	$\sqrt{2}$	$\sqrt{3-\tau}$	1	0
$2\cos 2\pi/p$	-2	-1	0	$\bar{\tau}$	1	$\sqrt{2}$	τ	$\sqrt{3}$	2

Here τ and $\bar{\tau}$ respectively denote the golden-section number ½$(\sqrt{5}+1)$ and its inverse ½$(\sqrt{5}-1)$; note that $\tau = 1 + \bar{\tau}$ so that $\tau^2 = \tau + 1$ and $\bar{\tau}^2 = 1 - \bar{\tau}$.

The *full apeirogonal* group $[\infty]$ is generated by reflections ρ_1 and ρ_2 in parallel lines of E^2 or H^2, satisfying the relations

$$\rho_1{}^2 = \rho_2{}^2 = (\rho_1\rho_2)^\infty = 1, \tag{11.2.3}$$

where the superfluous relation $(\rho_1\rho_2)^\infty = 1$ means that no finite power of the product $\rho_1\rho_2$ equals the identity. As an abstract group, $[\infty]$ is the infinite dihedral group D_∞. The *apeirogonal* group $[\infty]^+$ generated by the Euclidean translation or hyperbolic striation $\tau = \rho_1\rho_2$, with no relations, is the infinite cyclic group C_∞, the free group with one generator. The groups $[\infty]$ and $[\infty]^+$ are respectively the full symmetry group and the translation or striation group of a regular apeirogon $\{\infty\}$, either a partition of E^1 or an infinite hyperbolic polygon. Either of these groups can also be taken as operating on the Euclidean line or on a horocycle; in each case the mirror of a reflection is a point, together with the point at infinity of the line or the absolute point of the horocycle.

The reflections ρ_1 and ρ_2 may be represented by the transorthogonal reflection matrices

$$R_1 = \begin{pmatrix} -1 & 0 \\ 0 & 1 \end{pmatrix} \quad \text{and} \quad R_2 = \begin{pmatrix} -1 & 0 \\ 2\lambda & 1 \end{pmatrix}$$

and the Euclidean translation τ (or striation in H^2) by the translation matrix

$$T = R_1R_2 = \begin{pmatrix} 1 & 0 \\ 2\lambda & 1 \end{pmatrix}.$$

Here λ is the distance between the two mirrors along a Euclidean line.

The *full pseudogonal* group $[\pi i/\lambda]$ is generated by reflections ρ_1 and ρ_2 in diverging lines of H^2, satisfying the relations

$$\rho_1{}^2 = \rho_2{}^2 = (\rho_1\rho_2)^{\pi i/\lambda} = 1, \tag{11.2.4}$$

the parameter λ in the imaginary period of the product $\rho_1\rho_2$ being the distance between the lines. This is another representation of the infinite dihedral group D_∞, and the *pseudogonal* group $[\pi i/\lambda]^+$ generated by the hyperbolic translation $\upsilon = \rho_1\rho_2$ is again the infinite cyclic group C_∞. The groups $[\pi i/\lambda]$ and $[\pi i/\lambda]^+$ can also be taken as operating on the hyperbolic line, where they are respectively the full symmetry group and the translation group of a regular pseudogon $\{\pi i/\lambda\}$. The mirror of a reflection is now an ordinary point of H^1, together with the ideal point conjugate to it in the absolute involution.

The reflections ρ_1 and ρ_2 may be represented by the pseudo-orthogonal reflection matrices

$$R_1 = \begin{pmatrix} 1 & 0 \\ 0 & -1 \end{pmatrix} \quad \text{and} \quad R_2 = \begin{pmatrix} \cosh 2\lambda & \sinh 2\lambda \\ -\sinh 2\lambda & -\cosh 2\lambda \end{pmatrix}$$

(or their negatives) and the hyperbolic translation υ by the pseudo-orthogonal rotation matrix

$$U = R_1 R_2 = \begin{pmatrix} \cosh 2\lambda & \sinh 2\lambda \\ \sinh 2\lambda & \cosh 2\lambda \end{pmatrix},$$

with λ being the distance between the two mirrors. That $U^{\pi i/\lambda} = I$ follows from the identities $\cosh it = \cos t$ and $\sinh it = i \sin t$.

EXERCISES 11.2

1. Show that the full p-gonal group $[p] \cong \langle \rho_1, \rho_2 \rangle$, satisfying the relations (11.2.2), can be generated by the reflection $\rho = \rho_1$ and the rotation $\sigma = \rho_1 \rho_2$. What are the defining relations then?

2. The *Fibonacci sequence* $\{f\}$ is defined by the initial values $f_0 = 0$ and $f_1 = 1$ and the recursive rule $f_{n-1} + f_n = f_{n+1}$ for all $n \in \mathbb{Z}$. The limit as $n \to \infty$ of the quotient f_n/f_{n-1} of consecutive Fibonacci numbers is the golden ratio τ (the limit as $n \to -\infty$ is $-\bar{\tau}$). Show that $\tau^n = f_n\tau + f_{n-1}$ for all $n \in \mathbb{Z}$.

3. Show that the reflections ρ_1 and ρ_2 generating the full pseudogonal group $[\pi i/\lambda]$ and satisfying the relations (11.2.4) can be represented by the pseudo-orthogonal reflection matrices R_1 and R_2 given in this section.

11.3 PYRAMIDS, PRISMS, AND ANTIPRISMS

Finite groups operating in three-dimensional space include the identity group $[\,]^+ \cong C_1$ and the bilateral group $[\,] \cong D_1$. Other groups are associated with the symmetries of pyramids, prisms, antiprisms, and the regular polyhedra.

A rectangular *pyramid* $() \vee [\{\,\} \times \{\,\}]$ in Euclidean 3-space is the "dionic join" of a monon (), its *apex*, and a rectangle $\{\,\} \times \{\,\}$, its *base*; four isosceles triangles $()\vee\{\,\}$ surround the apex, each sharing one side with the base. The symmetry group of $()\vee[\{\,\}\times\{\,\}]$ is the *acrorectangular* group $[1, 2]$, whose generating reflections ρ_1 and ρ_2 in E^2 satisfy the relations (11.2.1). The direct subgroup, the *acrorhombic* group $[1,\ 2]^+$, is generated by the half-turn $\sigma = \rho_1\rho_2$, satisfying the relation $\sigma^2 = 1$. Thus $[1, 2]$ and $[1,\ 2]^+$ are respectively isomorphic to the rectangular group $[2] \cong D_2$ and the rhombic group $[2]^+ \cong C_2$.

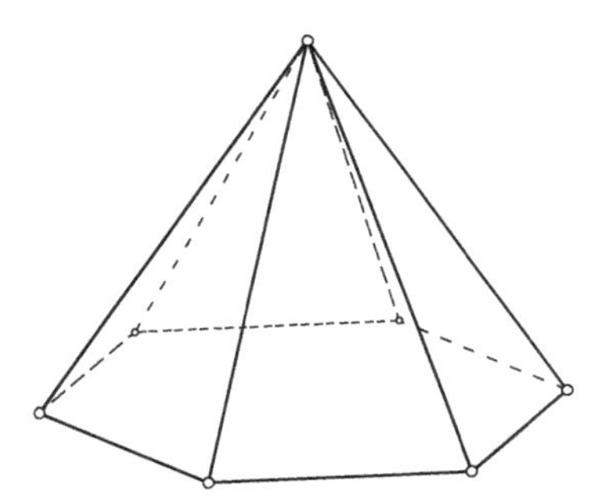

Figure 11.3a Pyramid $() \vee \{6\}$

Similarly, for $p \geq 3$, a p-gonal pyramid $() \vee \{p\}$ has a p-gonal base with p isosceles triangles surrounding the apex (see Figure 11.3a for $p = 6$). The symmetry group of $() \vee \{p\}$ is the *full acro-p-gonal* group $[1,\ p]$, whose generating reflections ρ_1 and ρ_2 satisfy (11.2.2). The direct subgroup, the *acro-p-gonal* group $[1,\ p]^+$, is generated by the rotation $\sigma = \rho_1\rho_2$, satisfying $\sigma^p = 1$. The groups $[1,\ p]$ and $[1,\ p]^+$ are respectively isomorphic to the full p-gonal group $[p] \cong D_p$ and the p-gonal group $[p]^+ \cong C_p$.

For $p \geq 2$, the reflections ρ_1 and ρ_2 are conveniently represented by the orthogonal reflection matrices

$$R_1 = \begin{pmatrix} 1 & 0 & 0 \\ 0 & -1 & 0 \\ 0 & 0 & 1 \end{pmatrix} \quad \text{and} \quad R_2 = \begin{pmatrix} \cos 2\pi/p & \sin 2\pi/p & 0 \\ \sin 2\pi/p & -\cos 2\pi/p & 0 \\ 0 & 0 & 1 \end{pmatrix}$$

and the rotation $\sigma = \rho_1\rho_2$ by the orthogonal rotation matrix

$$S = \begin{pmatrix} \cos 2\pi/p & \sin 2\pi/p & 0 \\ -\sin 2\pi/p & \cos 2\pi/p & 0 \\ 0 & 0 & 1 \end{pmatrix}.$$

A right rectangular *prism* $\{\} \times \{\} \times \{\}$ is the "rectangular product" of three orthogonal dions or (in three ways) of a dion $\{\}$ and a rectangle $\{\} \times \{\}$. One pair of opposite rectangles can be taken as *bases*, separated by a band of four lateral rectangles. If no two distances between pairs of opposite faces are equal, the symmetry group of $\{\} \times \{\} \times \{\}$ is the *orthorectangular* (or "full orthorhombic") group $[2, 2] \cong D_1 \times D_2$, of order 8, generated by three reflections ρ_0, ρ_1, and ρ_2 satisfying the relations

$$\rho_0{}^2 = \rho_1{}^2 = \rho_2{}^2 = (\rho_0\rho_1)^2 = (\rho_0\rho_2)^2 = (\rho_1\rho_2)^2 = 1. \qquad (11.3.1)$$

Since each commutator $(\rho_i\rho_j)^2$ is the identity, the generators commute, and the group is abelian. The direct subgroup is the *pararhombic* group $[2, 2]^+ \cong D_2$, of order 4, generated by the two half-turns $\sigma_1 = \sigma_{01} = \rho_0\rho_1$ and $\sigma_2 = \sigma_{12} = \rho_1\rho_2$, satisfying the relations

$$\sigma_1{}^2 = \sigma_2{}^2 = (\sigma_1\sigma_2)^2 = 1. \qquad (11.3.2)$$

The orthorectangular group $[2, 2]$ also has a "semidirect" subgroup of index 2, the *orthorhombic* group $[2, 2^+] \cong D_1 \times C_2$, of order 4, generated by the reflection ρ_0 and the half-turn $\sigma_{12} = \rho_1\rho_2$, satisfying the relations

$$\rho_0{}^2 = \sigma_{12}{}^2 = (\rho_0\sigma_{12})^2 = 1. \qquad (11.3.3)$$

There are also two isomorphic subgroups corresponding to other permutations of the reflections. The rectangular group [2], the acrorectangular group [1, 2], the pararhombic group $[2, 2]^+$, and the orthorhombic group $[2, 2^+]$ are different geometric representations of the same abstract dihedral group D_2.

The orthorectangular and orthorhombic groups have a subgroup of order 2, of index 4 in [2, 2] and of index 2 in $[2, 2^+]$, the *central* group $[2^+, 2^+] \cong 2$, generated by the *central inversion* (flip reflection) $\zeta = \rho_0\sigma_{12} = \rho_0\rho_1\rho_2$, satisfying the relation

$$\zeta^2 = 1. \tag{11.3.4}$$

The bilateral group [ ], the rhombic group $[2]^+$, and the central group $[2^+, 2^+]$ are different representations of the same abstract group $D_1 \cong C_2 \cong 2$.

For $p \geq 3$, a right p-gonal prism $\{\} \times \{p\}$ has two basal p-gons separated by a band of p lateral rectangles (see Figure 11.3b for $p = 6$). The symmetry group of $\{\} \times \{p\}$ is the *full ortho-p-gonal* group $[2, p] \cong D_1 \times D_p$, of order $4p$, generated by three reflections ρ_0, ρ_1, and ρ_2 satisfying the relations

$$\rho_0{}^2 = \rho_1{}^2 = \rho_2{}^2 = (\rho_0\rho_1)^2 = (\rho_0\rho_2)^2 = (\rho_1\rho_2)^p = 1. \tag{11.3.5}$$

The direct subgroup is the *para-p-gonal* group $[2, p]^+ \cong D_p$, of order $2p$, generated by the half-turn $\sigma_1 = \sigma_{01} = \rho_0\rho_1$ and the rotation $\sigma_2 = \sigma_{12} = \rho_1\rho_2$, satisfying the relations

$$\sigma_1{}^2 = \sigma_2{}^p = (\sigma_1\sigma_2)^2 = 1. \tag{11.3.6}$$

The full ortho-p-gonal group $[2, p]$ has another subgroup of index 2, the *ortho-p-gonal* group $[2, p^+] \cong D_1 \times C_p$, of order $2p$, generated by the reflection ρ_0 and the rotation $\sigma_{12} = \rho_1\rho_2$, which commute, satisfying

$$\rho_0{}^2 = \sigma_{12}{}^p = \rho_0\sigma_{12}{}^{-1}\rho_0\sigma_{12} = 1. \tag{11.3.7}$$

When p is even, the central group $[2^+, 2^+]$ is a subgroup of index p in $[2, p^+]$ and of index $2p$ in $[2, p]$, with $\zeta = \rho_0\sigma_{12}{}^{p/2} = \rho_0(\rho_1\rho_2)^{p/2}$.

For $p \geq 2$, the reflections ρ_0, ρ_1, and ρ_2 may be represented by the orthogonal reflection matrices

$$R_0 = \begin{pmatrix} 1 & 0 & 0 \\ 0 & 1 & 0 \\ 0 & 0 & -1 \end{pmatrix}, \quad R_1 = \begin{pmatrix} 1 & 0 & 0 \\ 0 & -1 & 0 \\ 0 & 0 & 1 \end{pmatrix},$$

$$R_2 = \begin{pmatrix} \cos 2\pi/p & \sin 2\pi/p & 0 \\ \sin 2\pi/p & -\cos 2\pi/p & 0 \\ 0 & 0 & 1 \end{pmatrix},$$

the half-turn $\sigma_1 = \rho_0\rho_1$ and the rotation $\sigma_2 = \rho_1\rho_2$ by the orthogonal rotation matrices

$$S_1 = \begin{pmatrix} 1 & 0 & 0 \\ 0 & -1 & 0 \\ 0 & 0 & -1 \end{pmatrix} \quad \text{and} \quad S_2 = \begin{pmatrix} \cos 2\pi/p & \sin 2\pi/p & 0 \\ -\sin 2\pi/p & \cos 2\pi/p & 0 \\ 0 & 0 & 1 \end{pmatrix},$$

and the central inversion $\zeta = \rho_0(\rho_1\rho_2)^{p/2}$ (p even) by the negation matrix

$$Z = \begin{pmatrix} -1 & 0 & 0 \\ 0 & -1 & 0 \\ 0 & 0 & -1 \end{pmatrix}.$$

The dual of a rectangle $\{\,\} \times \{\,\}$ is a rhombus $\{\,\} + \{\,\}$. The dual of a rectangular prism $\{\,\} \times \{\,\} \times \{\,\}$ is a rhombic *fusil* $\{\,\} + \{\,\} + \{\,\}$, the "rhombic sum" of three mutually orthogonal dions or (in three ways) of a dion $\{\,\}$ and a rhombus $\{\,\} + \{\,\}$. A rhombic fusil has six vertices and eight congruent triangular faces; each dion joins a pair of opposite vertices, or *apices*, with the other four vertices belonging to a mediary rhombus. The dual of a p-gonal prism $\{\,\} \times \{p\}$ is a p-gonal fusil $\{\,\} + \{p\}$, with an axial dion joining two apices and orthogonal to a mediary p-gon; it has $2p$ congruent isosceles triangular faces (see Figure 11.3c for $p = 6$). Dual figures have the same symmetry group.

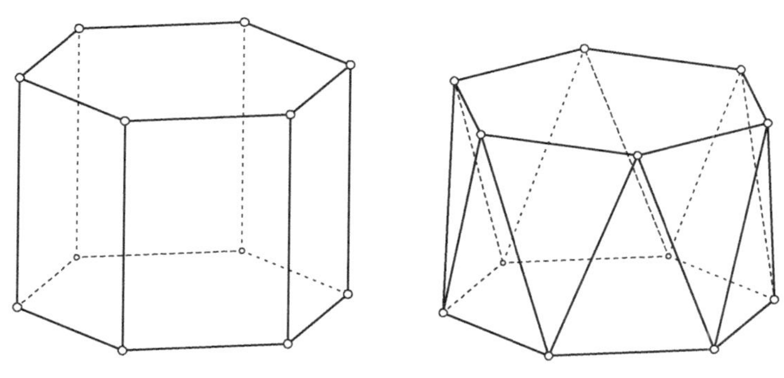

Figure 11.3b Prism { } × {6} and antiprism { } ⊗ {6}

For $p \geq 3$, a right p-gonal *antiprism* { } ⊗ {p}, the "skew-rectangular product" of a dion { } and a p-gon {p}, has two basal p-gons separated by a band of $2p$ isosceles triangles. The symmetry group of { } ⊗ {p} is a subgroup of index 2 in the full ortho-$2p$-gonal group $[2,\ 2p]$—namely, the *full gyro-p-gonal* group $[2^+,\ 2p] \cong \mathrm{D}_{2p}$, of order $4p$, generated by the half-turn $\sigma_{01} = \rho_0\rho_1$ and the reflection ρ_2, whose commutator $(\sigma_{01}\rho_2)^2$ is of period p:

$$\sigma_{01}{}^2 = \rho_2{}^2 = (\sigma_{01}\rho_2)^{2p} = 1. \tag{11.3.8}$$

The direct subgroup is the para-p-gonal group $[2^+,\ 2p]^+ \cong [2,\ p]^+ \cong \mathrm{D}_p$, of order $2p$.

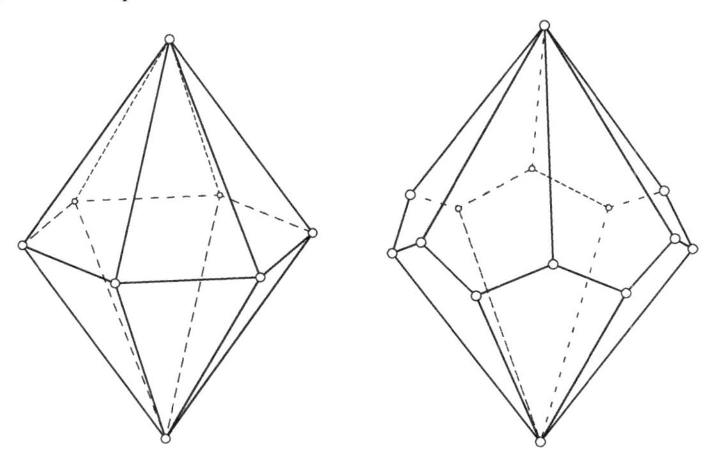

Figure 11.3c Fusil { } + {6} and antifusil { } ⊕ {6}

The groups $[2^+, 2p]$ and $[2, 2p^+]$ have a common subgroup, of index 2 in both—namely, the *gyro-p-gonal* group $[2^+, 2p^+] \cong C_{2p}$, of order $2p$, generated by the rotary reflection $\psi = \sigma_{01}\rho_2 = \rho_0\sigma_{12}$, satisfying the relation

$$\psi^{2p} = 1. \tag{11.3.9}$$

The $2p$-gonal group $[2p]^+$, the acro-$2p$-gonal group $[1, 2p]^+$, and the gyro-p-gonal group $[2^+, 2p^+]$ are different geometric representations of the same abstract group C_{2p}. The direct subgroup of $[2^+, 2p^+]$ is the acro-p-gonal group $[2^+, 2p^+]^+ \cong [1, p]^+ \cong C_p$, of order p, generated by the rotation $\psi^2 = (\rho_0\rho_1\rho_2)^2$.

When p is odd, the group $[2^+, 2^+]$ generated by the central inversion ζ is a subgroup of index p in $[2^+, 2p^+]$ and of index $2p$ in $[2^+, 2p]$ and $[2, 2p^+]$, with $\zeta = \psi^p = (\sigma_{01}\rho_2)^p = (\rho_0\sigma_{12})^p$.

When $p = 2$, we have the *gyrorectangular* (or "full gyrorhombic") group $[2^+, 4] \cong D_4$ and the *gyrorhombic* group $[2^+, 4^+] \cong C_4$. The respective direct subgroups are the pararhombic group $[2^+, 4]^+ \cong [2, 2]^+ \cong D_2$ and the acrorhombic group $[2^+, 4^+]^+ \cong [1, 2]^+ \cong C_2$. The group $[2^+, 4]$ is the symmetry group of a *tetragonal disphenoid* $\{\} \otimes \{2\}$, whose face polygons are four congruent isosceles triangles. The group $[2, 2]^+$ is the symmetry group of a *rhombic disphenoid*, whose face polygons are four congruent scalene triangles.

For $p \geq 2$, the reflections ρ_0 and ρ_1 and the half-turn $\sigma_{01} = \rho_0\rho_1$ can be represented by the matrices R_0, R_1, and S_1 given earlier, and the reflection ρ_2 and the rotation $\sigma_{12} = \rho_1\rho_2$ by the matrices R_2 and S_2 with $2\pi/p$ replaced by π/p. The rotary reflection $\psi = \sigma_{01}\rho_2 = \rho_0\sigma_{12}$ is then represented by the orthogonal matrix

$$Q = \begin{pmatrix} \cos \pi/p & \sin \pi/p & 0 \\ -\sin \pi/p & \cos \pi/p & 0 \\ 0 & 0 & -1 \end{pmatrix}.$$

It will be seen that $Q^{2p} = I$, with $Q^p = Z$ if p is odd.

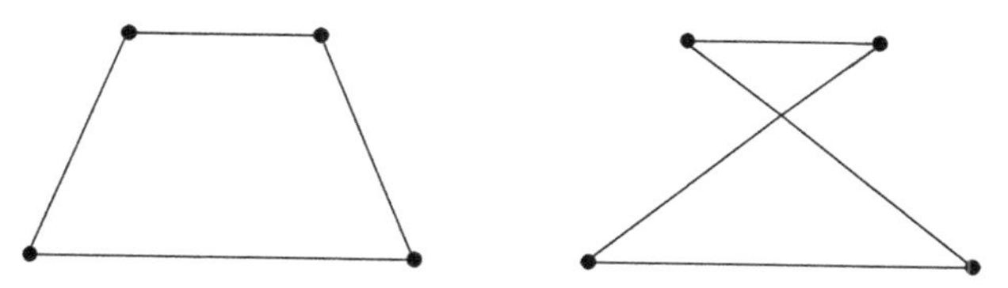

Figure 11.3d Convex and crossed trapezoids

A *trapezoid* is a quadrilateral with (at least) one pair of opposite sides parallel. Its dual is a *phylloid*, a quadrilateral one (at least) of whose diagonals bisects the other (a convex phylloid is commonly called a "kite" and a concave one a "dart"). These polygons are depicted in Figures 11.3d and 11.3e. Parallelograms are both trapezoids and phylloids. (These definitions apply equally well to affine quadrilaterals.) A Euclidean trapezoid or phylloid is *isosceles* if it has bilateral symmetry. Rectangles are doubly isosceles trapezoids, rhombuses are doubly isosceles phylloids, and squares are both.

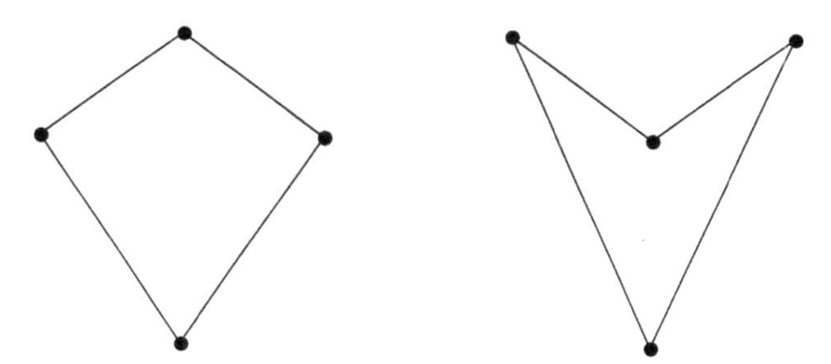

Figure 11.3e Convex and concave phylloids

A p-gonal antiprism $\{\,\} \otimes \{p\}$ has $2p$ vertices, with the vertex figures being isosceles trapezoids; its face polygons are two p-gonal bases and $2p$ isosceles triangles. Its dual is a p-gonal *antifusil* $\{\,\} \oplus \{p\}$, the "skew-rhombic sum" of a dion $\{\,\}$ and a p-gon $\{p\}$, which has two apices and $2p$ other vertices; its face polygons are $2p$ congruent isosceles phylloids.

When all of the isosceles triangular faces are equilateral, a triangular pyramid $() \vee \{3\}$ or a tetragonal disphenoid $\{\,\} \otimes \{2\} =$

{ } ⊕ {2} is a tetrahedron {3, 3}, while a rhombic fusil { } + { } + { }, a square fusil { } + {4}, or a triangular antiprism { } ⊗ {3} is an octahedron {3, 4}. Likewise, when all the face polygons are squares, a rectangular prism { } × { } × { }, a square prism { } × {4}, or a triangular antifusil { } ⊕ {3} is a cube {4, 3}. Each of these *regular* polyhedra has additional symmetries not included in the groups described in this section.

EXERCISES 11.3

1. Show that the reflections ρ_0, ρ_1, and ρ_2 generating the full ortho-p-gonal group [2, p] and satisfying the relations (11.3.5) can be represented by the matrices R_0, R_1, and R_2 given in this section.
2. Show that the reflection ρ_0 and the rotation σ_{12} generating the ortho-p-gonal group [2, p^+] and satisfying the relations (11.3.7) can be represented by the matrices R_0 and S_2 given in this section.
3. Show that the half-turn σ_{01} and the reflection ρ_2 generating the full gyro-p-gonal group [2^+, $2p$] and satisfying the relations (11.3.8) can be represented by the matrix S_1 given in this section and the matrix R_2 with $2\pi/p$ replaced by π/p.
4. Show that the rotary reflection ψ generating the gyro-p-gonal group [2^+, $2p^+$] and satisfying the relation (11.3.9) can be represented by the matrix Q given in this section.

11.4 POLYHEDRAL GROUPS

The five regular polyhedra described in Book XIII of Euclid's *Elements* (the "Platonic solids" of Figure 11.4a) belong to a family that includes the isomorphic tessellations of the sphere S^2, three regular tessellations of the Euclidean plane E^2, and infinitely many tessellations of the hyperbolic plane H^2. For p and q both greater than 2, $\{p, q\}$ is a spherical tessellation (or a polyhedron), a Euclidean tessellation (or a hyperbolic apeirohedron), or a hyperbolic tessellation according as $2(p+q)-pq$ is greater than, equal to, or less than zero.

The symmetry group $[p,\ q]$ of the regular polyhedron or tessellation $\{p,\ q\}$ is generated by three reflections ρ_0, ρ_1, and ρ_2, satisfying the relations

$$\rho_0{}^2 = \rho_1{}^2 = \rho_2{}^2 = (\rho_0\rho_1)^p = (\rho_0\rho_2)^2 = (\rho_1\rho_2)^q = 1. \qquad (11.4.1)$$

The direct subgroup $[p,\ q]^+$ is generated by the rotations $\sigma_1 = \sigma_{01} = \rho_0\rho_1$ and $\sigma_2 = \sigma_{12} = \rho_1\rho_2$, satisfying the relations

$$\sigma_1{}^p = \sigma_2{}^q = (\sigma_1\sigma_2)^2 = 1. \qquad (11.4.2)$$

The *tetrahedral* group $[3,\ 3]^+ \cong \mathrm{A}_4$ is of order 12, the *octahedral* (or *cubic*) group $[3,\ 4]^+ \cong \mathrm{S}_4$ of order 24, and the *icosahedral* (or *dodecahedral*) group $[3,\ 5]^+ \cong \mathrm{A}_5$ of order 60. Abstractly, these are alternating or symmetric groups of degree 4 or 5. The respective full symmetry groups are the *full tetrahedral* group $[3,\ 3] \cong \mathrm{S}_4$ of order 24, the *full octahedral* group $[3,\ 4] \cong 2 \times \mathrm{S}_4$ of order 48, and the *full icosahedral* group $[3,\ 5] \cong 2 \times \mathrm{A}_5$ of order 120.

For each group $[p,\ q]$, the *Coxeter number* h is the period of the “Petrie rotation” $\rho_0\rho_1\rho_2$ (after J. F. Petrie, 1907–1972); h is finite when the group is finite. The values of h for the groups [3, 3], [3, 4], and [3, 5] are respectively 4, 6, and 10, the number of sides of a skew *Petrie polygon* girdling a corresponding regular polyhedron. For [3, 4] and [3, 5], $(\rho_0\rho_1\rho_2)^{h/2}$ is the central inversion ζ (Coxeter 1940, p. 398).

Each generator of the direct subgroup W^+ of a Coxeter group W is a rotation or other direct isometry that is the product of two of the generating reflections of W. If the set of generating reflections of a Coxeter group W has a subset whose product with every other reflection is of even or infinite period, then W has an *ionic subgroup* generated by direct isometries that are products of two reflections in the subset, together with the other reflections and/or their conjugates by reflections in the subset. There is an ionic subgroup of index 2 for each subset of this kind, and there is an ionic subgroup of index 2^k for every k subsets that have no pairs of reflections in common. (Ionic subgroups, missing one or more of the original reflections, can be likened to atoms that have lost one or more electrons.)

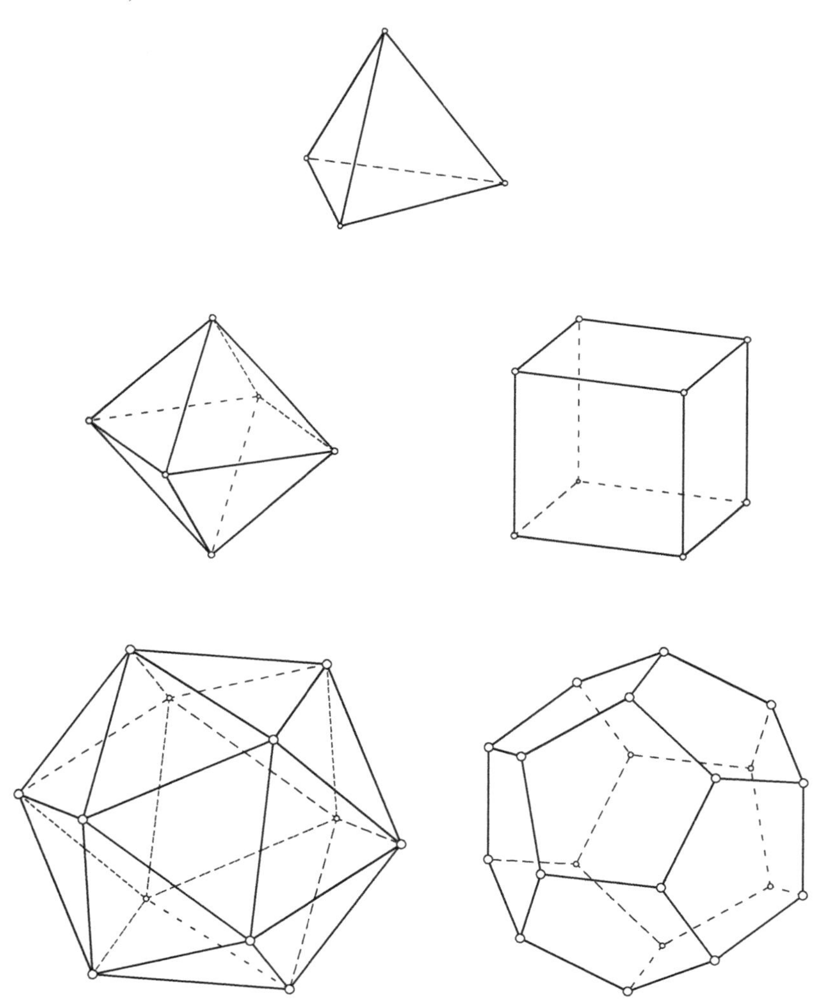

Figure 11.4a The Platonic solids

The commutator subgroup of a Coxeter group W is the normal subgroup generated by the commutators $(\rho_i\rho_j)^2$ of the generators and their conjugates. It is either the direct subgroup W^+ or the ionic subgroup W^{+c} of maximum index 2^c (Coxeter & Moser 1957, p. 126).

When p is even, the Coxeter group $[p, q]$ has a *halving subgroup* $[1^+, p, q]$ of index 2, also a Coxeter group, generated by the reflections ρ_1, ρ_2, and $\rho_{010} = \rho_0\rho_1\rho_0$, satisfying the relations

$$\rho_{010}{}^2 = \rho_1{}^2 = \rho_2{}^2 = (\rho_{010}\rho_1)^{p/2} = (\rho_{010}\rho_2)^q = (\rho_1\rho_2)^q = 1. \quad (11.4.3)$$

When q is even, $[p, q]$ has a *semidirect subgroup* $[p^+, q]$ of index 2 generated by the reflection ρ_2 and the rotation $\sigma_{01} = \rho_0\rho_1$, whose commutator is of period $q/2$:

$$\sigma_{02}{}^p = \rho_2{}^2 = (\sigma_{01}{}^{-1}\rho_2\sigma_{01}\rho_2)^{q/2} = 1. \quad (11.4.4)$$

In the case of the full octahedral or cubic group $[3, 4] \cong [4, 3] \cong 2 \times S_4$, the halving subgroup $[1^+, 4, 3]$ is the full tetrahedral group $[3, 3] \cong S_4$. A semidirect subgroup of $[3, 4]$ is the *pyritohedral* group $[3^+, 4] \cong 2 \times A_4$, with $(\sigma_{01}\rho_2)^3$ being the central inversion ζ. The commutator subgroup of both $[3, 4]$ and $[3, 3]$ is the tetrahedral group $[3^+, 4, 1^+] \cong [3, 3]^+ \cong A_4$, generated by two rotations σ_1 and σ_2 of period 3. The commutator subgroup of $[3, 3]^+$, of index 3, is the pararhombic group $[3, 3]^\Delta \cong [2, 2]^+ \cong D_2$, generated by the half-turns $\sigma_{12} = \sigma_1\sigma_2$ and $\sigma_{21} = \sigma_2\sigma_1$ (A_4 is the only alternating group with a proper normal subgroup). The group $[3, 3]^+ \cong A_4$ is a subgroup of index 5 in $[3, 5]^+ \cong A_5$, and $[3^+, 4] \cong 2 \times A_4$ is a subgroup of index 5 in $[3, 5] \cong 2 \times A_5$.

Relationships among the full polyhedral groups $[3, 3]$, $[3, 4]$, and $[3, 5]$ and their direct and ionic subgroups are shown in Figure 11.4b. When two groups are joined by a line, the one below is a subgroup of the one above (of index 2 unless otherwise indicated).

In addition, $[3, 3] \cong S_4$ contains the full acrotrigonal group $[1, 3] \cong D_3$ and the gyrorectangular group $[2^+, 4] \cong D_4$ as subgroups

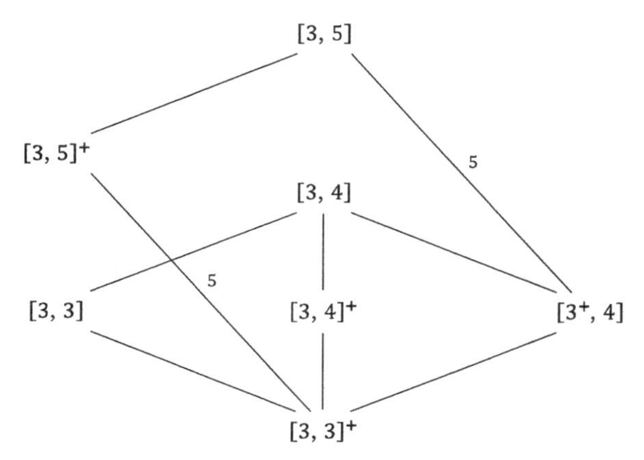

Figure 11.4b Polyhedral groups and subgroups

of index 4 and 3, respectively. Also, $[3, 4] \cong 2 \times S_4$ contains the full gyrotrigonal group $[2^+, 6] \cong 2 \times D_3$ and the full orthotetragonal group $[2, 4] \cong D_1 \times D_4$ as subgroups of index 4 and 3. The full icosahedral group $[3, 5] \cong 2 \times A_5$ contains the orthorectangular group $[2, 2] \cong D_1 \times D_2 \cong 2 \times D_2$ as a subgroup of index 15 and the full gyrotrigonal and gyropentagonal groups $[2^+, 6] \cong 2 \times D_3$ and $[2^+, 10] \cong 2 \times D_5$ as subgroups of index 10 and 6.

For the full tetrahedral group [3, 3], defined by (11.4.1) with $p = q = 3$, the reflections ρ_0, ρ_1, and ρ_2 are conveniently represented by the orthogonal reflection matrices

$$R_0 = \begin{pmatrix} 1 & 0 & 0 \\ 0 & 0 & -1 \\ 0 & -1 & 0 \end{pmatrix}, \quad R_1 = \begin{pmatrix} 1 & 0 & 0 \\ 0 & 0 & 1 \\ 0 & 1 & 0 \end{pmatrix}, \quad R_2 = \begin{pmatrix} 0 & 1 & 0 \\ 1 & 0 & 0 \\ 0 & 0 & 1 \end{pmatrix},$$

and the generating rotations $\sigma_1 = \sigma_{01} = \rho_0\rho_1$ and $\sigma_2 = \sigma_{12} = \rho_1\rho_2$ of the direct subgroup $[3, 3]^+$ by the orthogonal rotation matrices

$$S_1 = \begin{pmatrix} 1 & 0 & 0 \\ 0 & -1 & 0 \\ 0 & 0 & -1 \end{pmatrix} \quad \text{and} \quad S_2 = \begin{pmatrix} 0 & 1 & 0 \\ 0 & 0 & 1 \\ 1 & 0 & 0 \end{pmatrix}.$$

For the full octahedral or cubic group [4, 3], defined by (11.4.1) with $p = 4$ and $q = 3$, the reflections ρ_0, ρ_1, and ρ_2 are represented by the orthogonal reflection matrices

$$R_0 = \begin{pmatrix} 1 & 0 & 0 \\ 0 & 1 & 0 \\ 0 & 0 & -1 \end{pmatrix}, \quad R_1 = \begin{pmatrix} 1 & 0 & 0 \\ 0 & 0 & 1 \\ 0 & 1 & 0 \end{pmatrix}, \quad R_2 = \begin{pmatrix} 0 & 1 & 0 \\ 1 & 0 & 0 \\ 0 & 0 & 1 \end{pmatrix},$$

and the rotations $\sigma_1 = \sigma_{01} = \rho_0\rho_1$ and $\sigma_2 = \sigma_{12} = \rho_1\rho_2$ generating the direct subgroup $[4, 3]^+$ by the orthogonal rotation matrices

$$S_1 = \begin{pmatrix} 1 & 0 & 0 \\ 0 & 0 & 1 \\ 0 & -1 & 0 \end{pmatrix} \quad \text{and} \quad S_2 = \begin{pmatrix} 0 & 1 & 0 \\ 0 & 0 & 1 \\ 1 & 0 & 0 \end{pmatrix}.$$

The reflection ρ_0 and the rotation σ_{12} generate the pyritohedral group $[4, 3^+]$. The generating reflections $\rho_{010} = \rho_0\rho_1\rho_0$, ρ_1, and ρ_2 of the full tetrahedral subgroup $[3, 3] \cong [1^+, 4, 3]$ are represented by the matrices $R_{010} = R_0R_1R_0$, R_1, and R_2. The generating rotations $\sigma_{12} = \rho_1\rho_2$ and $\rho_0\sigma_{12}\rho_0 = \rho_{010}\rho_2$ of the tetrahedral subgroup $[3, 3]^+ \cong [1^+, 4, 3]^+$ are represented by the matrices $S_2 = R_1R_2$ and $R_0S_2R_0 = R_{010}R_2$.

For the full icosahedral or dodecahedral group [5, 3], defined by (11.4.1) with $p = 5$ and $q = 3$, it will be convenient to abbreviate $\cos 2\pi/5 = \tfrac{1}{2}\bar{\tau}$ and $\sin 2\pi/5 = \tfrac{1}{2}\sqrt{3 + \bar{\tau}}$ as $c(5)$ and $s(5)$. Then the reflections ρ_0, ρ_1, and ρ_2 are represented by the orthogonal reflection matrices

$$R_0 = \begin{pmatrix} 1 & 0 & 0 \\ 0 & 1 & 0 \\ 0 & 0 & -1 \end{pmatrix}, \quad R_1 = \begin{pmatrix} 1 & 0 & 0 \\ 0 & c(5) & s(5) \\ 0 & s(5) & -c(5) \end{pmatrix},$$

$$R_2 = \begin{pmatrix} \tfrac{1}{5}\sqrt{5} & -\tfrac{2}{5}\sqrt{5} & 0 \\ -\tfrac{2}{5}\sqrt{5} & -\tfrac{1}{5}\sqrt{5} & 0 \\ 0 & 0 & 1 \end{pmatrix}$$

and the rotations $\sigma_1 = \sigma_{01} = \rho_0\rho_1$ and $\sigma_2 = \sigma_{12} = \rho_1\rho_2$ generating the direct subgroup $[5,\ 3]^+$ by the orthogonal rotation matrices

$$S_1 = \begin{pmatrix} 1 & 0 & 0 \\ 0 & c(5) & s(5) \\ 0 & -s(5) & c(5) \end{pmatrix} \quad \text{and} \quad S_2 = \begin{pmatrix} \frac{1}{5}\sqrt{5} & -\frac{2}{5}\sqrt{5} & 0 \\ -\frac{2}{5}\sqrt{5}c(5) & -\frac{1}{5}\sqrt{5}c(5) & s(5) \\ -\frac{2}{5}\sqrt{5}s(5) & -\frac{1}{5}\sqrt{5}s(5) & -c(5) \end{pmatrix}.$$

The pyritohedral subgroup $[3^+,\ 4]$, which is generated by the reflection $\rho_{010} = \rho_0\rho_1\rho_0$ and the rotation $\sigma_{12} = \rho_1\rho_2$, is represented by the matrices $R_{010} = R_0R_1R_0$ and S_2. The tetrahedral subgroup $[3,\ 3]^+$, generated by rotations σ_{12} and $\rho_{010}\sigma_{12}\rho_{010}$, is represented by the matrices S_2 and $R_{010}S_2R_{010}$.

Every finite group of isometries in 3-space is either a group generated by three or fewer reflections, the direct subgroup of such a group, or one of the ionic subgroups $[2,\ p^+]$, $[2^+,\ 2p]$, $[2^+,\ 2p^+]$, or $[3^+,\ 4]$ (Coxeter 1985, p. 561). A summary of these groups, first described by J. F. Ch. Hessel in 1830, is given in Table 11.4. Each rotation group is the direct subgroup of one or more "extended" groups, as indicated (Klein 1884, part I, chap. 1; cf. Weyl 1952, pp. 77–80, 149–156).*

In this table it is to be assumed that $p \geq 2$. Recall that for $p = 2$, the suffix "-p-gonal" is replaced by "-rhombic" or, for the three "full" groups, by "-rectangular" (without "full").

EXERCISES 11.4

1. Show that the reflections ρ_0, ρ_1, and ρ_2 generating the full tetrahedral group $[3,\ 3]$, satisfying the relations (11.4.1) with $p = q = 3$, can be represented by the first set of matrices R_0, R_1, and R_2 given in this section.

2. Show that the reflections ρ_0, ρ_1, and ρ_2 generating the full octahedral group $[4,\ 3]$, satisfying the relations (11.4.1) with $p = 4$ and $q = 3$, can

* Conway & Smith (2003, pp. 25, 36) identify each of these groups by a symbol called its *signature*, based on William Thurston's theory of orbifolds, and also employ a somewhat different nomenclature. Their terms *pyramidal*, *prismatic*, and *antiprismatic* for types of axial symmetry correspond to our prefixes "acro-," "ortho-," and "gyro-," and their prefix "holo-" corresponds to our modifier *full*.

Table 11.4 *Finite Groups of Isometries in 3-Space*

Rotation Groups				Extended Groups			
Name	Symbol	Structure	Order	Name	Symbol	Structure	Order
Identity	$[\,]^+$	C_1	1	Bilateral	$[\,]$	D_1	2
				Central	$[2^+, 2^+]$	2	2
Acro-p-gonal	$[1, p]^+$	C_p	p	Full acro-p-gonal	$[1, p]$	D_p	$2p$
				Gyro-p-gonal	$[2^+, 2p^+]$	C_{2p}	$2p$
				Ortho-p-gonal	$[2, p^+]$	$D_1 \times C_p$	$2p$
Para-p-gonal	$[2, p]^+$	D_p	$2p$	Full gyro-p-gonal	$[2^+, 2p]$	D_{2p}	$4p$
				Full ortho-p-gonal	$[2, p]$	$D_1 \times D_p$	$4p$
Tetrahedral	$[3, 3]^+$	A_4	12	Full tetrahedral	$[3, 3]$	S_4	24
				Pyritohedral	$[3^+, 4]$	$2 \times A_4$	24
Octahedral	$[3, 4]^+$	S_4	24	Full octahedral	$[3, 4]$	$2 \times S_4$	48
Icosahedral	$[3, 5]^+$	A_5	60	Full icosahedral	$[3, 5]$	$2 \times A_5$	120

be represented by the second set of matrices R_0, R_1, and R_2 given in this section.

3. The matrices R_0 and $S_2 = R_1R_2$ generate the pyritohedral group $[4, 3^+]$ as a subgroup of [4, 3]. Show that these matrices satisfy the relations ${R_0}^2 = {S_2}^3 = ({S_2}^{-1}R_0S_2R_0)^2 = I$.

4. Show that the reflections ρ_0, ρ_1, and ρ_2 generating the full icosahedral group [5, 3], satisfying the relations (11.4.1) with $p = 5$ and $q = 3$, can be represented by the third set of matrices R_0, R_1, and R_2 given in this section.

5. The matrices $R_{010} = R_0R_1R_0$ and S_2 generate the pyritohedral group as a subgroup of [5, 3]. Find R_{010} and show that $({S_2}^{-1}R_{010}S_2R_{010})^2 = I$.

11.5 SPHERICAL COXETER GROUPS

The mirrors of a discrete group W generated by n reflections can be taken to be the n facets of a "Coxeter polytope" P in spherical, Euclidean, or hyperbolic space, the closure of which is the *fundamental region* for the group. In order for the group to be discrete, the dihedral angles of P must all be submultiples of π. When W is finite, P is a *Möbius simplex* in S^{n-1}.

In general, a Coxeter group $W \cong \langle \rho_0, \ldots, \rho_{n-1} \rangle$ generated by reflections satisfies the relations

$$(\rho_i\rho_j)^{p_{ij}} = 1 \quad (0 \le i \le j \le n-1,\ p_{ii} = 1) \tag{11.5.1}$$

(Coxeter 1934a, p. 588; Witt 1941, p. 294). Usually the exponents p_{ij} are positive integers, but infinite or imaginary exponents may be used to indicate that the product $\rho_i\rho_j$ does not have a finite period.

The discussion of Coxeter groups is greatly facilitated when a group W is represented by its *Coxeter diagram*, a labeled graph with n nodes representing the generators. If the period of the product of generators ρ_i and ρ_j is p_{ij}, the nodes are joined by a branch marked 'p_{ij}', except that marks '3' are customarily omitted and that when

$p_{ij} = 2$ the nodes are not joined. For example, the symmetry group $[p, q, \ldots, u, v]$ of the regular polytope or honeycomb $\{p, q, \ldots, u, v\}$ or its dual $\{v, u, \ldots, q, p\}$ is represented by the "string diagram"

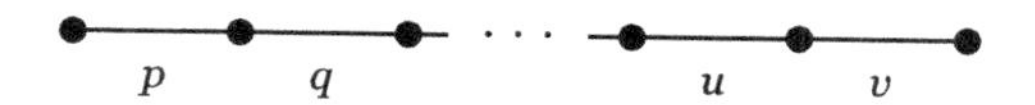

for which the Möbius simplex P is an *orthoscheme*, a simplex whose facets may be serially ordered so that any two that are not consecutive are orthogonal.

Since reflections are involutory, the relation $(\rho_i\rho_j)^2 = 1$ implies that generators ρ_i and ρ_j commute. The angle between facets of P corresponding to nodes i and j is π/p_{ij}, so that facets corresponding to nodes not joined by a branch are orthogonal. When the diagram of a Coxeter group W consists of two or more disconnected pieces, W is the direct product of the subgroups represented by the separate subdiagrams; otherwise, W is *irreducible.*

There is a relatively simple test for determining the nature of a Coxeter group W defined by (11.5.1) and acting in an $(n-1)$-dimensional space. Denote by P the simplex with (dihedral) angles $\alpha_{ij} = \alpha_{ji}$ $(1 \le i \le j \le n)$, and let A be the $n \times n$ *Gram matrix* with entries $a_{ij} = -\cos\alpha_{ij}$. The *Schläflian* σ of P is defined to be the determinant of the *Schläfli matrix* $2A$:

$$\sigma(\mathsf{P}) = \det 2A = 2^n \det A. \tag{11.5.2}$$

If $n = 0$, we take $\sigma(\mathsf{P})$ equal to unity. Then the simplex P, or the group generated by reflections in its facets, is spherical, Euclidean, or hyperbolic according as $\sigma(\mathsf{P})$ is greater than, equal to, or less than zero. We call this rule the Schläfli Criterion (Coxeter 1948, pp. 133–136; cf. Ratcliffe 1994/2006, §§7.2–7.3). The matrix of cosines, usually credited to J. P. Gram (1850–1916), was actually first employed by Schläfli (1858, §3).

Each irreducible Euclidean Coxeter group is an infinite extension of some *crystallographic* spherical Coxeter group (sometimes called a

"Weyl group"), with the exponents p_{ij} $(i \neq j)$ in (11.5.1) restricted to the values 2, 3, 4, and 6. Each such Euclidean group is associated with a simple *Lie group* (after Sophus Lie, 1842–1899), and the spherical subgroup is denoted by the corresponding *Cartan symbol*

$$\mathbf{A}_n\ (n \geq 0),\ \ \mathbf{B}_n \cong \mathbf{C}_n\ (n \geq 1),\ \ \mathbf{D}_n\ (n \geq 2),\ \ \mathbf{E}_n\ (n = 6, 7, 8),\ \ \mathbf{F}_4,\ \ \mathbf{G}_2.$$

Noncrystallographic spherical Coxeter groups have no such association but are often given analogous symbols $\mathbf{H}_n$ $(n = 2, 3, 4)$ or $\mathbf{I}_2(p)$ $(p > 6)$ (cf. Humphreys 1990, p. 32; Davis 2008, p. 104). A necessary and sufficient condition for a discrete group W generated by n reflections to be crystallographic is that, for each distinguished subgroup $W(P)$ generated by some subset P of the reflections, the Schläflian $\sigma(\mathsf{P})$ of the fundamental region P be an integer (Coxeter & Moser 1957, p. 131; Monson 1982; Humphreys 1990, pp. 135–137).

The Coxeter diagram for the trivial group $\mathbf{A}_0 \cong][$ is the empty graph with no nodes. The group $\mathbf{A}_1 \cong \mathbf{B}_1 \cong [\,]$ is denoted by a single node. The finite groups with two generators are the rectangular group $\mathbf{D}_2 \cong [2] \cong [\,] \times [\,]$ and the full polygonal groups

$$\mathbf{A}_2 \cong [3],\quad \mathbf{B}_2 \cong [4],\quad \mathbf{H}_2 \cong [5],\quad \mathbf{G}_2 \cong [6],\quad \text{and}\quad \mathbf{I}_2(p) \cong [p], \tag{11.5.3}$$

with the Coxeter diagrams

•—•   •—• (4)   •—• (5)   •—• (6)   and   •—• (p)

These are the respective symmetry groups of the regular polygons {3}, {4}, {5}, {6}, and $\{p\}$ $(p > 6)$.

The irreducible spherical groups with three generators are the full polyhedral groups

$$\mathbf{A}_3 \cong \mathbf{D}_3 \cong [3,\ 3],\qquad \mathbf{B}_3 \cong [3,\ 4],\qquad \mathbf{H}_3 \cong [3,\ 5], \tag{11.5.4}$$

of orders 24, 48, and 120, with Coxeter diagrams

These are the respective symmetry groups of the self-dual tetrahedron {3, 3}, the dual octahedron {3, 4} and cube {4, 3}, and the dual icosahedron {3, 5} and dodecahedron {5, 3}.

The groups with four generators are the *full polychoric* groups

$$\mathbf{A}_4 \cong [3, 3, 3], \qquad \mathbf{D}_4 \cong [3^{1,1,1}], \qquad \mathbf{B}_4 \cong [3, 3, 4],$$
$$\mathbf{F}_4 \cong [3, 4, 3], \qquad \mathbf{H}_4 \cong [3, 3, 5], \tag{11.5.5}$$

of orders 120, 192, 384, 1152, and 14400, with Coxeter diagrams

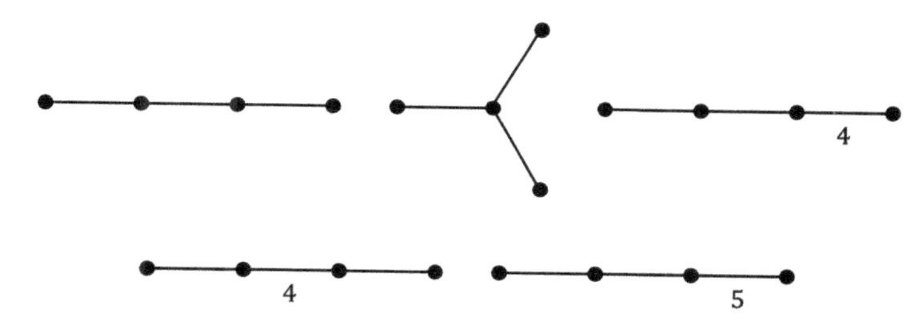

The four string diagrams represent the symmetry groups of the regular 4-polytopes, or *polychora*: the self-dual *pentachoron* ("5-cell") {3, 3, 3}, the dual *hexadecachoron* ("16-cell") {3, 3, 4} and *tesseract* ("hypercube") {4, 3, 3}, the self-dual *icositetrachoron* ("24-cell") {3, 4, 3}, and the dual *hexacosichoron* ("600-cell") {3, 3, 5}, and *dodecacontachoron* ("120-cell") {5, 3, 3}.

Abstractly, $\mathbf{A}_4$ is the symmetric group S_5, of order 5!; $\mathbf{B}_4$ is the "wreath product" $C_2 \wr S_4 \cong {}^2S_4$, of order $2^4 \cdot 4!$; $\mathbf{D}_4$ is its halving subgroup $\frac{1}{2} \cdot {}^2S_4$, of order $2^3 \cdot 4!$; $\mathbf{F}_4$ is the supergroup $3 \cdot {}^2S_4$, of order $3 \cdot 2^4 \cdot 4!$; and $\mathbf{H}_4$ is an extension $2 \cdot (A_5 \times A_5) \cdot 2$ of the direct product of two alternating groups A_5, of order $2^2 \cdot (\frac{1}{2} \cdot 5!)^2$. The groups $\mathbf{A}_4$, $\mathbf{B}_4$, $\mathbf{D}_4$, and $\mathbf{F}_4$, all of whose nontrivial rotations are of period 2, 3, or 4, are crystallographic.

Each Coxeter group in one of the three infinite families $\mathbf{A}_n$, $\mathbf{B}_n$, and $\mathbf{D}_n$ is an extension of the symmetric group S_n, of order $n!$. The *anasymmetric* group $(n+1) \cdot S_n \cong S_{n+1}$, of order $(n+1)!$ $(n \geq 0)$, is

$$\mathbf{A}_n \cong [3^{n-1}] \cong [3,\ 3, \ldots,\ 3,\ 3], \tag{11.5.6}$$

with Coxeter diagram

This "hypertetrahedral" group is the symmetry group of a regular n-dimensional *simplex* $\alpha_n = \{3,\ 3, \ldots,\ 3,\ 3\}$, the analogue of the triangle $\alpha_2 = \{3\}$ and the tetrahedron $\alpha_3 = \{3,\ 3\}$. The facet polytopes of α_n are $n + 1$ α_{n-1}'s.

The *bisymmetric* group $C_2 \wr S_n \cong {}^2S_n$, of order $2^n{\cdot}n!$ $(n \geq 1)$, is

$$\mathbf{B}_n \cong [3^{n-2},\ 4] \cong [3,\ 3, \ldots,\ 3,\ 4], \tag{11.5.7}$$

with Coxeter diagram

This "hyperoctahedral" group is the symmetry group of a regular n-dimensional *orthoplex* $\beta_n = \{3,\ 3, \ldots,\ 3,\ 4\}$ or *orthotope* $\gamma_n = \{4,\ 3, \ldots,\ 3,\ 3\}$, the analogues of the square $\beta_2 = \gamma_2 = \{4\}$ and the octahedron $\beta_3 = \{3,\ 4\}$ or cube $\gamma_3 = \{4,\ 3\}$ (Coxeter 1948, pp. 121–123; Conway & Smith 2003, p. 106). The facet polytopes of β_n are 2^n α_{n-1}'s, and those of γ_n are $2n$ γ_{n-1}'s.

The *demibisymmetric* group $\frac{1}{2}{\cdot}{}^2S_n$, of order $2^{n-1}{\cdot}n!$ $(n \geq 2)$, a halving subgroup of (11.5.7), is

$$\mathbf{D}_n \cong [3^{n-3,1,1}] \cong [3, \ldots,\ 3,\ 3^{1,1}]. \tag{11.5.8}$$

Its fundamental region is a *plagioscheme*, with a branching Coxeter diagram

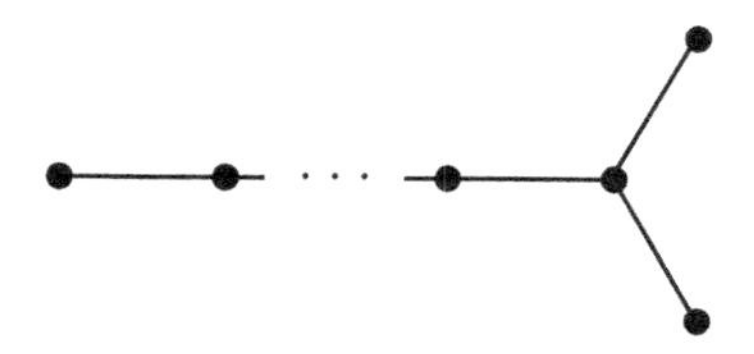

(The terms "orthoscheme" and "plagioscheme" are both due to Schläfli.) This is the symmetry group of a *demiorthotope* $h\gamma_n$, a "uniform" polytope with half the vertices of γ_n; its facet polytopes are 2^{n-1} α_{n-1}'s and $2n$ $h\gamma_{n-1}$'s (except that $h\gamma_2$ is a dion). The group $\mathbf{D}_3 \cong \frac{1}{2}\cdot{}^2\mathrm{S}_3$ is isomorphic to $\mathbf{A}_3 \cong \mathrm{S}_3$, and the "half-cube" $h\gamma_3$ is a tetrahedron α_3. Also, the "half-tesseract" $h\gamma_4$ is a hexadecachoron β_4.

For $n > 4$, α_n, β_n, and γ_n are the only regular n-polytopes. The regular n-simplex $\alpha_n = \{3^{n-1}\}$ is an n-dimensional pyramid

$$(n+1)\cdot\alpha_0 = (n+1)\cdot() = ()\vee()\vee\cdots\vee(),$$

the dionic join of $n+1$ mutually equidistant monons. The regular n-orthoplex (or "cross polytope") $\beta_n = \{3^{n-2}, 4\}$ is an n-dimensional fusil

$$n\beta_1 = n\{\,\} = \{\,\}+\{\,\}+\cdots+\{\,\},$$

the rhombic sum of n equal-length dions. The regular n-orthotope (or "block polytope") $\gamma_n = \{4, 3^{n-2}\}$ is an n-dimensional prism

$$\gamma_1^n = \{\,\}^n = \{\,\}\times\{\,\}\times\cdots\times\{\,\},$$

the rectangular product of n equal-length dions.

It should be noted that simplexes, orthoplexes, and orthotopes, the respective generalizations of triangles, rhombuses, and rectangles, do not have to be regular. Nor are these particular polytopes the only instances of pyramids, fusils, and prisms.

In general, if P and Q are disjoint polytopes of dimensions m and n in an $(m+n+1)$-dimensional space, a *pyramid* $\mathsf{P}\vee\mathsf{Q}$ is constructed by connecting each vertex of P to each vertex of Q. For example, $\{\,\}\vee\{\,\}$ is a disphenoid, $()\vee\{p\}$ is a p-gonal pyramid, and $\{\,\}\vee\{p\}$ is a dionic p-gonal pyramid (in 4-space) with $p+2$ vertices and $p+2$ cells.

Let P and Q be polytopes of dimensions m and n (each at least 1) lying in subspaces of an $(m+n)$-space that intersect in a single point contained in the body of each polytope. If each vertex of P is connected to each vertex of Q, the convex hull of the resulting figure defines a *fusil* $\mathsf{P}+\mathsf{Q}$. For example, $\{\,\}+\{p\}$ is a p-gonal fusil (or "bipyramid"), and $\{p\}+\{q\}$ is a four-dimensional

p-gonal q-gonal fusil with $p + q$ vertices and pq cells (a "double fusil" if $p = q$).

Again, let P and Q be polytopes of dimensions m and n (each at least 1) lying in subspaces of an $(m + n)$-space intersecting in a single point but with the point now a vertex of each polytope. If at each other vertex of P is erected a parallel copy of Q and vice versa, the resulting figure is a *prism* P × Q. For example, { } × {p} is a p-gonal prism, and {p} × {q} is a four-dimensional p-gonal q-gonal prism with pq vertices and $p + q$ cells (a "double prism" if $p = q$).

The join, sum, and product operations are associative and commutative, and a pyramid, fusil, or prism may have any number of components. The join of two or more simplexes is a simplex. The rhombic sum of two or more orthoplexes is an orthoplex. The rectangular product of two or more orthotopes is an orthotope.

The only other irreducible spherical Coxeter groups are three *exceptional* groups with 6, 7, and 8 generators, of respective orders 72·6!, 576·7!, and 17280·8!, namely,

$$\mathbf{E}_6 \cong [3^{2,2,1}], \qquad \mathbf{E}_7 \cong [3^{3,2,1}], \qquad \mathbf{E}_8 \cong [3^{4,2,1}], \qquad (11.5.9)$$

with Coxeter diagrams

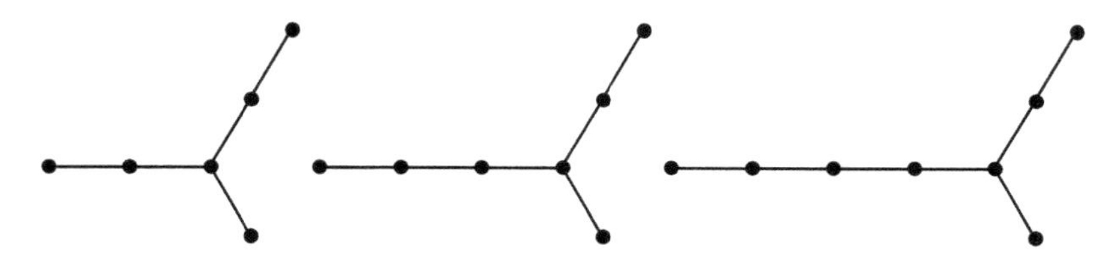

These are the symmetry groups of three "semiregular" n-polytopes $(n - 4)_{21}$ discovered by Thorold Gosset in 1900. The facet polytopes of $(n - 4)_{21}$ are simplexes α_{n-1} and orthoplexes β_{n-1}.

The group $\mathbf{E}_7 \cong [3^{3,2,1}]$ contains $\mathbf{A}_7 \cong [3^6]$ as a subgroup of index 72, while $\mathbf{E}_8 \cong [3^{4,2,1}]$ contains $\mathbf{D}_8 \cong [3^{5,1,1}]$ as a subgroup of index 135 and $\mathbf{A}_8 \cong [3^7]$ as a subgroup of index 1920 (Coxeter 1934b, table v).

The order of a finite Coxeter group may be determined from the content of the Möbius simplex forming its fundamental region (Coxeter 1935; Coxeter 1988, pp. 11–21), but this involves difficult integration formulas. McMullen (1991; cf. McMullen & Schulte 2002, pp. 83–94) has shown how to calculate the orders using only elementary methods.

When one Coxeter group W contains another Coxeter group V as a subgroup, the fundamental region for V can be dissected into c copies of the fundamental region for W, where $|W : V| = c$. In some cases, as with halving subgroups, the Coxeter diagram for V is symmetric, and the supergroup W is obtained from it as an extension by a group of automorphisms (reflections or other isometries) permuting the generators.

EXERCISES 11.5

1. The Schläflian of the circular arc $\mathsf{P} = (p)$ of angle π/p is

$$\sigma(\mathsf{P}) = 4 \sin^2 \pi/p = 2 - 2 \cos 2\pi/p.$$

The length of each part of a regular partition $\{p\}$ or the central angle subtended by each side of a regular polygon $\{p\}$ is $2\pi/p$. For each integer $p \geq 3$, the number $\xi_p = 2 \cos 2\pi/p$ is the largest zero of a *cyclotomic polynomial* of degree $\lfloor (p-1)/2 \rfloor$, whose other zeros are the numbers $\xi_{p/d} = 2 \cos 2d\pi/p$ $(1 < d < p/2)$. (Each of the zeros is an algebraic integer.) The polynomials for $p \leq 12$ are listed in the following table.

p	Polynomial	p	Polynomial
3	$x+1$	4	x
5	x^2+x-1	6	x^2-1
7	x^3+x^2-2x-1	8	x^3-2x
9	$x^4+x^3-3x^2-2x+1$	10	x^4-3x^2+1
11	$x^5+x^4-4x^3-3x^2+3x+1$	12	x^5-4x^3+3x

a. Find the zeros of the cyclotomic polynomials for p equal to 3, 4, 5, 6, 8, 10, and 12, and find $\sigma(\mathsf{P})$ in each case.

b. The coefficients of the terms in each polynomial can be expressed as binomial coefficients $\binom{n}{k}$. What is the general rule?

2. Denoting a triangle with angles π/p, π/q, π/r by $(p\ q\ r)$, find the Gram matrices for the spherical triangles $\mathsf{A}_3 = (3\ 3\ 2)$, $\mathsf{B}_3 = (4\ 3\ 2)$, and $\mathsf{H}_3 = (5\ 3\ 2)$. Then find the Schläflians $\sigma(\mathsf{A}_3)$, $\sigma(\mathsf{B}_3)$, and $\sigma(\mathsf{H}_3)$.

3. If A_n, B_n, and D_n are the simplexes that are fundamental regions for the Coxeter groups $\mathbf{A}_n \cong [3^{n-1}]$, $\mathbf{B}_n \cong [3^{n-2}, 4]$, and $\mathbf{D}_n \cong [3^{n-3,1,1}]$, use induction to show that $\sigma(\mathsf{A}_n) = n + 1$, $\sigma(\mathsf{B}_n) = 2$, and $\sigma(\mathsf{D}_n) = 4$. (Note that $\mathbf{D}_3 \cong \mathbf{A}_3$.)

11.6 SUBGROUPS AND EXTENSIONS

Each Coxeter group W generated by reflections has a direct subgroup W^+ generated by rotations or other direct isometries. It may also have various ionic subgroups corresponding to certain subsets of the generating reflections. Such subgroups are denoted by introducing superscript plus signs in the Coxeter symbol for W (cf. Coxeter & Moser 1957, pp. 124–129) and can be represented graphically by replacing solid nodes in the Coxeter diagram by open nodes (Johnson & Weiss 1999a, pp. 1309–1313).

A. *Polygonal groups.* The Coxeter diagram of the bilateral group $[\,] \cong \mathrm{D}_1$ generated by one reflection is a single solid node ●, with the identity group $[\,]^+ \cong \mathrm{C}_1$ being represented by an open node ○. The rectangular group $[2] \cong \mathrm{D}_2$ and the full p-gonal group $[p] \cong \mathrm{D}_p$ $(p \geq 3)$, generated by two reflections ρ_1 and ρ_2, have the Coxeter diagrams

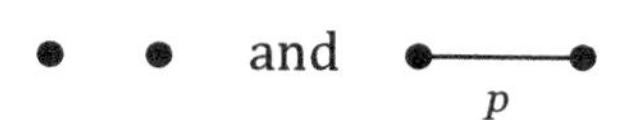

The respective direct subgroups $[2]^+ \cong \mathrm{C}_2$ and $[p]^+ \cong \mathrm{C}_p$ $(p \geq 3)$, generated by the rotation $\sigma_{12} = \rho_1\rho_2$, are represented by the diagrams

and

These are the respective rotation (direct symmetry) groups of a rhombus $\{\} + \{\}$ (or a rectangle) and a regular p-gon $\{p\}$.

Each of the even dihedral groups $[2p] \cong \mathrm{D}_{2p}$, with Coxeter diagram

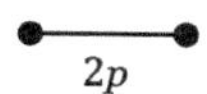

has three subgroups of index 2: a halving subgroup $[1^+, 2p] \cong \mathrm{D}_p$ generated by the reflections $\rho_{121} = \rho_1\rho_2\rho_1$ and ρ_2, the direct subgroup $[2p]^+ \cong \mathrm{C}_{2p}$ generated by the rotation $\sigma_{12} = \rho_1\rho_2$, and a halving subgroup $[2p, 1^+] \cong \mathrm{D}_p$ generated by the reflections ρ_1 and $\rho_{212} = \rho_2\rho_1\rho_2$. For $p \geq 2$, the respective diagrams are

The three groups have a common subgroup $[2p]^{+2} \cong \mathrm{C}_p$, the commutator subgroup of $[2p]$, of index 4, generated by the rotation $\sigma_{1212} = \rho_{121}\rho_2 = {\sigma_{12}}^2 = \rho_1\rho_{212}$. This can be variously described as $[1^+, 2p]^+$, $[1^+, 2p, 1^+]$, or $[2p, 1^+]^+$, corresponding to the diagram

As for supergroups, each group $[p]$ has a *double* group $[[p]] \cong \mathrm{D}_{2p}$, an extension obtained by adjoining to the generating reflections a "dualizing" automorphism that interchanges them. For $p \geq 3$, this is the symmetry group of a regular compound $\{\{p\}\}$ of two dual $\{p\}$'s (for example, $\{\{3\}\}$ is a regular *hexagram* or Star of David). It is also evident that every group $[cp]$ with $c > 1$ has $[p]$ as a subgroup of index c and that every group $[p]$ has extensions ${}^c[p] \cong \mathrm{D}_{cp}$ for all c.

B. *Pyramidal and prismatic groups.* In three dimensions, the pyramidal groups $[1, 2] \cong \mathrm{D}_2 \cong \mathrm{D}_1 \times \mathrm{D}_1$ and $[1, p] \cong \mathrm{D}_p$ $(p \geq 3)$, generated by two reflections, can be represented by the diagrams

○ ● ● and ○ ●—● (p)

obtained by adding a superfluous open node (the '1' in the Coxeter symbol) to the diagrams for the two-dimensional groups [2] and $[p]$. Diagrams for pyramidal subgroups can be constructed in the same manner.

The prismatic groups $[2,\ 2] \cong \mathrm{D}_1 \times \mathrm{D}_1 \times \mathrm{D}_1$ and $[2,\ p] \cong \mathrm{D}_1 \times \mathrm{D}_p$ $(p \geq 3)$ are generated by three reflections ρ_0, ρ_1, and ρ_2, with diagrams

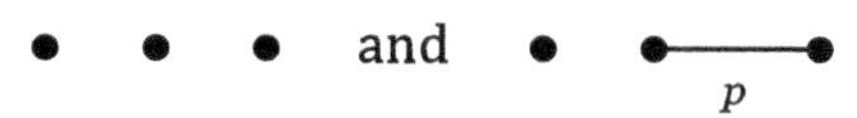

The respective halving subgroups $[1^+,\ 2,\ 2]$ and $[1^+,\ 2,\ p]$ are the pyramidal groups $[1, 2]$ and $[1,\ p]$. The direct subgroup $[2,\ 2]^+ \cong \mathrm{D}_2 \cong \mathrm{C}_2 \times \mathrm{C}_2$ is generated by any two of the three half-turns $\sigma_{01} = \rho_0\rho_1$, $\sigma_{02} = \rho_0\rho_2$, and $\sigma_{12} = \rho_1\rho_2$, while $[2,\ p]^+ \cong \mathrm{D}_p$ is generated by a half-turn σ_{01} or σ_{02} and the p-fold rotation σ_{12}. Diagrams for these groups are

o– –o– –o and o– –o——o (p)

A branch with a middle gap connecting a pair of open nodes (alternatively, a branch marked '2') represents a half-turn; it can be omitted if the nodes are linked by other branches.

The orthorectangular group [2, 2] has three other subgroups of index 2, each isomorphic to $\mathrm{D}_1 \times \mathrm{C}_2$. These are the orthorhombic groups $[2^+, 2]$, generated by the half-turn σ_{01} and the reflection ρ_2; $[(2,\ 2^+,\ 2)]$, generated by the half-turn σ_{02} and the reflection ρ_1; and $[2,\ 2^+]$, generated by the reflection ρ_0 and the half-turn σ_{12}. The respective diagrams are

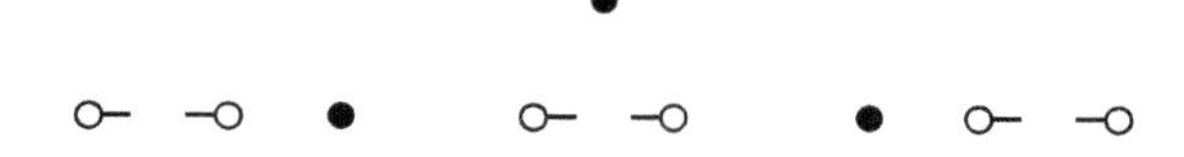

The three groups have a common subgroup, of index 2 in each of them and of index 4 in [2, 2]: the central group $[2^+,\ 2^+] \cong 2$, generated by the central inversion $\zeta = \sigma_{01}\rho_2 = \sigma_{02}\rho_1 = \rho_0\sigma_{12}$, with the diagram

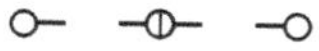

The full ortho-p-gonal group $[2,\ p]$ has an ionic subgroup of index 2, the ortho-p-gonal group $[2,\ p^+] \cong \mathrm{D}_1 \times \mathrm{C}_p$, generated by the reflection ρ_0 and the p-fold rotation σ_{12}. Its diagram is

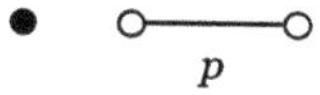

p

The direct subgroup is $[2,\ p^+]^+ \cong [1,\ p]^+ \cong \mathrm{C}_p$. When p is even, $[2,\ p^+]$ contains the central group $[2^+,\ 2^+]$ as a subgroup of index p.

The full ortho-$2p$-gonal group $[2,\ 2p] \cong \mathrm{D}_1 \times \mathrm{D}_{2p}$ $(p \geq 2)$ has another ionic subgroup of index 2, the full gyro-p-gonal group $[2^+,\ 2p] \cong \mathrm{D}_{2p}$, generated by the half-turn σ_{01} and the reflection ρ_2, with the diagram

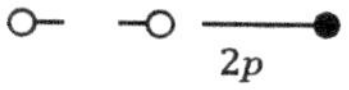

$2p$

The direct subgroup is $[2^+,\ 2p]^+ \cong [2,\ p]^+ \cong \mathrm{D}_p$. The two groups $[2^+,\ 2p]$ and $[2,\ 2p^+]$ have a common subgroup of index 2 in both and of index 4 in $[2,\ 2p]$: the gyro-p-gonal group $[2^+,\ 2p^+] \cong \mathrm{C}_{2p}$, generated by the rotary reflection $\phi_{012} = \sigma_{01}\rho_2 = \rho_0\sigma_{12}$, with the diagram

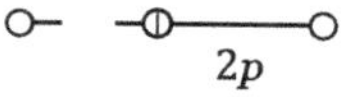

$2p$

The direct subgroup is $[2^+,\ 2p^+]^+ \cong [1,\ p]^+ \cong \mathrm{C}_p$. When p is odd, $[2^+,\ 2p^+]$ contains the central group $[2^+,\ 2^+]$ as a subgroup of index p.

For $p \geq 3$, the groups $[2,\ p]$ and $[2^+,\ 2p]$ are respectively the symmetry groups of a p-gonal prism $\{\ \} \times \{p\}$ and a p-gonal antiprism $\{\ \} \otimes \{p\}$.

C. *Polyhedral groups.* The full tetrahedral, octahedral, and icosahedral groups $[3,\ 3] \cong \mathrm{S}_4$, $[3,\ 4] \cong 2 \times \mathrm{S}_4$, and $[3,\ 5] \cong 2 \times \mathrm{A}_5$, each generated by three reflections ρ_0, ρ_1, and ρ_2, have the direct subgroups $[3,\ 3]^+ \cong \mathrm{A}_4$, $[3,\ 4]^+ \cong \mathrm{S}_4$, and $[3,\ 5]^+ \cong \mathrm{A}_5$, generated by two rotations $\sigma_{01} = \rho_0\rho_1$ and $\sigma_{12} = \rho_1\rho_2$. Diagrams for these subgroups are

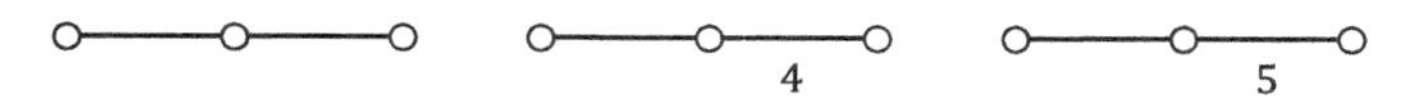

Their Coxeter diagrams having no branches with even marks, neither [3, 3] nor [3, 5] has any ionic subgroups. However, [3, 3] has two *trionic* subgroups: $[3,\ 3]^{\text{⅄}} \cong D_4$, of index 3, generated by the half-turn $\sigma_{02} = \rho_0\rho_2$ and the reflection ρ_1, and $[3,\ 3]^{\triangle} \cong D_2$, of index 6, generated by the half-turns σ_{02} and $\sigma_{1021} = \rho_1\sigma_{02}\rho_1$. The latter is also a subgroup of index 3 in $[3,\ 3]^+$. Diagrams for these groups are

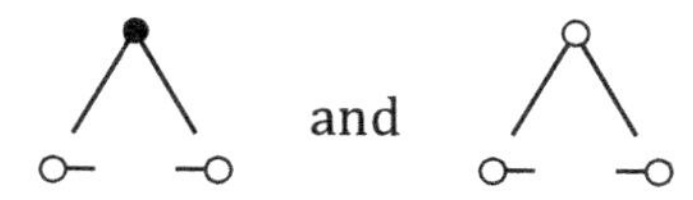

The group [3, 3] has a double group [[3, 3]], obtained by adjoining to the generating reflections ρ_0, ρ_1, and ρ_2 the automorphism that interchanges ρ_0 and ρ_2. This is a reflection ρ_3 in a mirror orthogonal to the mirror of ρ_1, and the extended group is generated by the reflectionsρ_1, ρ_2, and ρ_3, with $\rho_0 = \rho_3\rho_2\rho_3$. Since $(\rho_1\rho_2)^3 = (\rho_1\rho_3)^2 = (\rho_2\rho_3)^4 = 1$, we see that [[3, 3]] is the Coxeter group [3, 4].

With its generators taken as the three reflections ρ_0, ρ_1, and ρ_2, the full octahedral group $[3,\ 4] \cong 2\times S_4$ has three subgroups of index 2: the semidirect subgroup $[3^+,\ 4] \cong 2 \times A_4$, generated by the rotation σ_{01} and the reflection ρ_2; the direct subgroup $[3,\ 4]^+ \cong S_4$ already mentioned; and the halving subgroup $[3,\ 4,\ 1^+] \cong S_4$, generated by the reflections ρ_0, ρ_1, and $\rho_{212} = \rho_2\rho_1\rho_2$. The corresponding diagrams are

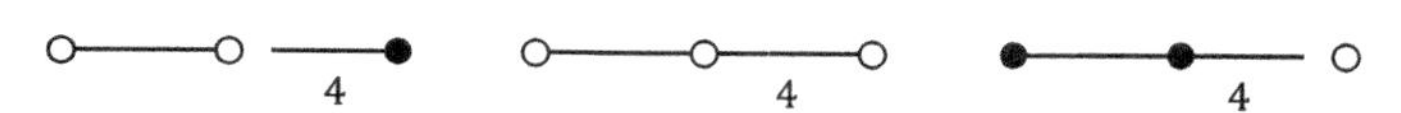

The first of these is known in crystallography as the *pyritohedral* group. Since $(\rho_0\rho_1)^3 = (\rho_1\rho_{212})^2 = (\rho_0\rho_{212})^3 = 1$, the last is just the Coxeter group [3, 3]. The three groups have a common subgroup of index 2, generated by the rotations $\sigma_{01} = \rho_0\rho_1$ and $\sigma_{0212} = \rho_2\sigma_{01}\rho_2 = \sigma_{01}{\sigma_{12}}^2 = \rho_0\rho_{212}$. This is the commutator subgroup $[3^+,\ 4,\ 1^+]$, of index 4 in [3, 4], which can also be denoted by $[3^+, 4]^+ \cong [3, 4]^{+2} \cong$ $[3, 4, 1^+]^+$, with the diagram

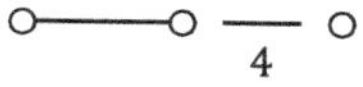

o———o $\underset{4}{—}$ o

It is, in fact, the tetrahedral group $[3, 3]^+ \cong A_4$.

A Coxeter group that has an ionic subgroup involving the removal of two or more reflections may also have a *radical subgroup* in which associated rotations are removed as well. The index of the subgroup is the order of the group generated by the removed reflections. Thus [3, 4] has the radical subgroup $[3^*, 4] \cong 2 \times D_2$, of index 3 in $[3^+, 4] \cong 2 \times A_4$ and of index 6 in [3, 4], generated by the reflections $\rho_{01210} = \sigma_{01}\rho_2\sigma_{01}^{-1} = \rho_0\rho_1\rho_2\rho_1\rho_0$ and ρ_2 and the central inversion $\zeta = (\sigma_{01}\rho_2)^3 = (\rho_0\rho_1\rho_2)^3$. The diagram for this group is

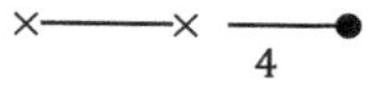

×———× $\underset{4}{—}$●

The intersection of the group $[3^*, 4]$ with the halving subgroup $[3, 4, 1^+]$ is the radical subgroup $[3^*, 4, 1^+] \cong D_2 \cong C_2 \times C_2$, of index 3 in $[3^+, 4, 1^+]$, of index 6 in $[3, 4, 1^+]$, and of index 12 in [3, 4], generated by the half-turns $\sigma_{012120} = \rho_{01210}\rho_2 = \rho_0\rho_1\rho_{212}\rho_0$ and $\sigma_{01201201} = \zeta\rho_2 = \rho_0\rho_1\rho_0\rho_{212}\rho_0\rho_1$. The corresponding diagram is

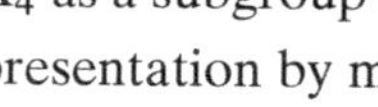

×———× $\underset{4}{—}$ o

The two subgroups $[3^*, 4]$ and $[3^*, 4, 1^+]$ can be recognized as the orthorectangular group [2, 2] and the pararhombic group $[2, 2]^+$.

The full icosahedral group $[3, 5] \cong 2 \times A_5$ contains the pyritohedral group $[3^+, 4] \cong 2 \times A_4$ as a subgroup of index 5, though not in a way that lends itself to representation by modified Coxeter diagrams. It follows that all the subgroups of $[3^+, 4]$ are also subgroups of [3, 5]; those consisting only of rotations are subgroups of $[3, 5]^+$ as well.

D. *Higher-dimensional groups.* Finite groups generated by four reflections include direct products like $[p] \times [q] \cong [p, 2, q]$ and

$[\,]\times[p, q] \cong [2, p, q]$ and irreducible groups $[p, q, r]$ and $[3^{1,1,1}]$. These groups, or their subgroups or extensions, are the symmetry groups of uniform prisms $\{p\} \times \{q\}$ and $\{\,\} \times \{p, q\}$ or of other regular or uniform figures.

Among the subgroups of the prismatic groups $[p, 2, q]$ and $[2, p, q]$ are their rotation groups $[p, 2, q]^+$ and $[2, p, q]^+$ and various ionic subgroups such as $[p^+, 2, q]$, $[p, 2, q^+]$, $[p^+, 2, q^+]$, and $[2, (p, q)^+]$ and still others when either p or q is even. We also note that the prismatic group $[p, 2, p]$ has an automorphism interchanging pairs of generators and hence a double group $[[p, 2, p]]$.

The groups $[3, 3, 3] \cong S_5$ and $[3^{1,1,1}] \cong \frac{1}{2}\cdot{}^2S_4$ have direct subgroups $[3, 3, 3]^+ \cong A_5$ and $[3^{1,1,1}]^+ \cong \frac{1}{2}\cdot{}^2A_4$, and $[3, 3, 5] \cong 2\cdot(A_5 \times A_5)\cdot 2$ has a direct subgroup $[3, 3, 5]^+ \cong 2\cdot(A_5 \times A_5)$. Having only odd numbers in their Coxeter symbols, none of these groups has any ionic subgroups. However, $[3^{1,1,1}]$ has trionic subgroups $[3^{1,1,1}]^{\curlywedge}$ of index 3 and $[3^{1,1,1}]^{\triangle} \cong 2\cdot(D_2 \times D_2)$ of index 6, the latter being the commutator subgroup of $[3^{1,1,1}]^+$.

The group $[3, 3, 4] \cong {}^2S_4$ has a direct subgroup $[3, 3, 4]^+ \cong {}^2A_4$, a semidirect subgroup $[(3, 3)^+, 4]$, and a halving subgroup $[3, 3, 4, 1^+] \cong [3^{1,1,1}]$; its commutator subgroup $[(3, 3)^+, 4, 1^+]$ is isomorphic to $[3^{1,1,1}]^+$. There are also the trionic subgroups $[(3, 3)^{\curlywedge}, 4]$ of index 3, $[(3, 3)^{\triangle}, 4]$ of index 6, $[(3, 3)^{\curlywedge}, 4, 1^+] \cong [3^{1,1,1}]^{\curlywedge}$ of index 6, and $[(3, 3)^{\triangle}, 4, 1^+] \cong [3^{1,1,1}]^{\triangle}$ of index 12.

The group $[3, 4, 3] \cong 3\cdot{}^2S_4$ has a direct subgroup $[3, 4, 3]^+ \cong 3\cdot{}^2A_4$ and two isomorphic semidirect subgroups $[3^+, 4, 3]$ and $[3, 4, 3^+]$; the commutator subgroup is $[3^+, 4, 3^+]$. The groups $[3, 4, 3]$ and $[3, 4, 3]^+$ respectively contain $[3, 3, 4]$ and $[3, 3, 4]^+$ as subgroups of index 3; their respective subgroups $[3^+, 4, 3]$ and $[3^+, 4, 3^+]$ (of index 2) contain $[(3, 3)^+, 4]$ and $[3^{1,1,1}]^+$ as subgroups of index 3. Thus there is a direct line of subgroups from $[3, 4, 3]$ to $[3, 3, 4]$ to $[3^{1,1,1}]$.

The group $[3, 3, 5]$ and its direct subgroup $[3, 3, 5]^+$ respectively contain $[3^+, 4, 3]$ and $[3^+, 4, 3^+]$ as subgroups of index 25, $[3^{1,1,1}]$

and $[3^{1,1,1}]^+$ as subgroups of index 75, and [3, 3, 3] and $[3, 3, 3]^+$ as subgroups of index 120.

The groups $\mathbf{A}_4 \cong [3, 3, 3]$ and $\mathbf{F}_4 \cong [3, 4, 3]$, whose Coxeter diagrams have bilateral symmetry, have extended groups [[3, 3, 3]] and [[3, 4, 3]] of twice the order, obtained by adjoining to the generators ρ_0, ρ_1, ρ_2, and ρ_3 an automorphism (a half-turn about the fundamental region's axis of symmetry) that interchanges ρ_i and ρ_{3-i}. These are the symmetry groups of the regular compounds {{3, 3, 3}} and {{3, 4, 3}} of two dual {3, 3, 3}'s or {3, 4, 3}'s.

Many of these connections are described by Conway & Smith (2003, pp. 45–48), who give different symbols for some of the groups. This is by no means an exhaustive account of finite groups of isometries operating in four-dimensional space (or on the 3-sphere S^3). However, it does cover the principal subgroups and supergroups of irreducible Coxeter groups and indicates how they are related to each other.

Groups generated by five or more reflections include direct products of lower-dimensional groups; the infinite families $\mathbf{A}_n \cong [3^{n-1}]$, $\mathbf{B}_n \cong [3^{n-2}, 4]$, and $\mathbf{D}_n \cong [3^{n-3,1,1}]$; and the three exceptional groups $\mathbf{E}_n$ ($n = 6, 7, 8$). All of these have direct subgroups, and $\mathbf{B}_n$ has ionic subgroups $[(3^{n-2})^+, 4]$, $[3^{n-2}, 4, 1^+] \cong [3^{n-3,1,1}]$, and $[(3^{n-2})^+, 4, 1^+] \cong [3^{n-3,1,1}]^+$. As noted in the previous section, $\mathbf{E}_7$ contains $\mathbf{A}_7$ as a subgroup of index 72, while $\mathbf{E}_8$ contains $\mathbf{D}_8$ as a subgroup of index 135 and $\mathbf{A}_8$ as a subgroup of index 1920. Each of the groups $\mathbf{A}_n$, $\mathbf{D}_n$, and $\mathbf{E}_6$ has a dualizing or bifurcating automorphism interchanging pairs of generators that can be adjoined to yield the double group $[[3^{n-1}]]$ or a *bipartite* group $\langle[3^{n-3,1,1}]\rangle \cong [3^{n-2}, 4]$ or $\langle[3^{2,2,1}]\rangle$.

EXERCISES 11.6

1. Given the generators ρ_0, ρ_1, and ρ_2 of the full ortho-2p-gonal group $[2, 2p]$, show that the subgroup $[2^+, 2p^+]$ generated by the rotary

reflection $\phi_{012} = \rho_0\rho_1\rho_2$ is a cyclic group of order $2p$. Also show that $\phi_{012}{}^p$ is the central inversion if and only if p is odd.

2. Show that the generating reflections ρ_1, ρ_2, and ρ_3 of the double group [[3, 3]] satisfy the relations $(\rho_1\rho_2)^3 = (\rho_1\rho_3)^2 = (\rho_2\rho_3)^4 = 1$.

3. Given the generating reflections ρ_0, ρ_1, and ρ_2 of the full octahedral group [3, 4], show that the generators ρ_0, ρ_1, and ρ_{212} of the halving subgroup [3, 4, 1⁺] satisfy the relations $(\rho_0\rho_1)^3 = (\rho_1\rho_{212})^2 = (\rho_0\rho_{212})^3 = 1$.

4. What are the commutator subgroups of the Coxeter groups $\mathbf{A}_n$, $\mathbf{B}_n$, and $\mathbf{D}_n$?

12

EUCLIDEAN SYMMETRY GROUPS

Repetitive patterns in the Euclidean plane have symmetry groups that include reflections and rotations of periods 2, 3, 4, and 6 but also involve aperiodic translations and transflections (glide reflections). There are just seven geometrically distinct "frieze patterns" and just seventeen "wallpaper patterns," all of which can be described in terms of Euclidean Coxeter groups. Generators and relations are given for each group, together with modified Coxeter diagrams. Certain patterns define *torohedral* quotient groups associated with finite Euclidean or projective planes. Irreducible Coxeter groups in higher space comprise four infinite families and five exceptional groups.

12.1 FRIEZE PATTERNS

A geometric figure in the Euclidean plane with a discrete symmetry group that includes translations along some line but not along any intersecting line is a *frieze pattern*. Symmetry groups of such patterns are extensions of the one-dimensional apeirogonal and full apeirogonal groups $[\infty]^+ \cong \mathrm{C}_\infty$ and $[\infty] \cong \mathrm{D}_\infty$, being the infinite analogues of the pyramidal, prismatic, and antiprismatic groups described in § 11.3. There are exactly seven geometrically distinct groups.

The *acro-apeiral* group $[1, \infty]^+ \cong C_\infty$ is generated by a translation τ, satisfying only the nominal relation

$$\tau^\infty = 1. \tag{12.1.1}$$

The *full acro-apeiral* group $[1, \infty] \cong D_\infty$ is generated by reflections ρ_1 and ρ_2 in parallel lines $\check{R}_1$ and $\check{R}_2$, whose product $\rho_1\rho_2$ is a translation through twice the distance $|\check{R}_1\check{R}_2|$, satisfying the relations

$$\rho_1{}^2 = \rho_2{}^2 = (\rho_1\rho_2)^\infty = 1. \tag{12.1.2}$$

Diagrams for the groups $[1, \infty]^+$ and $[1, \infty]$ are

○ ○—∞—○ and ○ ●—∞—●

Patterns with these symmetry groups are typified by the infinite strings

· · · b b b b b b b b · · ·
· · · b d b d b d b d · · ·

The *full ortho-apeiral* group $[2, \infty] \cong D_1 \times D_\infty$ is generated by reflections ρ_0, ρ_1, and ρ_2 in lines $\check{R}_0$, $\check{R}_1$, and $\check{R}_2$, with $\check{R}_1$ parallel to $\check{R}_2$ and with $\check{R}_0$ orthogonal to both. The products $\rho_0\rho_1$ and $\rho_0\rho_2$ are half-turns, and $\rho_1\rho_2$ is a translation. The generators satisfy the relations

$$\rho_0{}^2 = \rho_1{}^2 = \rho_2{}^2 = (\rho_0\rho_1)^2 = (\rho_0\rho_2)^2 = (\rho_1\rho_2)^\infty = 1. \tag{12.1.3}$$

The semidirect subgroup $[2, \infty^+] \cong D_1 \times C_\infty$ is the *ortho-apeiral* group, generated by the reflection $\rho = \rho_0$ and the translation $\tau = \tau_{12} = \rho_1\rho_2$, satisfying the relations

$$\rho_2 = \tau^\infty = 1, \quad \rho \rightleftarrows \tau, \tag{12.1.4}$$

where '$\rho \rightleftarrows \tau$' means that the generators ρ and τ commute. The direct subgroup $[2, \infty]^+ \cong D_\infty$ is the *para-apeiral* group, generated by the half-turn $\sigma = \sigma_{01} = \rho_0\rho_1$ and the translation $\tau = \tau_{12} = \rho_1\rho_2$, whose product $\sigma\tau$ is a half-turn, satisfying the relations

$$\sigma_2 = \tau^\infty = (\sigma\tau)^2 = 1. \tag{12.1.5}$$

Diagrams for the groups $[2,\ \infty]$, $[2,\ \infty^+]$, and $[2,\ \infty]^+$ are

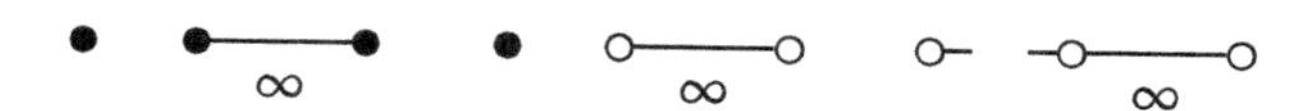

Typical patterns that have these symmetry groups are

$\cdots$ o o o o o o o o $\cdots$

$\cdots$ c c c c c c c c $\cdots$

$\cdots$ b q b q b q b q $\cdots$

Another semidirect subgroup of $[2,\ \infty]$ is the *full gyro-apeiral* group $[2^+,\ \infty] \cong \mathrm{D}_\infty$, generated by the half-turn $\sigma = \sigma_{01} = \rho_0\rho_1$ and the reflection $\rho = \rho_2$, whose product $\sigma\rho$ is a transflection (glide reflection), satisfying the relations

$$\sigma^2 = \rho^2 = (\sigma\rho)^\infty = 1. \tag{12.1.6}$$

The *gyro-apeiral* group $[2^+,\ \infty^+] \cong \mathrm{C}_\infty$ is generated by the transflection $\phi = \sigma_{01}\rho_2 = \rho_0\tau_{12} = \rho_0\rho_1\rho_2$, satisfying the relation

$$\phi^\infty = 1. \tag{12.1.7}$$

Diagrams for the groups $[2^+,\ \infty]$ and $[2^+,\ \infty^+]$ are

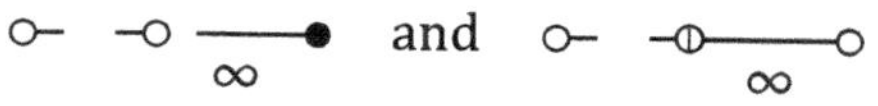

and

Typical patterns with these symmetry groups are

$\cdots$ b d p q b d p q $\cdots$

$\cdots$ b p b p b p b p $\cdots$

Every discrete group of isometries in E^2 that includes translations in only one or the other of two opposite directions is one of these seven groups. Each of the groups $[1,\ \infty]^+$ and $[2,\ \infty]^+$, which include only translations or half-turns, is the direct subgroup of two or more of the groups that include reflections or transflections. Table 12.1 shows how the seven groups are related.

Table 12.1 *Two-Dimensional Frieze-Pattern Groups*

Rotation Groups			Extended Groups		
Name	Symbol	Structure	Name	Symbol	Structure
Acro-apeiral	$[1, \infty]^+$	C_∞	Full acro-apeiral	$[1, \infty]$	D_∞
			Gyro-apeiral	$[2^+, \infty^+]$	C_∞
			Ortho-apeiral	$[2, \infty^+]$	$D_1 \times C_\infty$
Para-apeiral	$[2, \infty]^+$	D_∞	Full gyro-apeiral	$[2^+, \infty]$	D_∞
			Full ortho-apeiral	$[2, \infty]$	$D_1 \times D_\infty$

EXERCISES 12.1

1. Identify the groups of the frieze patterns formed by repetitions of each of the capital letters A, D, F, H, N, and X.

2. An infinite strip of rectangles $\{\ \} \times \{\ \}$ or isosceles triangles $() \vee \{\}$ between two parallel lines can be regarded as an apeirogonal prism $\{\ \} \times \{\infty\}$ or antiprism $\{\ \} \otimes \{\infty\}$. Describe the para-apeiral, gyro-apeiral, ortho-apeiral, full gyro-apeiral, and full ortho-apeiral groups in terms of symmetries of these figures.

12.2 LATTICE PATTERNS

A geometric figure in E^2 with a discrete symmetry group that includes two independent translations but no rotations other than half-turns is a *lattice pattern*. The symmetry group of such a pattern is an extension of the direct product of two apeirogonal groups $[\infty]^+ \cong C_\infty$. Each of the nine geometrically distinct groups has a standard abbreviated crystallographic symbol. Generators and relations for these groups are given by Coxeter & Moser (1957, pp. 40–48).

The simplest of the lattice-pattern groups is the *acroclinic* group $\mathbf{p1} \cong [\infty^+, 2, \infty^+] \cong C_\infty \times C_\infty$, generated by translations τ_1 and τ_2, satisfying the relations

$$\tau_1{}^\infty = \tau_2{}^\infty = 1, \quad \tau_1 \rightleftarrows \tau_2. \tag{12.2.1}$$

By adjoining a half-turn σ, whose product with either τ_1 or τ_2 is another half-turn, we obtain the *paraclinic* group **p2** $\cong [\infty, 2, \infty]^+ \cong 2 \cdot (\mathrm{C}_\infty \times C_\infty)$. The generators σ, τ_1, and τ_2 satisfy the relations

$$\sigma^2 = \tau_1{}^\infty = \tau_2{}^\infty = (\sigma\tau_1)^2 = (\sigma\tau_2)^2 = 1, \quad \tau_1 \rightleftarrows \tau_2. \tag{12.2.2}$$

Diagrams for the groups **p1** and **p2** are

∞ ∞ and ∞ ∞

The group **p1** is the direct subgroup of three different extended groups. When the translations τ_1 and τ_2 are in orthogonal directions, the *digyric* group **pg** $\cong [(\infty, 2)^+, \infty^+]$ is generated by two transflections ϕ_1 and ϕ_2 whose axes are parallel to the direction of τ_2, with $\phi_1\phi_2{}^{-1} = \tau_1$. The generators ϕ_1 and ϕ_2 satisfy the relations

$$\phi_1{}^\infty = \phi_2{}^\infty = 1, \quad \phi_1{}^2 = \phi_2{}^2. \tag{12.2.3}$$

If we replace the transflection ϕ_1 by the reflection $\rho = \phi_1\tau_2{}^{-1}$, we obtain the *gyro-orthic* group **cm** $\cong [\infty, 2^+, \infty^+]$, whose generators ρ and $\phi = \phi_2$ satisfy the relations

$$\rho^2 = \phi^\infty = 1, \quad \rho \rightleftarrows \phi^2. \tag{12.2.4}$$

Again, when the translations τ_1 and τ_2 of **p1** are in orthogonal directions, the *ortho-orthic* group **pm** $\cong [\infty, 2, \infty^+] \cong \mathrm{D}_\infty \times \mathrm{C}_\infty$, is generated by reflections ρ_1 and ρ_2 such that $\rho_1\rho_2 = \tau_1$, together with the translation $\tau = \tau_2$; the product of τ with either ρ_1 or ρ_2 is a transflection. The generators ρ_1, ρ_2, and τ satisfy the relations

$$\rho_1{}^2 = \rho_2{}^2 = \tau^\infty = (\rho_1\rho_2)^\infty = 1, \quad \rho_1 \rightleftarrows \tau, \ \rho_2 \rightleftarrows \tau. \tag{12.2.5}$$

Diagrams for the groups **pg**, **cm**, and **pm** are

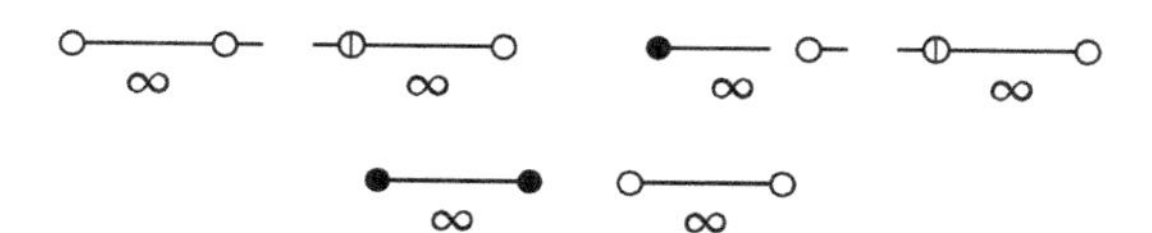

The group **p2** is the direct subgroup of four different groups. In contrast to the "bottom up" approach used for extensions of **p1**, extensions of **p2** are more conveniently derived "top down" as subgroups of the *full ortho-orthic* group **pmm** $\cong [\infty, 2, \infty] \cong [\infty] \times [\infty] \cong$ $D_\infty \times D_\infty$. The group **pmm** is generated by reflections $\rho_1, \rho_2, \rho_3, \rho_4$ in the sides of a rectangle, satisfying the relations

$$\begin{aligned}\rho_1{}^2 = \rho_2{}^2 = \rho_3{}^2 = \rho_4{}^2 = (\rho_1\rho_2)^\infty = (\rho_3\rho_4)^\infty \\ = (\rho_1\rho_3)^2 = (\rho_1\rho_4)^2 = (\rho_2\rho_3)^2 = (\rho_2\rho_4)^2 = 1.\end{aligned} \qquad (12.2.6)$$

The semidirect subgroup **cmm** $\cong [\infty, 2^+, \infty]$, the *full gyro-orthic* group, is generated by the reflections ρ_1 and ρ_4 and the half-turn $\sigma_{23} = \rho_2\rho_3$, satisfying the relations

$$\rho_1{}^2 = \sigma_{23}{}^2 = \rho_4{}^2 = (\rho_1\rho_4)^2 = (\rho_1\sigma_{23}\rho_4\sigma_{23})^2 = 1. \qquad (12.2.7)$$

The *full digyric* group **pmg** $\cong [(\infty, 2)^+, \infty]$ is generated by the half-turns $\sigma_{13} = \rho_1\rho_3$ and $\sigma_{23} = \rho_2\rho_3$ and the reflection ρ_4, satisfying the relations

$$\sigma_{13}{}^2 = \sigma_{23}{}^2 = \rho_4{}^2 = 1, \quad \sigma_{13}\rho_4\sigma_{13} = \sigma_{23}\rho_4\sigma_{23}. \qquad (12.2.8)$$

Diagrams for the groups **pmm**, **cmm**, and **pmg** are

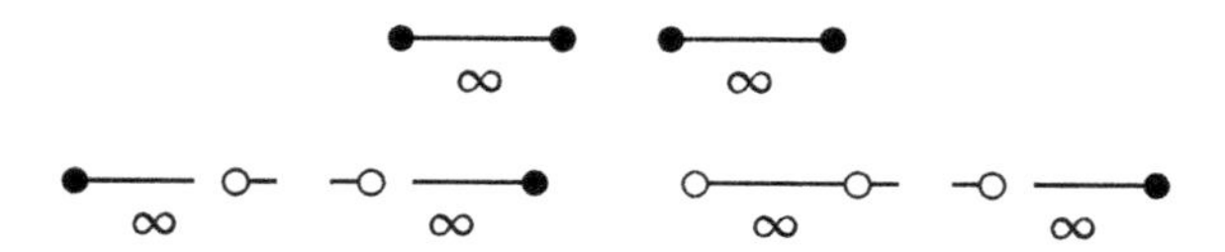

The *paragyric* group **pgg** $\cong [((\infty, 2)^+, (\infty, 2)^+)]$, a subgroup of each of these three groups, is generated by the orthogonal transflections $\phi_{123} = \rho_1\rho_2\rho_3$ and $\phi_{234} = \rho_2\rho_3\rho_4$, satisfying the relations

$$\phi_{123}{}^\infty = \phi_{234}{}^\infty = (\phi_{123}\phi_{234})^2 = (\phi_{123}{}^{-1}\phi_{234})^2 = 1. \qquad (12.2.9)$$

When the nodes corresponding to the generators of **pmm** are arranged in a circuit, in the order 1243, an appropriate diagram for **pgg** is

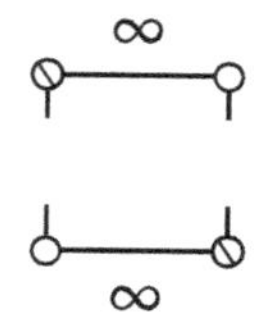

The direct subgroup **p2** common to the four groups is generated by the translations $\tau_1 = \rho_1\rho_2$ and $\tau_2 = \rho_3\rho_4$ and the half-turn $\sigma = \rho_2\rho_3$, satisfying the relations (12.2.2).

Generators for the group **p1** and its extensions can likewise be expressed in terms of the generating reflections $\rho_1, \rho_2, \rho_3, \rho_4$ of **pmm**. For **pm**, defined by (12.2.5), we have ρ_1, ρ_2, and $\tau = \tau_{34} = \rho_3\rho_4$. For **cm**, defined by (12.2.4), we have $\rho = \rho_1$ and $\phi = \phi_{234} = \rho_2\rho_3\rho_4$. For **pg**, defined by (12.2.3), we have $\phi_1 = \phi_{134} = \rho_1\rho_3\rho_4$ and $\phi_2 = \phi_{234} = \rho_2\rho_3\rho_4$. Finally, for **p1**, defined by (12.2.1), we have $\tau_1 = \rho_1\rho_2$ and $\tau_2 = \rho_3\rho_4$.

Table 12.2 *Two-Dimensional Lattice-Pattern Groups*

Rotation Groups			Extended Groups		
Group	Symbol	Generators	Group	Symbol	Generators
p1	$[\infty^+, 2, \infty^+]$	τ_1, τ_2	**pg**	$[(\infty, 2)^+, \infty^+]$	ϕ_{134}, ϕ_{234}
			cm	$[\infty, 2^+, \infty^+]$	ρ_1, ϕ_{234}
			pm	$[\infty, 2, \infty^+]$	$\rho_1, \rho_2, \tau_{34}$
p2	$[\infty, 2, \infty]^+$	σ, τ_1, τ_2	**pgg**	$[((\infty, 2)^+, (\infty, 2)^+)]$	ϕ_{123}, ϕ_{234}
			pmg	$[(\infty, 2)^+, \infty]$	$\sigma_{13}, \sigma_{23}, \rho_4$
			cmm	$[\infty, 2^+, \infty]$	$\rho_1, \sigma_{23}, \rho_4$
			pmm	$[\infty, 2, \infty]$	$\rho_1, \rho_2, \rho_3, \rho_4$

Table 12.2 shows how the symmetry groups of the nine lattice patterns are related. Examples of each pattern are shown in Figure 12.2a (the portion shown is assumed to be indefinitely repeated both horizontally and vertically).

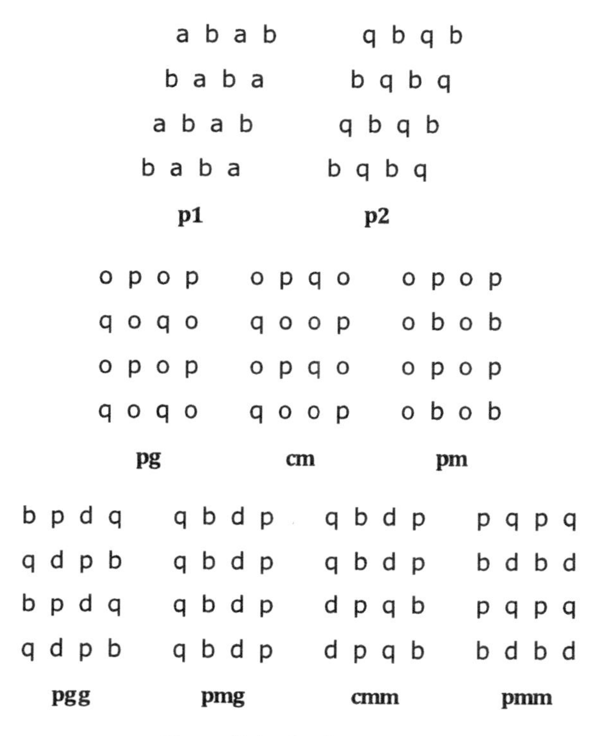

Figure 12.2a Lattice patterns

EXERCISES 12.2

1. By expressing the generators of the one in terms of the generators of the other, show that the groups **pg**, **cm**, and **pm** are subgroups of **pmg**, **cmm**, and **pmm**, respectively.

2. Show that **cmm** is a subgroup of **pmm** and that **pgg** is a subgroup of **cmm**.

3. To show that **pgg** is a subgroup of **pmg**, let the latter group be generated by the half-turns σ_{14} and σ_{24} and the reflection ρ_3, satisfying the relations (12.2.8) with subscripts 3 and 4 switched. Then express the generators ϕ_{123} and ϕ_{234} of **pgg** in terms of the generators of **pmg** and show that they satisfy the relations (12.2.9).

4. If the portions of the lattice patterns shown in Figure 12.2a are repeated only in the horizontal direction, what are the symmetry groups of the resulting frieze patterns?

5. ⊢ If orthogonal reflections are replaced by affine reflections, each of the nine lattice-pattern groups can be taken as operating in the affine plane.

12.3 APEIROHEDRAL GROUPS

The nine lattice-pattern groups just presented and eight other *apeirohedral groups* comprise the seventeen "wallpaper pattern" groups first determined by Fedorov (1891; cf. Fricke & Klein 1897, pp. 226–234; Pólya 1924; Niggli 1924). For an extensive treatment of the plane symmetry groups, see Schattschneider 1978. The eight groups described in this section are infinite analogues of the polyhedral groups discussed in § 11.4 (cf. Coxeter & Moser 1957, pp. 46–51).

If we denote a triangle with angles π/p, π/q, and π/r by $(p\ q\ r)$, then an equilateral triangle $(3\ 3\ 3)$ and the right triangles $(4\ 4\ 2)$ and $(6\ 3\ 2)$ are the respective fundamental regions of the Coxeter groups $[3^{[3]}]$, $[4, 4]$, and $[3, 6]$. The *full trigonohedral* group **p3m1** $\cong [3^{[3]}]$ is generated by reflections ρ_0, ρ_1, and ρ_2 satisfying the relations

$$\rho_0{}^2 = \rho_1{}^2 = \rho_2{}^2 = (\rho_0\rho_1)^3 = (\rho_0\rho_2)^3 = (\rho_1\rho_2)^3 = 1. \qquad (12.3.1)$$

For the *full tetragonohedral* group **p4m** $\cong [4, 4]$, the relations are those of (11.4.1) with $p = q = 4$, i.e.,

$$\rho_0{}^2 = \rho_1{}^2 = \rho_2{}^2 = (\rho_0\rho_1)^4 = (\rho_0\rho_2)^2 = (\rho_1\rho_2)^4 = 1. \qquad (12.3.2)$$

For the *full hexagonohedral* group **p6m** $\cong [3, 6]$, we take $p = 3$ and $q = 6$ (or vice versa):

$$\rho_0{}^2 = \rho_1{}^2 = \rho_2{}^2 = (\rho_0\rho_1)^3 = (\rho_0\rho_2)^2 = (\rho_1\rho_2)^6 = 1. \qquad (12.3.3)$$

The groups $[4, 4]$ and $[3, 6]$ are the respective symmetry groups of the self-dual square tessellation $\{4, 4\}$ and of the dual triangular and hexagonal tessellations {3, 6} and {6, 3}. Diagrams for the groups **p3m1**, **p4m**, and **p6m** are

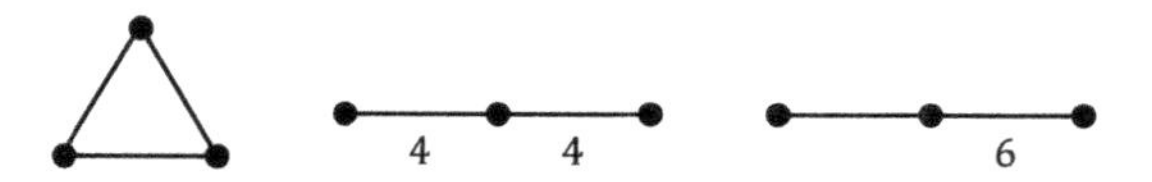

The direct subgroup of each of these reflection groups is generated by the rotations $\sigma_1 = \sigma_{01} = \rho_0\rho_1$ and $\sigma_2 = \sigma_{12} = \rho_1\rho_2$. For the *trigonohedral* group **p3** $\cong [3^{[3]}]^+$, the generators satisfy the relations

$$\sigma_1{}^3 = \sigma_2{}^3 = (\sigma_1\sigma_2)^3 = 1. \tag{12.3.4}$$

The generators of the *tetragonohedral* group **p4** $\cong$ $[4,\ 4]^+$ satisfy (11.4.2) with $p = q = 4$, i.e.,

$$\sigma_1{}^4 = \sigma_2{}^4 = (\sigma_1\sigma_2)^2 = 1, \tag{12.3.5}$$

while for the *hexagonohedral* group **p6** $\cong [3,\ 6]^+$ we take $p = 3$ and $q = 6$:

$$\sigma_1{}^3 = \sigma_2{}^6 = (\sigma_1\sigma_2)^2 = 1. \tag{12.3.6}$$

Diagrams for the groups **p3**, **p4**, and **p6** are

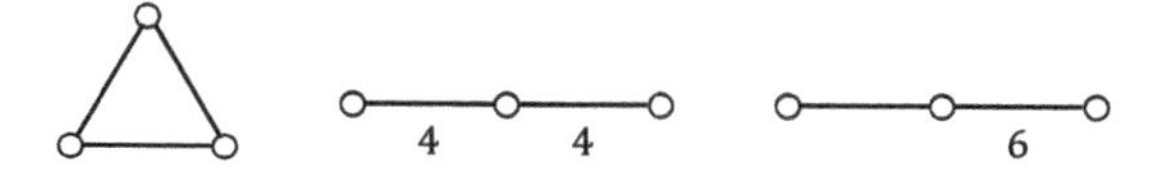

The group **p4m** $\cong$ [4, 4] has a halving subgroup $[1^+,\ 4,\ 4]$, generated by the reflections ρ_1, ρ_2, and $\rho_{010} = \rho_0\rho_1\rho_0$, and satisfying the relations (11.4.3) with $p = q = 4$. An isomorphic subgroup $[4,\ 4,\ 1^+]$ is generated by the reflections ρ_0, ρ_1, and $\rho_{212} = \rho_2\rho_1\rho_2$. Each of these subgroups is isomorphic to [4, 4] but with a larger fundamental region. Likewise, **p6m** $\cong$ [3, 6] has a halving subgroup $[3,\ 6,\ 1^+] \cong [3^{[3]}]$, defined by (11.4.3) with $p = 6$ and $q = 3$.

The group [4, 4] has a semidirect subgroup, the *gyro-tetragonohedral* group **p4g** $\cong$ $[4^+,\ 4]$, generated by the rotation

$\sigma_{01} = \rho_0\rho_1$ and the reflection ρ_2, satisfying the relations (11.4.4) with $p = q = 4$, i.e.,

$$\sigma_{01}{}^4 = \rho_2{}^2 = (\sigma_{01}{}^{-1}\rho_2\sigma_{01}\rho_2)^2 = 1. \tag{12.3.7}$$

An isomorphic subgroup $[4,\ 4^+]$ is generated by the reflection ρ_0 and the rotation $\sigma_{12} = \rho_1\rho_2$. Similarly, $[3, 6]$ has a semidirect subgroup, the *gyro-trigonohedral* group **p31m** $\cong [3^+,\ 6]$, with the rotation σ_{01} and the reflection ρ_2 satisfying (11.4.4) with $p = 3$ and $q = 6$, i.e.,

$$\sigma_{01}{}^3 = \rho_2{}^2 = (\sigma_{01}{}^{-1}\rho_2\sigma_{01}\rho_2)^3 = 1. \tag{12.3.8}$$

Diagrams for the groups **p4g** and **p31m** are

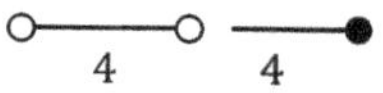

4 4 and 6

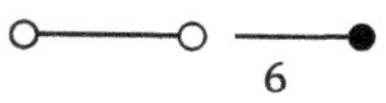

Table 12.3 shows how the eight apeirohedral groups are related. The groups $[3^{[3]}]$, $[4, 4]$, and $[3, 6]$ are generated by reflections ρ_i in the sides of a triangle $(p\ q\ r)$. The subgroups $[3^{[3]}]^+$, $[4,\ 4]^+$, and $[3,\ 6]^+$ are generated by the rotations $\sigma_1 = \rho_0\rho_1$ and $\sigma_2 = \rho_1\rho_2$. The groups $[4^+,\ 4]$ and $[3^+,\ 6]$ are each generated by a rotation σ_{01} and a reflection ρ_2 with parameters $(p,\ q)$. The symmetry group of any repeating pattern filling the Euclidean plane is one of the seventeen lattice-pattern or apeirohedral groups.

Table 12.3 *Apeirohedral Groups*

Rotation Groups			Extended Groups		
Group	Symbol	Generators	Group	Symbol	Generators
p3	$[3^{[3]}]^+$	σ_1, σ_2 (3 3 3)	**p3m1**	$[3^{[3]}]$	ρ_0, ρ_1, ρ_2 (3 3 3)
			p31m	$[3^+,\ 6]$	σ_{01}, ρ_2 (3, 6)
p4	$[4,\ 4]^+$	σ_1, σ_2 (4 4 2)	**p4m**	$[4, 4]$	ρ_0, ρ_1, ρ_2 (4 4 2)
			p4g	$[4^+,\ 4]$	σ_{01}, ρ_2 (4, 4)
p6	$[3,\ 6]^+$	σ_1, σ_2 (6 3 2)	**p6m**	$[3, 6]$	ρ_0, ρ_1, ρ_2 (6 3 2)

Euclidean symmetry groups are not only infinite but have the property of containing scaled-up copies of themselves as subgroups. Thus $[4, 4] \cong$ **p4m** has halving subgroups $[1^+, 4, 4]$ and $[4, 4, 1^+]$ isomorphic to itself. The semidirect subgroups $[4, 4^+]$ and $[4^+, 4]$ are both isomorphic to **p4g**. The common direct subgroup of $[1^+, 4, 4]$ and $[4, 4^+]$ is $[1^+, 4, 4^+]$, while that of $[4^+, 4]$ and $[4, 4, 1^+]$ is $[4^+, 4, 1^+]$; both of these are subgroups of and isomorphic to $[4, 4]^+ \cong$ **p4**. The groups $[1^+, 4, 4]$ and $[4, 4, 1^+]$ have a common subgroup $[1^+, 4, 4, 1^+] \cong [\infty, 2, \infty] \cong$ **pmm**, and the groups $[4, 4^+]$ and $[4^+, 4]$ a common subgroup $[4^+, 4^+] \cong [((\infty, 2)^+, (\infty, 2)^+)] \cong$ **pgg**. A third halving subgroup of $[4, 4]$ is $[4, 1^+, 4] \cong [\infty, 2, \infty] \cong$ **pmm**. The ionic subgroup common to $[1^+, 4, 4]$ and $[4, 1^+, 4]$ is $[1^+, 4, 1^+, 4]$, and the one common to $[4, 1^+, 4]$ and $[4, 4, 1^+]$ is $[4, 1^+, 4, 1^+]$; each of these is isomorphic to $[\infty, 2^+, \infty] \cong$ **cmm**. The two groups $[1^+, 4, 1^+, 4]$ and $[4, 1^+, 4, 1^+]$, together with $[1^+, 4, 4, 1^+]$ and $[4^+, 4^+]$, have the same direct subgroup $[1^+, 4, 1^+, 4, 1^+] \cong [4^+, 4^+]^+ \cong [\infty, 2, \infty]^+ \cong$ **p2**, the commutator subgroup of $[4, 4]$.

Besides its direct subgroup, two semidirect subgroups, and three halving subgroups, the group $[4, 4]$ has still another subgroup of index 2. Given the generating reflections ρ_0, ρ_1, and ρ_2 for $[4, 4]$, the half-turn $\sigma_{02} = \rho_0\rho_2$, the reflection ρ_1, and either of the reflections $\rho_{010} = \rho_0\rho_1\rho_0$ or $\rho_{212} = \rho_2\rho_1\rho_2$ generate a subgroup $[(4, 2^+, 4)] \cong [\infty, 2^+, \infty] \cong$ **cmm**. The direct subgroup $[(4, 2^+, 4)]^+$, generated by the three half-turns σ_{02}, $\sigma_{0101} = \rho_{010}\rho_1$, and $\sigma_{1212} = \rho_1\rho_{212}$, is isomorphic to **p2**.

The group $[3, 6] \cong$ **p6m** has three subgroups of index 2: the direct subgroup $[3, 6]^+ \cong$ **p6**, a semidirect subgroup $[3^+, 6] \cong$ **p31m**, and a halving subgroup $[3, 6, 1^+] \cong [3^{[3]}] \cong$ **p3m1**. Their common subgroup, the commutator subgroup of both $[3, 6]$ and $[3^{[3]}]$, is $[3^+, 6, 1^+] \cong [3^{[3]}]^+ \cong$ **p3**.

The fundamental region for the group $[4, 4]$ is an isosceles right triangle $(4\ 4\ 2)$, which can be divided into two similar triangles, with sides shorter by a factor of $\sqrt{2}$. Either of the smaller triangles may be taken as the fundamental region for an isomorphic group ${}^{\sqrt{2}}[4,\ 4] \cong [[4, 4]]$, in which the original [4, 4] is a subgroup of index 2. The process can be repeated, with [4, 4] being a subgroup of index c^2 or $2c^2$ in an isomorphic group ${}^{c}[4,\ 4]$ or ${}^{c\sqrt{2}}[4,\ 4]$.

In the same manner, the fundamental region for the group [3, 6], a right triangle (6 3 2), can be divided into three or four similar triangles. Repeated subdivision exhibits [3, 6] as a subgroup of index c^2 or $3c^2$ in an isomorphic group ${}^{c}[3,\ 6]$ or ${}^{c\sqrt{3}}[3,\ 6]$. The fundamental region for

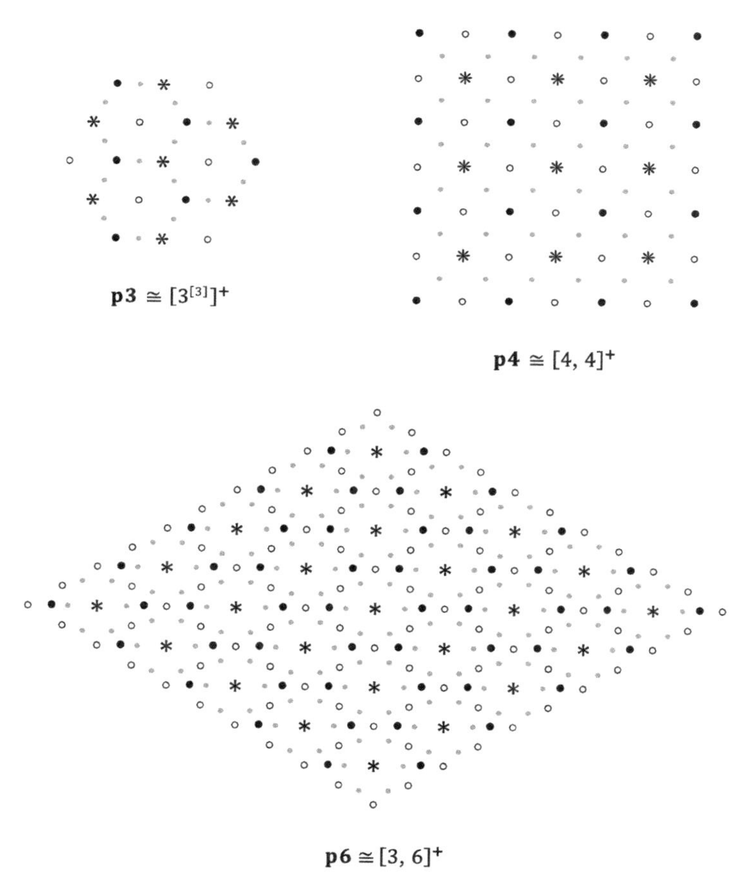

Figure 12.3a Apeirohedral patterns: rotation groups

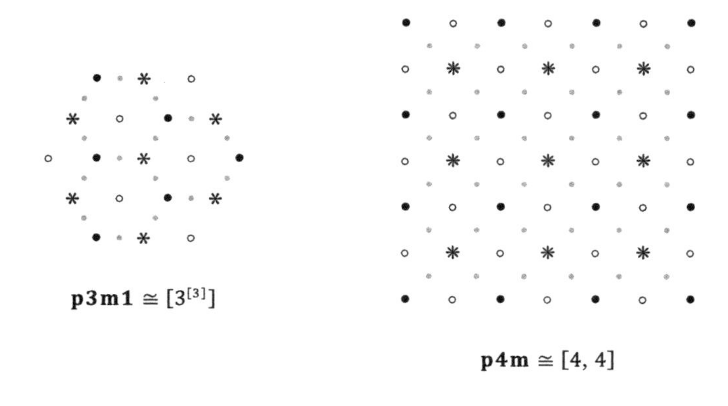

p3m1 $\cong [3^{[3]}]$

p4m $\cong [4, 4]$

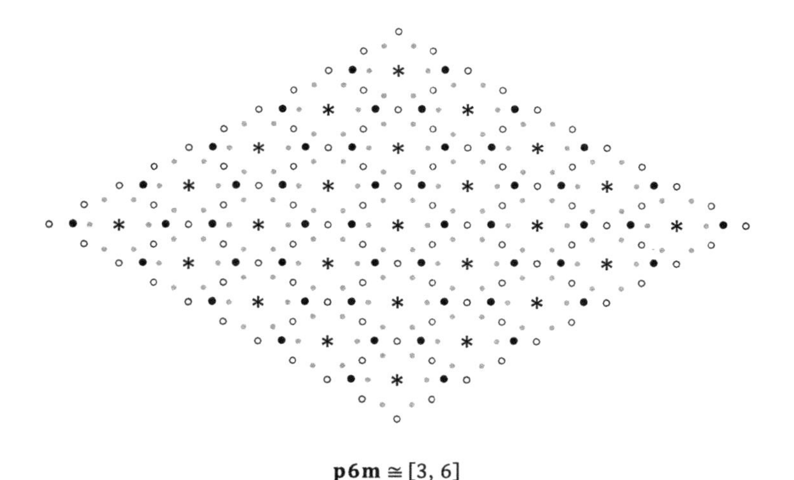

p6m $\cong [3, 6]$

Figure 12.3b Apeirohedral patterns: full symmetry groups

the group $[3^{[3]}]$ is an equilateral triangle (3 3 3), which can be divided into four similar triangles or into two or six triangles (6 3 2). Thus $[3^{[3]}]$ is a subgroup of index c^2 in an isomorphic group ${}^{c}[3^{[3]}]$ and a subgroup of index $2c^2$ or $6c^2$ in the group [3, 6].

Each of these relationships can be turned around. For example, we can say that the group [4, 4] contains an isomorphic subgroup ${}^{1/\sqrt{2}}[4, 4]$ (a halving subgroup). The various subgroups of [4, 4], [3, 6],

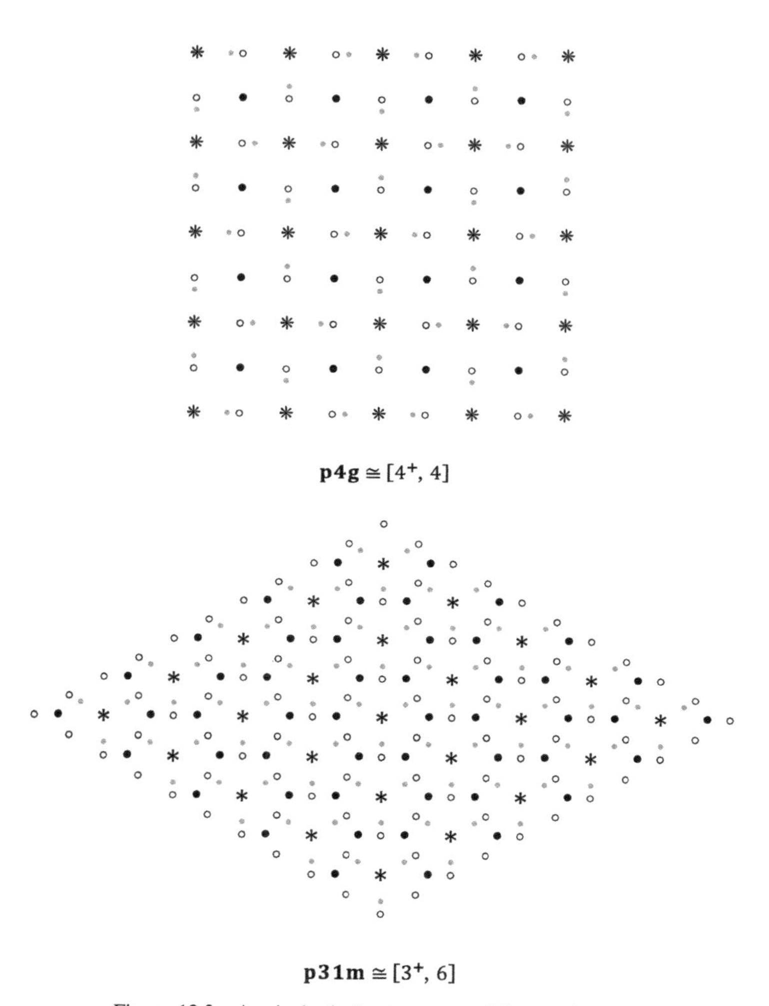

Figure 12.3c Apeirohedral patterns: semidirect subgroups

and $[3^{[3]}]$ can be combined with their larger or smaller isomorphs to give infinite sequences of groups and subgroups.

Examples of each of the eight apeirohedral patterns are shown in Figures 12.3a, 12.3b, and 12.3c. In each case, an arrangement of asterisks and black and white dots is superimposed on a background of gray dots that, continued indefinitely in every direction, are the vertices of some uniform tiling of the plane. Each tiling is a covering of the plane by regular polygons—triangles, squares, or hexagons—with the symmetry group of the pattern being transitive on the vertices of the tiling.

EXERCISES 12.3

1. If the reflections ρ_0, ρ_1, and ρ_2 satisfy the relations (12.3.2) for the group [4, 4], show that the rotation $\sigma_{01} = \rho_0\rho_1$ and the reflection ρ_2 satisfy the relations (12.3.7) for the group $[4^+, 4]$.

2. If the reflections ρ_0, ρ_1, and ρ_2 satisfy the relations (12.3.3) for the group [3, 6], show that the rotation $\sigma_{01} = \rho_0\rho_1$ and the reflection ρ_2 satisfy the relations (12.3.8) for the group $[3^+, 6]$.

12.4 TOROHEDRAL GROUPS

Any parallelogram or centrally symmetric hexagon can be taken as the fundamental region for a lattice group **p1** $\cong C_\infty \times C_\infty$ generated by two translations τ_1 and τ_2 satisfying the relations (12.2.1) or by three translations τ_1, τ_2, and τ_3 satisfying the relations

$$\tau_1{}^\infty = \tau_2{}^\infty = \tau_3{}^\infty = \tau_1\tau_2\tau_3 = \tau_3\tau_2\tau_1 = 1 \qquad (12.4.1)$$

(Coxeter & Moser 1957, pp. 40–41). By identifying opposite sides of the parallelogram or hexagon, we convert the Euclidean plane E^2 into a *torus* $S^1 \times S^1 = E^2/(C_\infty \times C_\infty)$.

A. *Finite Euclidean planes.* The fundamental region of a group $[3^{[3]}]$, [4, 4], or [3, 6] generated by three reflections is the closure of a triangle (3 3 3), (4 4 2), or (6 3 2) in the Euclidean plane, and the mirrors and virtual mirrors fall into three, four, or six equivalence classes (i.e., directions) of parallel lines. For each of the groups, the images of the vertex (or one of the vertices) of the fundamental region at which two sides make the smallest angle—respectively $\pi/3$, $\pi/4$, or $\pi/6$—are the points of a lattice; these are the points marked with an asterisk in Figure 12.3b. When certain sets of points are identified, the lattice becomes a finite arrangement of points on a torus.

Starting at any lattice point, proceed in a straight line toward one of the others closest to it to the bth lattice point in that direction, then turn to the left through an appropriate angle—$\pi/2$ for the group $[4, 4]$, $\pi/3$ for the group $[3^{[3]}]$ or $[3, 6]$—and proceed to the cth lattice point in the new direction. Either b or c, but not both, may be zero. If we identify all the points that can be reached from a given lattice point by repeated applications of this recipe, then there are only finitely many equivalence classes of lattice points. For the group $[4, 4]$, the number is $b^2 + c^2$; for $[3^{[3]}]$ or $[3, 6]$, it is $b^2 + bc + c^2$ (Coxeter & Moser 1957, pp. 103–109).

If the images of the other vertices of the fundamental region—the black and white dots of Figure 12.3b—retain their relative positions, the effect of the procedure described above is to map a Euclidean tessellation of triangles $(3\ 3\ 3)$, $(4\ 4\ 2)$, or $(6\ 3\ 2)$ onto a torus. The Euclidean tessellation, with symmetry group $[3^{[3]}]$, $[4, 4]$, or $[3, 6]$, is reflexible. The toroidal tessellation is reflexible if and only if $bc(b - c) = 0$, and its symmetry group is then the *torohedral* quotient group

$$[3^{[3]}]_{(b,c)}, \quad [4, 4]_{(b,c)}, \quad \text{or} \quad [3, 6]_{(b,c)}.$$

The order of the group is the number of toroidal triangles, respectively $6(b^2 + bc + c^2)$, $8(b^2 + c^2)$, or $12(b^2 + bc + c^2)$. When $bc(b - c) \neq 0$, the toroidal tessellation is chiral, with only rotational symmetry, the group being

$$[3^{[3]}]^+_{(b,c)}, \quad [4, 4]^+_{(b,c)}, \quad \text{or} \quad [3, 6]^+_{(b,c)},$$

of half the order, i.e., $3(b^2 + bc + c^2)$, $4(b^2 + c^2)$, or $6(b^2 + bc + c^2)$.

If the apeirohedral patterns of Figure 12.3b are taken to be confined to the hexagonal, square, or rhombic regions shown, with opposite sides of the boundary polygon identified, they represent the torohedral groups

$$[3^{[3]}]_{(2,0)}, \quad [4,\ 4]_{(3,0)}, \quad \text{and} \quad [3,\ 6]_{(5,0)}.$$

These three patterns are of particular interest as Euclidean models for the finite affine planes

$$\mathbb{F}_2\mathrm{A}^2, \qquad \mathbb{F}_3\mathrm{A}^2, \quad \text{and} \quad \mathbb{F}_5\mathrm{A}^2.$$

A finite affine plane is a configuration $((q^2)_{q+1},\ (q^2+q)_q)$ of q^2 points and $q^2 + q$ lines ($q > 1$), with $q + 1$ lines through each point and q points on each line. In all known cases, q is a prime or a prime power. With q equal to 2, 3, or 5, this is a configuration $(4_3,\ 6_2)$, $(9_4,\ 12_3)$, or $(25_6,\ 30_5)$. The first of these is simply a complete quadrangle. The second is the *Hessian configuration* (after Otto Hesse, 1811–1874), which also exists in any projective plane over any field (e.g., $\mathbb{C}$) that contains two cube roots of unity different from 1, say ω and ω^2 (Coxeter & Moser 1957, p. 98; Coxeter 1974, p. 123). In such a plane, the cross ratio $\| UV,\ \check{W}\check{Z} \|$ of two points and two lines of the configuration can take on only the two values $-\omega$ and $-\omega^2$.

Figures 12.4a, 12.4b, and 12.4c illustrate the three cases. Points are denoted by capital letters, replacing the asterisks of Figure 12.3b, and lines are indicated by sequences of three or more dots between letters. Repeated occurrences of each of the 4, 9, or 25 points and of each of the 6, 12, or 30 lines are to be identified. Any two points lie on a unique line, and any two lines either meet in a unique point or are parallel. The lines comprise 3, 4, or 6 sets of parallel lines, with 2, 3, or 5 lines in each direction. Because of their toroidal realizations, each of these affine geometries also has a metric structure and may be regarded as a finite *Euclidean* plane $\mathbb{F}_2\mathrm{E}^2$, $\mathbb{F}_3\mathrm{E}^2$, or $\mathbb{F}_5\mathrm{E}^2$.

Every affinity of the plane $\mathbb{F}_2\mathrm{A}^2$—i.e., every permutation of the four points—is an isometry of $\mathbb{F}_2\mathrm{E}^2$, and the isometry group $\mathrm{E}_2(\mathbb{F}_2) \cong [3^{[3]}]_{(2,0)}$ is the symmetric group S_4, of order 24. The subgroup $\mathrm{E}_2^+(\mathbb{F}_2)$

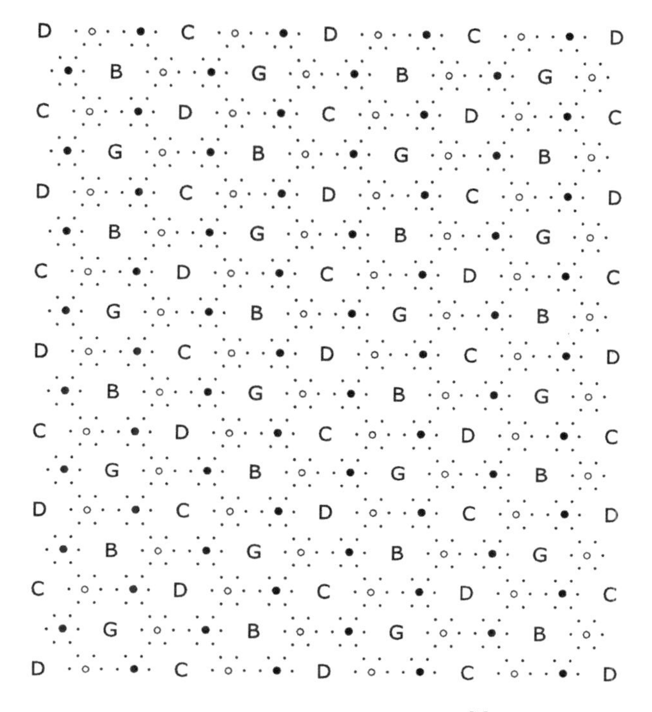

Figure 12.4a Torohedral pattern for $[3^{[3]}]_{(2,0)}$

of direct isometries is the alternating group A_4, of order 12. All triangles in $\mathbb{F}_2E^2$ are equilateral.

Of the 432 affinities of $\mathbb{F}_3A^2$, only those that permute the 72 copies of the fundamental region are isometries of $\mathbb{F}_3E^2$. The isometry group $E_2(\mathbb{F}_3) \cong [4, 4]_{(3,0)}$, of order 72, is generated by reflections ρ_0, ρ_1, and ρ_2, satisfying the relations (12.3.2) with the additional relation

$$(\rho_0\rho_1\rho_2\rho_1)^3 = 1. \tag{12.4.2}$$

The direct subgroup $E_2^+(\mathbb{F}_3)$ is of order 36. All triangles are isosceles right triangles (4 4 2) but come in two sizes, with areas in the ratio of 1 to 2 (mod 3). The similarity and direct similarity groups $\tilde{E}_2(\mathbb{F}_3)$ and $\tilde{E}_2^+(\mathbb{F}_3)$ are of orders 144 and 72.

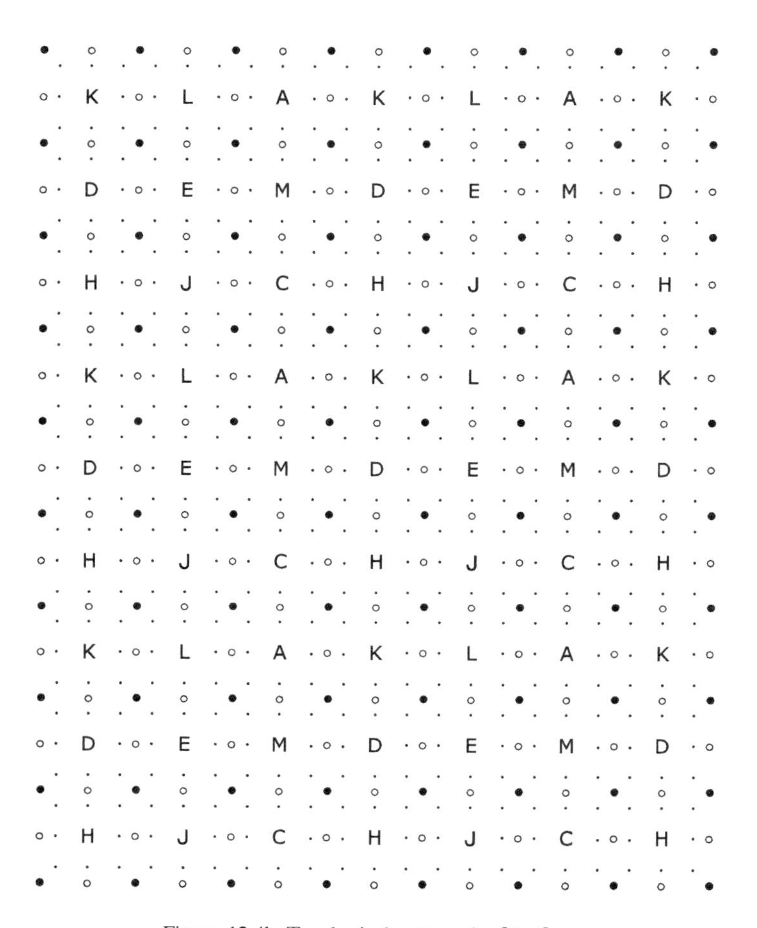

Figure 12.4b Torohedral pattern for $[4, 4]_{(3,0)}$

Likewise, of the 12000 affinities of $\mathbb{F}_5\mathrm{A}^2$, only those that permute the 300 copies of the fundamental region are isometries of $\mathbb{F}_5\mathrm{E}^2$. The isometry group $\mathrm{E}_2(\mathbb{F}_5) \cong [3, 6]_{(5,0)}$ is of order 300, generated by reflections ρ_0, ρ_1, and ρ_2, satisfying the relations (12.3.3) with the additional relation

$$(\rho_0\rho_1\rho_2\rho_1\rho_2\rho_1)^5 = 1. \tag{12.4.3}$$

The direct subgroup $\mathrm{E}_2^+(\mathbb{F}_5)$ is of order 150. Triangles come in three shapes, namely, (6 3 2), (3 3 3), and (6 6 3/2), and four sizes, with

Figure 12.4c Torohedral pattern for $[3, 6]_{(5,0)}$

areas proportional to 1, 2, 3, and 4 (mod 5). The similarity and direct similarity groups $\tilde{\mathrm{E}}_2(\mathbb{F}_5)$ and $\tilde{\mathrm{E}}_2^+(\mathbb{F}_5)$ have orders 1200 and 600. The finite geometry $\mathbb{F}_5\mathrm{E}^2$ is treated at length by O'Beirne (1965, pp. 77–98).

B. *Finite projective planes.* By adjoining a "point at infinity" to each line, any finite affine plane $\mathbb{F}_q\mathrm{A}^2$ can be extended to a finite projective plane $\mathbb{F}_q\mathrm{P}^2$, which is a self-dual configuration $(q^2 + q + 1)_{q+1}$ of

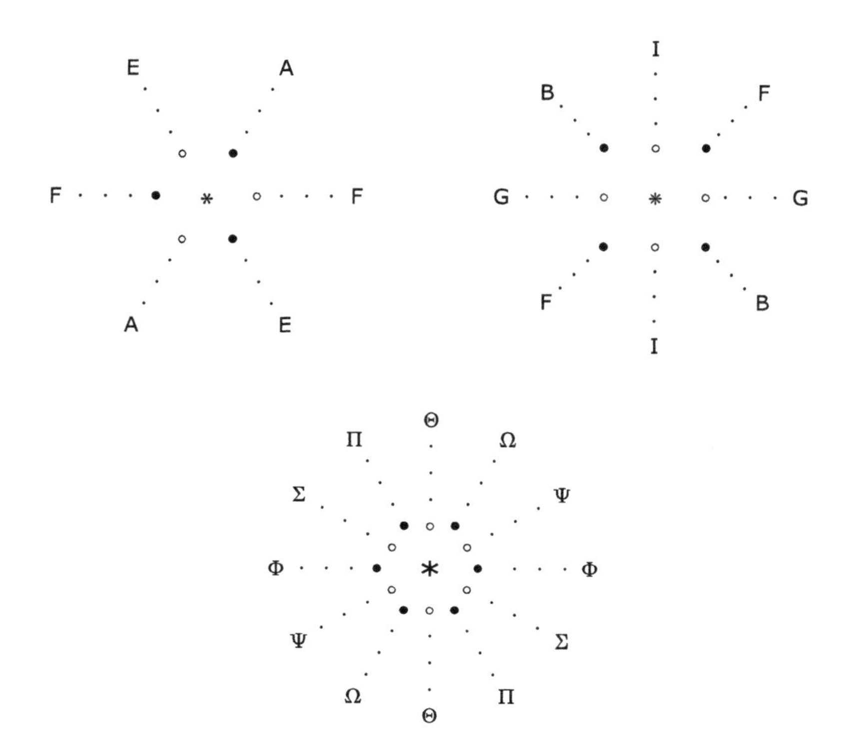

Figure 12.4d Points at infinity for torohedral patterns

q^2+q+1 points and q^2+q+1 lines, with $q+1$ lines through each point and $q+1$ points on each line. With q equal to 2, 3, or 5, this is a regular configuration 7_3, 13_4, or 31_6, the first of which will be recognized as that of Fano.

The directions of the additional three, four, or six points at infinity required for the projective planes $\mathbb{F}_2\mathrm{P}^2$, $\mathbb{F}_3\mathrm{P}^2$, and $\mathbb{F}_5\mathrm{P}^2$ are indicated in Figure 12.4d. These extra points are denoted either by other capital letters or, in the case where twenty-five Latin letters have already been used, by letters from the Greek alphabet. It is convenient here to denote lines by corresponding lowercase letters, with the three projective planes having the following incidence tables, showing which points lie on each line.

$\mathbb{F}_2\mathrm{P}^2 = 7_3$

a	b	c	d	e	f	g
A	G	F	E	D	C	B
B	A	G	F	E	D	C
D	C	B	A	G	F	E

$\mathbb{F}_3\mathrm{P}^2 = 13_4$

a	b	c	d	e	f	g	h	i	j	k	l	m
A	M	L	K	J	I	H	G	F	E	D	C	B
B	A	M	L	K	J	I	H	G	F	E	D	C
D	C	B	A	M	L	K	J	I	H	G	F	E
J	I	H	G	F	E	D	C	B	A	M	L	K

$\mathbb{F}_5\mathrm{P}^2 = 31_6$

a	b	c	d	θ	e	f	g	h	i	j	k	l	m	n	o	π	p	q	r	σ	s	t	u	v	x	y	φ	z	ψ	ω
A	Ω	Ψ	Z	Φ	Y	X	V	U	T	S	Σ	R	Q	P	Π	O	N	M	L	K	J	I	H	G	F	E	Θ	D	C	B
Θ	D	C	B	A	Ω	Ψ	Z	Φ	Y	X	V	U	T	S	Σ	R	Q	P	Π	O	N	M	L	K	J	I	H	G	F	E
K	J	I	H	G	F	E	Θ	D	C	B	A	Ω	Ψ	Z	Φ	Y	X	V	U	T	S	Σ	R	Q	P	Π	O	N	M	L
M	L	K	J	I	H	G	F	E	Θ	D	C	B	A	Ω	Ψ	Z	Φ	Y	X	V	U	T	S	Σ	R	Q	P	Π	O	N
N	M	L	K	J	I	H	G	F	E	Θ	D	C	B	A	Ω	Ψ	Z	Φ	Y	X	V	U	T	S	Σ	R	Q	P	Π	O
R	Q	P	Π	O	N	M	L	K	J	I	H	G	F	E	Θ	D	C	B	A	Ω	Ψ	Z	Φ	Y	X	V	U	T	S	Σ

By interchanging capital and lowercase letters, the same tables can be used to tell which lines pass through each point.

C. *Coordinates and matrices.* The points and lines of each finite projective plane can be assigned homogeneous coordinates in the finite field $\mathbb{F}_2$, $\mathbb{F}_3$, or $\mathbb{F}_5$. With the line at infinity d, i, or o taken as the unit line [1, 1, 1], compatible choices for the reference points

$$(1,\ 0,\ 0), \quad (0,\ 1,\ 0), \quad (0,\ 0,\ 1), \quad \text{and} \quad (1,\ 1,\ 1)$$

are B, C, G, and D for $\mathbb{F}_2\mathrm{P}^2$; E, K, M, and I for $\mathbb{F}_3\mathrm{P}^2$; and N, P, Z, and O for $\mathbb{F}_5\mathrm{P}^2$. Point and line coordinates for each of the planes are listed below. A point with coordinates (x_1, x_2, x_3) lies on a line with coordinates $[u_1, u_2, u_3]$ if and only if $x_1u_1 + x_2u_2 + x_3u_3$ is congruent to zero modulo 2, 3, or 5, respectively.

Any 3×3 invertible matrix over $\mathbb{F}_2$, $\mathbb{F}_3$, or $\mathbb{F}_5$ defines a projective collineation of $\mathbb{F}_2\mathrm{P}^2$, $\mathbb{F}_3\mathrm{P}^2$, or $\mathbb{F}_5\mathrm{P}^2$. Those collineations that take the unit line [1, 1, 1] into itself are affinities of the associated affine plane. Among the affinities, those that transform the reference triangle BCG, EKM, or NPZ into a triangle of the same size and shape are isometries of one of the finite Euclidean planes $\mathbb{F}_2\mathrm{E}^2$, $\mathbb{F}_3\mathrm{E}^2$, or $\mathbb{F}_5\mathrm{E}^2$.

The group $\mathrm{E}_2(\mathbb{F}_q)$ of isometries of a finite plane $\mathbb{F}_q\mathrm{E}^2$ is the extension of the dihedral group $[q+1] \cong D_{q+1}$ of rotations and reflections fixing one point by the abelian group $\mathrm{T}^2(\mathbb{F}_q) \cong \mathrm{C}_q \times \mathrm{C}_q$ of translations, i.e., the semidirect product $\mathrm{T}^2(\mathbb{F}_q) \rtimes [q+1]$, of order $2(q+1)q^2$. If $[q+1]$ is replaced by the dihedral group $[q^2-1]$ of dilative rotations and dilative reflections fixing a point, we have the group $\tilde{\mathrm{E}}_2(\mathbb{F}_q) \cong \mathrm{T}^2(\mathbb{F}_q) \rtimes [q^2-1]$ of similarities, of order $2(q^2-1)q^2$. When $q = 2$, every similarity is an isometry.

For the plane $\mathbb{F}_2\mathrm{E}^2$, the group $\mathrm{E}_2^+(\mathbb{F}_2)$ of direct isometries is generated by a rotation σ : DB $\mapsto$ DC of period 3 about the point

$\mathbb{F}_2\mathrm{P}^2$

Points	Lines
A (0, 1, 1)	a [0, 1, 1]
B (1, 0, 0)	b [1, 0, 0]
C (0, 1, 0)	c [0, 1, 0]
D (1, 1, 1)	d [1, 1, 1]
E (1, 1, 0)	e [1, 1, 0]
F (1, 0, 1)	f [1, 0, 1]
G (0, 0, 1)	g [0, 0, 1]

$\mathbb{F}_3\mathrm{P}^2$

Points	Lines
A (1, 1, 2)	a [1, 1, 2]
B (1, 2, 0)	b [1, 2, 0]
C (1, 1, 0)	c [1, 1, 0]
D (1, 0, 1)	d [1, 0, 1]
E (1, 0, 0)	e [1, 0, 0]
F (0, 1, 2)	f [0, 1, 2]
G (2, 0, 1)	g [2, 0, 1]
H (1, 2, 1)	h [1, 2, 1]
I (1, 1, 1)	i [1, 1, 1]
J (0, 1, 1)	j [0, 1, 1]
K (0, 1, 0)	k [0, 1, 0]
L (2, 1, 1)	l [2, 1, 1]
M (0, 0, 1)	m [0, 0, 1]

$\mathbb{F}_5\mathrm{P}^2$

Points	Lines
A (0, 1, 2)	a [0, 1, 2]
B (4, 1, 1)	b [4, 1, 1]
C (2, 0, 1)	c [2, 0, 1]
D (1, 1, 0)	d [1, 1, 0]
Θ (1, 3, 1)	θ [1, 3, 1]
E (0, 1, 1)	e [0, 1, 1]
F (1, 3, 2)	f [1, 3, 2]
G (2, 1, 0)	g [2, 1, 0]
H (1, 4, 1)	h [1, 4, 1]
I (1, 2, 3)	i [1, 2, 3]
J (3, 2, 1)	j [3, 2, 1]
K (2, 3, 1)	k [2, 3, 1]
L (2, 1, 1)	l [2, 1, 1]
M (3, 1, 2)	m [3, 1, 2]
N (1, 0, 0)	n [1, 0, 0]
O (1, 1, 1)	o [1, 1, 1]
Π (1, 4, 0)	π [1, 4, 0]
P (0, 1, 0)	p [0, 1, 0]
Q (1, 0, 1)	q [1, 0, 1]
R (1, 1, 2)	r [1, 1, 2]
Σ (3, 1, 1)	σ [3, 1, 1]
S (0, 2, 1)	s [0, 2, 1]
T (1, 2, 0)	t [1, 2, 0]
U (1, 2, 1)	u [1, 2, 1]
V (2, 1, 3)	v [2, 1, 3]
X (1, 0, 2)	x [1, 0, 2]
Y (1, 1, 4)	y [1, 1, 4]
Φ (4, 0, 1)	ϕ [4, 0, 1]
Z (0, 0, 1)	z [0, 0, 1]
Ψ (1, 1, 3)	ψ [1, 1, 3]
Ω (0, 1, 4)	ω [0, 1, 4]

D and two translations $\tau_1 : \mathsf{D} \mapsto \mathsf{B}$ and $\tau_2 : \mathsf{D} \mapsto \mathsf{C}$, both of period 2, defined by the matrices

$$S = \begin{pmatrix} 0 & 1 & 0 \\ 0 & 0 & 1 \\ 1 & 0 & 0 \end{pmatrix}, \quad T_1 = \begin{pmatrix} 1 & 1 & 1 \\ 0 & 0 & 1 \\ 0 & 1 & 0 \end{pmatrix}, \quad T_2 = \begin{pmatrix} 0 & 0 & 1 \\ 1 & 1 & 1 \\ 1 & 0 & 0 \end{pmatrix}.$$

If we adjoin the reflection ρ in the line DC, defined by the matrix

$$R = \begin{pmatrix} 0 & 1 & 0 \\ 1 & 0 & 0 \\ 0 & 0 & 1 \end{pmatrix},$$

we obtain the full isometry group $\mathrm{E}_2(\mathbb{F}_2) \cong \mathrm{T}^2(\mathbb{F}_2) \rtimes [3] \cong [3^{[3]}]_{(2,0)}$, of order 24.

The group $\mathrm{E}_2^+(\mathbb{F}_3)$ of direct isometries of the plane $\mathbb{F}_3\mathrm{E}^2$ is generated by a quarter-turn $\sigma : \mathsf{DE} \mapsto \mathsf{DK}$ about the point D and two translations $\tau_1 : \mathsf{D} \mapsto \mathsf{E}$ and $\tau_2 : \mathsf{D} \mapsto \mathsf{K}$, both of period 3, defined by the matrices

$$S = \begin{pmatrix} 0 & 1 & 0 \\ 0 & 0 & 1 \\ 1 & 2 & 1 \end{pmatrix}, \quad T_1 = \begin{pmatrix} 0 & 0 & 1 \\ 2 & 1 & 1 \\ 2 & 0 & 2 \end{pmatrix}, \quad T_2 = \begin{pmatrix} 2 & 1 & 1 \\ 1 & 2 & 1 \\ 1 & 1 & 2 \end{pmatrix}.$$

If we adjoin the reflection ρ in the line DK, defined by the matrix

$$R = \begin{pmatrix} 0 & 0 & 1 \\ 0 & 1 & 0 \\ 1 & 0 & 0 \end{pmatrix},$$

we obtain the full isometry group $\mathrm{E}_2(\mathbb{F}_3) \cong \mathrm{T}^2(\mathbb{F}_3) \rtimes [4] \cong [4,\ 4]_{(3,0)}$, of order 72. If the rotation σ is replaced by the dilative rotation $\tilde{\sigma} : \mathsf{DE} \mapsto \mathsf{DL}$, of period 8, defined by the matrix

$$\tilde{S} = \begin{pmatrix} 2 & 1 & 1 \\ 1 & 1 & 2 \\ 2 & 2 & 0 \end{pmatrix},$$

we have the similarity group $\tilde{E}_2(\mathbb{F}_3) \cong T^2(\mathbb{F}_3) \rtimes [8] \cong [[4,\ 4]]_{(3,0)}$, of order 144. As elements of $\mathbb{F}_3$, the entries of any matrix can all be multiplied by 2, thus interchanging 1's and 2's.

The group $E_2^+(\mathbb{F}_5)$ of direct isometries of the plane $\mathbb{F}_5E^2$ is generated by a rotation σ : ON $\mapsto$ OR of period 6 about the point O and two translations τ_1 : O $\mapsto$ N and τ_2 : O $\mapsto$ P, both of period 5, defined by the matrices

$$S = \begin{pmatrix} 1 & 1 & 2 \\ 2 & 1 & 1 \\ 1 & 2 & 1 \end{pmatrix}, \quad T_1 = \begin{pmatrix} 0 & 1 & 1 \\ 3 & 3 & 1 \\ 3 & 1 & 3 \end{pmatrix}, \quad T_2 = \begin{pmatrix} 3 & 3 & 1 \\ 1 & 0 & 1 \\ 1 & 3 & 3 \end{pmatrix}.$$

If we adjoin the reflection ρ in the line OP, defined by the matrix

$$R = \begin{pmatrix} 0 & 0 & 1 \\ 0 & 1 & 0 \\ 1 & 0 & 0 \end{pmatrix},$$

we obtain the full isometry group $E_2(\mathbb{F}_5) \cong T^2(\mathbb{F}_3) \rtimes [6] \cong [3,\ 6]_{(5,0)}$, of order 300. If the rotation σ is replaced by the dilative rotation $\tilde{\sigma}$: ON $\mapsto$ OJ, of period 24, defined by the matrix

$$\tilde{S} = \begin{pmatrix} 3 & 2 & 1 \\ 1 & 3 & 2 \\ 2 & 1 & 1 \end{pmatrix},$$

we have the similarity group $\tilde{E}_2(\mathbb{F}_5) \cong T^2(\mathbb{F}_5) \rtimes [24] \cong {}^2[3,\ 6]_{(5,0)}$, of order 1200. The entries of any matrix can all be multiplied by any nonzero element of $\mathbb{F}_5$.

EXERCISES 12.4

1. Show that in the complex projective plane $\mathbb{C}P^2$ the nine points

$$\begin{array}{lll} (0,\ 1,\ -1), & (0,\ 1,\ -\omega), & (0,\ 1,\ -\omega^2), \\ (-1,\ 0,\ 1), & (-\omega,\ 0,\ 1), & (-\omega^2,\ 0,\ 1), \\ (1,\ -1,\ 0), & (1,\ -\omega,\ 0), & (1,\ -\omega^2,\ 0), \end{array}$$

where $\omega = -\frac{1}{2} + \frac{1}{2}\sqrt{3}\mathrm{i}$ and $\omega^2 = -\frac{1}{2} - \frac{1}{2}\sqrt{3}\mathrm{i}$ are primitive cube roots of unity, are the vertices of a configuration $(9_4, 12_3)$. Find coordinates for the twelve lines.

2. If the reflections ρ_0, ρ_1, and ρ_2 satisfy the relations (12.3.2) for the group [4, 4], describe the isometry $\rho_0\rho_1\rho_2\rho_1$. Show that the extra relation (12.4.2) can be interpreted as the identification of opposite sides of the square region in the pattern for [4, 4] in Figure 12.3b.

3. If the reflections ρ_0, ρ_1, and ρ_2 satisfy the relations (12.3.3) for the group [3, 6], describe the isometry $\rho_0\rho_1\rho_2\rho_1\rho_2\rho_1$. Show that the extra relation (12.4.3) can be interpreted as the identification of opposite sides of the rhombic region in the pattern for [3, 6] in Figure 12.3b.

4. Verify that the first set of matrices S, T_1, and T_2 in this section correspond in the finite plane $\mathbb{F}_2\mathrm{E}^2$ to the rotation $\sigma : \mathsf{DB} \mapsto \mathsf{DC}$ about the point D and the translations $\tau_1 : \mathsf{D} \mapsto \mathsf{B}$ and $\tau_2 : \mathsf{D} \mapsto \mathsf{C}$.

5. Verify that the second set of matrices S, T_1, and T_2 in this section correspond in the finite plane $\mathbb{F}_3\mathrm{E}^2$ to the rotation $\sigma : \mathsf{DE} \mapsto \mathsf{DK}$ about the point D and the translations $\tau_1 : \mathsf{D} \mapsto \mathsf{E}$ and $\tau_2 : \mathsf{D} \mapsto \mathsf{K}$.

6. Verify that the third set of matrices S, T_1, and T_2 in this section correspond in the finite plane $\mathbb{F}_5\mathrm{E}^2$ to the rotation $\sigma : \mathsf{ON} \mapsto \mathsf{OR}$ about the point O and the translations $\tau_1 : \mathsf{O} \mapsto \mathsf{N}$ and $\tau_2 : \mathsf{O} \mapsto \mathsf{P}$.

12.5 EUCLIDEAN COXETER GROUPS

The fundamental region of a finite Coxeter group $[p]$ may be taken to be an arc of a circle or, in the Euclidean plane, an angular region bounded by two rays. In general, a finite group generated by n reflections may be realized geometrically either on the $(n-1)$-sphere S^{n-1}, where its fundamental region is the closure of a Möbius simplex bounded by parts of n great hyperspheres, or in Euclidean n-space E^n, where parts of n hyperplanes bound an infinite fundamental region. In the latter case, whenever the group is crystallographic, it is always possible to find an additional hyperplane, not concurrent with the others, whose angle of intersection with each of the other hyperplanes is a submultiple of π. The $n+1$ hyperplanes (the last one being the "roof") then bound a *Cartan simplex*, reflections in the facets of which generate an infinite (Euclidean) Coxeter group. In some cases there is more than one way to choose the final hyperplane. In other cases more than one additional hyperplane may be adjoined to form a prism, which becomes the fundamental region for a reducible group.

The group $\mathbf{A}_1 \cong \mathbf{B}_1 \cong \mathbf{C}_1 \cong [\ ]$ is generated by a reflection in a point of the Euclidean line E^1. Reflections in two points of E^1, the ends of a straight dion (∞) of length λ, generate the full apeirogonal group

$$\tilde{\mathbf{A}}_1 \cong \tilde{\mathbf{B}}_1 \cong \tilde{\mathbf{C}}_1 \cong [\infty], \tag{12.5.1}$$

defined by (11.2.3), with the Coxeter diagram

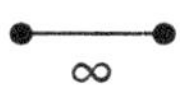

This is the symmetry group of a regular apeirogon $\{\infty\}$, partitioning E^1 into segments of length 2λ.

Extending the crystallographic polygonal groups $\mathbf{A}_2 \cong [3]$, $\mathbf{B}_2 \cong \mathbf{C}_2 \cong [4]$, and $\mathbf{G}_2 \cong [6]$, we obtain the irreducible Euclidean groups with three generators. These are the full apeirohedral groups

$$\tilde{\mathbf{A}}_2 \cong [3^{[3]}], \qquad \tilde{\mathbf{B}}_2 \cong \tilde{\mathbf{C}}_2 \cong [4,\ 4], \qquad \tilde{\mathbf{G}}_2 \cong [3,\ 6] \tag{12.5.2}$$

defined in § 12.3, with triangular fundamental regions corresponding to the Coxeter diagrams

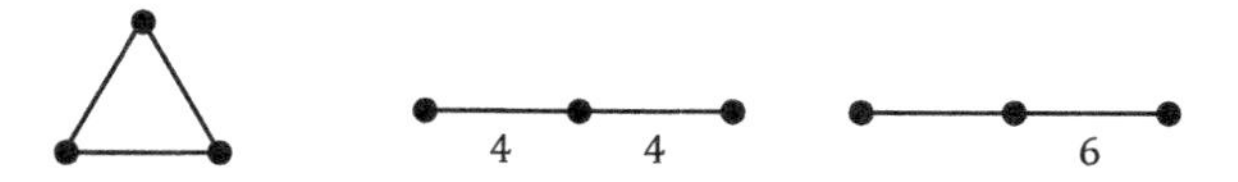

The rectangular group $\mathbf{D}_2 \cong [2] \cong [\ ] \times [\ ]$ is generated by reflections in two perpendicular lines, and its Euclidean extension $\tilde{\mathbf{D}}_2 \cong [2^{[4]}] \cong [\infty,\ 2,\ \infty] \cong [\infty] \times [\infty]$ is generated by reflections in the sides of a rectangle (2 2 2 2).

From the crystallographic polyhedral groups $\mathbf{A}_3 \cong \mathbf{D}_3 \cong [3, 3]$ and $\mathbf{B}_3 \cong \mathbf{C}_3 \cong [3, 4]$ we obtain the three-dimensional Euclidean groups

$$\tilde{\mathbf{A}}_3 \cong \tilde{\mathbf{D}}_3 \cong [3^{[4]}], \quad \tilde{\mathbf{B}}_3 \cong [4, 3^{1,1}], \quad \tilde{\mathbf{C}}_3 \cong [4, 3, 4], \tag{12.5.3}$$

whose tetrahedral fundamental regions correspond to the Coxeter diagrams

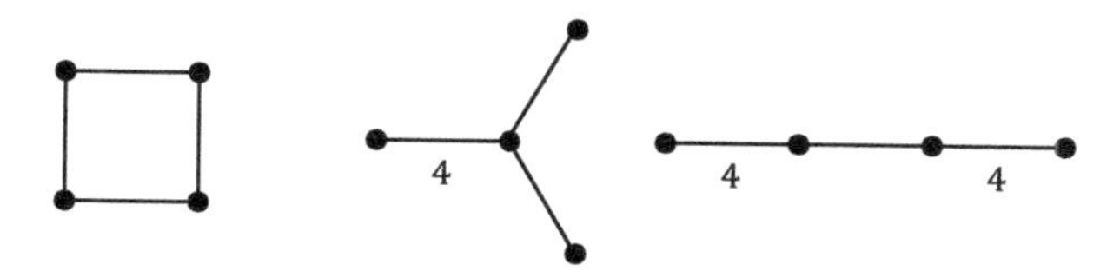

The group $\tilde{\mathbf{C}}_3 \cong [4, 3, 4]$ is the symmetry group of the self-dual regular cubic cellulation $\{4, 3, 4\}$. The group $\tilde{\mathbf{D}}_3 \cong [3^{[4]}]$ occurs as a subgroup of index 2 in $\tilde{\mathbf{B}}_3 \cong [4, 3^{1,1}]$, which in turn is a subgroup of index 2 in $\tilde{\mathbf{C}}_3 \cong [4, 3, 4]$; in addition, $\tilde{\mathbf{C}}_3$ is a subgroup of index 4 in another group $\tilde{\mathbf{B}}_3$.

The crystallographic polychoric groups $\mathbf{A}_4 \cong [3, 3, 3]$, $\mathbf{B}_4 \cong \mathbf{C}_4 \cong [3, 3, 4]$, $\mathbf{D}_4 \cong [3^{1,1,1}]$, and $\mathbf{F}_4 \cong [3, 4, 3]$ extend to the four-dimensional Euclidean groups

$$\tilde{\mathbf{A}}_4 \cong [3^{[5]}], \quad \tilde{\mathbf{B}}_4 \cong [4, 3, 3^{1,1}], \quad \tilde{\mathbf{C}}_4 \cong [4, 3, 3, 4],$$
$$\tilde{\mathbf{D}}_4 \cong [3^{1,1,1,1}], \quad \tilde{\mathbf{F}}_4 \cong [3, 3, 4, 3], \tag{12.5.4}$$

with pentachoric fundamental regions corresponding to the Coxeter diagrams

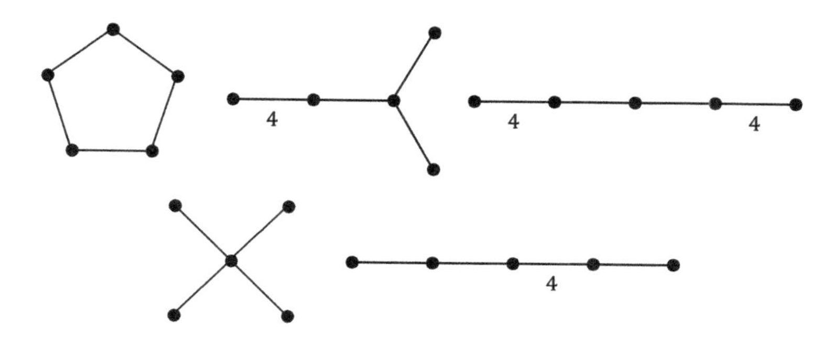

The group $\tilde{\mathbf{C}}_4 \cong [4, 3, 3, 4]$ is the symmetry group of the self-dual regular tesseractic honeycomb $\{4, 3, 3, 4\}$. The group $\tilde{\mathbf{F}}_4 \cong [3, 3, 4, 3]$ is the symmetry group of the dual regular honeycombs $\{3, 3, 4, 3\}$ and $\{3, 4, 3, 3\}$ of 16-cells $\{3, 3, 4\}$ or 24-cells $\{3, 4, 3\}$. The group $\tilde{\mathbf{D}}_4 \cong [3^{1,1,1,1}]$ is a subgroup of index 2 in $\tilde{\mathbf{B}}_4 \cong [4, 3, 3^{1,1}]$, which is itself a subgroup of index 2 in $\tilde{\mathbf{C}}_4 \cong [4, 3, 3, 4]$. Also, $\tilde{\mathbf{C}}_4$ is a subgroup of index 6 in $\tilde{\mathbf{F}}_4$ and of index 8 in another group $\tilde{\mathbf{B}}_4$, while $\tilde{\mathbf{F}}_4$ is a subgroup of index 4 in a larger group $\tilde{\mathbf{F}}_4$.

There are four infinite families of irreducible Euclidean groups—$\tilde{\mathbf{A}}_n$, $\tilde{\mathbf{B}}_n$, $\tilde{\mathbf{C}}_n$, and $\tilde{\mathbf{D}}_n$. The *cyclosymmetric* group of dimension n, an infinite extension $\mathrm{S}_{[n+1]}$ of the symmetric group $\mathrm{S}_{n+1} \cong (n+1) \cdot \mathrm{S}_n$, includes $\tilde{\mathbf{A}}_1 \cong [\infty]$ and

$$\tilde{\mathbf{A}}_n \cong [3^{[n+1]}] \quad (n \geq 2). \tag{12.5.5}$$

For $n \geq 2$, the fundamental region for $\tilde{\mathbf{A}}_n$ is a *cycloscheme*, a simplex whose facets may be cyclically ordered so that any two that are not consecutive are orthogonal. The Coxeter diagram is a circuit of $n+1$ nodes joined by $n+1$ unmarked branches:

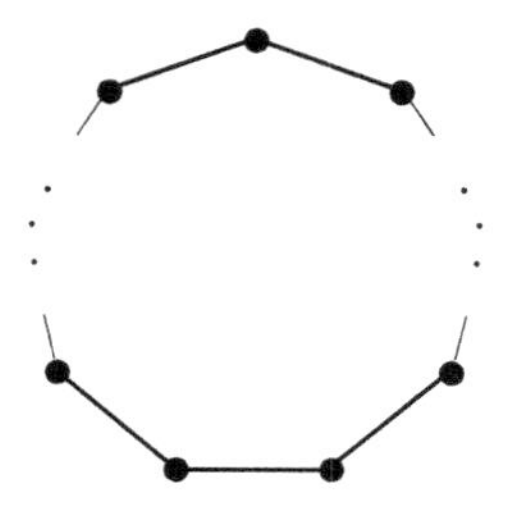

The *orthobisymmetric* group, an infinite extension of the bisymmetric group ${}^2\mathrm{S}_n$, includes $\tilde{\mathbf{C}}_1 \cong [\infty]$, $\tilde{\mathbf{C}}_2 \cong [4, 4]$, and

$$\tilde{\mathbf{C}}_n \cong [4, 3^{n-2}, 4] \quad (n \geq 3). \tag{12.5.6}$$

For $n \geq 2$, the fundamental region for $\tilde{\mathbf{C}}_n$ is an orthoscheme, and the Coxeter diagram is a string graph of $n+1$ nodes and n branches:

The group (12.5.6) is the symmetry group of a regular n-dimensional *orthocomb* $\delta_{n+1} = \{4, 3^{n-2}, 4\}$, the analogue of the apeirogon $\{\infty\}$ and the square tessellation $\{4, 4\}$.

The *metabisymmetric* group, another extension of ${}^2\mathrm{S}_n$ and a halving subgroup of (12.5.6), includes $\tilde{\mathbf{B}}_1 \cong [\infty]$, $\tilde{\mathbf{B}}_2 \cong [4^{1,1}]$, $\tilde{\mathbf{B}}_3 \cong [4, 3^{1,1}]$, and

$$\tilde{\mathbf{B}}_n \cong [4, 3^{n-3}, 3^{1,1}] \quad (n \geq 4). \tag{12.5.7}$$

For $n \geq 3$, the fundamental region for $\tilde{\mathbf{B}}_n$ is a plagioscheme, and the Coxeter diagram is a bifurcated graph of $n + 1$ nodes and n branches:

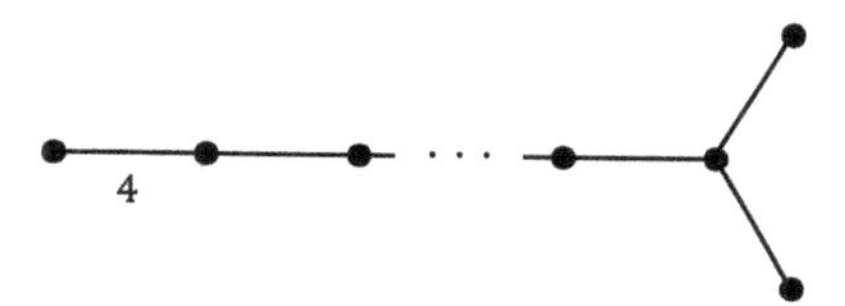

The *metademibisymmetric* group, an infinite extension of ½·${}^2\mathrm{S}_n$ and a halving subgroup of (12.5.7), includes $\tilde{\mathbf{D}}_2 \cong [2^{[4]}]$, $\tilde{\mathbf{D}}_3 \cong [3^{[4]}]$, $\tilde{\mathbf{D}}_4 \cong [3^{1,1,1,1}]$, and

$$\tilde{\mathbf{D}}_n \cong [3^{1,1}, 3^{n-4}, 3^{1,1}] \quad (n \geq 5). \tag{12.5.8}$$

For $n \geq 4$, the fundamental region for $\tilde{\mathbf{D}}_n$ is a plagioscheme, and the Coxeter diagram is a doubly bifurcated graph of $n + 1$ nodes and n unmarked branches:

The group (12.5.7) is the symmetry group of a *demiorthocomb* $\mathrm{h}\delta_{n+1}$, which has half the vertices of δ_{n+1}; its cellular polytopes are β_n's and $\mathrm{h}\gamma_n$'s. The "half-square tessellation" $\mathrm{h}\delta_3$ is another δ_3, and the "half-tesseractic honeycomb" $\mathrm{h}\delta_5$ is the regular 16-cell honeycomb $\{3, 3, 4, 3\}$. An extension of the group (12.5.8) is the symmetry group of a "quarter-hypercubic honeycomb" $\mathrm{k}\delta_{n+1}$ with half the vertices of $\mathrm{h}\delta_{n+1}$.

For $n > 4$, δ_{n+1} is the only regular n-honeycomb. The regular n-orthocomb (or "grid honeycomb") $\delta_{n+1} = \{4, 3^{n-2}, 4\}$ is an n-dimensional *prismatic honeycomb*

$$\delta_2{}^n = \{\infty\}^n = \{\infty\} \times \{\infty\} \times \cdots \times \{\infty\},$$

the rectangular product of n apeirogons of equal partition length.

Each of the three exceptional groups $\mathbf{E}_6 \cong [3^{2,2,1}]$, $\mathbf{E}_7 \cong [3^{3,2,1}]$, and $\mathbf{E}_8 \cong [3^{4,2,1}]$ can be extended to give a corresponding Euclidean group

$$\tilde{\mathbf{E}}_6 \cong [3^{2,2,2}], \qquad \tilde{\mathbf{E}}_7 \cong [3^{3,3,1}], \qquad \tilde{\mathbf{E}}_8 \cong [3^{5,2,1}]. \quad (12.5.9)$$

The respective Coxeter diagrams are

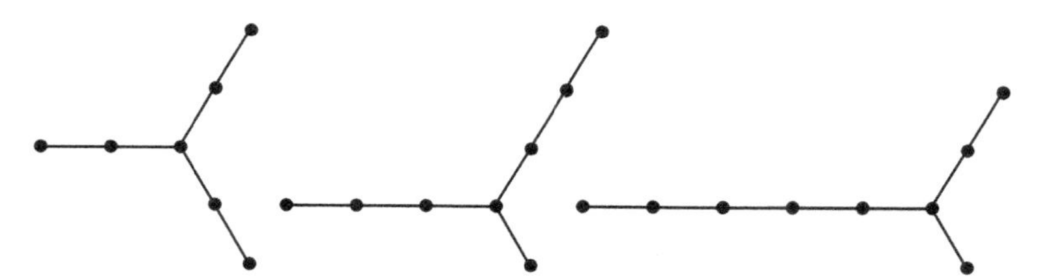

The group $\tilde{\mathbf{E}}_7 \cong [3^{3,3,1}]$ contains $\tilde{\mathbf{A}}_7 \cong [3^{[8]}]$ as a subgroup of index 144, and $\tilde{\mathbf{E}}_8 \cong [3^{5,2,1}]$ contains $\tilde{\mathbf{D}}_8 \cong [3^{1,1}, 3^4, 3^{1,1}]$ as a subgroup of index 270 and $\tilde{\mathbf{A}}_8 \cong [3^{[9]}]$ as a subgroup of index 5760.

For every integer $c > 1$, any irreducible Euclidean group W generated by $n + 1$ reflections is a subgroup of index c^n in an isomorphic "c-plex" group ${}^cW$, with a similar but smaller fundamental region, the edge lengths being in the ratio $c : 1$. Alternatively, we can write W for the larger group and ${}^{1/c}W$ for the subgroup. In a few cases, c can also have an irrational value: the group $[4, 4]$ is a subgroup of index 2 in the isomorphic ${}^{\sqrt{2}}[4, 4] \cong [[4, 4]]$, the group $[3, 6]$ is a subgroup of index 3 in ${}^{\sqrt{3}}[3, 6]$, and $[3, 3, 4, 3]$ is a subgroup of index 4 in ${}^{\sqrt{2}}[3, 3, 4, 3]$.

Up to isomorphism, the metabisymmetric and orthobisymmetric groups $\tilde{\mathbf{B}}_n \cong [4, 3^{n-3}, 3^{1,1}]$ and $\tilde{\mathbf{C}}_n \cong [4, 3^{n-2}, 4]$ have the same subgroups, for $\tilde{\mathbf{C}}_n$ not only contains $\tilde{\mathbf{B}}_n$ as a subgroup of index 2 but is itself a subgroup of the duplex group ${}^2\tilde{\mathbf{B}}_n$ with index 2^{n-1} (Coxeter 1934b, § 12).

The anasymmetric group $\mathbf{A}_{n-1} \cong [3^{n-2}]$ is a subgroup of the cyclosymmetric group $\tilde{\mathbf{A}}_{n-1} \cong [3^{[n]}]$ for all $n \geq 3$. Also, $\tilde{\mathbf{A}}_{n-1} \cong [3^{[n]}]$ is a subgroup of $\tilde{\mathbf{C}}_n \cong [4, 3^{n-2}, 4]$ for all $n \geq 3$ (Schoute 1908). However, $\tilde{\mathbf{A}}_{n-1}$ is a subgroup of $\tilde{\mathbf{C}}_{n-1} \cong [4, 3^{n-3}, 4]$ only for $n = 4$. A regular simplex $\alpha_{n-1} = \{3^{n-2}\}$ can be inscribed in a regular $(n-1)$-orthocomb $\delta_n = \{4, 3^{n-3}, 4\}$ if n is an odd square, a multiple of 4, or the sum of two odd squares (Schoenberg 1937, p. 49), but this fact does not imply that $\mathbf{A}_{n-1}$ is a subgroup of $\tilde{\mathbf{C}}_{n-1}$ except for $n = 4$.

EXERCISES 12.5

1. Find the Gram matrix for the Euclidean line segment $\mathsf{W} = (\infty)$ of angle zero, and show that $\sigma(\mathsf{W}) = 0$.

2. Find the Gram matrices for the Euclidean triangles $\mathsf{P}_3 = (3\ 3\ 3)$, $\mathsf{R}_3 = (4\ 4\ 2)$, and $\mathsf{V}_3 = (6\ 3\ 2)$. Show that the Schläflians $\sigma(\mathsf{P}_3)$, $\sigma(\mathsf{R}_3)$, and $\sigma(\mathsf{V}_3)$ are all zero.

3. By expressing the generators of one group in terms of the generators of the other, show that $\tilde{\mathbf{B}}_3$ is a subgroup of $\tilde{\mathbf{C}}_3$ and that $\tilde{\mathbf{D}}_3$ is a subgroup of $\tilde{\mathbf{B}}_3$.

12.6 OTHER NOTATIONS

In the course of investigating semisimple continuous groups, Hermann Weyl (1925, 1926a,b) was led to consider those finite groups generated by reflections that satisfy the Crystallographic Restriction (i.e., the product of any two distinct reflections is of period 2, 3, 4, or 6); finite crystallographic reflection groups consequently became known as *Weyl groups*. Élie Cartan (1927, 1928) showed that each simple Lie group is related to an infinite (Euclidean) group generated by reflections. Moreover, each such "affine Weyl group" is an extension of one of the finite Weyl groups. This connection was rigorously established by Eduard Stiefel (1942). The symbols introduced by Cartan (1927, pp. 218–224) for families of simple Lie groups provide a concise alternative notation for the corresponding crystallographic Coxeter groups.

A slightly different approach was taken by Ernst Witt (1941, p. 301), who gave each group generated by n reflections in a Euclidean space of dimension n or $n-1$ its own distinctive symbol, using letters from the beginning or the end of the alphabet for finite or infinite groups, respectively. Coxeter (1948, p. 297) adopted Witt's group symbols as names for the simplexes that serve as fundamental regions for the groups (cf. McMullen & Schulte 2002, pp. 72–73).

Witt's notation for the finite (spherical) groups is consistent with that of Cartan in denoting $[3^{n-1}]$ by $\mathbf{A}_n$, $[3^{n-2}, 4]$ by $\mathbf{C}_n$, $[3^{n-4,2,1}]$ $(n = 6, 7, 8)$ by $\mathbf{E}_n$, and $[3, 4, 3]$ by $\mathbf{F}_4$. However, Witt's use of $\mathbf{B}_n$ for $[3^{n-3,1,1}]$, of $\mathbf{D}_2{}^p$ for $[p]$, and of $\mathbf{G}_3$ and $\mathbf{G}_4$ for $[3, 5]$ and $[3, 3, 5]$ obviously conflicts with the Cartan symbols for other groups. The Cartan notation for Lie groups is firmly established, and the corresponding Cartan symbols for finite Coxeter groups are the ones now used by most authors.

Nonetheless, Witt's notation for the infinite groups has certain advantages when used in conjunction with the Cartan symbols for the finite groups. First, each spherical or Euclidean group generated by n reflections has a simple literal symbol, with subscript n. Second, it makes possible a notation for *hyperbolic* groups generated by $n+1$ reflections analogous to the extended Cartan symbols for Euclidean groups generated by $n+1$ reflections.

Each Euclidean group belonging to one of the families (12.5.5), (12.5.8), (12.5.6), and (12.5.7) has a *Witt symbol* as follows (Coxeter 1948, p. 194):

$$\mathbf{W}_2 \cong [\infty], \quad \mathbf{P}_n \cong [3^{[n]}] \quad (n \geq 3),$$
$$\mathbf{Q}_4 \cong [3^{[4]}], \quad \mathbf{Q}_n \cong [3^{1,1},\ 3^{n-5},\ 3^{1,1}] \quad (n \geq 5),$$
$$\mathbf{R}_3 \cong [4,\ 4], \quad \mathbf{R}_n \cong [4,\ 3^{n-3},\ 4] \quad (n \geq 4), \tag{12.6.1}$$
$$\mathbf{S}_3 \cong [4^{1,1}], \quad \mathbf{S}_n \cong [4,\ 3^{n-4},\ 3^{1,1}] \quad (n \geq 3).$$

There are also the five exceptional groups

$$\mathbf{T}_7 \cong [3^{2,2,2}], \quad \mathbf{T}_8 \cong [3^{3,3,1}], \quad \mathbf{T}_9 \cong [3^{5,2,1}],$$
$$\mathbf{U}_5 \cong [3,\ 3,\ 4,\ 3], \tag{12.6.2}$$
$$\mathbf{V}_3 \cong [3,\ 6].$$

We shall regard these as useful alternatives to the extended Cartan symbols employed in the preceding section.

Each hyperbolic Coxeter group is an extension of one or another spherical or Euclidean group and has an appropriate "extended" Witt symbol. In order to distinguish different extensions of the same spherical or Euclidean group, synonyms are used for the names of a few groups. In particular, we have the identifications

$$\mathbf{H}_3 \cong \mathbf{J}_3 \cong \mathbf{K}_3 \cong [3,\ 5], \quad \mathbf{H}_4 \cong \mathbf{K}_4 \cong [3,\ 3,\ 5],$$
$$\mathbf{M}_2 \cong \mathbf{N}_2 \cong \mathbf{O}_2 \cong \mathbf{P}_2 \cong \mathbf{Q}_2 \cong \mathbf{W}_2 \cong [\infty],$$
$$\mathbf{M}_3 \cong \mathbf{N}_3 \cong \mathbf{O}_3 \cong \mathbf{R}_3 \cong \mathbf{S}_3 \cong [4,\ 4], \quad \mathbf{V}_3 \cong \mathbf{Y}_3 \cong \mathbf{Z}_3 \cong [3,\ 6],$$
$$\mathbf{M}_4 \cong \mathbf{N}_4 \cong \mathbf{O}_4 \cong \mathbf{S}_4 \cong [4,\ 3^{1,1}], \tag{12.6.3}$$
$$\mathbf{L}_5 \cong \mathbf{Q}_5 \cong [3^{1,1,1,1}], \quad \mathbf{M}_5 \cong \mathbf{N}_5 \cong \mathbf{O}_5 \cong \mathbf{S}_5 \cong [4,\ 3,\ 3^{1,1}],$$
$$\mathbf{U}_5 \cong \mathbf{X}_5 \cong [3,\ 3,\ 4,\ 3].$$

The Coxeter diagram for a discrete group generated by n reflections consists of n nodes, one for each generator, with two nodes being joined by a branch if the period of the product of the corresponding generators is greater than 2. The branch is unmarked if the period is 3 and is marked 'p' if the period p is greater than 3. Witt (1941, p. 301) proposed the substitution of a $(p - 2)$-tuple branch for a branch marked 'p'. Thus a single branch would indicate a period of 3, a double branch a period of 4, a triple branch a period of 5, and so on.

A still neater version of this scheme, especially well suited to the representation of simple Lie groups and crystallographic Coxeter groups, was devised by E. B. Dynkin (1946, p. 349). In independently inventing what is now often called the *Coxeter-Dynkin diagram*, Dynkin used single, double, and triple branches to denote periods of 3, 4, and 6. The general rule is that two nodes are joined by a branch of multiplicity $4\cos^2 \pi/p = 4 - \sigma(p)$, where p is the period of the product of the generators and $\sigma(p) = 4\sin^2 \pi/p$ is the Schläflian (11.5.2) of the group $[p]$. Nodes corresponding to parallel mirrors $(p = \infty)$ are joined by a quadruple branch. Any other value of p is indicated by a branch marked 'p', as in the equivalent Coxeter diagram.

13

HYPERBOLIC COXETER GROUPS

The fundamental region for an irreducible spherical or Euclidean Coxeter group is the closure of a simplex all of whose dihedral angles are submultiples of π. A discrete group generated by reflections in hyperbolic space whose fundamental region has finite content may be *compact*, *paracompact*, or *hypercompact*, according as the region is the closure of an ordinary simplex, an asymptotic simplex (with one or more absolute vertices), or some other polytope. As first shown by Poincaré, the infinitely many such groups that operate in the hyperbolic plane have important applications to complex analysis. Numerous subgroup relationships hold between hyperbolic Coxeter groups; some of these are quite remarkable. In contrast to the infinite families of crystallographic spherical and Euclidean groups, finitary hyperbolic groups become progressively scarcer in higher dimensions, eventually disappearing altogether.

13.1 PSEUDOHEDRAL GROUPS

Coxeter groups generated by two reflections include the finite dihedral groups $\mathbf{I}_2(p) \cong [p]$, of order $2p$, and the infinite dihedral group $\tilde{\mathbf{A}}_1 \cong \mathbf{W}_2 \cong [\infty]$. On the hyperbolic line H^1, the full pseudogonal group $\tilde{\mathbf{A}}_1(\lambda) \cong [\pi i/\lambda]$, the symmetry group of a regular pseudogon $\{\pi i/\lambda\}$, is generated by reflections in the ends of a dion of length λ.

When $n = 3$ in (11.5.1), a Coxeter group $[(p, q, r)]$, with p, q, and r all greater than or equal to 2, is generated by reflections in the sides of a triangle $(p\ q\ r)$, with angles π/p, π/q, π/r. If $p = q = r > 2$, the group is denoted more concisely by $[p^{[3]}]$. The triangle $(p\ q\ 2)$ is an orthoscheme (right triangle) with Coxeter diagram

and the group is denoted by $[p, q]$ or $[q, p]$.

The group $[(p, q, r)]$ is spherical, Euclidean, or hyperbolic according as the sum of the angles of the triangle $(p\ q\ r)$ is greater than, equal to, or less than π, i.e., according as $p^{-1} + q^{-1} + r^{-1} - 1$ is positive, zero, or negative. As we have seen, the only spherical groups are the orthorectangular group $[2, 2]$, the full ortho-p-gonal groups $[2, p]$ $(p \geq 3)$, and the full polyhedral groups $[3, 3]$, $[3, 4]$, and $[3, 5]$. The only Euclidean groups are the full ortho-apeiral group $[2, \infty]$ and the full apeirohedral groups $[3^{[3]}]$, $[4, 4]$, and $[3, 6]$.

All other groups $[(p, q, r)]$ with p, q, and r finite are generated by reflections in the sides of an ordinary hyperbolic triangle $(p\ q\ r)$. We may also consider groups whose fundamental regions are *asymptotic* (or "ideal") triangles, with one, two, or all three vertices being absolute points, or "cusps." Groups $[(p, q, \infty)]$ $(pq > p + q)$, $[(p, \infty, \infty)]$ $(p \geq 2)$, and $[\infty^{[3]}]$ are generated by reflections in the sides of unicuspal, bicuspal, or tricuspal (i.e., "singly," "doubly," or "trebly" asymptotic) triangles $(p\ q\ \infty)$, $(p\ \infty\ \infty)$, and $(\infty\ \infty\ \infty)$. We note especially the groups $[p, q]$ with $pq > 2(p + q)$, $[p, \infty]$ with $p > 2$, and $[\infty, \infty]$, which are the symmetry groups of regular tessellations of the hyperbolic plane. The area of any ordinary or asymptotic triangle $(p\ q\ r)$ is finite, being equal to its "hyperbolic defect" $(1 - p^{-1} - q^{-1} - r^{-1})\pi$.

When the finite group $\mathbf{I}_2(p) \cong [p]$ $(p \geq 3)$ is generated by reflections in two intersecting lines of the hyperbolic plane, a third line perpendicular to the first and making an angle of π/q with the second exists for all values of q such that $pq > 2(p + q)$. Reflections in the three lines then generate the hyperbolic group $\bar{\mathbf{I}}_2(p, q) \cong [p, q]$. Likewise, if $pqr > pq + pr + qr$, a third line can be chosen that makes angles of π/q and π/r with the other two, and the group $\bar{\mathbf{I}}_2(p, q, r) \cong [(p, q, r)]$ is generated by reflections in the three lines.

In the same fashion, when the Euclidean group $\mathbf{P}_2 \cong \mathbf{W}_2 \cong [\infty]$ is generated by reflections in two parallel lines of H^2, a third line may be drawn making angles of π/p and π/q with the first two, where $p \geq 2$ and $q \geq 3$, forming a unicuspal triangle $(p\ q\ \infty)$. Reflections in the three lines then generate one of the hyperbolic groups $\bar{\mathbf{P}}_2 \cong [3, \infty]$, $\bar{\mathbf{O}}_2 \cong [(3, 3, \infty)]$, $\bar{\mathbf{W}}_2(p) \cong [p, \infty]$, or $\bar{\mathbf{W}}_2(p, q) \cong [(p, q, \infty)]$.

If the third line is parallel to the second and makes an angle of π/p with the first, the three lines form a bicuspal triangle $(p\ \infty\ \infty)$, and the group is $\bar{\mathbf{N}}_2 \cong [\infty, \infty]$, $\hat{\mathbf{P}}_2 \cong [(3, \infty, \infty)]$, or $\hat{\mathbf{W}}_2(p) \cong [(p, \infty, \infty)]$ for some $p > 3$. The group $\bar{\mathbf{M}}_2 \cong [\infty^{[3]}]$ is generated by reflections in the sides of a tricuspal triangle $(\infty\ \infty\ \infty)$.

Among hyperbolic groups generated by reflections in the sides of an ordinary triangle, we find that $[3, 2p]$ $(p > 3)$ contains $[p, 2p]$ as a subgroup of index 3 and $[p^{[3]}]$ as a subgroup of index 6, that $[3, 8]$ also contains $[8, 8]$ as a subgroup of index 6, and that $[3, 7]$ contains $[7^{[3]}]$ as a subgroup of index 24 (Coxeter 1964, pp. 154–155, 157, 162). The group $[4, 2p]$ contains $[(2p, 2p, p)]$ as a subgroup of index 4, $[3, 3p]$ contains $[(3p, p, 3)]$ as a subgroup of index 4, and $[3, 4p]$ contains $[(4p, 4p, p)]$ as a subgroup of index 6. In addition, $[3, 8]$ contains $[(8, 8, 4)]$ as a subgroup of index 12.

Figure 13.1a (cf. Johnson & Weiss 1999a, p. 1317) shows relationships among the groups $\bar{\mathbf{P}}_2 \cong [3, \infty]$, $\bar{\mathbf{Q}}_2 \cong [4, \infty]$, and $\overline{\mathbf{GP}}_2 \cong [6, \infty]$ and certain other groups generated by reflections in the sides of asymptotic triangles. We shall encounter these groups again in the next chapter.

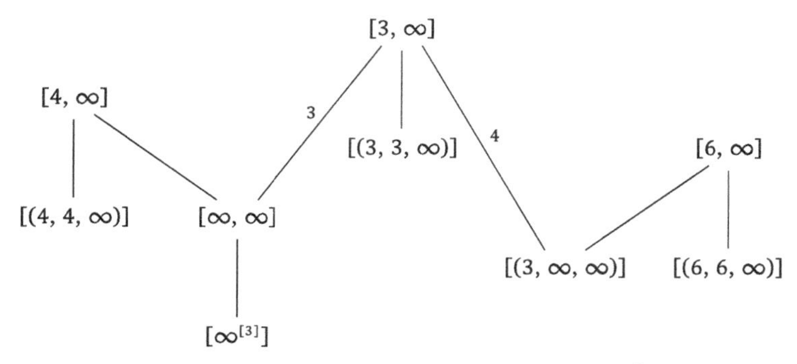

Figure 13.1a Subgroups of reflection groups in H^2

For $p \geq 3$, the hyperbolic Coxeter group $[p, \infty]$ is generated by reflections ρ_0, ρ_1, and ρ_2 in the sides of a unicuspal right triangle with acute angle π/p, satisfying the relations

$$\rho_0{}^2 = \rho_1{}^2 = \rho_2{}^2 = (\rho_0\rho_1)^p = (\rho_0\rho_2)^2 = (\rho_1\rho_2)^\infty = 1. \qquad (13.1.1)$$

The direct subgroup $[p, \infty]^+$ is generated by the rotation $\sigma_1 = \rho_0\rho_1$ and the striation $\sigma_2 = \rho_1\rho_2$, satisfying

$$\sigma_1{}^p = \sigma_2{}^\infty = (\sigma_1\sigma_2)^2 = 1. \qquad (13.1.2)$$

The reflections ρ_0, ρ_1, and ρ_2 may be represented by linear fractional transformations $\cdot\langle R_0\rangle$, $\cdot\langle R_1\rangle$, and $\cdot\langle R_2\rangle$, defined by the matrices

$$R_0 = \begin{pmatrix} 0 & 1 \\ 1 & 0 \end{pmatrix}, \quad R_1 = \begin{pmatrix} -1 & 0 \\ 2\cos\pi/p & 1 \end{pmatrix}, \quad R_2 = \begin{pmatrix} -1 & 0 \\ 0 & 1 \end{pmatrix}.$$

The corresponding direct isometries σ_1 and σ_2 may likewise be represented by linear fractional transformations $\cdot\langle S_1\rangle$ and $\cdot\langle S_2\rangle$, defined by

$$S_1 = \begin{pmatrix} 2\cos\pi/p & 1 \\ -1 & 0 \end{pmatrix} \quad \text{and} \quad S_2 = \begin{pmatrix} 1 & 0 \\ 2\cos\pi/p & 1 \end{pmatrix}.$$

The group $[\infty, \infty]$ is generated by reflections ρ_0, ρ_1, and ρ_2 in the sides of a bicuspal right triangle, satisfying the relations

$$\rho_0{}^2 = \rho_1{}^2 = \rho_2{}^2 = (\rho_0\rho_1)^\infty = (\rho_0\rho_2)^2 = (\rho_1\rho_2)^\infty = 1, \qquad (13.1.3)$$

with the direct subgroup $[\infty, \infty]^+$ being generated by the striations $\sigma_1 = \rho_0\rho_1$ and $\sigma_2 = \rho_1\rho_2$, satisfying

$$\sigma_1{}^\infty = \sigma_2{}^\infty = (\sigma_1\sigma_2)^2 = 1. \tag{13.1.4}$$

The generators may be represented by linear fractional transformations defined by the matrices

$$R_0 = \begin{pmatrix} 0 & 1 \\ 1 & 0 \end{pmatrix}, \quad R_1 = \begin{pmatrix} -1 & 0 \\ 2 & 1 \end{pmatrix}, \quad R_2 = \begin{pmatrix} -1 & 0 \\ 0 & 1 \end{pmatrix},$$

$$S_1 = \begin{pmatrix} 2 & 1 \\ -1 & 0 \end{pmatrix} \quad \text{and} \quad S_2 = \begin{pmatrix} 1 & 0 \\ -2 & 1 \end{pmatrix}.$$

In addition to the infinitely many Coxeter groups generated by reflections in the sides of triangles, there are uncountably many others whose fundamental regions are the closures of hyperbolic polygons with more than three sides. In H^2, polygons exist with any number of sides, or even infinitely many sides, having any specified set of acute angles, in any order, provided only that the angle sum is less than the Euclidean value (Beardon 1979).

If the angles of a convex hyperbolic polygon P are all submultiples of π (or zero), the group generated by reflections in the sides of P will be discrete. Furthermore, since nonadjacent sides of P lie on diverging lines, the removal of any side leaves an unbounded region of infinite area that is the fundamental region for a group generated by reflections in the remaining sides. Discrete groups generated by products of reflections in adjacent sides of such a polygon are what Poincaré called *Fuchsian groups* (after I. L. Fuchs, 1833–1902).

EXERCISES 13.1

1. Find the Gram matrix for the hyperbolic line segment $\mathsf{L} = (\pi i/\lambda)$ of length λ. Show that $\sigma(\mathsf{L}) = -4\sinh^2\lambda$. (Hint: $\cos it = \cosh t$.)

2. The interior angle of a Euclidean pentagon $\{5\}$ is $108°$. Find the interior angle of a face polygon in the spherical tessellation $\{5, 3\}$ and in the hyperbolic tessellations $\{5, 4\}$ and $\{5, 5\}$.

3. Show that the reflections ρ_0, ρ_1, and ρ_2 generating the hyperbolic Coxeter group $\bar{\mathbf{W}}_2(p) \cong [p, \infty]$, satisfying the relations (13.1.1), can be represented by the first set of matrices R_0, R_1, and R_2 given in this section. (Note that both the identity matrix I and the negation matrix $-I$ represent the identity transformation.)

4. Show that the reflections ρ_0, ρ_1, and ρ_2 generating the hyperbolic Coxeter group $\bar{\mathbf{N}}_2 \cong [\infty, \infty]$, satisfying the relations (13.1.3), can be represented by the second set of matrices R_0, R_1, and R_2 given in this section.

5. Show that the group $\bar{\mathbf{N}}_2 \cong [\infty, \infty]$ is a subgroup of both $\bar{\mathbf{P}}_2 \cong [3, \infty]$ and $\bar{\mathbf{Q}}_2 \cong [4, \infty]$. (Hint: Express the generators of $\bar{\mathbf{N}}_2$ in terms of the generators of $\bar{\mathbf{P}}_2$ or $\bar{\mathbf{Q}}_2$.)

6. By subdividing the triangle $(3\ \infty\ \infty)$ into four copies of the triangle $(2\ 3\ \infty)$, show that the group $\hat{\mathbf{P}}_2 \cong [(3, \infty, \infty)]$ is a subgroup of index 4 in the group $\bar{\mathbf{P}}_2 \cong [3, \infty]$.

13.2 COMPACT HYPERBOLIC GROUPS

A Coxeter group W will be called *finitary* if it is generated by reflections in the n facets of a convex polytope P and if its fundamental region (the closure of P) has finite content. A finitary group is also said to be "of finite covolume." If in addition each subgroup generated by $n - 1$ of the reflections is spherical, we say that W is *compact* (cf. Humphreys 1990, pp. 138–144). If W is finitary and if each subgroup generated by $n - 1$ of the reflections is either spherical or Euclidean, including at least one of the latter, we say that W is *paracompact.* If W is finitary but at least one subgroup generated by $n - 1$ of the reflections is hyperbolic, we say that W is *hypercompact.*

A hyperbolic group generated by n reflections in the facets of a convex polytope P is compact if P is an ordinary simplex, paracompact if P is an asymptotic simplex, or hypercompact if P is not a simplex. We distinguish the three cases by calling P a *Lannér simplex*, a *Koszul simplex*, or a *Vinberg polytope*. Compact hyperbolic groups exist only for $n \leq 5$ and paracompact groups only for $n \leq 10$. There are 9 compact and 23 paracompact hyperbolic groups generated by four reflections, 5 compact and 9 paracompact groups generated by five reflections, 12 paracompact groups generated by six reflections, 3 by seven reflections, 4 by eight reflections, 4 by nine reflections, and 3 by ten reflections (Lannér 1950, pp. 51–56; Coxeter & Whitrow 1950, pp. 430–431; Koszul 1968; Chein 1969; cf. Humphreys 1990, pp. 141–144; Davis 2008, p. 105).

The nine compact groups generated by four reflections in hyperbolic 3-space are extensions of the spherical groups $\mathbf{A}_3 \cong \mathbf{D}_3 \cong [3, 3]$, $\mathbf{B}_3 \cong [3, 4]$, and $\mathbf{H}_3 \cong \mathbf{J}_3 \cong \mathbf{K}_3 \cong [3, 5]$. They include the groups

$$\bar{\mathbf{J}}_3 \cong [3, 5, 3], \qquad \bar{\mathbf{K}}_3 \cong [5, 3, 5],$$

$$\overline{\mathbf{BH}}_3 \cong [4, 3, 5], \tag{13.2.1}$$

whose fundamental regions are orthoschemes, with Coxeter diagrams

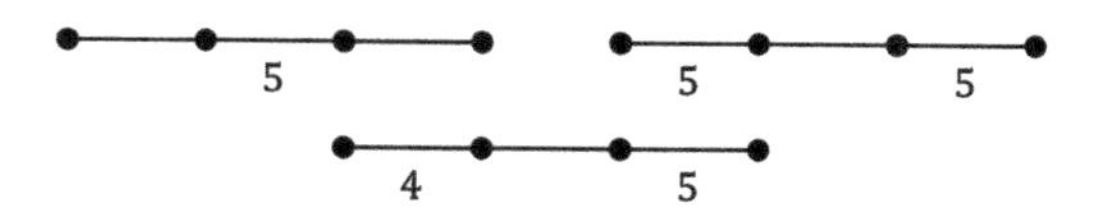

These are the symmetry groups of the self-dual regular cellulations {3, 5, 3} and {5, 3, 5} and the dual cellulations {4, 3, 5} and {5, 3, 4}.

The group $\overline{\mathbf{BH}}_3$ has a halving subgroup

$$\overline{\mathbf{DH}}_3 \cong [5, 3^{1,1}], \tag{13.2.2}$$

whose fundamental region is a plagioscheme, represented by the Coxeter diagram

The fundamental regions of five other groups

$$\widehat{\mathbf{AB}}_3 \cong [(3^3, 4)], \qquad \widehat{\mathbf{BB}}_3 \cong [(3, 4)^{[2]}], \tag{13.2.3}$$
$$\widehat{\mathbf{AH}}_3 \cong [(3^3, 5)], \quad \widehat{\mathbf{BH}}_3 \cong [(3, 4, 3, 5)], \quad \widehat{\mathbf{HH}}_3 \cong [(3, 5)^{[2]}]$$

are cycloschemes, represented by the Coxeter diagrams

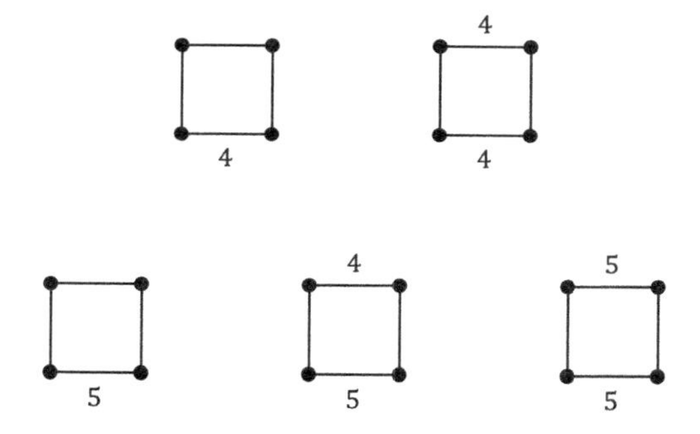

The compact groups generated by reflections in hyperbolic 4-space are extensions of the spherical groups $\mathbf{A}_4 \cong [3, 3, 3]$, $\mathbf{B}_4 \cong [3, 3, 4]$, $\mathbf{D}_4 \cong [3^{1,1,1}]$, $\mathbf{F}_4 \cong [3, 4, 3]$, and $\mathbf{H}_4 \cong \mathbf{K}_4 \cong [3, 3, 5]$. They include the groups

$$\bar{\mathbf{H}}_4 \cong [3, 3, 3, 5], \qquad \bar{\mathbf{K}}_4 \cong [5, 3, 3, 5],$$
$$\overline{\mathbf{BH}}_4 \cong [4, 3, 3, 5], \tag{13.2.4}$$

whose fundamental regions are orthoschemes, with Coxeter diagrams

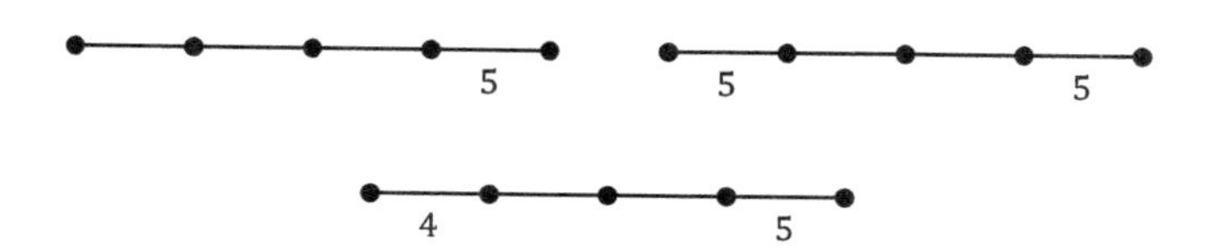

These are the symmetry groups of the dual regular honeycombs {3, 3, 3, 5} and {5, 3, 3, 3}, the self-dual honeycomb {5, 3, 3, 5}, and the dual honeycombs {4, 3, 3, 5} and {5, 3, 3, 4}.

The group $\mathbf{B\overline{H}}_4$ has a halving subgroup

$$\mathbf{D\overline{H}}_4 \cong [5,\ 3,\ 3^{1,1}], \qquad (13.2.5)$$

whose fundamental region is a plagioscheme, represented by the Coxeter diagram

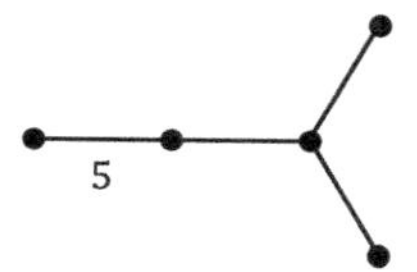

The only other compact group is

$$\mathbf{\widehat{AF}}_4 \cong [(3^4,\ 4)], \qquad (13.2.6)$$

whose fundamental region is a cycloscheme, with Coxeter diagram

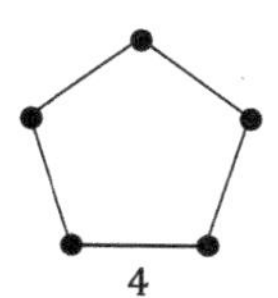

The description of compact hyperbolic groups ends here. For $m \geq 5$ there are no discrete groups generated by reflections in the facets of an ordinary simplex in H^m.

As with direct products of Euclidean groups, hyperbolic Coxeter groups may be generated by reflections in the facets of simplicial prisms (Kaplinskaya 1974; Esselmann 1996), but many other combinatorial types are possible. However, as the number of dimensions increases, fewer and fewer polytopes whose dihedral angles are submultiples of π will fit in hyperbolic space. Coxeter groups with bounded fundamental regions are known to exist in H^m for $m \leq 8$ and are known not to exist for $m \geq 30$ (Vinberg 1984). Other finitary groups are known to exist for $m \leq 21$; there are no finitary groups in H^m for m sufficiently large (Prokhorov 1986). Fuller descriptions of hyperbolic reflection groups are given by Vinberg (1985) and Vinberg & Shvartsman (1988, chaps. 5–6).

Im Hof (1985, 1990) has enumerated all Coxeter groups defined by simplexes or simplexes truncated by one or two hyperplanes whose facets can be serially or cyclically ordered so that any two nonadjacent ones are orthogonal. Groups of this kind exist in H^m for $m \leq 9$. Ruth Kellerhals (1989, pp. 550–568; 1991, pp. 206–210) has calculated the content of the fundamental regions of some of them. The sizes of other Lannér and Koszul simplexes have been determined by Kellerhals (1992, 1995), Ratcliffe & Tschantz (1997, pp. 64–68), and Johnson, Kellerhals, Ratcliffe & Tschantz (1999).

EXERCISES 13.2

1. The dihedral angle of a Euclidean dodecahedron {5, 3} is $\pi - \tan^{-1} 2$, about $116°34'$. Find the dihedral angle of a cell polyhedron in the spherical cellulation {5, 3, 3} and in the hyperbolic cellulations {5, 3, 4} and {5, 3, 5}.

2. The dihedral angle of a Euclidean 120-cell {5, 3, 3} is 144°. Find the dihedral angle of a cellular polychoron in each of the hyperbolic honeycombs {5, 3, 3, 3}, {5, 3, 3, 4}, and {5, 3, 3, 5}.

3. A Coxeter group generated by n reflections is compact if and only if each distinguished subgroup generated by $n - 1$ of the reflections is spherical; the fundamental region of a compact Coxeter group is the closure of a simplex. Use this fact to show that there are no compact hyperbolic Coxeter groups generated by more than five reflections.

13.3 PARACOMPACT GROUPS IN H^3

There are twenty-three discrete groups generated by reflections in the faces of asymptotic tetrahedra in hyperbolic 3-space. Many of the groups are extensions of one of the polyhedral groups $\mathbf{A}_3 \cong \mathbf{D}_3 \cong [3, 3]$, $\mathbf{B}_3 \cong \mathbf{C}_3 \cong [3, 4]$, $\mathbf{H}_3 \cong [3, 5]$, and all are extensions of one of the apeirohedral groups $\mathbf{P}_3 \cong [3^{[3]}]$, $\mathbf{M}_3 \cong \mathbf{N}_3 \cong \mathbf{O}_3 \cong \mathbf{R}_3 \cong [4, 4]$, or $\mathbf{V}_3 \cong \mathbf{Y}_3 \cong \mathbf{Z}_3 \cong [3, 6]$.

Seven such paracompact groups, namely,

$$\bar{\mathbf{R}}_3 \cong [3, 4, 4], \tag{13.3.1}$$

$$\bar{\mathbf{V}}_3 \cong [3, 3, 6], \qquad \overline{\mathbf{BV}}_3 \cong [4, 3, 6], \qquad \overline{\mathbf{HV}}_3 \cong [5, 3, 6],$$

$$\bar{\mathbf{Y}}_3 \cong [3, 6, 3], \qquad \bar{\mathbf{N}}_3 \cong [4, 4, 4], \qquad \bar{\mathbf{Z}}_3 \cong [6, 3, 6],$$

have fundamental regions that are orthoschemes, with Coxeter diagrams

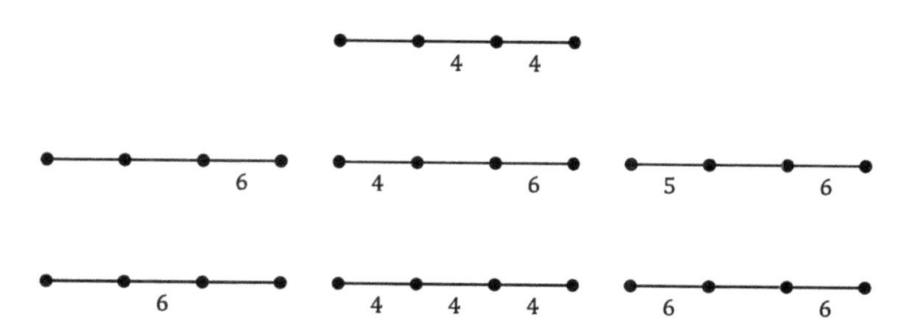

These are the symmetry groups of the regular cellulations

$$\{3, 4, 4\}, \quad \{3, 3, 6\}, \quad \{4, 3, 6\}, \quad \{5, 3, 6\}$$

of asymptotic octahedra, tetrahedra, hexahedra, or dodecahedra; the dual apeirohedral cellulations

$$\{4, 4, 3\}, \quad \{6, 3, 3\}, \quad \{6, 3, 4\}, \quad \{6, 3, 5\}$$

with hexahedral, tetrahedral, octahedral, or icosahedral vertex figures; and the self-dual apeiroasymptotic cellulations

$$\{3, 6, 3\}, \quad \{4, 4, 4\}, \quad \{6, 3, 6\}.$$

All of the groups (13.3.1) except $\bar{\mathbf{Y}}_3$ have at least one halving subgroup, and $\mathbf{B}\bar{\mathbf{V}}_3$ has two different ones. These are the seven groups

$$\begin{aligned} &\bar{\mathbf{O}}_3 \cong [3,\ 4^{1,1}], \\ \bar{\mathbf{P}}_3 \cong [3,\ 3^{[3]}],\ &\mathbf{B}\bar{\mathbf{P}}_3 \cong [4,\ 3^{[3]}],\ \mathbf{H}\bar{\mathbf{P}}_3 \cong [5,\ 3^{[3]}], \\ \mathbf{D}\bar{\mathbf{V}}_3 \cong [6,\ 3^{1,1}],\ &\bar{\mathbf{M}}_3 \cong [4^{1,1,1}],\ \ \mathbf{V}\bar{\mathbf{P}}_3 \cong [6,\ 3^{[3]}], \end{aligned} \tag{13.3.2}$$

with Coxeter diagrams

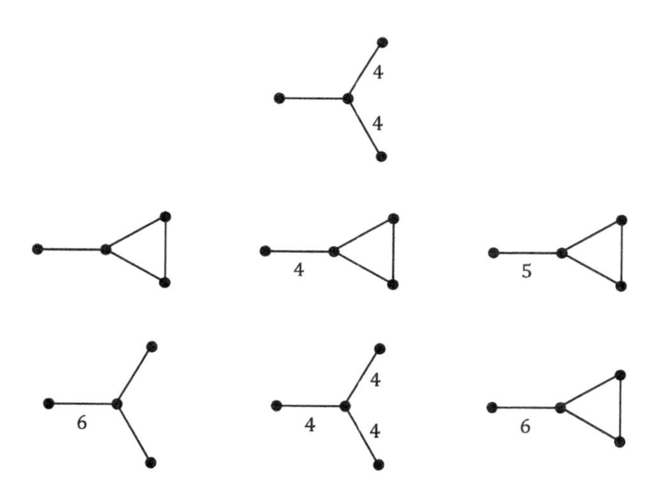

The fundamental regions of seven other groups,

$$\begin{aligned} \mathbf{B}\widehat{\mathbf{R}}_3 \cong [(3^2,\ 4^2)], \quad &\widehat{\mathbf{C}\mathbf{R}}_3 \cong [(3,\ 4^3)], \quad \widehat{\mathbf{R}\mathbf{R}}_3 \cong [4^{[4]}], \\ &\widehat{\mathbf{A}\mathbf{V}}_3 \cong [(3^3,\ 6)],\ \mathbf{B}\widehat{\mathbf{V}}_3 \cong [(3,\ 4,\ 3,\ 6)], \\ &\widehat{\mathbf{H}\mathbf{V}}_3 \cong [(3,\ 5,\ 3,\ 6)],\ \widehat{\mathbf{V}\mathbf{V}}_3 \cong [(3,\ 6)^{[2]}], \end{aligned} \tag{13.3.3}$$

are cycloschemes, with Coxeter diagrams

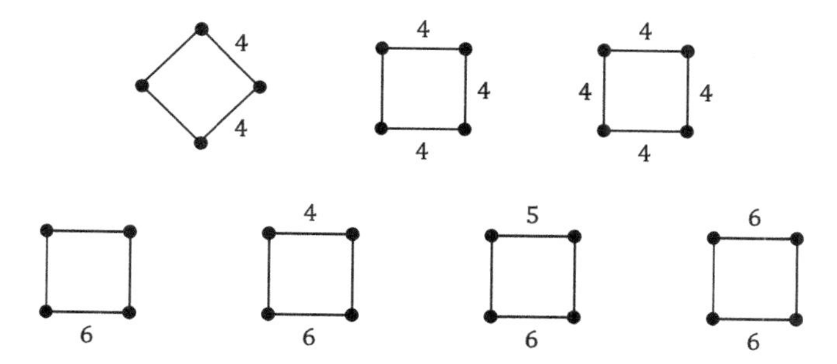

The two remaining groups

$$\overline{\mathbf{DP}}_3 \cong [3^{[\,]\times[\,]}] \quad \text{and} \quad \widehat{\mathbf{PP}}_3 \cong [3^{[3,3]}] \qquad (13.3.4)$$

have fundamental tetrahedra with five or all six dihedral angles equal to $\pi/3$, as represented by the Coxeter diagrams

With four exceptions ($\widehat{\mathbf{CR}}_3$, $\widehat{\mathbf{AV}}_3$, $\widehat{\mathbf{BV}}_3$, $\widehat{\mathbf{HV}}_3$), each of the groups described above is a subgroup or supergroup of one or more others. The group $\overline{\mathbf{HV}}_3 \cong [5, 3, 6]$ contains $\overline{\mathbf{HP}}_3 \cong [5, 3^{[3]}]$ as a subgroup of index 2. The following diagrams (Figure 13.3a) show the relationships among subgroups of $\bar{\mathbf{R}}_3 \cong [3, 4, 4]$ and $\bar{\mathbf{N}}_3 \cong [4, 4, 4]$ and among subgroups of $\bar{\mathbf{V}}_3 \cong [3, 3, 6]$ and $\overline{\mathbf{BV}}_3 \cong [4, 3, 6]$ (cf. Johnson & Weiss 1999a; Felikson 2002).

As noted in § 13.1, there exist hyperbolic polygons with any number of sides, or even infinitely many sides, and any specified set of acute angles. Consequently, the Coxeter groups generated by three reflections do not by any means account for all the discrete groups generated by reflections in H^2.

A somewhat similar situation obtains in three-dimensional space. There exist hyperbolic polyhedra with arbitrarily many faces, all of

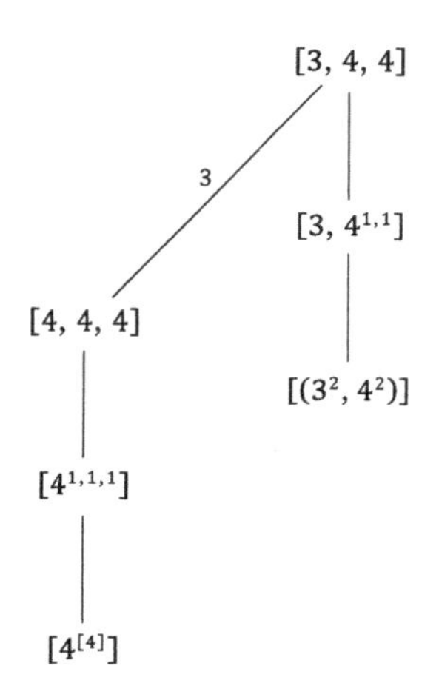

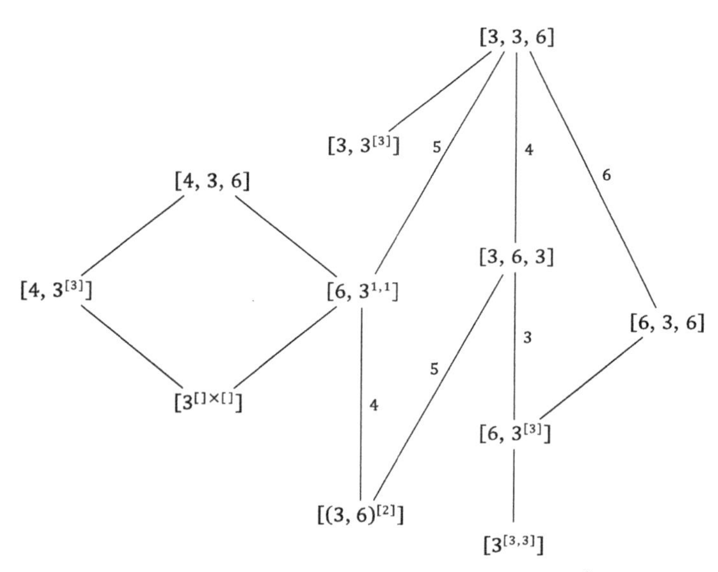

Figure 13.3a Subgroups of reflection groups in H^3

whose dihedral angles are submultiples of π (Andreev 1970a; cf. Davis 2008, pp. 106–107). It follows that, in addition to the thirty-two compact and paracompact Coxeter groups generated by four reflections in H^3, there are infinitely many other discrete groups generated by more than four reflections.

In one essential respect, however, such hypercompact Coxeter groups in a hyperbolic space of three or more dimensions differ from their counterparts in H^2. Two hyperbolic n-gons ($n > 3$) having the same angles, in the same cyclic order, may have sides of different lengths and so need not be congruent. But the shape of a convex hyperbolic polyhedron or higher-dimensional polytope is completely determined by its dihedral angles. This fact is usually called the Mostow Rigidity Theorem (Mostow 1968; cf. Vinberg & Shvartsman 1988, chap. 7, § 1; Ratcliffe 1994/2006, § 11.8). Thus if two groups generated by reflections in H^m ($m \geq 3$) are isomorphic, their fundamental regions are congruent.

EXERCISES 13.3

1. Show that the group $\bar{\mathbf{N}}_3 \cong [4, 4, 4]$ is a subgroup of $\bar{\mathbf{R}}_3 \cong [3, 4, 4]$.
2. Show that the groups $\bar{\mathbf{Y}}_3 \cong [3, 6, 3]$ and $\bar{\mathbf{Z}}_3 \cong [6, 3, 6]$ are both subgroups of $\bar{\mathbf{V}}_3 \cong [3, 3, 6]$.

13.4 PARACOMPACT GROUPS IN H^4 AND H^5

In addition to the five compact Coxeter groups in hyperbolic 4-space, there are nine groups generated by reflections in the five facets of an asymptotic simplex. Only one group in H^4 has an orthoscheme for a fundamental region; this is

$$\bar{\mathbf{R}}_4 \cong [3, 4, 3, 4], \tag{13.4.1}$$

with the Coxeter diagram

This is the symmetry group of the dual asymptotic and apeirohedral regular honeycombs {3, 4, 3, 4} and {4, 3, 4, 3}.

Four groups have fundamental regions that are plagioschemes. These are

$$\bar{\mathbf{S}}_4 \cong [4, 3^{2,1}], \quad \bar{\mathbf{N}}_4 \cong [4\urcorner\, 3\ulcorner\, 3, 4], \quad \bar{\mathbf{M}}_4 \cong [4, 3^{1,1,1}], \quad \bar{\mathbf{O}}_4 \cong [3, 4, 3^{1,1}], \tag{13.4.2}$$

with corresponding Coxeter diagrams

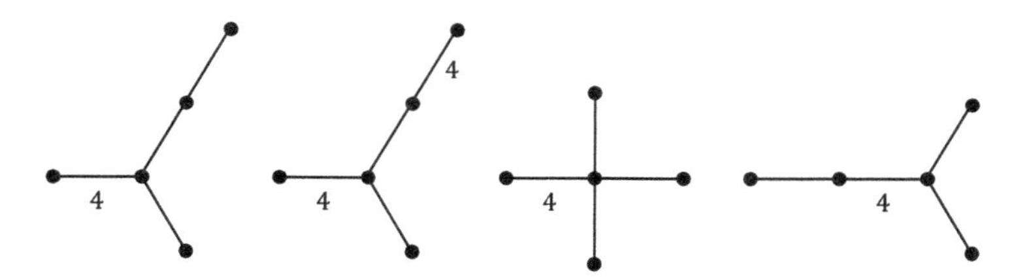

The group $\bar{\mathbf{M}}_4$ is a halving subgroup of $\bar{\mathbf{N}}_4$, and $\bar{\mathbf{O}}_4$ is a halving subgroup of $\bar{\mathbf{R}}_4$.

Four other groups have Coxeter diagrams with circuits:

$$\bar{\mathbf{P}}_4 \cong [3, 3^{[4]}], \; \overline{\mathbf{BP}}_4 \cong [4, 3^{[4]}], \; \overline{\mathbf{DP}}_4 \cong [3^{[3]\times[\,]}], \; \widehat{\mathbf{FR}}_4 \cong [(3^2, 4, 3, 4)], \tag{13.4.3}$$

represented by

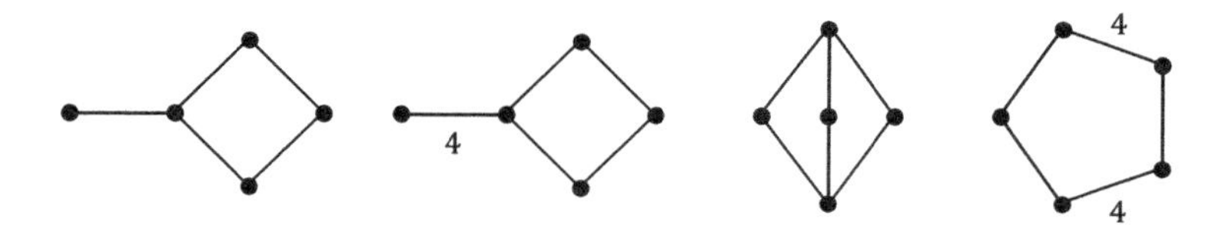

The group $\bar{\mathbf{P}}_4$ is a halving subgroup of $\bar{\mathbf{S}}_4$, $\overline{\mathbf{BP}}_4$ a halving subgroup of $\bar{\mathbf{N}}_4$, and $\overline{\mathbf{DP}}_4$ a halving subgroup of both $\bar{\mathbf{M}}_4$ and $\overline{\mathbf{BP}}_4$.

In hyperbolic 5-space we find twelve groups generated by reflections in the six facets of an asymptotic simplex. Three groups have orthoschemes for fundamental regions; these are

$$\bar{\mathbf{U}}_5 \cong [3, 3, 3, 4, 3], \tag{13.4.4}$$

$$\bar{\mathbf{X}}_5 \cong [3, 3, 4, 3, 3], \qquad \bar{\mathbf{R}}_5 \cong [3, 4, 3, 3, 4].$$

Their Coxeter diagrams are

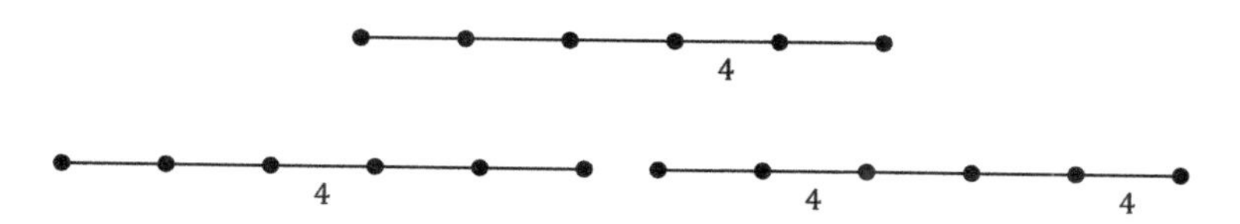

These are the symmetry groups of the dual asymptotic and apeirohedral regular honeycombs

$$\{3, 3, 3, 4, 3\} \quad \text{and} \quad \{3, 4, 3, 3, 3\},$$

the self-dual apeiroasymptotic honeycomb

$$\{3, 3, 4, 3, 3\},$$

and the dual apeiroasymptotic honeycombs

$$\{3, 4, 3, 3, 4\} \quad \text{and} \quad \{4, 3, 3, 4, 3\}.$$

Another six groups, namely,

$$\bar{\mathbf{S}}_5 \cong [4, 3, 3^{2,1}], \quad \bar{\mathbf{N}}_5 \cong [4, 3_{\rceil}\, 3_{\lceil}\, 3, 4], \quad \bar{\mathbf{O}}_5 \cong [3, 4, 3, 3^{1,1}],$$
$$\bar{\mathbf{Q}}_5 \cong [3^{2,1,1,1}], \quad \bar{\mathbf{M}}_5 \cong [4, 3, 3^{1,1,1}], \quad \bar{\mathbf{L}}_5 \cong [3^{1,1,1,1,1}] \tag{13.4.5}$$

have fundamental regions that are plagioschemes, with Coxeter diagrams

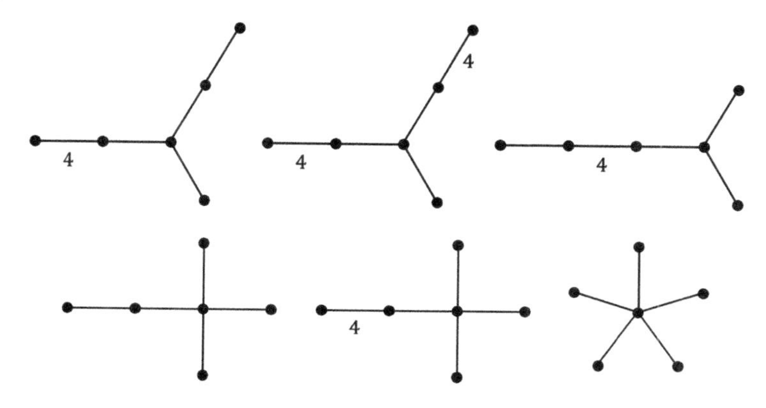

The group $\bar{\mathbf{L}}_5$ is a halving subgroup of $\bar{\mathbf{M}}_5$, and $\bar{\mathbf{M}}_5$ a halving subgroup of $\bar{\mathbf{N}}_5$; $\bar{\mathbf{O}}_5$ is a halving subgroup of $\bar{\mathbf{R}}_5$, and $\bar{\mathbf{Q}}_5$ is a halving subgroup of $\bar{\mathbf{S}}_5$.

Three other groups have Coxeter diagrams with circuits:

$$\bar{\mathbf{P}}_5 \cong [3, 3^{[5]}], \quad \widehat{\mathbf{AU}}_5 \cong [(3^5, 4)], \quad \widehat{\mathbf{UR}}_5 \cong [(3^2, 4)^{[2]}], \qquad (13.4.6)$$

represented by

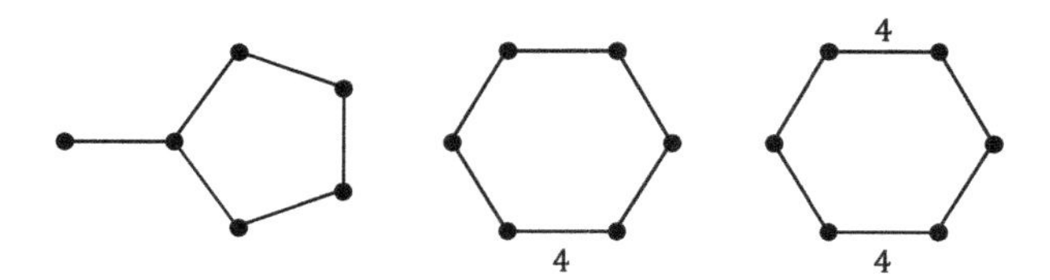

Subgroup relationships among paracompact groups in H^4 and H^5 have been determined by Johnson & Weiss (1999b) and Felikson (2002) and are shown in Figures 13.4a and 13.4b. As before, the lower of two groups joined by a line is a subgroup of the upper, of

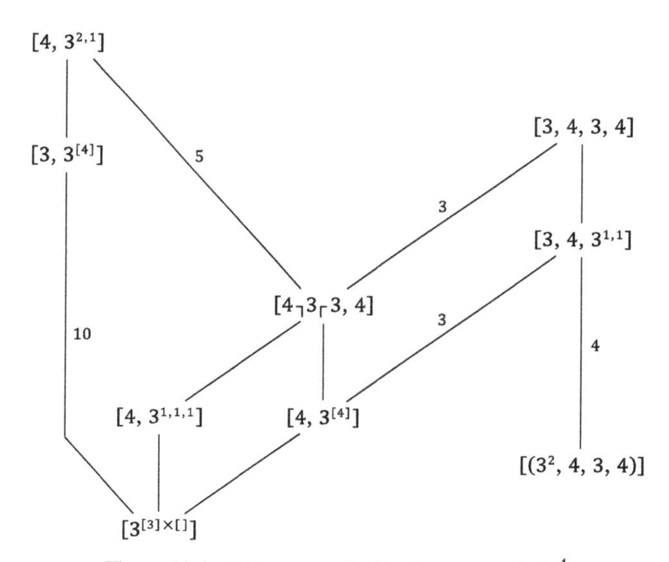

Figure 13.4a Subgroups of reflection groups in H^4

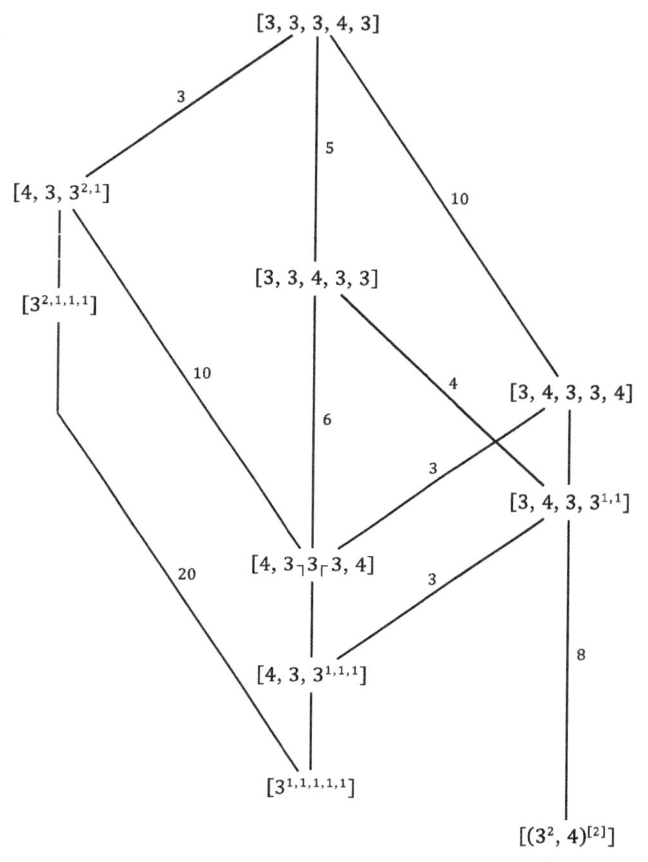

Figure 13.4b Subgroups of reflection groups in H^5

index 2 unless otherwise indicated. All nine of the four-dimensional groups and all but two ($\bar{\mathbf{P}}_5$ and $\widehat{\mathbf{AU}}_5$) of the twelve five-dimensional ones are related to each other.

EXERCISES 13.4

1. Show that $\bar{\mathbf{N}}_4 \cong [4_\urcorner\, 3_\ulcorner\, 3, 4]$ is a subgroup of $\bar{\mathbf{R}}_4 \cong [3, 4, 3, 4]$.
2. Show that $\bar{\mathbf{N}}_5 \cong [4, 3_\urcorner 3_\ulcorner\, 3, 4]$ is a subgroup of $\bar{\mathbf{R}}_5 \cong [3, 4, 3, 3, 4]$.

3. Show that $\bar{\mathbf{X}}_5 \cong [3, 3, 4, 3, 3]$ and $\bar{\mathbf{R}}_5 \cong [3, 4, 3, 3, 4]$ are both subgroups of $\bar{\mathbf{U}}_5 \cong [3, 3, 3, 4, 3]$.

13.5 PARACOMPACT GROUPS IN HIGHER SPACE

The number of Coxeter groups in hyperbolic n-space with simplicial fundamental regions drops sharply when n is greater than 5. We find no more orthoschemes and hence no more regular honeycombs. For $6 \leq n \leq 9$ there are only three or four paracompact groups, and there are none at all for $n > 9$.

In hyperbolic 6-space we have just the three groups

$$\bar{\mathbf{S}}_6 \cong [4, 3^2, 3^{2,1}], \qquad \bar{\mathbf{Q}}_6 \cong [3^{1,1}, 3, 3^{2,1}], \qquad \bar{\mathbf{P}}_6 \cong [3, 3^{[6]}], \tag{13.5.1}$$

with the respective Coxeter diagrams

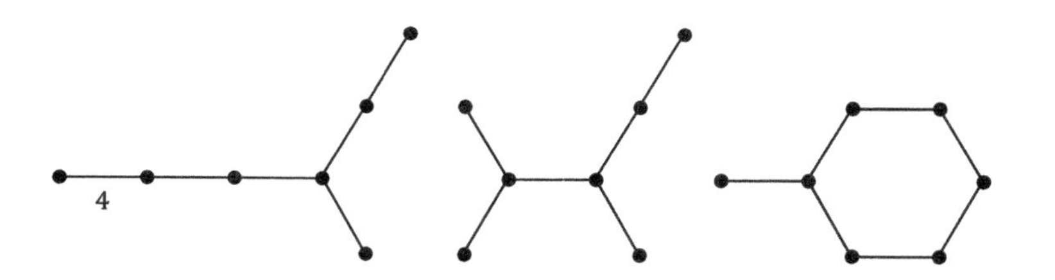

The group $\bar{\mathbf{Q}}_6$ is a halving subgroup of $\bar{\mathbf{S}}_6$.

In hyperbolic 7-space there are four groups, including an extension of the exceptional Euclidean group $\tilde{\mathbf{E}}_6 \cong \mathbf{T}_7 \cong [3^{2,2,2}]$. These are

$$\bar{\mathbf{T}}_7 \cong [3^{3,2,2}],$$
$$\bar{\mathbf{S}}_7 \cong [4, 3^3, 3^{2,1}], \tag{13.5.2}$$
$$\bar{\mathbf{Q}}_7 \cong [3^{1,1}, 3^2, 3^{2,1}], \qquad \bar{\mathbf{P}}_7 \cong [3, 3^{[7]}],$$

with Coxeter diagrams

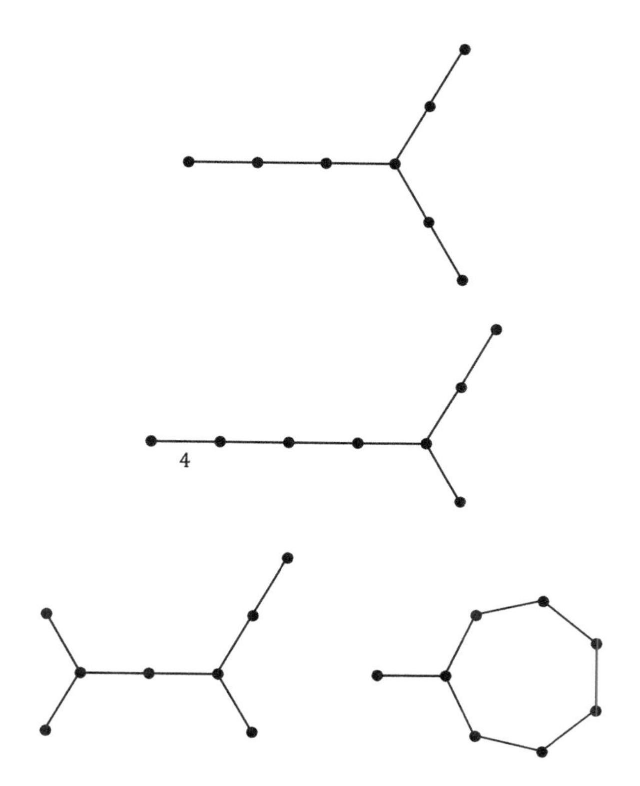

The group $\bar{\mathbf{Q}}_7$ is a halving subgroup of $\bar{\mathbf{S}}_7$.

The four analogous groups in hyperbolic 8-space include an extension of the exceptional Euclidean group $\tilde{\mathbf{E}}_7 \cong \mathbf{T}_8 \cong [3^{3,3,1}]$. These are

$$\bar{\mathbf{T}}_8 \cong [3^{4,3,1}],$$

$$\bar{\mathbf{S}}_8 \cong [4,\ 3^4,\ 3^{2,1}], \tag{13.5.3}$$

$$\bar{\mathbf{Q}}_8 \cong [3^{1,1},\ 3^3,\ 3^{2,1}], \qquad \bar{\mathbf{P}}_8 \cong [3,\ 3^{[8]}],$$

with Coxeter diagrams

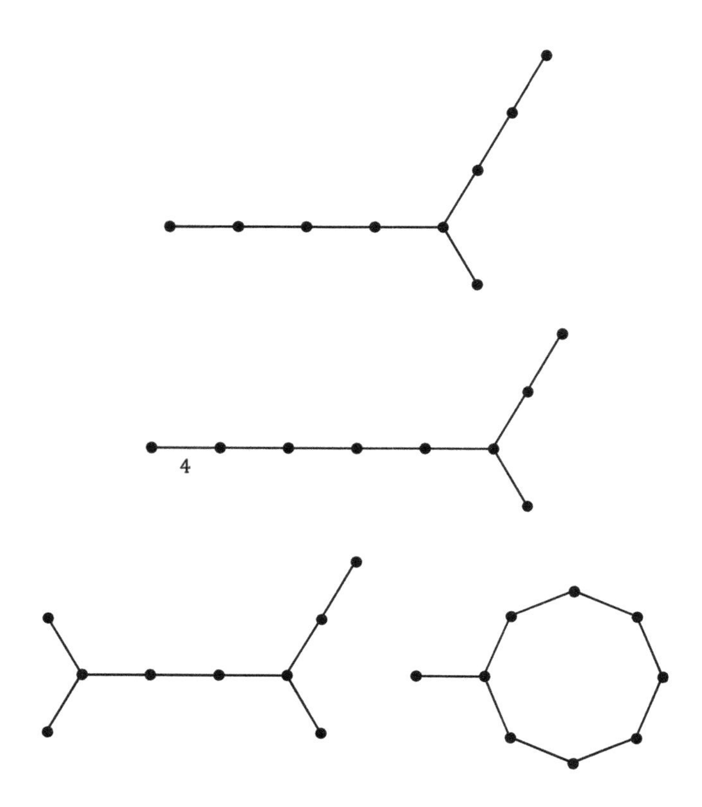

The group $\bar{\mathbf{Q}}_8$ is a halving subgroup of $\bar{\mathbf{S}}_8$. More remarkably, as will be shown in the next section, $\bar{\mathbf{P}}_8$ is a subgroup of index 272 in $\bar{\mathbf{T}}_8$.

There are only three groups in hyperbolic 9-space, including an extension of the exceptional Euclidean group $\tilde{\mathbf{E}}_8 \cong \mathbf{T}_9 \cong [3^{5,2,1}]$:

$$\begin{aligned} \bar{\mathbf{T}}_9 &\cong [3^{6,2,1}], \\ \bar{\mathbf{S}}_9 &\cong [4,\ 3^5,\ 3^{2,1}], \\ \bar{\mathbf{Q}}_9 &\cong [3^{1,1},\ 3^4,\ 3^{2,1}]. \end{aligned} \tag{13.5.4}$$

The respective Coxeter diagrams are

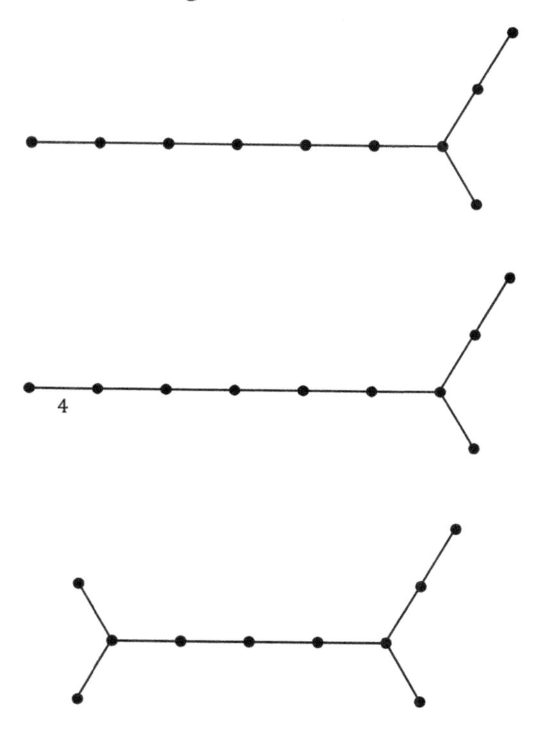

All three groups are related: $\bar{\mathbf{Q}}_9$ is a halving subgroup of $\bar{\mathbf{S}}_9$ and, as we shall see, a subgroup of index 527 in $\bar{\mathbf{T}}_9$.

EXERCISES 13.5

1. Show that the fundamental region of a hyperbolic Coxeter group with more than six generators cannot be an orthoscheme.

2. A Coxeter group generated by n reflections is paracompact if and only if each distinguished subgroup generated by $n-1$ of the reflections is either spherical or Euclidean, including at least one of the latter; the fundamental region of a paracompact hyperbolic Coxeter group is the closure of a simplex. Show that there are no hyperbolic Coxeter groups with simplicial fundamental regions when $n > 10$.

13.6 LORENTZIAN LATTICES

In order to determine how certain hyperbolic Coxeter groups are related, we utilize the connection between hyperbolic and Lorentzian geometry. The metric of Lorentzian $(n + 1)$-space $\mathrm{E}^{n,1}$ is defined by the dual bilinear forms

$$(x\ddot{}y) = x_1y_1 + \cdots + x_ny_n - x_0y_0 \quad \text{and} \quad [u\ddot{}v] = u_1v_1 + \cdots + u_nv_n - u_0v_0.$$

Vectors $(x) = (x_1, \ldots, x_n \mid x_0)$ may be regarded as directed line segments joining points (events) in $\mathrm{E}^{n,1}$; nonzero covectors $[u] = [u_1, \ldots, u_n \mid u_0]$ correspond to (homogeneous) direction numbers for pencils of parallel hyperplanes $[[u], r]$. A Lorentzian vector (x) or covector $[u]$ is said to be

timelike	if	$(x\ddot{}x) < 0$	or	$[u\ddot{}u] > 0$,
lightlike	if	$(x\ddot{}x) = 0$	or	$[u\ddot{}u] = 0$,
spacelike	if	$(x\ddot{}x) > 0$	or	$[u\ddot{}u] < 0$.

The Lorentzian length of a vector (x) is given by

$$|\dot{}(x)\dot{}| = \sqrt{(x\ddot{}x)} = [x_1{}^2 + \cdots + x_n{}^2 - x_0{}^2]^{1/2}. \tag{13.6.1}$$

The length is real and positive if (x) is spacelike, zero if (x) is lightlike, pure imaginary if (x) is timelike. The Lorentzian angle between two spacelike vectors (x) and (y) is given by

$$|(x)\ddot{}(y)| = \cos^{-1} \frac{|(x\ddot{}y)|}{\sqrt{(x\ddot{}x)}\sqrt{(y\ddot{}y)}}. \tag{13.6.2}$$

and that between two timelike covectors $[u]$ and $[v]$ by

$$|[u]\ddot{}[v]| = \cos^{-1} \frac{|(u\ddot{}v)|}{\sqrt{(u\ddot{}u)}\sqrt{(u\ddot{}v)}}. \tag{13.6.3}$$

The vectors or covectors are (pseudo-) orthogonal if and only if $(x\ddot{}y) = 0$ or $[u\ddot{}v] = 0$.

The space at infinity of $\mathrm{E}^{n,1}$ is complete hyperbolic n-space hP^n. Parallel projection takes Lorentzian lines $\langle(x)\rangle$ to hyperbolic points X and pencils of parallel Lorentzian hyperplanes $\langle[u]\rangle$ to hyperbolic hyperplanes $\check{U}$. A point X and a hyperplane $\check{U}$ are incident if and only if the corresponding vector (x) and covector $[u]$ each belong to the annihilator of the other, i.e., if and only if $(x)[u] = 0$. The images of timelike vectors and covectors of $E^{n,1}$ lie in the ordinary region H^n, the images of lightlike ones on the absolute hypersphere, and the images of spacelike ones in the ideal region $\check{H}^n$.

The absolute polar of the ideal point X of hP^n associated with a spacelike vector (x) of $E^{n,1}$ is an ordinary hyperplane $\check{X}$, and the angle $|(x)\ddot{\ }(y)|$ between two spacelike vectors is equal to the angle between the corresponding hyperplanes in H^n. It is also true that the absolute pole of the ideal hyperplane $\check{U}$ of hP^n associated with a spacelike covector $[u]$ of $E^{n,1}$ is an ordinary point U; the imaginary angle between two covectors defines the distance between the corresponding points.

A. *Paracompact groups in hyperbolic 9-space.* Lattices in Lorentzian $(n+1)$-space may be connected with discrete groups generated by reflections in hyperbolic n-space. In particular, in ten-dimensional Lorentzian space $E^{9,1}$, let spacelike vectors $\mathbf{0}$, $\mathbf{1}, \ldots,$ $\mathbf{9}$, and $\bar{\mathbf{8}}$, each of length $\sqrt{2}$, be defined as follows:

$$
\begin{aligned}
\mathbf{0} &= (\ \ 1,\ -1,\ \ 0,\ \ 0,\ \ 0,\ \ 0,\ \ 0,\ \ 0,\ \ 0 \mid 0\ ),\\
\mathbf{1} &= (\ \ 0,\ \ 1,\ -1,\ \ 0,\ \ 0,\ \ 0,\ \ 0,\ \ 0,\ \ 0 \mid 0\ ),\\
\mathbf{2} &= (\ \ 0,\ \ 0,\ \ 1,\ -1,\ \ 0,\ \ 0,\ \ 0,\ \ 0,\ \ 0 \mid 0\ ),\\
\mathbf{3} &= (\ \ 0,\ \ 0,\ \ 0,\ \ 1,\ -1,\ \ 0,\ \ 0,\ \ 0,\ \ 0 \mid 0\ ),\\
\mathbf{4} &= (\ \ 0,\ \ 0,\ \ 0,\ \ 0,\ \ 1,\ -1,\ \ 0,\ \ 0,\ \ 0 \mid 0\ ),\\
\mathbf{5} &= (\ \ 0,\ \ 0,\ \ 0,\ \ 0,\ \ 0,\ \ 1,\ -1,\ \ 0,\ \ 0 \mid 0\ ),\\
\mathbf{6} &= (\ \ 0,\ \ 0,\ \ 0,\ \ 0,\ \ 0,\ \ 0,\ \ 1,\ -1,\ \ 0 \mid 0\ ),\\
\mathbf{7} &= (\ \ 0,\ \ 0,\ \ 0,\ \ 0,\ \ 0,\ \ 0,\ \ 0,\ \ 1,\ -1 \mid 0\ ),\\
\mathbf{8} &= (\ \tfrac{1}{2},\ \tfrac{1}{2},\ \tfrac{1}{2},\ \tfrac{1}{2},\ \tfrac{1}{2},\ \tfrac{1}{2},\ \tfrac{1}{2},\ \tfrac{1}{2},\ \tfrac{1}{2} \mid \tfrac{1}{2}\ ),\\
\mathbf{9} &= (-1,\ -1,\ \ 0,\ \ 0,\ \ 0,\ \ 0,\ \ 0,\ \ 0,\ \ 0 \mid 0\ ),\\
\bar{\mathbf{8}} &= (\ \ 0,\ \ 0,\ \ 0,\ \ 0,\ \ 0,\ \ 0,\ \ 1,\ \ 1,\ \ 1 \mid 1\ ).
\end{aligned}
\tag{13.6.4}
$$

These may be taken as coordinates of ideal points in hyperbolic 9-space absolutely polar to the mirrors of reflections ρ_0, $\rho_1, \ldots,$ ρ_9, and $\rho_{\bar{8}}$, related as in the Coxeter diagrams

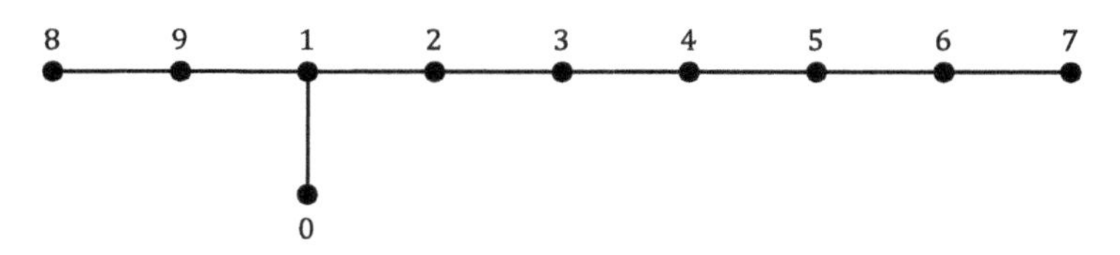

and

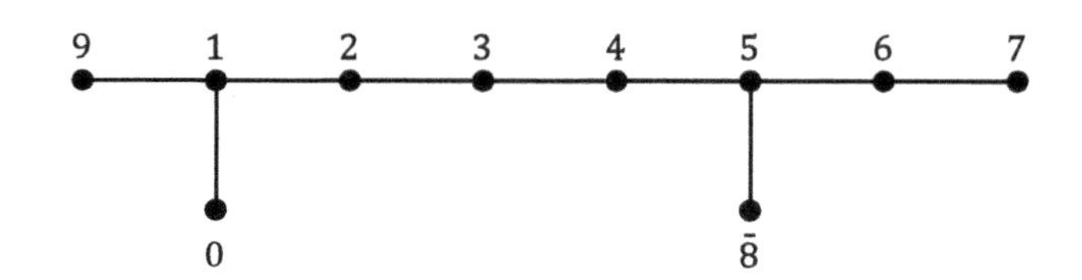

for the nine-dimensional hyperbolic groups $\bar{\mathbf{T}}_9$ and $\bar{\mathbf{Q}}_9$. (Vectors $\mathbf{8}$ and $\bar{\mathbf{8}}$ also make an angle of $\pi/3$.)

John Conway (1983, p. 160) has shown that vectors $\mathbf{0}$ through $\mathbf{9}$ span the even unimodular Lorentzian lattice $\mathrm{II}_{9,1}$. We see that vector $\bar{\mathbf{8}}$ is also in the lattice, since it can be expressed as an integral linear combination of the others:

$$\bar{\mathbf{8}} = 2(\mathbf{8}) + 3(\mathbf{9}) + 2(\mathbf{0}) + 4(\mathbf{1}) + 3(\mathbf{2}) + 2(\mathbf{3}) + (\mathbf{4}). \qquad (13.6.5)$$

Inasmuch as vectors **5**, **6**, and **7** do not appear on the right side of this equation, it follows that the spherical Coxeter groups

$$\mathbf{A}_0 \times \mathbf{D}_6 \cong \langle \rho_{\bar{8}}, \rho_9, \rho_0, \rho_1, \rho_2, \rho_3, \rho_4 \rangle \quad \text{and}$$

$$\mathbf{D}_8 \cong \langle \rho_{\bar{8}}, \rho_9, \rho_0, \rho_1, \rho_2, \rho_3, \rho_4, \rho_5 \rangle$$

are subgroups of

$$\mathbf{E}_7 \cong \langle \rho_8, \rho_9, \rho_0, \rho_1, \rho_2, \rho_3, \rho_4 \rangle \quad \text{and}$$

$$\mathbf{E}_8 \cong \langle \rho_8, \rho_9, \rho_0, \rho_1, \rho_2, \rho_3, \rho_4, \rho_5 \rangle$$

(cf. Coxeter 1934b, table VI); the respective indices are 63 and 135. Likewise, the Euclidean group

$$\tilde{\mathbf{D}}_8 \cong \langle \rho_{\bar{8}}, \rho_9, \rho_0, \rho_1, \rho_2, \rho_3, \rho_4, \rho_5, \rho_6 \rangle \cong \mathbf{Q}_9$$

is a subgroup of

$$\tilde{\mathbf{E}}_8 \cong \langle \rho_8, \rho_9, \rho_0, \rho_1, \rho_2, \rho_3, \rho_4, \rho_5, \rho_6 \rangle \cong \mathbf{T}_9$$

(the index is 270). Finally, the hyperbolic group

$$\bar{\mathbf{Q}}_9 \cong \langle \rho_{\bar{8}}, \rho_9, \rho_0, \rho_1, \rho_2, \rho_3, \rho_4, \rho_5, \rho_6, \rho_7 \rangle$$

is a subgroup of

$$\bar{\mathbf{T}}_9 \cong \langle \rho_8, \rho_9, \rho_0, \rho_1, \rho_2, \rho_3, \rho_4, \rho_5, \rho_6, \rho_7 \rangle$$

(the index is 527).

The index of each subgroup can be determined by dissecting its fundamental region into copies of the fundamental region for the supergroup. The fundamental 9-simplex T for the hyperbolic group $\bar{\mathbf{T}}_9$ is unicuspal, with the absolute vertex section being the fundamental 8-simplex for the Euclidean group $\tilde{\mathbf{E}}_8$. The fundamental 9-simplex Q for $\bar{\mathbf{Q}}_9$ is tricuspal, each absolute vertex section being the fundamental region for a subgroup of $\tilde{\mathbf{E}}_8$, namely, $\tilde{\mathbf{D}}_8$ of index 270, the "semiplex" group ${}^{1/2}\tilde{\mathbf{E}}_8$ of index 256, and $\tilde{\mathbf{E}}_8$ itself of index 1. Together they account for all the absolute vertices of the 527 copies of T into which Q can be dissected.

B. *Paracompact groups in hyperbolic 8-space.* In similar fashion, let vectors $\mathbf{0}$, $\mathbf{1}, \ldots,$ $\mathbf{9}$, each of length $\sqrt{2}$, be defined by

$$
\begin{aligned}
\mathbf{0} &= (\ \tfrac{1}{2},\ \tfrac{1}{2},\ \tfrac{1}{2},\ \tfrac{1}{2},\ -\tfrac{1}{2},\ -\tfrac{1}{2},\ -\tfrac{1}{2},\ -\tfrac{1}{2},\ 0 \mid 0\ ),\\
\mathbf{1} &= (\ -1,\ 1,\ 0,\ 0,\ 0,\ 0,\ 0,\ 0,\ 0 \mid 0\ ),\\
\mathbf{2} &= (\ 0,\ -1,\ 1,\ 0,\ 0,\ 0,\ 0,\ 0,\ 0 \mid 0\ ),\\
\mathbf{3} &= (\ 0,\ 0,\ -1,\ 1,\ 0,\ 0,\ 0,\ 0,\ 0 \mid 0\ ),\\
\mathbf{4} &= (\ 0,\ 0,\ 0,\ -1,\ 1,\ 0,\ 0,\ 0,\ 0 \mid 0\ ),\\
\mathbf{5} &= (\ 0,\ 0,\ 0,\ 0,\ -1,\ 1,\ 0,\ 0,\ 0 \mid 0\ ),\\
\mathbf{6} &= (\ 0,\ 0,\ 0,\ 0,\ 0,\ -1,\ 1,\ 0,\ 0 \mid 0\ ),\\
\mathbf{7} &= (\ 0,\ 0,\ 0,\ 0,\ 0,\ 0,\ -1,\ 1,\ 1 \mid -1\ ),\\
\mathbf{8} &= (\ \tfrac{1}{2},\ \tfrac{1}{2},\ \tfrac{1}{2},\ \tfrac{1}{2},\ \tfrac{1}{2},\ \tfrac{1}{2},\ \tfrac{1}{2},\ \tfrac{1}{2},\ \tfrac{1}{2} \mid \tfrac{1}{2}\ ),\\
\mathbf{9} &= (\ 1,\ 0,\ 0,\ 0,\ 0,\ 0,\ 0,\ -1,\ 0 \mid 0\ ).
\end{aligned}
\tag{13.6.6}
$$

Each of these spacelike vectors belongs to the annihilator of the timelike covector

$$[1,\ 1,\ 1,\ 1,\ 1,\ 1,\ 1,\ 1,\ -4 \mid -4],$$

which is a nine-dimensional subspace of the ten-dimensional vector space. The vectors correspond to ideal points in hyperbolic 8-space whose absolute polars are the mirrors of reflections $\rho_0,\ \rho_1, \ldots,\ \rho_9$, relations among which can be seen in the Coxeter diagrams

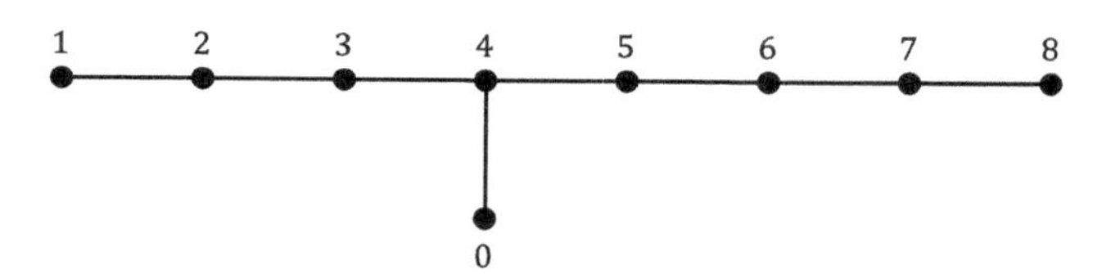

and

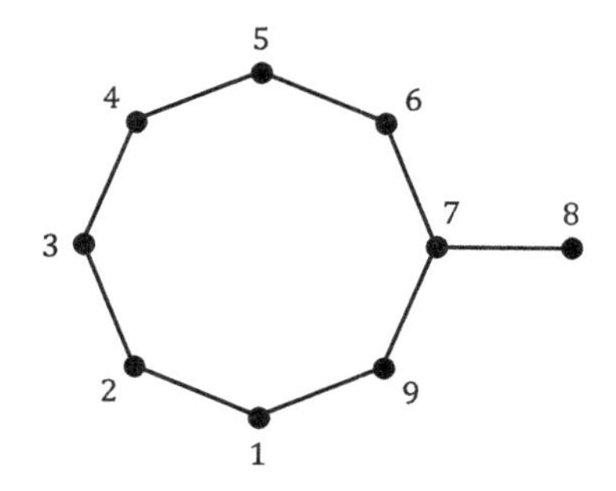

for the eight-dimensional hyperbolic groups $\bar{\mathbf{T}}_8$ and $\bar{\mathbf{P}}_8$. (Vectors **0** and **9** also make an angle of $\pi/3$.)

Now, vector **9** is in the lattice spanned by vectors **0** through **8**, since

$$\mathbf{9} = 2(\mathbf{0}) + (\mathbf{2}) + 2(\mathbf{3}) + 3(\mathbf{4}) + 2(\mathbf{5}) + (\mathbf{6}). \tag{13.6.7}$$

Noting that vectors **1**, **7**, and **8** do not appear on the right side of this equation, we see that the spherical Coxeter groups

$$\mathbf{A}_0 \times \mathbf{A}_5 \cong \langle \rho_9, \rho_2, \rho_3, \rho_4, \rho_5, \rho_6 \rangle \quad \text{and} \quad \mathbf{A}_7 \cong \langle \rho_9, \rho_1, \rho_2, \rho_3, \rho_4, \rho_5, \rho_6 \rangle$$

are subgroups of

$$\mathbf{E}_6 \cong \langle \rho_0, \rho_2, \rho_3, \rho_4, \rho_5, \rho_6 \rangle \quad \text{and} \quad \mathbf{E}_7 \cong \langle \rho_0, \rho_1, \rho_2, \rho_3, \rho_4, \rho_5, \rho_6 \rangle$$

(cf. Coxeter 1934b, table VI); the respective indices are 36 and 72. Likewise, the Euclidean group

$$\tilde{\mathbf{A}}_7 \cong \langle \rho_9, \rho_1, \rho_2, \rho_3, \rho_4, \rho_5, \rho_6, \rho_7 \rangle \cong \mathbf{P}_8$$

is a subgroup of

$$\tilde{\mathbf{E}}_7 \cong \langle \rho_0, \rho_1, \rho_2, \rho_3, \rho_4, \rho_5, \rho_6, \rho_7 \rangle \cong \mathbf{T}_8$$

(the index is 144). Finally, the hyperbolic group

$$\bar{\mathbf{P}}_8 \cong \langle \rho_9, \rho_1, \rho_2, \rho_3, \rho_4, \rho_5, \rho_6, \rho_7, \rho_8 \rangle$$

is a subgroup of

$$\bar{\mathbf{T}}_8 \cong \langle \rho_0, \rho_1, \rho_2, \rho_3, \rho_4, \rho_5, \rho_6, \rho_7, \rho_8 \rangle$$

(the index is 272).

Again, the index of each subgroup can be determined by dissecting its fundamental region. The fundamental 8-simplex T for the hyperbolic group $\bar{\mathbf{T}}_8$ is unicuspal, the absolute vertex section being the fundamental 7-simplex for the Euclidean group $\tilde{\mathbf{E}}_7$. The fundamental 8-simplex P for $\bar{\mathbf{P}}_8$ is bicuspal, corresponding to the group $\tilde{\mathbf{A}}_7$ and the "semiplex" group ${}^{1/2}\tilde{\mathbf{E}}_7$, which are subgroups of

$\tilde{\mathbf{E}}_7$ of index 144 and 128, respectively, thereby accounting for all the absolute vertices of the 272 copies of T into which P can be dissected.

An equivalent presentation of the groups $\bar{\mathbf{T}}_8$ and $\bar{\mathbf{P}}_8$, involving vectors in $\mathrm{E}^{8,1}$ rather than $\mathrm{E}^{9,1}$, is given by Johnson, Kellerhals, Ratcliffe & Tschantz (2002, pp. 141–143). In an earlier paper (1999), the same authors determined the size of the fundamental region of each compact or paracompact hyperbolic Coxeter group (see Table B). In agreement with the foregoing results, the size of the Koszul simplex for $\bar{\mathbf{P}}_8$ is 272 times the size of that for $\bar{\mathbf{T}}_8$, and the Koszul simplex for $\bar{\mathbf{Q}}_9$ is 527 times the size of that for $\bar{\mathbf{T}}_9$.

However, the mere fact that their fundamental regions have commensurable sizes does not imply a subgroup relationship between two groups. For example, the sizes of the Lannér simplexes for the compact groups $\bar{\mathbf{K}}_4 \cong [5, 3, 3, 5]$, $\overline{\mathbf{DH}}_4 \cong [5, 3, 3^{1,1}]$, and $\bar{\mathbf{H}}_4 \cong [3, 3, 3, 5]$ are in the ratio 26 : 17 : 1, suggesting that $\bar{\mathbf{K}}_4$ and $\overline{\mathbf{DH}}_4$ could be subgroups of $\bar{\mathbf{H}}_4$, but this is not the case. In fact, Brent Everitt and Colin Maclachlan have established that $\bar{\mathbf{H}}_4$ has no subgroups whatever of index 26 or 17.

EXERCISES 13.6

1. Use Formula (13.6.1) to show that the length of each of the vectors (13.6.4) is $\sqrt{2}$.
2. Use Formula (13.6.2) to verify that the angle between consecutive vectors (13.6.4) is either $\pi/2$ or $\pi/3$, as indicated in the Coxeter diagrams for the groups $\bar{\mathbf{T}}_9$ and $\bar{\mathbf{Q}}_9$.
3. Show that the length of each of the vectors (13.6.6) is $\sqrt{2}$.
4. Verify that the angle between consecutive vectors (13.6.6) is either $\pi/2$ or $\pi/3$, as indicated in the Coxeter diagrams for the groups $\bar{\mathbf{T}}_8$ and $\bar{\mathbf{P}}_8$.

5. The Euclidean group $\tilde{\mathbf{F}}_4$ contains $\tilde{\mathbf{C}}_4$ as a subgroup of index 6 and an isomorphic group ${}^{1/\sqrt{2}}\tilde{\mathbf{F}}_4$ as a subgroup of index 4. Use this fact to account for all the absolute vertices of the five or ten copies of the fundamental region for the hyperbolic group $\bar{\mathbf{U}}_5$ into which the fundamental region for the subgroup $\bar{\mathbf{X}}_5$ or $\bar{\mathbf{R}}_5$ can be dissected.

14

MODULAR TRANSFORMATIONS

LINEAR FRACTIONAL TRANSFORMATIONS over various rings of real, complex, or quaternionic integers turn out to be closely related to certain paracompact or hypercompact Coxeter groups operating in hyperbolic n-space H^n ($2 \leq n \leq 5$). Some of these correspondences were established by Klein, Poincaré, and others in the late nineteenth century, but other connections have been discovered only recently. The discussion in this chapter and the next draws extensively on collaborative work with Asia Ivić Weiss (Johnson & Weiss 1999a,b). The key concept is that of a *modular group*, in which the linear fractional coefficients are integers of a specified type.

14.1 REAL MODULAR GROUPS

We observed in § 9.3 that isometries of the hyperbolic plane H^2 can be represented as linear fractional transformations $\cdot\langle A\rangle : \mathbb{C} \cup \{\infty\} \to \mathbb{C} \cup \{\infty\}$, where A is a real invertible 2×2 matrix. The group $\mathrm{PO}_{2,1}$ of all isometries of H^2 is isomorphic to the real projective general linear group PGL_2, and the subgroup $\mathrm{P}^+\mathrm{O}_{2,1}$ of direct isometries, with $\det A > 0$, is isomorphic to the real projective special linear group PSL_2.

A. *The rational modular group*. The (integral) *special linear* group $\mathrm{SL}_2(\mathbb{Z})$ of 2×2 integer matrices of determinant 1 is generated by the matrices

$$A = \begin{pmatrix} 0 & 1 \\ -1 & 0 \end{pmatrix} \quad \text{and} \quad B = \begin{pmatrix} 1 & 0 \\ 1 & 1 \end{pmatrix}.$$

(Equivalently, the group is generated by the linear transformations $\cdot A$ and $\cdot B$.) The unit linear group $\bar{\mathrm{S}}\mathrm{L}_2(\mathbb{Z})$ of 2×2 integer matrices of determinant ± 1, which is the same as the general linear group $\mathrm{GL}_2(\mathbb{Z})$ of invertible matrices with integer entries, is generated by the three matrices $A,\ B$, and $L = \backslash -1,\ 1\backslash$.

Various other representations of both groups are possible (e.g., see Coxeter & Moser 1957, pp. 83–88; Magnus 1974, pp. 107–111). In particular, if we let $R_0 = AL,\ R_1 = LB$, and $R_2 = L$, then $\bar{\mathrm{S}}\mathrm{L}_2(\mathbb{Z})$ is generated by the matrices

$$R_0 = \begin{pmatrix} 0 & 1 \\ 1 & 0 \end{pmatrix},\ R_1 = \begin{pmatrix} -1 & 0 \\ 1 & 1 \end{pmatrix},\ R_2 = \begin{pmatrix} -1 & 0 \\ 0 & 1 \end{pmatrix}$$

Likewise, with $S_1 = R_0R_1 = AB$ and $S_2 = R_1R_2 = B^{-1}$, $\mathrm{SL}_2(\mathbb{Z})$ is generated by

$$S_1 = \begin{pmatrix} 1 & 1 \\ -1 & 0 \end{pmatrix} \quad \text{and} \quad S_2 = \begin{pmatrix} 1 & 0 \\ -1 & 1 \end{pmatrix}.$$

The two groups have a common commutator subgroup $\mathrm{SL}_2'(\mathbb{Z})$, of index 2 in $\mathrm{SL}_2(\mathbb{Z})$ and of index 4 in $\bar{\mathrm{S}}\mathrm{L}_2(\mathbb{Z})$, generated by the matrices $S = S_1$ and $W = S_2{}^{-1}S_1S_2$. Each of these matrix groups may be regarded as a group of linear transformations of the lattice $\mathbb{Z}^2$ of points with integer coordinates (x_1, x_2).

The generators ρ_0, ρ_1, ρ_2 of the *projective unit linear* group $\mathrm{P}\bar{\mathrm{S}}\mathrm{L}_2(\mathbb{Z}) \cong \bar{\mathrm{S}}\mathrm{L}_2(\mathbb{Z})/\mathrm{SZ}_2(\mathbb{Z})$, where $\mathrm{SZ}_2(\mathbb{Z}) \cong \{\pm I\}$, satisfy the relations

$$\rho_0{}^2 = \rho_1{}^2 = \rho_2{}^2 = (\rho_0\rho_1)^3 = (\rho_0\rho_2)^2 = (\rho_1\rho_2)^\infty = 1. \qquad (14.1.1)$$

The *projective special linear* group $\mathrm{PSL}_2(\mathbb{Z}) \cong \mathrm{SL}_2(\mathbb{Z})/\mathrm{SZ}_2(\mathbb{Z})$ is equivalent to the (rational) *modular group* of linear fractional transformations

$$\cdot\langle M\rangle : \mathbb{C} \cup \{\infty\} \to \mathbb{C} \cup \{\infty\}, \text{ with } z \mapsto \frac{az+c}{bz+d}, \begin{pmatrix} a & b \\ c & d \end{pmatrix} \in \mathrm{SL}_2(\mathbb{Z}). \tag{14.1.2}$$

As a linear fractional group, $\mathrm{P\bar{S}L}_2(\mathbb{Z})$ is the *extended modular group*. The generators $\sigma_1 = \cdot\langle S_1\rangle$ and $\sigma_2 = \cdot\langle S_2\rangle$ of the modular group $\mathrm{PSL}_2(\mathbb{Z})$ satisfy

$$\sigma_1{}^3 = \sigma_2{}^\infty = (\sigma_1\sigma_2)^2 = 1 \tag{14.1.3}$$

(Klein 1879, pp. 120–121). The modular group's commutator subgroup $\mathrm{PSL}_2{}'(\mathbb{Z}) \cong \mathrm{SL}_2{}'(\mathbb{Z})/\mathrm{SZ}_2(\mathbb{Z})$ is generated by $\sigma = \sigma_1$ and $\omega = \sigma_2{}^{-1}\sigma_1\sigma_2$, satisfying

$$\sigma^3 = \omega^3 = (\sigma\omega)^\infty = 1 \tag{14.1.4}$$

(Coxeter & Moser 1957, p. 86).

The relations (14.1.1) define the paracompact Coxeter group $\bar{\mathbf{P}}_2 \cong [3, \infty]$, which is the symmetry group of the regular hyperbolic tessellation $\{3, \infty\}$ of asymptotic triangles (see Figure 14.1a). The modular group, defined by (14.1.3), is the direct subgroup $\bar{\mathbf{P}}_2^+ \cong [3, \infty]^+$, the rotation group of $\{3, \infty\}$. The commutator subgroup of both $\bar{\mathbf{P}}_2$ and $\bar{\mathbf{P}}_2^+$, defined by (14.1.4), is $\bar{\mathbf{O}}_2^+ \cong [(3, 3, \infty)]^+ \cong [3^+, \infty, 1^+]$.

Besides the halving subgroup $\bar{\mathbf{O}}_2 \cong [(3, 3, \infty)] \cong [3, \infty, 1^+]$, the extended modular group $\bar{\mathbf{P}}_2 \cong [3, \infty]$ has three other subgroups, namely, $\bar{\mathbf{N}}_2 \cong [\infty, \infty], \hat{\mathbf{P}}_2 \cong [(3, \infty, \infty)]$, and $\bar{\mathbf{M}}_2 \cong [\infty^{[3]}]$, of respective indices 3, 4, and 6, that are themselves Coxeter groups. The subgroup $\bar{\mathbf{M}}_2^+ \cong [\infty^{[3]}]^+$, generated by the striations $\sigma\omega$ and $\omega\sigma$, is the free group with two generators.

Linear fractional transformations can be defined not only over the ring $\mathbb{Z}$ of rational integers but also over certain other rings of algebraic integers, which are similarly related to paracompact groups operating in the hyperbolic plane.

Figure 14.1a The modular tessellation $\{3, \infty\}$

B. *Semiquadratic modular groups.* For any square-free integer $d \neq 1$, the *quadratic field* $\mathbb{Q}(\sqrt{d})$ has elements $r + s\sqrt{d}$, where r and s belong to the rational field $\mathbb{Q}$. The *conjugate* of $a = r + s\sqrt{d}$ is $\tilde{a} = r - s\sqrt{d}$, its *trace* tr a is $a + \tilde{a} = 2r$, and its *norm* $N(a)$ is $a\tilde{a} = r^2 - s^2 d$.

A *quadratic integer* is a root of a monic quadratic equation with integer coefficients. For $d \equiv 2$ or $d \equiv 3 \pmod 4$, $r + s\sqrt{d}$ is a quadratic integer if and only if r and s are both rational integers; for $d \equiv 1 \pmod 4$, r and s may be both integers or both halves of odd integers. The quadratic integers of $\mathbb{Q}(\sqrt{d})$ form an integral domain, a two-dimensional algebra $\mathbb{Z}^2(d)$ over $\mathbb{Z}$, whose invertible elements, or *units*, have norm ± 1.

When d is negative, the invertible elements of $\mathbb{Z}^2(d)$ are complex numbers of absolute value 1, and there are only finitely many units: four if $d = -1$, six if $d = -3$, and two in all other cases. When d is positive, $\mathbb{Z}^2(d)$ has an infinite number of units. It may be noted that $\mathbb{Z}^2(d)$ is just the ring $\mathbb{Z}[\delta]$ of numbers of the form $m + n\delta$ (m and n in $\mathbb{Z}$), where $\delta = \sqrt{d}$ if $d \equiv 2$ or $d \equiv 3 \pmod 4$ and $\delta = -\frac{1}{2} + \frac{1}{2}\sqrt{d}$ if $d \equiv 1 \pmod 4$; in the latter case, $\mathbb{Z}^2(d)$ contains $\mathbb{Z}[\sqrt{d}]$ as a proper subdomain.

The group $\bar{\mathrm{S}}\mathrm{L}_2(\mathbb{Z}^2(d))$ of 2×2 invertible matrices over $\mathbb{Z}^2(d)$ has two discrete subgroups analogous to the groups $\bar{\mathrm{S}}\mathrm{L}_2(\mathbb{Z})$ and $\mathrm{SL}_2(\mathbb{Z})$ discussed earlier. In both cases, each matrix entry is either a rational integer or an integral multiple of $\sqrt{d}$; entries of the form $r + s\sqrt{d}$ with $rs \neq 0$ do not occur. The *semiquadratic unit linear* group $\bar{\mathrm{S}}\mathrm{L}_{1+1}(\mathbb{Z}[\sqrt{d}])$ is generated by the matrices

$$R_0 = \begin{pmatrix} 0 & 1 \\ 1 & 0 \end{pmatrix}, \quad R_1 = \begin{pmatrix} -1 & 0 \\ \sqrt{d} & 1 \end{pmatrix}, \quad R_2 = \begin{pmatrix} -1 & 0 \\ 0 & 1 \end{pmatrix},$$

and the *semiquadratic special linear* group $\mathrm{SL}_{1+1}(\mathbb{Z}[\sqrt{d}])$ is generated by

$$S_1 = R_0 R_1 = \begin{pmatrix} \sqrt{d} & 1 \\ -1 & 0 \end{pmatrix} \quad \text{and} \quad S_2 = R_1 R_2 = \begin{pmatrix} 1 & 0 \\ -\sqrt{d} & 1 \end{pmatrix}.$$

If we factor out the center (the subgroup $\mathrm{SZ}_2(\mathbb{Z}[\sqrt{d}]) \cong \{\pm I\}$ if $d \neq -1$) we obtain the *semiquadratic extended modular group* $\mathrm{P}\bar{\mathrm{S}}\mathrm{L}_{1+1}(\mathbb{Z}[\sqrt{d}])$, generated by $\rho_0 = \cdot\langle R_0\rangle$, $\rho_1 = \cdot\langle R_1\rangle$, and $\rho_2 = \cdot\langle R_2\rangle$, and the *semiquadratic modular group* $\mathrm{PSL}_{1+1}(\mathbb{Z}[\sqrt{d}])$, with generators $\sigma_1 = \rho_0\rho_1$ and $\sigma_2 = \rho_1\rho_2$ (Johnson & Weiss 1999a, p. 1316).

The period of the matrix S_1 is finite when d has one of the values 0, 1, 2, or 3, with $S_1{}^p = -I$ for p equal to 2, 3, 4, or 6, respectively. Thus for d equal to 2 or 3, the generators of the extended modular and modular groups $\mathrm{P}\bar{\mathrm{S}}\mathrm{L}_{1+1}(\mathbb{Z}[\sqrt{d}])$ and $\mathrm{PSL}_{1+1}(\mathbb{Z}[\sqrt{d}])$ satisfy the respective relations

$$\rho_0{}^2 = \rho_1{}^2 = \rho_2{}^2 = (\rho_0\rho_1)^p = (\rho_0\rho_2)^2 = (\rho_1\rho_2)^\infty = 1 \qquad (14.1.5)$$

and

$$\sigma_1{}^p = \sigma_2{}^\infty = (\sigma_1\sigma_2)^2 = 1, \qquad (14.1.6)$$

with p equal to 4 or 6. As seen in § 13.1, these relations define the Coxeter group $[p, \infty]$ and its direct subgroup $[p, \infty]^+$. Hence, for d equal to 2 or 3, $\mathrm{P\bar{S}L}_{1+1}(\mathbb{Z}[\sqrt{d}])$ is the group $\bar{\mathbf{Q}}_2 \cong [4, \infty]$ or $\bar{\mathbf{GP}}_2 \cong [6, \infty]$, the symmetry group of the regular hyperbolic tessellation $\{4, \infty\}$ or $\{6, \infty\}$, and $\mathrm{PSL}_{1+1}(\mathbb{Z}[\sqrt{d}])$ is the rotation group $\bar{\mathbf{Q}}_2^+ \cong [4, \infty]^+$ or $\bar{\mathbf{GP}}_2^+ \cong [6, \infty]^+$.

The group $[4, \infty]$ contains two other Coxeter groups as halving subgroups, namely, $\bar{\mathbf{W}}_2(4,4) \cong [(4, 4, \infty)] \cong [4, \infty, 1^+]$ and $\bar{\mathbf{N}}_2 \cong [\infty, \infty] \cong [1^+, 4, \infty]$. The latter group has a subgroup $\bar{\mathbf{M}}_2 \cong [\infty^{[3]}]$, of index 4 in $[4, \infty]$. The commutator subgroup of $[4, \infty]$ and $[4, \infty]^+$, of index 8 in the former and of index 4 in the latter, is $[1^+, 4, 1^+, \infty, 1^+] \cong [4, \infty]^{+3}$.

Likewise, the group $[6, \infty]$ contains the halving subgroups $\bar{\mathbf{W}}_2(6,6) \cong [(6, 6, \infty)] \cong [6, \infty, 1^+]$ and $\hat{\mathbf{P}}_2 \cong [(3, \infty, \infty)] \cong [1^+, 6, \infty]$. The commutator subgroup of $[6, \infty]$ and $[6, \infty]^+$, of index 8 in the former and of index 4 in the latter, is $[1^+, 6, 1^+, \infty, 1^+] \cong [6, \infty]^{+3}$.

The quadratic integral domain $\mathbb{Z}^2(5) = \mathbb{Z}[\tau] = \mathbb{Z}[\bar{\tau}]$ comprises numbers of the form $m + n\tau = (m + n) + n\bar{\tau}$ (m and n in $\mathbb{Z}$), where $\tau = ½(\sqrt{5} + 1)$ and $\bar{\tau} = ½(\sqrt{5} - 1)$. The unit linear group $\mathrm{\bar{S}L}_2(\mathbb{Z}^2(5))$ of 2×2 invertible matrices over $\mathbb{Z}^2(5)$ has a semiquadratic unit linear subgroup $\mathrm{\bar{S}L}_{1+1}(\mathbb{Z}[\tau])$, generated by the matrices

$$R_0 = \begin{pmatrix} 0 & 1 \\ 1 & 0 \end{pmatrix}, \quad R_1 = \begin{pmatrix} -1 & 0 \\ \tau & 1 \end{pmatrix}, \quad R_2 = \begin{pmatrix} -1 & 0 \\ 0 & 1 \end{pmatrix},$$

and a semiquadratic special linear subgroup $\mathrm{SL}_{1+1}(\mathbb{Z}[\tau])$, generated by

$$S_1 = \begin{pmatrix} \tau & 1 \\ -1 & 0 \end{pmatrix} \quad \text{and} \quad S_2 = \begin{pmatrix} 1 & 0 \\ -\tau & 1 \end{pmatrix}.$$

It also has the isomorphic subgroups $\mathrm{\bar{S}L}_{1+1}(\mathbb{Z}[\bar{\tau}])$ and $\mathrm{SL}_{1+1}(\mathbb{Z}[\bar{\tau}])$, defined similarly except with the matrix entry τ replaced by $-\bar{\tau}$ and $-\tau$ by $\bar{\tau}$. The generators ρ_0, ρ_1, ρ_2 of the semiquadratic extended modular group $\mathrm{P\bar{S}L}_{1+1}(\mathbb{Z}[\tau]) \cong \mathrm{P\bar{S}L}_{1+1}(\mathbb{Z}[\bar{\tau}])$ satisfy (14.1.5), and the generators σ_1 and σ_2 of the semiquadratic modular group

$\mathrm{PSL}_{1+1}(\mathbb{Z}[\tau]) \cong \mathrm{PSL}_{1+1}(\mathbb{Z}[\bar{\tau}])$ satisfy (14.1.6), with $p = 5$ in both cases. It follows that $\mathrm{P\bar{S}L}_{1+1}(\mathbb{Z}[\tau])$ is isomorphic to the Coxeter group $\bar{\mathbf{HP}}_2 \cong [5,\infty]$, with $\mathrm{PSL}_{1+1}(\mathbb{Z}[\tau])$ being the rotation group $\bar{\mathbf{HP}}_2^+ \cong [5,\infty]^+$.

Thus each of the regular hyperbolic tessellations $\{p,\infty\}$ ($3 \leq p \leq 6$) is associated with a group of rational or semiquadratic modular transformations. The above matrix representations of this section agree with those given in § 13.1, with the respective values of $1, \sqrt{2}, \tau$, and $\sqrt{3}$ for $2\cos\pi/p$. The latter expression is neither a rational integer nor a quadratic integer when $p > 6$.

In addition to its representation as $\mathrm{P\bar{S}L}_{1+1}(\mathbb{Z}[\sqrt{2}])$, the group $[4,\infty]$ is isomorphic to the (integral) *projective pseudo-orthogonal* group $\mathrm{PO}_{2,1}(\mathbb{Z})$, the central quotient group of the group $\mathrm{O}_{2,1}(\mathbb{Z})$ of 3×3 pseudo-orthogonal matrices with integer entries (Coxeter & Whitrow 1950, pp. 423–424; cf. Ratcliffe 1994/2006, § 7.3).

EXERCISES 14.1

1. Show that matrices A and B define transformations of $\mathbb{C}\cup\{\infty\}$ respectively taking z to $-1/z$ and z to $z+1$.

2. Show that the linear fractional transformations ρ_0, ρ_1, and ρ_2 represented by the first set of matrices R_0, R_1, and R_2 given in this section satisfy the relations (14.1.1), i.e., that the extended modular group $\mathrm{P\bar{S}L}_2(\mathbb{Z})$ is isomorphic to the Coxeter group $\bar{\mathbf{P}}_2 \cong [3,\infty]$.

3. Show that the linear fractional transformations σ_1 and σ_2 represented by the first set of matrices S_1 and S_2 satisfy the relations (14.1.3), i.e., that the modular group $\mathrm{PSL}_2(\mathbb{Z})$ is isomorphic to the rotation group $\bar{\mathbf{P}}_2^+ \cong [3,\infty]^+$.

4. Show that the linear fractional transformations ρ_0, ρ_1, and ρ_2 represented by the second set of matrices R_0, R_1, and R_2 given in this section with d equal to 2 or 3 satisfy the relations (14.1.5) with p equal to 4 or 6.

5. Show that the linear fractional transformations ρ_0, ρ_1, and ρ_2 represented by the third set of matrices R_0, R_1, and R_2 given in this section satisfy the relations (14.1.5) with p equal to 5.

14.2 THE GAUSSIAN MODULAR GROUP

The integral domain $\mathbb{G} = \mathbb{Z}[\mathrm{i}] = \mathbb{Z}^2(-1)$ of *Gaussian integers* comprises the complex numbers $g = g_0 + g_1\mathrm{i}$, where $(g_0, g_1) \in \mathbb{Z}^2$ and $\mathrm{i} = \sqrt{-1}$. This system was first described by C. F. Gauss (1777–1855). Each Gaussian integer g has a *norm* $N(g) = |g|^2 = {g_0}^2 + {g_1}^2$. The units of $\mathbb{G}$ are the four numbers with norm 1, namely, ± 1 and $\pm\mathrm{i}$, which form the group $\bar{\mathrm{S}}\mathrm{L}(\mathbb{G}) \cong \mathrm{C}_4(\mathbb{C})$, with the proper subgroup $\mathrm{C}_2(\mathbb{C}) \cong \{\pm 1\}$.

The *special linear* group $\mathrm{SL}_2(\mathbb{G})$ of 2×2 Gaussian integer matrices of determinant 1 is generated by the matrices

$$A = \begin{pmatrix} 0 & 1 \\ -1 & 0 \end{pmatrix}, \quad B = \begin{pmatrix} 1 & 0 \\ 1 & 1 \end{pmatrix}, \quad C = \begin{pmatrix} 1 & 0 \\ \mathrm{i} & 1 \end{pmatrix}$$

(Bianchi 1891, p. 314). The *semispecial linear* group $\mathrm{S}_2\mathrm{L}_2(\mathbb{G})$ of 2×2 matrices S over $\mathbb{G}$ with $(\det S)^2 = 1$ is generated by A, B, C, and $L = \backslash 1, -1\backslash$. The *unit linear* group $\bar{\mathrm{S}}\mathrm{L}_2(\mathbb{G})$ of matrices U with $|\det U| = 1$ is generated by A, B, and $M = \backslash 1, \mathrm{i}\backslash$ (Schulte & Weiss 1994, pp. 230–231). The center of $\mathrm{SL}_2(\mathbb{G})$ is the *special scalar* group $\mathrm{SZ}_2(\mathbb{G}) \cong \{\pm I\}$, and the center of both $\bar{\mathrm{S}}\mathrm{L}_2(\mathbb{G})$ and $\mathrm{S}_2\mathrm{L}_2(\mathbb{G})$ is the *unit scalar* group $\bar{\mathrm{S}}\mathrm{Z}(\mathbb{G}) \cong \{\pm I, \pm\mathrm{i}I\}$.

The *Gaussian modular group*

$$\mathrm{PSL}_2(\mathbb{G}) \cong \mathrm{SL}_2(\mathbb{G})/\mathrm{SZ}_2(\mathbb{G}) \cong \mathrm{S}_2\mathrm{L}_2(\mathbb{G})/\bar{\mathrm{S}}\mathrm{Z}(\mathbb{G}),$$

constructed by Émile Picard (1884) and generally known as the "Picard group," is generated in H^3 by the half-turn $\alpha = \cdot\langle A\rangle$ and the striations $\beta = \cdot\langle B\rangle$ and $\gamma = \cdot\langle C\rangle$. The *Gaussian extended modular group*

$$\mathrm{P}\bar{\mathrm{S}}\mathrm{L}_2(\mathbb{G}) \cong \bar{\mathrm{S}}\mathrm{L}_2(\mathbb{G})/\bar{\mathrm{S}}\mathrm{Z}(\mathbb{G})$$

is likewise generated by the half-turn α, the striation β, and the quarter-turn $\mu = \cdot\langle M\rangle$.

When the complex field $\mathbb{C}$ is regarded as a two-dimensional vector space over $\mathbb{R}$, the Gaussian integers constitute a two-dimensional lattice C_2 spanned by the units 1 and i, as shown in Figure 14.2a.

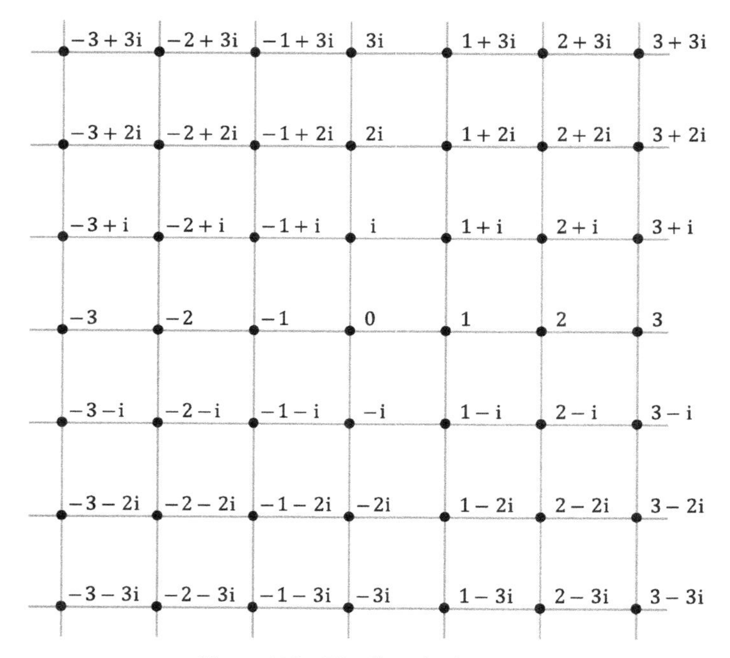

Figure 14.2a The Gaussian integers

The points of C_2 are the vertices of a regular tessellation {4, 4} of the Euclidean plane E^2, whose symmetry group [4, 4] is generated by three reflections ρ_1, ρ_2, ρ_3, satisfying the relations

$$\rho_1{}^2 = \rho_2{}^2 = \rho_3{}^2 = (\rho_1\rho_2)^4 = (\rho_1\rho_3)^2 = (\rho_2\rho_3)^4 = 1. \qquad (14.2.1)$$

As a horospherical tessellation, {4, 4} is the vertex figure of a regular cellulation {3, 4, 4} of hyperbolic 3-space H^3, the cell polyhedra of which are regular octahedra {3, 4} whose vertices all lie on the absolute sphere.

The symmetry group [3, 4, 4] of the cellulation {3, 4, 4} is generated by four reflections $\rho_0, \rho_1, \rho_2, \rho_3$, satisfying (14.2.1) as well as

$$\rho_0{}^2 = (\rho_0\rho_1)^3 = (\rho_0\rho_2)^2 = (\rho_1\rho_2)^2 = 1. \qquad (14.2.2)$$

The combined relations (14.2.1) and (14.2.2) are indicated in the Coxeter diagram

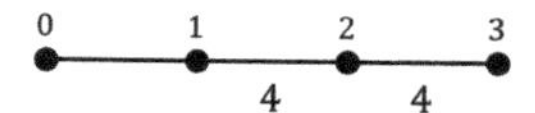

The generators $\rho_0, \rho_1, \rho_2, \rho_3$ can be represented by antilinear fractional transformations $\odot\langle R_0\rangle$, $\odot\langle R_1\rangle$, $\odot\langle R_2\rangle$, $\odot\langle R_3\rangle$, determined by the matrices

$$R_0 = \begin{pmatrix} 0 & 1 \\ 1 & 0 \end{pmatrix}, \ R_1 = \begin{pmatrix} -1 & 0 \\ 1 & 1 \end{pmatrix}, \ R_2 = \begin{pmatrix} \mathrm{i} & 0 \\ 0 & 1 \end{pmatrix}, \ R_3 = \begin{pmatrix} 1 & 0 \\ 0 & 1 \end{pmatrix}.$$

The direct subgroup $[3,4,4]^+$ is generated by three rotations $\sigma_1 = \rho_0\rho_1, \sigma_2 = \rho_1\rho_2, \sigma_3 = \rho_2\rho_3$, with the defining relations

$$\sigma_1{}^3 = \sigma_2{}^4 = \sigma_3{}^4 = (\sigma_1\sigma_2)^2 = (\sigma_2\sigma_3)^2 = (\sigma_1\sigma_2\sigma_3)^2 = 1. \qquad (14.2.3)$$

The generators $\sigma_1, \sigma_2, \sigma_3$ can be represented by linear fractional transformations $\cdot\langle s_1\rangle, \cdot\langle s_2\rangle, \cdot\langle S_3\rangle$, corresponding to the unit matrices

$$S_1 = \begin{pmatrix} 1 & 1 \\ -1 & 0 \end{pmatrix}, \ S_2 = \begin{pmatrix} -\mathrm{i} & 0 \\ \mathrm{i} & 1 \end{pmatrix}, \ S_3 = \begin{pmatrix} -\mathrm{i} & 0 \\ 0 & 1 \end{pmatrix}.$$

Fricke & Klein (1897, pp. 76–93; cf. Magnus 1974, pp. 60, 196) identified the Gaussian modular group $\mathrm{PSL}_2(\mathbb{G})$ with a subgroup of the rotation group of the cellulation {3, 4, 4}. Schulte & Weiss (1994, pp. 235–236) showed that it is a subgroup of index 2 in $[3,4,4]^+$, and Monson & Weiss (1995, pp. 188–189) exhibited it as a subgroup of index 2 in the hypercompact Coxeter group $[\infty,3,3,\infty]$. Johnson & Weiss (1999a, pp. 1319–1323) found a presentation for $\mathrm{PSL}_2(\mathbb{G})$ as an ionic subgroup of $\bar{\mathbf{R}}_3 \cong [3,4,4]$.

The group [3, 4, 4] has a halving subgroup $[3,4,1^+,4] \cong [\infty,3,3,\infty]$, generated by the reflections ρ_0, ρ_1, ρ_3, $\rho_{212} = \rho_2\rho_1\rho_2$, and $\rho_{232} = \rho_2\rho_3\rho_2$, satisfying the relations indicated in the diagram

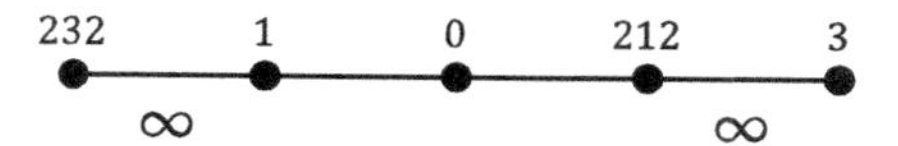

The five mirrors are the bounding planes of a quadrangular pyramid whose apex lies on the absolute sphere of H^3. The two groups $[3,4,4]^+$ and $[3,4,1^+,4]$ have a common subgroup $[3,4,1^+,4]^+ \cong [\infty,3,3,\infty]^+$, of index 2 in both and of index 4 in $[3,4,4]$, generated by the striations β and γ and the rotations σ and ϕ, where

$$\beta = \sigma_3{\sigma_2}^{-1}, \quad \gamma = {\sigma_3}^{-1}\sigma_2, \quad \sigma = \sigma_1, \quad \text{and} \quad \phi = \sigma_1{\sigma_2}^2.$$

Since the corresponding matrices

$$B = S_3S_2^{-1}, \qquad C = {S_3}^{-1}S_2, \quad S = S_1, \quad \text{and} \quad U = -\mathrm{i}S_1{S_2}^2$$

all belong to the Gaussian special linear group $\mathrm{SL}_2(\mathbb{G}) \cong \langle A, B, C\rangle$ and since $A = SB^{-1}$, it follows that $\mathrm{SL}_2(\mathbb{G})$ is generated by the matrices

$$B = \begin{pmatrix} 1 & 0 \\ 1 & 1 \end{pmatrix}, \; C = \begin{pmatrix} 1 & 0 \\ \mathrm{i} & 1 \end{pmatrix}, \; S = \begin{pmatrix} 1 & 1 \\ -1 & 0 \end{pmatrix}.$$

Hence the Gaussian modular group $\mathrm{PSL}_2(\mathbb{G}) \cong \langle \alpha, \beta, \gamma\rangle$ is generated by the corresponding isometries β, γ, and σ. The matrix U and the rotation ϕ are superfluous, since $U = C^{-1}ACA^{-1} = C^{-1}SCS^{-1}$, so that $\phi = \gamma^{-1}\alpha\gamma\alpha = \gamma^{-1}\sigma\gamma\sigma$. Thus the group $[3,4,1^+,4]^+$, actually generated by just β, γ, and σ, is isomorphic to $\mathrm{PSL}_2(\mathbb{G})$.

The group $[\infty,(3,3)^+,\infty]^+ \cong [3^+,4,1^+,4,1^+] \cong [3,4,4]^{+3} \cong \mathrm{P\bar{S}L}_2'(\mathbb{G})$ is the commutator subgroup of [3, 4, 4] and $[3,4,4]^+ \cong \mathrm{P\bar{S}L}_2(\mathbb{G})$, of index 4 in $\mathrm{P\bar{S}L}_2(\mathbb{G})$ and of index 2 in $\mathrm{PSL}_2(\mathbb{G})$. It is generated by the rotations $\sigma = \sigma_1$, $\tau = {\sigma_3}^2$, and $\phi = \sigma_1{\sigma_2}^2$, satisfying the relations

$$\sigma^3 = \tau^2 = \phi^3 = (\sigma^{-1}\phi)^2 = (\sigma^{-1}\tau\phi\tau)^2 = 1. \tag{14.2.4}$$

The corresponding matrices are $S = S_1$, $T = \mathrm{i}{S_3}^2$, and $U = -\mathrm{i}S_1{S_2}^2$.

The hypercompact group $[3,4,1^+,4] \cong [\infty,3,3,\infty]$ and its subgroups $[\infty,3,3,\infty]^+ \cong \mathrm{PSL}_2(\mathbb{G})$ and $[\infty,(3,3)^+,\infty]^+ \cong \mathrm{P\bar{S}L}_2'(\mathbb{G})$ have a common commutator subgroup $[1^+,\infty,(3,3)^+,\infty,1^+] \cong [\infty,3,3,\infty]^{+3} \cong \mathrm{PSL}_2'(\mathbb{G}) \cong \mathrm{P\bar{S}L}_2''(\mathbb{G})$. This group, of index 4 in

$\mathrm{PSL}_2(\mathbb{G})$ and of index 2 in $\mathrm{P\bar{S}L}_2{}'(\mathbb{G})$, is generated by the rotations

$$\sigma = \sigma_1, \quad \phi = \sigma_1\sigma_2{}^2, \quad \psi = \sigma_3{}^{-1}\sigma_1\sigma_3, \quad \omega = \sigma_3\sigma_1\sigma_2{}^2\sigma_3{}^{-1},$$

satisfying the relations

$$\sigma^3 = \phi^3 = \psi^3 = \omega^3 = (\sigma^{-1}\phi)^2 = (\sigma^{-1}\psi)^2 = (\phi^{-1}\omega)^2 = (\psi^{-1}\omega)^2 = 1. \tag{14.2.5}$$

The corresponding matrices are

$$S = S_1, \quad U = -\mathrm{i}S_1S_2{}^2, \quad V = S_3{}^{-1}S_1S_3, \quad W = -\mathrm{i}S_3S_1S_2{}^2S_3{}^{-1}.$$

That is,

$$S = \begin{pmatrix} 0 & 1 \\ -1 & 0 \end{pmatrix}, \; U = \begin{pmatrix} 1 & -\mathrm{i} \\ -\mathrm{i} & 0 \end{pmatrix}, \; V = \begin{pmatrix} 1 & \mathrm{i} \\ \mathrm{i} & 0 \end{pmatrix}, \; W = \begin{pmatrix} 1 & -1 \\ 1 & 0 \end{pmatrix}.$$

The group $\bar{\mathbf{R}}_3 \cong [3, 4, 4]$ has several subgroups that are themselves Coxeter groups. For instance, the subgroup $\bar{\mathbf{N}}_3 \cong [4, 4, 4]$, of index 3, is generated by the reflections ρ_0, ρ_2, ρ_3, and $\rho_{121} = \rho_1\rho_2\rho_1$, satisfying the relations indicated in the Coxeter diagram

0 —4— 121 —4— 3 —4— 2

This is the symmetry group of a self-dual regular cellulation $\{4, 4, 4\}$.

Groups involving antilinear or antilinear fractional transformations can also be represented by subgroups of $\bar{\mathrm{S}}\mathrm{L}_4(\mathbb{Z})$ or $\mathrm{P\bar{S}L}_4(\mathbb{Z})$. In particular, the group [3, 4, 4] is isomorphic to $\mathrm{PO}_{3,1}(\mathbb{Z})$, the central quotient group of the group $\mathrm{O}_{3,1}(\mathbb{Z})$ of 4×4 pseudo-orthogonal matrices with integral entries (Coxeter & Whitrow 1950, pp. 428–429; cf. Ratcliffe 1994/2006, § 7.3).

EXERCISES 14.2

1. Show that the antilinear fractional transformations ρ_0, ρ_1, ρ_2, and ρ_3 represented by the matrices R_0, R_1, R_2, and R_3 given in this section satisfy the relations (14.2.1) and (14.2.2) that define the Coxeter group $\bar{\mathbf{R}}_3 \cong [3, 4, 4]$.

2. Show that the linear fractional transformations σ_1, σ_2, and σ_3 represented by the matrices S_1, S_2, and S_3 satisfy the relations (14.2.3) that define the rotation group $\bar{\mathbf{R}}_3^+ \cong [3, 4, 4]^+$.
3. Assuming the relations for the Coxeter group [3, 4, 4], show that the reflections $\rho_0, \rho_1, \rho_3, \rho_{212}$, and ρ_{232} satisfy the relations indicated in the diagram for the halving subgroup $[3, 4, 1^+, 4] \cong [\infty, 3, 3, \infty]$.
4. Show that the rotations σ, ϕ, ψ, and ω satisfy the relations (14.2.5) that define the group $[\infty, 3, 3, \infty]^{+3} \cong \mathrm{PSL}_2{}'(\mathbb{G})$.

14.3 THE EISENSTEIN MODULAR GROUP

The integral domain $\mathbb{E} = \mathbb{Z}[\omega] = \mathbb{Z}^2(-3)$ of *Eisenstein integers* comprises the complex numbers $e = e_0 + e_1\omega$, where $(e_0, e_1) \in \mathbb{Z}^2$ and $\omega = -½ + ½\sqrt{-3}$ is a primitive cube root of unity, so that $\omega^2 + \omega + 1 = 0$. Quadratic integers of this type were investigated by Gotthold Eisenstein (1823–1852). Each Eisenstein integer e has a *norm* $N(e) = |e|^2 = e_0{}^2 - e_0e_1 + e_1{}^2$. The units of $\mathbb{E}$ are the six numbers with norm 1, namely, ±1, $\pm\omega$, $\pm\omega^2$, which form the group $\bar{\mathrm{S}}\mathrm{L}(\mathbb{E}) \cong \mathrm{C}_6(\mathbb{C})$, with the proper subgroups $\mathrm{C}_3(\mathbb{C}) \cong \{1, \omega, \omega^2\}$ and $\mathrm{C}_2(\mathbb{C}) \cong \{\pm1\}$.

The *special linear* group $\mathrm{SL}_2(\mathbb{E})$ of 2×2 Eisenstein integer matrices of determinant 1 is generated by the matrices

$$A = \begin{pmatrix} 0 & 1 \\ -1 & 0 \end{pmatrix}, \quad B = \begin{pmatrix} 1 & 0 \\ 1 & 1 \end{pmatrix}, \quad C = \begin{pmatrix} 1 & 0 \\ \omega & 1 \end{pmatrix}$$

(Bianchi 1891, p. 316). The *semispecial linear* group $\mathrm{S}_2\mathrm{L}_2(\mathbb{E})$ of 2×2 matrices S over $\mathbb{E}$ with $(\det S)^2 = 1$ is generated by A, B, C, and $L = \backslash 1, -1\backslash$ (Schulte & Weiss 1994, p. 231). The *ternispecial linear* group $\mathrm{S}_3\mathrm{L}_2(\mathbb{E})$ of matrices T with $(\det T)^3 = 1$ is generated by A, B, and $M = \backslash 1, \omega\backslash$. The *unit linear* group $\bar{\mathrm{S}}\mathrm{L}_2(\mathbb{E})$ of matrices U with $|\det U| = 1$ is generated by A, B, and $N = \backslash 1, -\omega\backslash$. The center of both $\mathrm{SL}_2(\mathbb{E})$ and $\mathrm{S}_2\mathrm{L}_2(\mathbb{E})$ is the *special scalar* group $\mathrm{SZ}_2(\mathbb{E}) \cong \{\pm I\}$, and the center of both $\bar{\mathrm{S}}\mathrm{L}_2(\mathbb{E})$ and $\mathrm{S}_3\mathrm{L}_2(\mathbb{E})$ is the *unit scalar* group $\bar{\mathrm{S}}\mathrm{Z}(\mathbb{E}) \cong \{\pm I, \pm\omega I, \pm\omega^2 I\}$.

The *Eisenstein modular group*

$$\mathrm{PSL}_2(\mathbb{E}) \cong \mathrm{SL}_2(\mathbb{E})/\mathrm{SZ}_2(\mathbb{E}) \cong \mathrm{S}_3\mathrm{L}_2(\mathbb{E})/\bar{\mathrm{S}}\mathrm{Z}(\mathbb{E})$$

is generated in H^3 by the half-turn $\alpha = \cdot\langle A\rangle$ and the striations $\beta = \cdot\langle B\rangle$ and $\gamma = \cdot\langle C\rangle$; alternatively, it is also generated by the half-turn α, the striation β, and the rotation $\mu = \cdot\langle M\rangle$ (of period 3). The *Eisenstein extended modular group*

$$\mathrm{P}\bar{\mathrm{S}}\mathrm{L}_2(\mathbb{E}) \cong \bar{\mathrm{S}}\mathrm{L}_2(\mathbb{E})/\bar{\mathrm{S}}\mathrm{Z}(\mathbb{E}) \cong \mathrm{S}_2\mathrm{L}_2(\mathbb{E})/\mathrm{SZ}_2(\mathbb{E})$$

is similarly generated either by the half-turn α, the striation β, and the rotation $\nu = \cdot\langle N\rangle$ (of period 6) or by the half-turn α, the striations β and γ, and the half-turn $\lambda = \cdot\langle L\rangle$.

When the complex field $\mathbb{C}$ is regarded as a two-dimensional vector space over $\mathbb{R}$, the Eisenstein integers constitute a two-dimensional lattice A_2 spanned by the units 1 and ω, as shown in Figure 14.3a. The points of A_2 are the vertices of a regular tessellation {3, 6} of the Euclidean plane E^2, whose symmetry group [3, 6] is generated by three reflections ρ_1, ρ_2, ρ_3, satisfying the relations

$$\rho_1{}^2 = \rho_2{}^2 = \rho_3{}^2 = (\rho_1\rho_2)^3 = (\rho_1\rho_3)^2 = (\rho_2\rho_3)^6 = 1. \tag{14.3.1}$$

The (horospherical) tessellation {3, 6} is the vertex figure of a regular cellulation {3, 3, 6} of hyperbolic 3-space H^3, the cell polyhedra of which are regular tetrahedra {3, 3} whose vertices all lie on the absolute sphere.

The symmetry group [3, 3, 6] of the cellulation {3, 3, 6} is generated by four reflections $\rho_0, \rho_1, \rho_2, \rho_3$, satisfying (14.3.1) as well as

$$\rho_0{}^2 = (\rho_0\rho_1)^3 = (\rho_0\rho_2)^2 = (\rho_1\rho_2)^2 = 1. \tag{14.3.2}$$

The combined relations (14.3.1) and (14.3.2) are indicated in the Coxeter diagram

The generators $\rho_0, \rho_1, \rho_2, \rho_3$ can be represented by antilinear fractional transformations $\odot\langle R_0\rangle$, $\odot\langle R_1\rangle$, $\odot\langle R_2\rangle$, $\odot\langle R_3\rangle$, determined by the matrices

$$R_0 = \begin{pmatrix} 0 & 1 \\ 1 & 0 \end{pmatrix}, \ R_1 = \begin{pmatrix} -1 & 0 \\ 1 & 1 \end{pmatrix}, \ R_2 = \begin{pmatrix} -\omega & 0 \\ 0 & \omega^2 \end{pmatrix}, \ R_3 = \begin{pmatrix} -1 & 0 \\ 0 & -1 \end{pmatrix}.$$

The direct subgroup $[3, 3, 6]^+$ is generated by three rotations $\sigma_1 = \rho_0\rho_1, \sigma_2 = \rho_1\rho_2, \sigma_3 = \rho_2\rho_3$, with the defining relations

$$\sigma_1{}^3 = \sigma_2{}^3 = \sigma_3{}^6 = (\sigma_1\sigma_2)^2 = (\sigma_2\sigma_3)^2 = (\sigma_1\sigma_2\sigma_3)^2 = 1. \qquad (14.3.3)$$

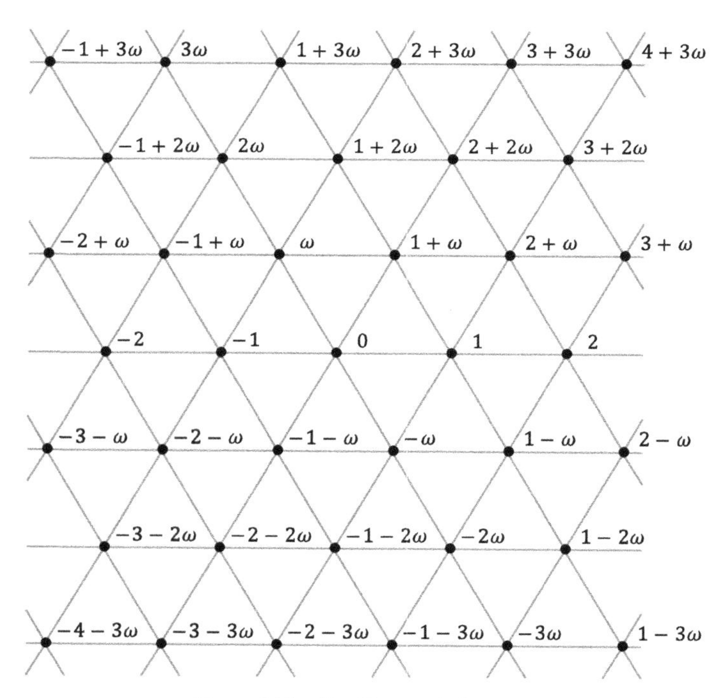

Figure 14.3a The Eisenstein integers

The generators $\sigma_1, \sigma_2, \sigma_3$ can be represented by linear fractional transformations $\cdot\langle S_1\rangle$, $\cdot\langle S_2\rangle$, $\cdot\langle S_3\rangle$, corresponding to the unit matrices

$$S_1 = \begin{pmatrix} 1 & 1 \\ -1 & 0 \end{pmatrix}, \quad S_2 = \begin{pmatrix} \omega & 0 \\ -\omega & \omega^2 \end{pmatrix}, \quad S_3 = \begin{pmatrix} \omega^2 & 0 \\ 0 & -\omega \end{pmatrix}.$$

Luigi Bianchi (1891, 1892) showed that if D is an imaginary quadratic integral domain, the group $\mathrm{PSL}_2(\mathsf{D})$ acts discontinuously on hyperbolic 3- space. As Fricke and Klein had done a century before with $\mathrm{PSL}_2(\mathbb{G})$ and the honeycomb {3, 4, 4}, Schulte & Weiss (1994, pp. 234–236; cf. Monson & Weiss 1995, pp. 188–189, and 1997, pp. 102–103) connected the Eisenstein modular group $\mathrm{PSL}_2(\mathbb{E})$ with the honeycomb {3, 3, 6}. They proved that $\mathrm{PSL}_2(\mathbb{E})$ is isomorphic to a subgroup of $[3,3,6]^+$, which Johnson & Weiss (1999a, pp. 1328–1331) subsequently showed to be the commutator subgroup of $\bar{\mathbf{V}}_3 \cong [3,3,6]$.

The matrices S_1, S_2, S_3 not only belong to but generate the semi-special linear group $\mathrm{S_2L_2}(\mathbb{E}) \cong \langle A, B, C, L\rangle$, since

$$A = S_1S_2S_3{}^{-2}, \quad B = S_3{}^2S_2{}^{-1}, \quad C = S_2S_3{}^2S_2, \quad \text{and} \quad L = S_3{}^3.$$

Thus the group $[3,3,6]^+$, generated by σ_1, σ_2, and σ_3, is the Eisenstein extended modular group $\mathrm{P\bar{S}L}_2(\mathbb{E}) \cong \langle \alpha, \beta, \gamma, \lambda\rangle$.

The group [3, 3, 6] has a halving subgroup $[3,3,6,1^+] \cong [3,3^{[3]}]$, generated by the reflections ρ_0, ρ_1, ρ_2, and $\rho_{323} = \rho_3\rho_2\rho_3$, satisfying the relations indicated in the diagram

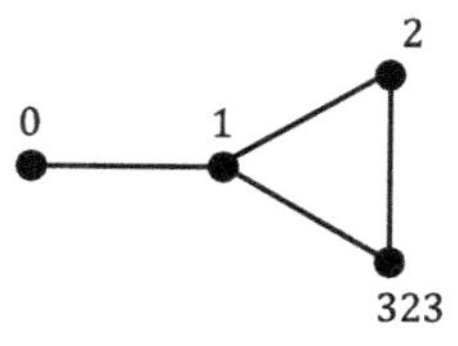

There is also a semidirect subgroup $[(3,3)^+,6]$, generated by the rotations $\sigma_1 = \rho_0\rho_1$ and $\sigma_2 = \rho_1\rho_2$ and the reflection ρ_3. The four groups $[3,3,6]$, $[3,3,6]^+$, $[3,3,6,1^+]$, and $[(3,3)^+,6]$ have a common commutator subgroup $[3,3,6,1^+]^+ \cong [(3,3)^+,6,1^+] \cong [3,3,6]^{+2}$, of index 4 in $[3,3,6]$ and of index 2 in the others, generated by the three rotations σ_1, σ_2, and $\sigma_{33} = \sigma_3{}^2$. Defining relations for the group $[(3,3)^+, 6, 1^+]$ are

$$\sigma_1{}^3 = \sigma_2{}^3 = \sigma_{33}{}^3 = (\sigma_1\sigma_2)^2 = (\sigma_2\sigma_{33})^3 = (\sigma_1\sigma_2\sigma_{33})^2 = 1. \quad (14.3.4)$$

Since the corresponding matrices S_1, S_2, and $S_{33} = S_3{}^2$ all belong to the Eisenstein special linear group $\mathrm{SL}_2(\mathbb{E}) \cong \langle A,\ B,\ C\rangle$, and since

$$A = S_1 S_2 S_{33}{}^{-1}, \quad B = S_{33} S_2{}^{-1}, \quad C = S_2 S_{33} S_2,$$

it follows that $\mathrm{SL}_2(\mathbb{E})$ is generated by the matrices

$$S_1 = \begin{pmatrix} 1 & 1 \\ -1 & 0 \end{pmatrix}, \ S_2 = \begin{pmatrix} \omega & 0 \\ -\omega & \omega^2 \end{pmatrix}, \ S_{33} = \begin{pmatrix} \omega & 0 \\ 0 & \omega^2 \end{pmatrix}.$$

Hence the Eisenstein modular group $\mathrm{PSL}_2(\mathbb{E}) \cong \langle \alpha,\ \beta,\ \gamma\rangle$ is generated by the corresponding isometries $\sigma_1,\ \sigma_2$, and σ_{33}. That is $\mathrm{PSL}_2(\mathbb{E})$ is isomorphic to the group $[(3,3)^+, 6, 1^+] \cong \langle \sigma_1,\ \sigma_2,\ \sigma_{33}\rangle$.

The "trionic" subgroup $[(3,3)^\triangle,\ 6, 1^+] \cong \mathrm{PSL}_2{}'(\mathbb{E})$ is the commutator subgroup of $[(3,3)^+, 6, 1^+] \cong \mathrm{PSL}_2(\mathbb{E})$, of index 3. It is generated by the four half-turns

$$\sigma_{12} = \sigma_1\sigma_2, \quad \sigma_{21} = \sigma_2\sigma_1, \quad \bar{\sigma}_{12} = \sigma_1\sigma_2\sigma_{33}, \quad \bar{\sigma}_{21} = \sigma_2\sigma_{33}\sigma_1,$$

satisfying the relations

$$\begin{aligned} \sigma_{12}{}^2 = \sigma_{21}{}^2 = \bar{\sigma}_{12}{}^2 = \bar{\sigma}_{21}{}^2 &= (\sigma_{12}\sigma_{21})^2 = (\bar{\sigma}_{12}\bar{\sigma}_{21})^2 \\ &= (\sigma_{12}\bar{\sigma}_{12})^3 = (\sigma_{21}\bar{\sigma}_{21})^3 = (\sigma_{12}\sigma_{21}\bar{\sigma}_{12}\bar{\sigma}_{21})^3 = 1. \quad (14.3.5) \end{aligned}$$

The corresponding matrices are

$$S_{12} = S_1 S_2, \quad S_{21} = S_2 S_1, \quad \bar{S}_{12} = S_1 S_2 S_{33}, \quad \bar{S}_{21} = S_2 S_{33} S_1,$$

which evaluate as

$$S_{12} = \begin{pmatrix} 0 & \omega^2 \\ -\omega & 0 \end{pmatrix}, \ S_{21} = \begin{pmatrix} \omega & \omega \\ 1 & -\omega \end{pmatrix},$$

$$\bar{S}_{12} = \begin{pmatrix} 0 & \omega \\ -\omega^2 & 0 \end{pmatrix}, \ \bar{S}_{21} = \begin{pmatrix} \omega^2 & \omega^2 \\ 1 & -\omega^2 \end{pmatrix}.$$

Besides the halving subgroup $\bar{\mathbf{P}}_3 \cong [3, 3^{[3]}]$, the group $\bar{\mathbf{V}}_3 \cong [3, 3, 6]$ has several other subgroups that are themselves Coxeter groups. The subgroup $\bar{\mathbf{Y}}_3 \cong [3, 6, 3]$, of index 4, is generated by the reflections ρ_0, ρ_1, ρ_3, and $\rho_{232} = \rho_2\rho_3\rho_2$. The subgroup $\overline{\mathbf{DV}}_3 \cong [6, 3^{1,1}]$, of index 5, is generated by the reflections ρ_0, ρ_1, ρ_{232}, and $\rho_{323} = \rho_3\rho_2\rho_3$. The subgroup $\bar{\mathbf{Z}}_3 \cong [6, 3, 6]$, of index 6, is generated by the reflections ρ_0, ρ_2, ρ_3, and $\rho_{12321} = \rho_1\rho_{232}\rho_1$. The three groups $[3, 6, 3], [6, 3^{1,1}]$, and [6, 3, 6] have the respective Coxeter diagrams

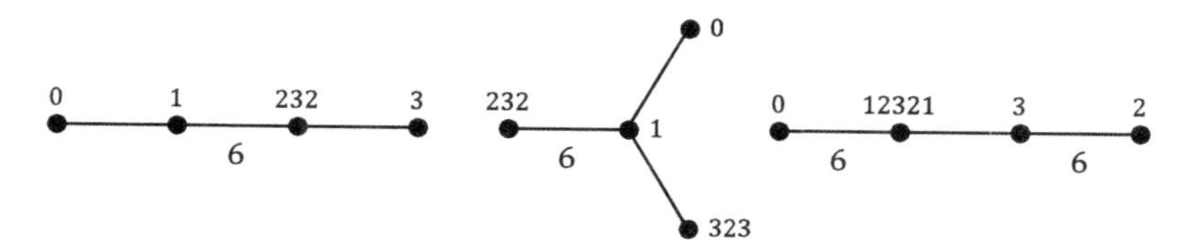

As evidenced by the bilateral symmetry of their graphs, each of these groups has an involutory automorphism. For the group $\overline{\mathbf{DV}}_3 \cong [6, 3^{1,1}]$, this is conjugation by a reflection ρ_{-1} in a plane bisecting the fundamental region. Augmenting $[6, 3^{1,1}]$ by this automorphism, we get another Coxeter group $\overline{\mathbf{BV}}_3 \cong [4, 3, 6]$, generated by the reflections ρ_{-1}, ρ_0, ρ_1, and ρ_{232} (with $\rho_{323} = \rho_{-1}\rho_0\rho_{-1}$), as in the diagram

The generators $\rho_{-1}, \rho_0, \rho_1, \rho_{232}$ can be represented by antilinear fractional transformations $\odot\langle R_{-1}\rangle$, $\odot\langle R_0\rangle$, $\odot\langle R_1\rangle$, $\odot\langle R_{232}\rangle$, determined by the matrices

$$R_{-1} = \frac{1}{\sqrt{2}}\begin{pmatrix} -\omega^2 & 1 \\ 1 & \omega \end{pmatrix}, \quad R_0 = \begin{pmatrix} 0 & 1 \\ 1 & 0 \end{pmatrix}, \quad R_1 = \begin{pmatrix} -1 & 0 \\ 1 & 1 \end{pmatrix},$$

$$R_{232} = \begin{pmatrix} -\omega^2 & 0 \\ 0 & -\omega \end{pmatrix}.$$

EXERCISES 14.3

1. Show that the antilinear fractional transformations ρ_0, ρ_1, ρ_2, and ρ_3 represented by the matrices R_0, R_1, R_2, and R_3 given in this section satisfy the relations (14.3.1) and (14.3.2) that define the Coxeter group $\bar{\mathbf{V}}_3 \cong [3,3,6]$.

2. Show that the linear fractional transformations σ_1, σ_2, and σ_3 represented by the matrices S_1, S_2, and S_3 satisfy the relations (14.3.3) that define the rotation group $\bar{\mathbf{V}}_3^+ \cong [3,3,6]^+$.

3. Assuming the relations for the Coxeter group $[3,3,6]$, show that the reflections ρ_0, ρ_1, ρ_2, and ρ_{323} satisfy the relations indicated in the diagram for the halving subgroup $[3,3,6,1^+] \cong [3,3^{[3]}]$.

4. Show that the half-turns $\sigma_{12}, \sigma_{21}, \bar{\sigma}_{12}$, and $\bar{\sigma}_{21}$ satisfy the relations (14.3.5) that define the group $[(3,3)^{\triangle},6,1^+] \cong \mathrm{PSL}_2{}'(\mathbb{E})$.

15

QUATERNIONIC MODULAR GROUPS

In each of the normed division algebras over the real field $\mathbb{R}$—namely, $\mathbb{R}$ itself, the complex numbers $\mathbb{C}$, the quaternions $\mathbb{H}$, and the octonions $\mathbb{O}$—certain elements can be characterized as *integers*. An integer of norm 1 is a *unit*. In a *basic system* of integers, the units span a 1-, 2-, 4-, or 8-dimensional lattice. Each of the three basic systems of integral quaternions defines a modular group of linear fractional transformations corresponding to a subgroup of a Coxeter group operating in H^5, as well as a "pseudo-modular" group of transformations corresponding to isometries of H^4. While there are no linear fractional transformations or modular groups over nonassociative subrings of $\mathbb{O}$, each of the four basic systems of integral octonions defines an *octonionic modular loop*.

15.1 INTEGRAL QUATERNIONS

The division ring $\mathbb{H}$ of quatemions was discovered in 1843 by William Rowan Hamilton. In 1856, at the instigation of his friend John Graves, Hamilton devised a system of quaternionic integers, but his notes remained unpublished for more than a century before being appended to his *Mathematical Papers* (Hamilton 1967). An extensive theory along the same lines was subsequently developed by Rudolph Lipschitz (1886). Whereas both Hamilton and Lipschitz defined an integral quaternion as one whose components are rational integers,

Adolf Hurwitz (1896) constructed a system with nicer arithmetic properties by allowing the components to be either all integers or all halves of odd integers.

In defining a *basic system* of quadratic integral quaternions, a reasonable approach is to require (1) that each element be a zero of a monic quadratic polynomial with coefficients in $\mathbb{Z}$, (2) that the elements form a subring of $\mathbb{H}$ with the units forming a subgroup of the group $\mathrm{SL}(\mathbb{H})$ of unit quaternions, and (3) that when $\mathbb{H}$ is taken as a four-dimensional vector space over $\mathbb{R}$, the elements are the points of a four-dimensional lattice spanned by the units. As we shall see, there are exactly three such systems.

The first condition is satisfied by any quaternion $\mathrm{Q} = q_0 + q_1\mathrm{i} + q_2\mathrm{j} + q_3\mathrm{k}$ such that q_0 is an integer or half an odd integer and $q_0{}^2 + q_1{}^2 + q_2{}^2 + q_3{}^2$ is an integer (Dickson 1923b p. 111).

Du Val (1964, pp. 49–54) described all the finite groups of quaternions, which operate as four-dimensional rotation groups. Abstractly, such a group is either a cyclic group C_p, of order p $(p \geq 1)$, or an extension of one of the *polyhedral* groups $(p, q, 2) \cong [p, q]^+$, which are the rotation groups of the regular polyhedra $\{p, q\}$ and the right prisms $\{p\} \times \{\}$. The *binary polyhedral* group $((p, q, 2))$, of order $8pq/(2p + 2q - pq)$, is defined by the relations

$$\sigma_1{}^p = \sigma_2{}^q = \sigma_3{}^2 = \sigma_1\sigma_2\sigma_3, \tag{15.1.1}$$

where $p \geq 2$, $q \geq 2$, and $pq < 2(p + q)$ (Coxeter & Moser 1957, pp. 67–69). The center of $((p, q, 2))$ is the group 2 generated by $\zeta = \sigma_1\sigma_2\sigma_3$, and the central quotient group $((p, q, 2))/2$ is $(p, q, 2)$. The actual cases are the *binary dihedral* (or "dicyclic") groups $((p, 2, 2)) \cong 2\mathrm{D}_p$ $(p \geq 2)$, the *binary tetrahedral* group $((3, 3, 2)) \cong 2\mathrm{A}_4 \cong \mathrm{SL}_2(\mathbb{F}_3)$, the *binary octahedral* group $((4, 3, 2)) \cong 2\mathrm{S}_4$, and the *binary icosahedral* group $((5, 3, 2)) \cong 2\mathrm{A}_5 \cong \mathrm{SL}_2(\mathbb{F}_5)$, of respective orders $4p$, 24, 48, and 120 (Coxeter 1974 pp. 74–80; cf. Conway & Smith 2003 p. 33).

As a subgroup of $\mathrm{SL}(\mathbb{H})$, C_p is a two-dimensional rotation group generated by a single quaternion in the conjugacy class of $\cos 2\pi/p +$

i sin $2\pi/p$. The group $2D_p$, generators for which are $\cos\pi/p + \mathrm{i}\sin\pi/p$ and j, contains C_{2p} as a subgroup of index 2. When $p = 2$, this is the *Hamilton group* generated by i and j, whose eight elements are

$$\pm 1, \quad \pm\mathrm{i}, \quad \pm\mathrm{j}, \quad \pm\mathrm{k}.$$

When $p = 3$, it is a "hybrid" group of order 12 generated by the quaternions $\omega = -½ + ½\sqrt{3}\mathrm{i}$ (of period 3) and j (of period 4). Although every binary dihedral group $2D_p$ can be extended to a quaternion ring, the ring elements are the points of a four-dimensional lattice only when p equals 2 or 3.

The binary tetrahedral group $2A_4$ contains $2D_2$ as subgroup of index 3. As a subgroup of SL($\mathbb{H}$), it is the *Hurwitz group*, generated by the quaternions

$$u = \tfrac{1}{2} - \tfrac{1}{2}\mathrm{i} - \tfrac{1}{2}\mathrm{j} + \tfrac{1}{2}\mathrm{k} \quad \text{and} \quad v = \tfrac{1}{2} + \tfrac{1}{2}\mathrm{i} - \tfrac{1}{2}\mathrm{j} - \tfrac{1}{2}\mathrm{k}$$

(both of period 6), consisting of the eight Hamilton units together with the sixteen additional elements

$$\pm\tfrac{1}{2} \pm \tfrac{1}{2}\mathrm{i} \pm \tfrac{1}{2}\mathrm{j} \pm \tfrac{1}{2}\mathrm{k}.$$

The group $2A_4$ is a subgroup of index 2 in $2S_4$ and a subgroup of index 5 in $2A_5$. As a subgroup of SL($\mathbb{H}$), the binary octahedral group $2S_4$ is the *octian group*, generated by u, v, and $\eta = ½\sqrt{2} + ½\sqrt{2}\mathrm{i}$ (of period 8), consisting of the twenty-four Hurwitz units together with twenty-four others obtained by multiplying each of them by η, i.e., quaternions whose coordinates are permutations of $(\pm½\sqrt{2}, \pm½\sqrt{2}, 0, 0)$. Likewise, the binary icosahedral group $2A_5$ is the *icosian group* (cf. Conway & Sloane 1988, pp. 207–210), which consists of the twenty-four Hurwitz units and ninety-six others (each of period 5) whose coordinates are even permutations of $(\pm½\tau, \pm½\bar{\tau}, \pm½, 0)$, where $\tau = ½(\sqrt{5} + 1)$ and $\bar{\tau} = ½(\sqrt{5} - 1)$.

Johnson & Weiss (1999b, pp. 167–168) proved that the only basic systems of quadratic integral quaternions are the rings associated with the binary dihedral groups $2D_2$ and $2D_3$ and the binary tetrahedral group $2A_4$. The lattice of *Hamilton integers* $\mathbb{H}\mathsf{am} = \mathbb{H}_{22} = \mathbb{Z}[\mathrm{i}, \mathrm{j}]$ derived from the Hamilton group $2D_2$ is spanned by the four unit quaternions 1, i, j, and k = ij. The lattice of *hybrid integers* $\mathbb{H}\mathsf{yb} = \mathbb{H}_{32} = \mathbb{Z}[\omega, \mathrm{j}]$ derived from the hybrid group $2D_3$ is spanned by 1, ω, j, and ωj. The lattice of *Hurwitz integers* $\mathbb{H}\mathsf{ur} = \mathbb{H}_{33} = \mathbb{Z}[u, v]$ derived from the Hurwitz group $2A_4$ is spanned by $1, u, v$, and $w = (uv)^{-1}$. All three systems are of interest; there is no need to decide which is the "best."

Corresponding to the occurrence of $2D_2$ as a subgroup of $2A_4$, $\mathbb{H}\mathsf{am}$ is a subring of $\mathbb{H}\mathsf{ur}$, so that $\mathbb{H}\mathsf{am}$ is not a "maximal" system. Although neither the ring $\mathbb{H}_{43}$ of octians nor the ring $\mathbb{H}_{53}$ of icosians has a representation as a four-dimensional lattice, both can be realized as lattices in $\mathbb{R}^8$. As quaternion rings, $\mathbb{H}_{43}$ and $\mathbb{H}_{53}$ both contain $\mathbb{H}\mathsf{am}$ and $\mathbb{H}\mathsf{ur}$ as subrings.

As points of Euclidean 4-space, the elements of $\mathbb{H}\mathsf{am}$ are the vertices of the regular honeycomb $\{4, 3, 3, 4\}$ of tesseracts $\{4, 3, 3\}$. The elements of $\mathbb{H}\mathsf{ur}$ are the vertices of the regular honeycomb $\{3, 3, 4, 3\}$ of 16-cells $\{3, 3, 4\}$. The elements of $\mathbb{H}\mathsf{yb}$ are the vertices of a uniform honeycomb $\{3, 6\} \times \{3, 6\}$, all of whose cellular polychora are double prisms $\{3\} \times \{3\}$.

The rings $\mathbb{H}\mathsf{am}$ and $\mathbb{H}\mathsf{ur}$ of Hamilton and Hurwitz integers may be regarded as quaternionic analogues of the rings $\mathbb{G}$ and $\mathbb{E}$ of Gaussian and Eisenstein integers. The subrings $\mathbb{Z}[\mathrm{i}]$, $\mathbb{Z}[\mathrm{j}]$, and $\mathbb{Z}[\mathrm{k}]$ of either are isomorphic to $\mathbb{G}$, and the subrings $\mathbb{Z}[u]$, $\mathbb{Z}[v]$, and $\mathbb{Z}[w]$ of $\mathbb{H}\mathsf{ur}$ are isomorphic to $\mathbb{E}$. The ring $\mathbb{H}\mathsf{yb}$ of hybrid integers has a subring $\mathbb{Z}[\omega]$ isomorphic to $\mathbb{E}$ and subrings $\mathbb{Z}[\mathrm{j}]$, $\mathbb{Z}[\omega\mathrm{j}]$, and $\mathbb{Z}[\omega^2\mathrm{j}]$ isomorphic to $\mathbb{G}$.

The elements of $\mathbb{H}\mathsf{am}$ are the quaternions $\mathrm{G} = g_0 + g_1\mathrm{i} + g_2\mathrm{j} + g_3\mathrm{k}$, where i, j, and k satisfy Hamilton's relations

$$\mathrm{i}^2 = \mathrm{j}^2 = \mathrm{k}^2 = \mathrm{ijk} = -1 \qquad (15.1.2)$$

and $(g_0, g_1, g_2, g_3) \in \mathbb{Z}^4$. The elements of $\mathbb{H}\mathsf{yb}$ are quaternions of the form $\mathrm{E} = e_0 + e_1\omega + e_2\mathrm{j} + e_3\omega\mathrm{j}$, where ω and j satisfy the relations

$$\omega + \omega^2 = \mathrm{j}^2 = (\omega\mathrm{j})^2 = (\omega^2\mathrm{j})^2 = -1 \tag{15.1.3}$$

and $(e_0, e_1, e_2, e_3) \in \mathbb{Z}^4$. The elements of $\mathbb{H}\mathsf{ur}$ are quaternions of the form $\mathrm{H} = h_0 + h_1 u + h_2 v + h_3 w$, where u, v, and w satisfy the relations

$$u - u^2 = v - v^2 = w - w^2 = uvw = 1 \tag{15.1.4}$$

and $(h_0, h_1, h_2, h_3) \in \mathbb{Z}^4$.

To facilitate our treatment of integral quaternion matrices, we take the twenty-four Hurwitz units to be the four Hamilton units

$$\begin{aligned}
1 &= 1 && = 1,\\
\mathrm{i} &= 1 - u \qquad - w && = vu^{-1},\\
\mathrm{j} &= 1 - u - v && = u^{-1}vuv^{-1},\\
\mathrm{k} &= 1 \qquad - v - w && = v^{-1}u
\end{aligned} \tag{15.1.5}$$

and their negatives, together with the eight auxiliary numbers

$$\begin{aligned}
l &= -\tfrac{1}{2} + \tfrac{1}{2}\mathrm{i} + \tfrac{1}{2}\mathrm{j} + \tfrac{1}{2}\mathrm{k} = \ \ 1 - u - v - w = -vu,\\
p &= -\tfrac{1}{2} - \tfrac{1}{2}\mathrm{i} + \tfrac{1}{2}\mathrm{j} - \tfrac{1}{2}\mathrm{k} = -1 \qquad\qquad + w = -uv,\\
q &= -\tfrac{1}{2} - \tfrac{1}{2}\mathrm{i} - \tfrac{1}{2}\mathrm{j} + \tfrac{1}{2}\mathrm{k} = -1 + u \qquad\qquad = -u^{-1},\\
r &= -\tfrac{1}{2} + \tfrac{1}{2}\mathrm{i} - \tfrac{1}{2}\mathrm{j} - \tfrac{1}{2}\mathrm{k} = -1 \quad + v \qquad = -v^{-1},\\
s &= -\tfrac{1}{2} - \tfrac{1}{2}\mathrm{i} - \tfrac{1}{2}\mathrm{j} - \tfrac{1}{2}\mathrm{k} = -2 + u + v + w = -u^{-1}v^{-1},\\
u &= \ \ \tfrac{1}{2} - \tfrac{1}{2}\mathrm{i} - \tfrac{1}{2}\mathrm{j} + \tfrac{1}{2}\mathrm{k} = \qquad u \qquad\qquad = \ u,\\
v &= \ \ \tfrac{1}{2} + \tfrac{1}{2}\mathrm{i} - \tfrac{1}{2}\mathrm{j} - \tfrac{1}{2}\mathrm{k} = \qquad\quad v \qquad = \ v,\\
w &= \ \ \tfrac{1}{2} - \tfrac{1}{2}\mathrm{i} + \tfrac{1}{2}\mathrm{j} - \tfrac{1}{2}\mathrm{k} = \qquad\qquad w = \ v^{-1}u^{-1}
\end{aligned} \tag{15.1.6}$$

and their negatives. The unit -1 is of period 2, eight units $(l,\ p,\ q,\ r,\ s, -u, -v, -w)$ are of period 3, eight $(-l, -p, -q, -r, -s,$ $u,\ v,\ w)$ are of period 6, and six $(\pm\mathrm{i},\ \pm\mathrm{j},\ \pm\mathrm{k})$ are of period 4.

The twelve hybrid units are the quaternions $1,\ \omega,\ \omega^2,\ \mathrm{j},\ \omega\mathrm{j},\ \omega^2\mathrm{j}$ and their negatives. The unit -1 is of period 2, two units (ω and ω^2) are of period 3, two ($-\omega$ and $-\omega^2$) are of period 6, and six ($\pm\mathrm{j}, \pm\omega\mathrm{j}, \pm\omega^2\mathrm{j}$) are of period 4.

The entries of the matrices we shall be considering are not necessarily restricted to quaternionic integers but in some cases belong to the ring of octians. The following list of values will be useful:

$$\begin{aligned}
\eta &= \tfrac{1}{2}\sqrt{2}+\tfrac{1}{2}\sqrt{2}\mathrm{i}, & l\eta &= -\tfrac{1}{2}\sqrt{2}+\tfrac{1}{2}\sqrt{2}\mathrm{j}, & s\eta &= -\tfrac{1}{2}\sqrt{2}\mathrm{i}-\tfrac{1}{2}\sqrt{2}\mathrm{j},\\
\mathrm{i}\eta &= -\tfrac{1}{2}\sqrt{2}+\tfrac{1}{2}\sqrt{2}\mathrm{i}, & p\eta &= -\tfrac{1}{2}\sqrt{2}\mathrm{i}-\tfrac{1}{2}\sqrt{2}\mathrm{k}, & u\eta &= \tfrac{1}{2}\sqrt{2}+\tfrac{1}{2}\sqrt{2}\mathrm{k},\\
\mathrm{j}\eta &= \tfrac{1}{2}\sqrt{2}\mathrm{j}-\tfrac{1}{2}\sqrt{2}\mathrm{k}, & q\eta &= -\tfrac{1}{2}\sqrt{2}\mathrm{i}+\tfrac{1}{2}\sqrt{2}\mathrm{k}, & v\eta &= \tfrac{1}{2}\sqrt{2}\mathrm{i}-\tfrac{1}{2}\sqrt{2}\mathrm{j},\\
\mathrm{k}\eta &= \tfrac{1}{2}\sqrt{2}\mathrm{j}+\tfrac{1}{2}\sqrt{2}\mathrm{k}, & r\eta &= -\tfrac{1}{2}\sqrt{2}-\tfrac{1}{2}\sqrt{2}\mathrm{j}, & w\eta &= \tfrac{1}{2}\sqrt{2}-\tfrac{1}{2}\sqrt{2}\mathrm{k}.
\end{aligned} \tag{15.1.7}$$

The twelve Hurwitz units (15.1.5) and (15.1.6) and their negatives, together with these twelve octians and their negatives, are the forty-eight units of the ring $\mathbb{H}_{43}$.

EXERCISES 15.1

1. Verify the relations (15.1.2), (15.1.3), and (15.1.4).
2. Construct a multiplication table for the binary dihedral group $2\mathrm{D}_2$, taking as elements the eight Hamilton units ± 1, $\pm\mathrm{i}$, $\pm\mathrm{j}$, $\pm\mathrm{k}$. (A table giving products of elements $1, \mathrm{i}, \mathrm{j}, \mathrm{k}$ will suffice.)
3. Show that the four complex matrices

$$\mathbf{1}=\begin{pmatrix}1&0\\0&1\end{pmatrix},\quad \mathbf{i}=\begin{pmatrix}\mathrm{i}&0\\0&-\mathrm{i}\end{pmatrix},\quad \mathbf{j}=\begin{pmatrix}0&1\\-1&0\end{pmatrix},\quad \mathbf{k}=\begin{pmatrix}0&\mathrm{i}\\\mathrm{i}&0\end{pmatrix}$$

generate a multiplicative group isomorphic to $2\mathrm{D}_2$ and span a lattice isomorphic to $\mathbb{H}\mathrm{am}$.

4. The identity matrix I_4 and the skew-symmetric matrices J_4, J_4', and J_4'' were used in Chapters 5 and 6 to define the orthogonal and symplectic groups O_4, Sp_4, Sp_4', and Sp_4''. Show that these four matrices and their negatives form a multiplicative group isomorphic to $2\mathrm{D}_2$.

5. Construct a multiplication table for the binary dihedral group $2D_3$, taking as elements the twelve hybrid units $\pm 1, \pm\omega, \pm\omega^2, \pm j, \pm\omega j, \pm\omega^2 j$.

6. Construct a multiplication table for the binary tetrahedral group $2A_4$, taking as elements the twenty-four Hurwitz units $\pm 1, \pm i, \pm j, \pm k, \pm l, \pm p, \pm q, \pm r, \pm s, \pm u, \pm v, \pm w$.

15.2 PSEUDO-MODULAR GROUPS

The division ring $\mathbb{H}$ of quaternions can be identified with the Clifford algebra $\mathbb{C}_2 = \mathbb{R}(i, j)$, with $i^2 = j^2 = -1, ji = -ij$. As in §10.4, we define the *reverse* of a quaternion

$$Q = q_0 + q_1 i + q_2 j + q_3 ij = q_0 + q_1 i + q_2 j + q_3 k$$

to be the quaternion

$$Q^* = q_0 + q_1 i + q_2 j + q_3 ji = q_0 + q_1 i + q_2 j - q_3 k.$$

We say that Q is *primitive* (a "Clifford vector") if $Q^* = Q$, i.e., if $q_3 = 0$.

Every 2×2 quaternionic matrix $M = [(A, B), (C, D)]$ has a *pseudo-determinant* $\Delta(M) = AD^* - BC^*$. A matrix M is *reversible* if $AB^* = BA^*$ and $CD^* = DC^*$. The set of reversible 2×2 matrices with real pseudo-determinants is closed under multiplication, forming the *pseudo-linear* semigroup ${}^*L_2(\mathbb{H})$.

A. *The Hamilton pseudo-modular group.* The skew-domain $\mathbb{H}\mathsf{am} = \mathbb{Z}[i, j]$ of Hamilton integers comprises the quaternions $G = g_0 + g_1 i + g_2 j + g_3 k$, where $(g_0, g_1, g_2, g_3) \in \mathbb{Z}^4$. Each Hamilton integer G has a norm

$$N(G) = |G|^2 = g_0{}^2 + g_1{}^2 + g_2{}^2 + g_3{}^2 \qquad (15.2.1)$$

The units of $\mathbb{H}\mathsf{am}$ are the eight numbers with norm 1, namely, $\pm 1, \pm i, \pm j, \pm k$. These form the Hamilton *special scalar* group

$SL(\mathbb{H}am) \cong 2D_2 \cong \langle i, j\rangle$, with proper subgroups C_4 generated by i, j, or k and center $SZ(\mathbb{H}am) \cong C_2 \cong \{\pm 1\}$.

The Hamilton *specialized linear* group $S^*L_2(\mathbb{H}am)$ consists of the reversible 2×2 Hamilton integer matrices of pseudo-determinant 1. Combining the results of Maclachlan, Waterman & Wielenberg (1989) and Johnson & Weiss (1999b, pp. 173–174), it can be established—as will subsequently be verified below—that $S^*L_2(\mathbb{H}am)$ is generated by the matrices

$$A = \begin{pmatrix} 0 & 1 \\ -1 & 0 \end{pmatrix}, \; B = \begin{pmatrix} 1 & 0 \\ 1 & 1 \end{pmatrix}, \; C = \begin{pmatrix} 1 & 0 \\ i & 1 \end{pmatrix}, \; D = \begin{pmatrix} 1 & 0 \\ j & 1 \end{pmatrix}.$$

The center of $S^*L_2(\mathbb{H}am)$ is the *special central scalar* group $SZ_2(\mathbb{H}am) \cong \{\pm I\}$. It follows that the *Hamilton pseudo-modular group*

$$PS^*L_2(\mathbb{H}am) \cong S^*L_2(\mathbb{H}am)/SZ_2(\mathbb{H}am)$$

is generated in H^4 by the half-turn $\alpha = \cdot\langle A\rangle$ and the striations $\beta = \cdot\langle B\rangle$, $\gamma = \cdot\langle C\rangle$, and $\delta = \cdot\langle D\rangle$.

When the set of primitive quaternions is taken as a three-dimensional vector space over $\mathbb{R}$, the primitive Hamilton integers $g_0 + g_1 i + g_2 j$ form a three-dimensional lattice C_3. The points of C_3 are the vertices of a regular cellulation $\{4, 3, 4\}$ of Euclidean 3-space E^3, whose symmetry group $[4, 3, 4]$ is generated by four reflections $\rho_1, \rho_2, \rho_3, \rho_4$ satisfying the relations

$$\rho_1{}^2 = \rho_2{}^2 = \rho_3{}^2 = \rho_4{}^2 = (\rho_1\rho_2)^4 = (\rho_2\rho_3)^3 = (\rho_3\rho_4)^4 = 1,$$
$$\rho_1 \rightleftarrows \rho_3, \; \rho_1 \rightleftarrows \rho_4, \; \rho_2 \rightleftarrows \rho_4. \qquad (15.2.2)$$

This cellulation is the vertex figure of a regular honeycomb $\{3, 4, 3, 4\}$ of hyperbolic 4-space H^4, the cellular polychora of which are asymptotic 24-cells $\{3, 4, 3\}$.

The symmetry group of the honeycomb $\{3, 4, 3, 4\}$ is a paracompact Coxeter group $\bar{\mathbf{R}}_4 \cong [3, 4, 3, 4]$ generated by five reflections $\rho_0, \rho_1, \rho_2, \rho_3, \rho_4$, satisfying (15.2.2) and

$$\rho_0{}^2 = (\rho_0\rho_1)^3 = (\rho_0\rho_2)^2 = (\rho_0\rho_3)^2 = (\rho_0\rho_4)^2 = 1. \qquad (15.2.3)$$

The combined relations (15.2.2) and (15.2.3) are indicated in the Coxeter diagram

0 —— 1 —4— 2 —— 3 —4— 4

The generators ρ_0, ρ_1, ρ_2, ρ_3, ρ_4 can be represented by antilinear fractional transformations $\odot\langle R_0\rangle$, $\odot\langle R_1\rangle$, $\odot\langle R_2\rangle$, $\odot\langle R_3\rangle$, $\odot\langle R_4\rangle$, determined by the unit matrices

$$R_0 = \begin{pmatrix} 0 & 1 \\ 1 & 0 \end{pmatrix}, \; R_1 = \begin{pmatrix} -1 & 0 \\ 1 & 1 \end{pmatrix}, \; R_2 = \begin{pmatrix} -\mathrm{i}\eta & 0 \\ 0 & -\eta \end{pmatrix},$$

$$R_3 = \begin{pmatrix} \nu\eta & 0 \\ 0 & \nu\eta \end{pmatrix}, \; R_4 = \begin{pmatrix} -\mathrm{j} & 0 \\ 0 & -\mathrm{j} \end{pmatrix},$$

where the octians appearing as entries of R_2 and R_3 are defined by (15.1.7).

The direct subgroup $[3,4,3,4]^+$ is generated by four rotations $\sigma_1 = \rho_0\rho_1, \sigma_2 = \rho_1\rho_2, \sigma_3 = \rho_2\rho_3, \sigma_4 = \rho_3\rho_4$, with the defining relations

$$\sigma_1{}^3 = \sigma_2{}^4 = \sigma_3{}^3 = \sigma_4{}^4 = (\sigma_1\sigma_2)^2 = (\sigma_2\sigma_3)^2 = (\sigma_3\sigma_4)^2$$
$$= (\sigma_1\sigma_2\sigma_3)^2 = (\sigma_2\sigma_3\sigma_4)^2 = (\sigma_1\sigma_2\sigma_3\sigma_4)^2 = 1. \qquad (15.2.4)$$

The generators σ_1, σ_2, σ_3, σ_4 can be represented by linear fractional transformations $\cdot\langle S_1\rangle$, $\cdot\langle S_2\rangle$, $\cdot\langle S_3\rangle$, $\cdot\langle S_4\rangle$, respectively corresponding to the unit matrices

$$S_1 = \begin{pmatrix} 1 & 1 \\ -1 & 0 \end{pmatrix}, \quad S_2 = \begin{pmatrix} \mathrm{i}\eta & 0 \\ -\mathrm{i}\eta & -\eta \end{pmatrix},$$

$$S_3 = \begin{pmatrix} r & 0 \\ 0 & p \end{pmatrix}, \quad S_4 = \begin{pmatrix} u\eta & 0 \\ 0 & u\eta \end{pmatrix},$$

where the entries of S_2, S_3, and S_4 are defined by (15.1.6) and (15.1.7). Since each of these matrices is reversible and has pseudo-determinant 1, we see that $[3,4,3,4]^+$ is a subgroup of $\mathrm{PS^*L_2}(\mathbb{H})$—as it should be, since $\mathrm{PS^*L_2}(\mathbb{H}) \cong \mathrm{P^+O_{4,1}}$ is the group of direct isometries of H^4.

The group [3, 4, 3, 4] has an ionic subgroup $[3,4,3^+,4]$, of index 2, generated by the reflections ρ_0, ρ_1, and ρ_4 and the rotation σ_3. The groups $[3,4,3,4]^+$ and $[3,4,3^+,4]$ have a common subgroup $[3,4,3^+,4]^+$, of index 2 in both and of index 4 in $[3,4,3,4]$, generated by the rotations σ_1, σ_2, and $\sigma_{1234} = \sigma_1\sigma_2\sigma_3\sigma_4 = \rho_0\rho_4$, satisfying the relations

$$\begin{aligned}\sigma_1{}^3 = \sigma_3{}^3 = \sigma_{1234}{}^2 = (\sigma_1\sigma_{1234})^2 = (\sigma_1\sigma_3{}^{-1}\sigma_1{}^{-1}\sigma_3)^2 \\ = (\sigma_1\sigma_3{}^{-1}\sigma_{1234}\sigma_3)^2 = (\sigma_{1234}\sigma_3{}^{-1}\sigma_{1234}\sigma_3)^2 = 1. \qquad (15.2.5)\end{aligned}$$

The corresponding matrices are S_1, S_2, and $S_{1234} = -S_1S_2S_3S_4$.

The group [3, 4, 3, 4] has a hypercompact subgroup $[\infty,3,3,4,4]$, of index 3, generated by the six reflections $\rho_0, \rho_1, \rho_2, \rho_4, \rho_{343} = \rho_3\rho_4\rho_3$, and $\rho_{32123} = \rho_3\rho_2\rho_1\rho_2\rho_3$, satisfying the relations indicated in the diagram

This group has a halving subgroup

$$[3,4,3^*,4] \cong [\infty,3,3,4,1^+,4] \cong [(3,\infty)^{1,1,1}],$$

a "radical" subgroup of index 6 in $[3,4,3,4]$, as well as a subgroup of index 3 in $[3,4,3^+,4]$, generated by the reflections ρ_0, ρ_1, and ρ_4 and their conjugates

$$\rho_{212} = \rho_2\rho_1\rho_2 = \sigma_3\rho_1\sigma_3{}^{-1}, \quad \rho_{343} = \rho_3\rho_4\rho_3 = \sigma_3{}^{-1}\sigma_4\sigma_3,$$
$$\rho_{32123} = \rho_3\rho_{212}\rho_3 = \sigma_3{}^{-1}\rho_1\sigma_3, \quad \rho_{23432} = \rho_2\rho_{343}\rho_2 = \sigma_3\rho_4\sigma_3{}^{-1},$$

satisfying the relations indicated in the diagram

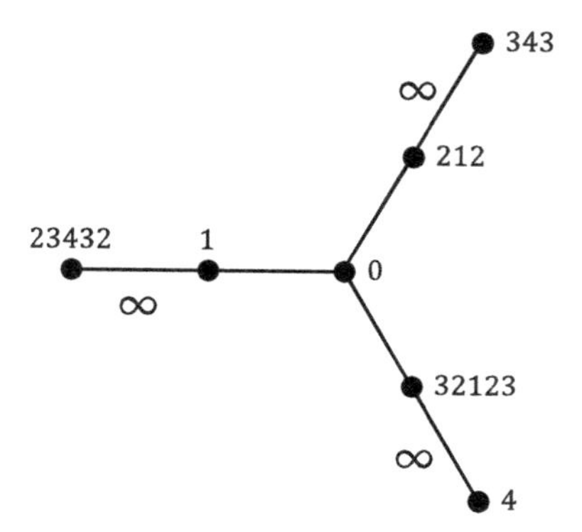

The latter group has an automorphism group D_3, permuting the pairs of generators (ρ_1, ρ_{23432}), (ρ_{212}, ρ_{343}), and (ρ_{32123}, ρ_4).

The foregoing groups have a common subgroup $[3,4,3^*,4]^+$, of index 2 in $[3,4,3^*,4]$, of index 4 in $[\infty,3,3,4,4]$, of index 6 in $[3,4,3^+,4]$ and $[3,4,3,4]^+$, and of index 12 in $[3,4,3,4]$, generated by the striations β, γ, and δ and the rotations σ, ϕ, and ψ, where

$$\beta = \rho_{23432}\rho_1 = \sigma_3\sigma_4{}^{-1}\sigma_2{}^{-1}, \quad \sigma = \rho_0\rho_1 = \sigma_1,$$

$$\gamma = \rho_{343}\rho_{212} = \sigma_4\sigma_3{}^{-1}\sigma_2, \quad \phi = \rho_0\rho_{212} = \sigma_1\sigma_2{}^2,$$

$$\delta = \rho_4\rho_{23123} = \sigma_4{}^{-1}\sigma_2{}^{-1}\sigma_3, \quad \psi = \rho_0\rho_{32123} = \sigma_3{}^{-1}\sigma_1\sigma_3.$$

The corresponding matrices

$$B = S_3S_4{}^{-1}S_2{}^{-1}, \quad S = S_1,$$

$$C = S_4S_3{}^{-1}S_2, \quad U = S_1S_2{}^2,$$

$$D = S_4{}^{-1}S_2{}^{-1}S_3, \quad V = -S_3{}^{-1}S_1S_3$$

are easily determined to be

$$B = \begin{pmatrix} 1 & 0 \\ 1 & 1 \end{pmatrix}, \quad C = \begin{pmatrix} 1 & 0 \\ \mathrm{i} & 1 \end{pmatrix}, \quad D = \begin{pmatrix} 1 & 0 \\ \mathrm{j} & 1 \end{pmatrix},$$

$$S = \begin{pmatrix} 1 & 1 \\ -1 & 0 \end{pmatrix}, \quad U = \begin{pmatrix} -1 & \mathrm{i} \\ \mathrm{i} & 0 \end{pmatrix}, \quad V = \begin{pmatrix} -1 & \mathrm{j} \\ \mathrm{j} & 0 \end{pmatrix}.$$

The matrix $A = SB^{-1}$ belongs to the corresponding matrix group, which can then be generated by just the matrices A, B, C, and D, since $S = AB$, $U = C^{-1}ACA$, and $V = D^{-1}ADA$. It follows that the half-turn $\alpha = \sigma\beta^{-1}$ and the striations β, γ, and δ suffice to generate $[3, 4, 3^*, 4]^+$.

These matrices all belong to the specialized linear group $\mathrm{S^*L_2}(\mathbb{H}\mathrm{am})$, so that $[3, 4, 3^*, 4]^+ \cong \langle\alpha, \beta, \gamma, \delta\rangle$ is a subgroup of $\mathrm{PS^*L_2}(\mathbb{H}\mathrm{am})$. Maclachlan, Waterman & Wielenberg (1989, pp. 748–751) present the group $\mathrm{PS^*L_2}(\mathbb{H}\mathrm{am})$ as a "graph amalgamation product" of eight finite groups generated by subsets of the half-turns α, $\phi\gamma^{-1}$, and $\psi\delta^{-1}$ and the rotations σ, ϕ, and ψ (each of period 3). The Hamilton pseudo-modular group $\mathrm{PS^*L_2}(\mathbb{H}\mathrm{am})$ is therefore isomorphic to $[3, 4, 3^*, 4]^+$, a subgroup of index 6 in the rotation group of the regular honeycomb {3, 4, 3, 4}. Applying these results to the corresponding matrix group, we confirm that the Hamilton specialized linear group $\mathrm{S^*L_2}(\mathbb{H}\mathrm{am})$ is generated by the matrices A, B, C, and D.

B. *The Hurwitz pseudo-modular group*. The skew-domain $\mathbb{H}\mathrm{ur} = \mathbb{Z}[u, v]$ of Hurwitz integers comprises the quaternions $\mathrm{H} = h_0 + h_1u + h_2v + h_3w$, where $(h_0, h_1, h_2, h_3) \in \mathbb{Z}^4$. Each Hurwitz integer H has a norm

$$N(\mathrm{H}) = |\mathrm{H}|^2 = h_0{}^2 + h_1{}^2 + h_2{}^2 + h_3{}^2 + h_0h_1 + h_0h_2 + h_0h_3. \quad (15.2.6)$$

The units of $\mathbb{H}\mathrm{ur}$ are the twenty-four numbers with norm 1, namely, the eight Hamilton units and sixteen others defined by (15.1.6). These form the Hurwitz *special scalar* group $\mathrm{SL}(\mathbb{H}\mathrm{ur}) \cong 2\mathrm{A}_4 \cong \langle u, v\rangle$, with proper subgroups $2\mathrm{D}_2 \cong \langle \mathrm{i}, \mathrm{j}\rangle$; C_6 generated by u, v, or w; C_4 generated by i, j, or k; C_3 generated by $-u, -v$, or $-w$; and center $\mathrm{SZ}(\mathbb{H}\mathrm{ur}) \cong \mathrm{C}_2 \cong \{\pm 1\}$.

The group $[3, 4, 3^*, 4]^+ \cong [(3, \infty)^{1,1,1}]^+$ has an automorphism of period 3, conjugation by σ_3, cyclically permuting the pairs of generators $(\beta, \sigma), (\gamma, \phi)$, and (δ, ψ). Adjoining σ_3, we obtain the group $[3, 4, 3^+, 4]^+$, which is thus a semidirect product of $[3, 4, 3^*, 4]^+$ and the automorphism group C_3 generated by σ_3. The automorphism transforms Hamilton integers into Hurwitz integers, so that $[3,4,3^+,4]^+$ is the *Hurwitz pseudo-modular group*

$$\mathrm{PS^*L_2(\mathbb{H}ur)} \cong \mathrm{S^*L_2(\mathbb{H}ur)/SZ_2(\mathbb{H}ur)}$$

(Maclachlan, Waterman & Wielenberg, 1989 p. 751). Since $[3,4,3^+,4]^+$ is generated by the rotations σ_1, σ_3, and σ_{1234}, it follows that the Hurwitz *specialized linear* group $\mathrm{S^*L_2(\mathbb{H}ur)}$ is generated by the matrices

$$S_1 = \begin{pmatrix} 1 & 1 \\ -1 & 0 \end{pmatrix}, \quad S_3 = \begin{pmatrix} r & 0 \\ 0 & p \end{pmatrix}, \quad S_{1234} = \begin{pmatrix} 0 & \mathrm{j} \\ \mathrm{j} & 0 \end{pmatrix},$$

where $p = -uv$ and $r = -v^{-1}$.

C. *The hybrid pseudo-modular group.* The skew-domain $\mathbb{H}\mathrm{yb} = \mathbb{Z}[\omega, \mathrm{j}]$ of hybrid integers comprises the quaternions $\mathrm{E} = e_0 + e_1\omega + e_2\mathrm{j} + e_3\omega\mathrm{j}$, where $\omega = -½ + ½\sqrt{3}\mathrm{i}$ and $(e_0, e_1, e_2, e_3) \in \mathbb{Z}^4$. Each hybrid integer E has a norm

$$N(\mathrm{E}) = |\mathrm{E}|^2 = e_0{}^2 - e_0e_1 + e_1{}^2 + e_2{}^2 - e_2e_3 + e_3{}^2. \qquad (15.2.7)$$

The units of $\mathbb{H}\mathrm{yb}$ are the twelve numbers with norm 1: ± 1, $\pm\omega$, $\pm\omega^2$, $\pm\mathrm{j}$, $\pm\omega\mathrm{j}$, $\pm\omega^2\mathrm{j}$. These form the hybrid *special scalar* group $\mathrm{SL(\mathbb{H}yb)} \cong 2\mathrm{D}_3 \cong \langle\omega, \mathrm{j}\rangle$, with proper subgroups C_6 generated by $-\omega$; C_4 generated by j, ωj, or ω^2j; C_3 generated by ω; and center $\mathrm{SZ(\mathbb{H}yb)} \cong \mathrm{C}_2 \cong \{\pm 1\}$.

Like $\mathrm{S^*L_2(\mathbb{H}am)}$ with i replaced by ω, the hybrid *specialized linear* group $\mathrm{S^*L_2(\mathbb{H}yb)}$ is generated by the matrices

$$A = \begin{pmatrix} 0 & 1 \\ -1 & 0 \end{pmatrix}, \quad B = \begin{pmatrix} 1 & 0 \\ 1 & 1 \end{pmatrix}, \quad C = \begin{pmatrix} 1 & 0 \\ \omega & 1 \end{pmatrix}, \quad D = \begin{pmatrix} 1 & 0 \\ \mathrm{j} & 1 \end{pmatrix}.$$

As we shall see, the generators are also related to hyperbolic Coxeter groups.

The primitive hybrid integers $e_0 + e_1\omega + e_2\mathrm{j}$ form a three-dimensional lattice $\mathsf{A}_2 \oplus \mathsf{A}_1$. The points of $\mathsf{A}_2 \oplus \mathsf{A}_1$ are the vertices of a uniform prismatic cellulation $\{3,6\} \times \{\infty\}$ of Euclidean 3-space E^3, the cell polyhedra being triangular prisms $\{3\} \times \{\ \}$. The symmetry group $[3,6] \times [\infty]$ of the cellulation, generated by five reflections $\rho_1, \rho_2, \rho_3, \rho_{\bar{1}}$, and $\rho_{\bar{2}}$, can be realized in hyperbolic 4-space H^4. We can then adjoin another reflection ρ_0 to generate a hypercompact Coxeter group $[\infty,3,3,3,6]$, whose fundamental region is the closure of a prismatic pyramid with its apex on the absolute hyper-sphere. Defining relations for this group are implicit in the diagram

The generators can be represented by antilinear fractional transformations $\odot\langle R_0\rangle$, $\odot\langle R_1\rangle$, $\odot\langle R_2\rangle$, $\odot\langle R_3\rangle$, $\odot\langle R_{\bar{1}}\rangle$, $\langle\odot R_{\bar{2}}\rangle$, defined by the unit matrices

$$R_0 = \begin{pmatrix} 0 & 1 \\ 1 & 0 \end{pmatrix}, \quad R_1 = \begin{pmatrix} -1 & 0 \\ 1 & 1 \end{pmatrix}, \quad R_2 = \begin{pmatrix} -\omega & 0 \\ 0 & \omega^2 \end{pmatrix},$$

$$R_3 = \begin{pmatrix} -\mathrm{i} & 0 \\ 0 & -\mathrm{i} \end{pmatrix}, \quad R_{\bar{1}} = \begin{pmatrix} \mathrm{j} & 0 \\ 1 & \mathrm{j} \end{pmatrix}, \quad R_{\bar{2}} = \begin{pmatrix} \mathrm{j} & 0 \\ 0 & \mathrm{j} \end{pmatrix}.$$

The generators of the direct subgroup $[\infty,3,3,3,6]^+$ are the four rotations $\sigma_1 = \rho_0\rho_1$, $\sigma_2 = \rho_1\rho_2$, $\sigma_3 = \rho_2\rho_3$, and $\acute{\sigma}_1 = \rho_0\rho_{\bar{1}}$, of respective periods 3, 3, 6, and 3, and the striation $\acute{\sigma}_2 = \rho_{\bar{1}}\rho_{\bar{2}}$. These generators can be represented by linear fractional transformations $\cdot\langle S_1\rangle$, $\cdot\langle S_2\rangle$, $\cdot\langle S_3\rangle$, $\cdot\langle \acute{S}_1\rangle$, $\cdot\langle \acute{S}_2\rangle$, corresponding to the unit matrices

$$S_1 = \begin{pmatrix} 1 & 1 \\ -1 & 0 \end{pmatrix}, \quad S_2 = \begin{pmatrix} \omega & 0 \\ -\omega & \omega^2 \end{pmatrix}, \quad S_3 = \begin{pmatrix} \omega^2\mathrm{i} & 0 \\ 0 & -\omega\mathrm{i} \end{pmatrix},$$

$$\acute{S}_1 = \begin{pmatrix} 1 & \mathrm{j} \\ \mathrm{j} & 0 \end{pmatrix}, \quad \acute{S}_2 = \begin{pmatrix} 1 & 0 \\ \mathrm{j} & 1 \end{pmatrix}.$$

The group $[\infty,3,3,3,6]$ has a halving subgroup

$$[\infty,3,3,3,6,1^+] \cong [\infty,3,3,3^{[3]}],$$

generated by the reflections $\rho_{\bar{2}}$, $\rho_{\bar{1}}$, ρ_0, ρ_1, ρ_2, and $\rho_{323} = \rho_3\rho_2\rho_3$, satisfying the relations indicated in the diagram

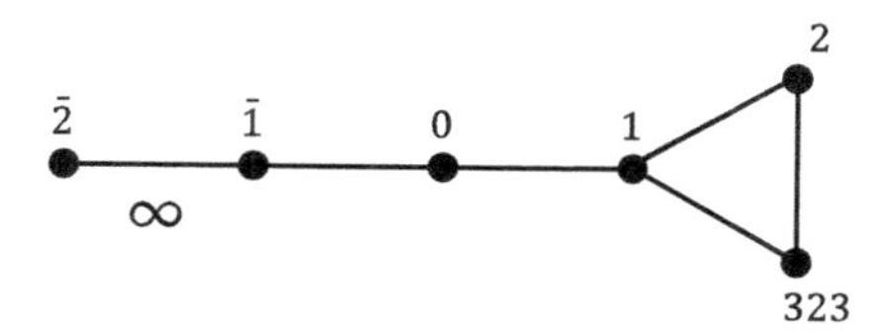

The groups $[\infty, 3, 3, 3, 6]^+$ and $[\infty, 3, 3, 3, 6, 1^+]$ have a common subgroup $[\infty, 3, 3, 3, 6, 1^+]^+ \cong [\infty, 3, 3, 3^{[3]}]^+$, of index 4 in $[\infty, 3, 3, 3, 6]$ and of index 2 in the others, generated by the striation $\acute{\sigma}_2$ and the four rotations $\acute{\sigma}_1$, σ_1, σ_2, and $\sigma_{33} = {\sigma_3}^2 = (\rho_2\rho_3)^2 = \rho_2\rho_{323}$, all of period 3 (only three of the rotations are actually needed).

The corresponding matrices $\acute{S}_2$, $\acute{S}_1$, S_1, S_2, and $S_{33} = -{S_3}^2 = \backslash\omega,\ \omega^2\backslash$ all belong to the hybrid specialized linear group $\mathrm{S^*L_2}(\mathbb{H}\mathrm{yb})$. The group $\mathrm{SL_2}(\mathbb{E})$ generated by S_1, S_2, and S_{33} contains the matrices A, B, and C, since

$$A = S_1S_2{S_{33}}^{-1}, \quad B = S_{33}{S_2}^{-1}, \quad C = S_2S_{33}S_2.$$

We also observe that $D = \acute{S}_2$. Thus the group generated by S_1, S_2, S_{33}, and $\acute{S}_2$ is $\mathrm{S^*L_2}(\mathbb{H}\mathrm{yb}) \cong \langle A, B,\ C,\ D\rangle$. It follows that $[\infty, 3, 3, 3, 6, 1^+]^+$, generated by the three rotations σ_1, σ_2, and σ_{33} and the striation $\acute{\sigma}_2$, is the *hybrid pseudo-modular group*

$$\mathrm{PS^*L_2}(\mathbb{H}\mathrm{yb}) \cong \mathrm{S^*L_2}(\mathbb{H}\mathrm{yb})/\mathrm{SZ_2}(\mathbb{H}\mathrm{yb}).$$

EXERCISES 15.2

1. Show that the antilinear fractional transformations $\rho_0, \rho_1, \rho_2, \rho_3$, and ρ_4 represented by the matrices R_0, R_1, R_2, R_3, and R_4 given in this section satisfy the relations (15.2.2) and (15.2.3) that define the Coxeter group $\bar{\mathbf{R}}_4 \cong [3, 4, 3, 4]$.

2. Show that each of the matrices S_1, S_2, S_3, and S_4 whose associated rotations generate the group $[3,4,3,4]^+$ is reversible and has pseudo-determinant 1.

3. Assuming the relations for the Coxeter group [3, 4, 3, 4], show that the reflections ρ_0, ρ_1, ρ_{23432}, ρ_{212}, ρ_{343}, ρ_{32123}, and ρ_4 satisfy the relations indicated in the diagram for the hypercompact subgroup $[3,4,3^*,4] \cong [(3,\infty)^{1,1,1}]$.

4. Show that the three pairs of generators $(\beta,\ \sigma), (\gamma,\ \phi)$, and $(\delta,\ \psi)$ of the group $[3,4,3^*,4]^+ \cong [(3,\infty)^{1,1,1}]^+$ are cyclically permuted by the automorphism $\theta \mapsto \sigma_3\theta\sigma_3{}^{-1}$.

5. Show that the antilinear fractional transformations ρ_0, ρ_1, ρ_2, ρ_3, $\rho_{\bar{1}}$, $\rho_{\bar{2}}$ represented by the matrices $R_0, R_1, R_2, R_3, R_{\bar{1}}, R_{\bar{2}}$ given in this section satisfy the relations indicated in the diagram for the hypercompact group $[\infty,3,3,3,6]$.

15.3 THE HAMILTON MODULAR GROUP

The Hamilton *special linear* group $\mathrm{SL}_2(\mathbb{H}\mathrm{am})$ consists of all 2×2 matrices over the ring $\mathbb{H}\mathrm{am}$ of determinant 1. The center of $\mathrm{SL}_2(\mathbb{H}\mathrm{am})$ is the group $\mathrm{SZ}_2(\mathbb{H}\mathrm{am}) \cong \{\pm I\}$, and the central quotient group $\mathrm{SL}_2(\mathbb{H}\mathrm{am})/\mathrm{SZ}_2(\mathbb{H}\mathrm{am})$ is the *Hamilton modular group* $\mathrm{PSL}_2(\mathbb{H}\mathrm{am})$.

Unlike either of the other quaternionic systems, the Hamilton integers admit only a restricted Euclidean division algorithm on the norm. Given any Hamilton integer G and any positive odd integer m, one can find a Hamilton integer L such that $|\mathrm{G} - m\mathrm{L}|^2 < m^2$, but this cannot always be done when m is even. In spite of the restriction, the algorithm still suffices (see Johnson & Weiss 1999b, pp. 177–179) to show that $\mathrm{SL}_2(\mathbb{H}\mathrm{am})$ is generated by the matrices

$$A = \begin{pmatrix} 0 & 1 \\ -1 & 0 \end{pmatrix},\ B = \begin{pmatrix} 1 & 0 \\ 1 & 1 \end{pmatrix},\ M = \begin{pmatrix} 1 & 0 \\ 0 & \mathrm{i} \end{pmatrix},\ N = \begin{pmatrix} 1 & 0 \\ 0 & \mathrm{j} \end{pmatrix}.$$

It follows that $\mathrm{PSL}_2(\mathbb{H}\mathrm{am})$ is generated by the corresponding linear fractional transformations

$$\alpha = \cdot\langle A\rangle, \quad \beta = \cdot\langle B\rangle, \quad \mu = \cdot\langle M\rangle, \quad \nu = \cdot\langle N\rangle.$$

We note that $A^2 = -I$ and that $M^4 = N^4 = I$. As isometries of hyperbolic 5-space H^5, α is a half-turn, β is a striation, and μ and ν are quarter-turns.

When the skew-field $\mathbb{H}$ of quaternions is taken as a four-dimensional vector space over $\mathbb{R}$, the Hamilton integers form a four-dimensional lattice C_4. The points of C_4 are the vertices of a regular honeycomb {4, 3, 3, 4} of Euclidean 4-space E^4, whose symmetry group [4, 3, 3, 4] is generated by five reflections $\rho_1, \rho_2, \rho_3, \rho_4, \rho_5$, satisfying the relations

$$\begin{aligned}\rho_1{}^2 = \rho_2{}^2 = \rho_3{}^2 = \rho_4{}^2 = \rho_5{}^2 = (\rho_1\rho_2)^4 = (\rho_2\rho_3)^3 = (\rho_3\rho_4)^3 \\ = (\rho_4\rho_5)^4 = 1, \quad \rho_1 \rightleftarrows \rho_3, \ \rho_1 \rightleftarrows \rho_4, \ \rho_1 \rightleftarrows \rho_5, \ \rho_2 \rightleftarrows \rho_4, \\ \rho_2 \rightleftarrows \rho_5, \ \rho_3 \rightleftarrows \rho_5. \qquad (15.3.1)\end{aligned}$$

This honeycomb is the vertex figure of a regular honeycomb {3, 4, 3, 3, 4} of hyperbolic 5-space H^5, the cellular polytopes of which are asymptotic apeirotopes {3, 4, 3, 3}.

The symmetry group $\bar{\mathbf{R}}_5 \cong$ [3, 4, 3, 3, 4] of the honeycomb {3, 4, 3, 3, 4} is generated by six reflections $\rho_0, \rho_1, \rho_2, \rho_3, \rho_4, \rho_5$, satisfying (15.3.1) and

$$\rho_0{}^2 = (\rho_0\rho_1)^3 = (\rho_0\rho_2)^2 = (\rho_0\rho_3)^2 = (\rho_0\rho_4)^2 = (\rho_0\rho_5)^2 = 1. \qquad (15.3.2)$$

The combined relations (15.3.1) and (15.3.2) are indicated in the Coxeter diagram

The generators $\rho_0, \rho_1, \rho_2, \rho_3, \rho_4, \rho_5$ can be represented by antilinear fractional transformations $\odot\langle R_0\rangle, \odot\langle R_1\rangle, \odot\langle R_2\rangle, \odot\langle R_3\rangle, \odot\langle R_4\rangle, \odot\langle R_5\rangle$, defined by the unit matrices

$$R_0 = \begin{pmatrix} 0 & 1 \\ 1 & 0 \end{pmatrix}, \quad R_1 = \begin{pmatrix} -1 & 0 \\ 1 & 1 \end{pmatrix}, \quad R_2 = \begin{pmatrix} -\mathrm{i}\eta & 0 \\ 0 & -\eta \end{pmatrix},$$

$$R_3 = \begin{pmatrix} \nu\eta & 0 \\ 0 & \nu\eta \end{pmatrix}, \quad R_4 = \begin{pmatrix} \mathrm{j}\eta & 0 \\ 0 & \mathrm{j}\eta \end{pmatrix}, \quad R_5 = \begin{pmatrix} \mathrm{k} & 0 \\ 0 & \mathrm{k} \end{pmatrix},$$

where the entries of R_2, R_3, and R_4 are defined by (15.1.7).

The direct subgroup $[3, 4, 3, 3, 4]^+$ is generated by the five rotations $\sigma_1 = \rho_0\rho_1, \sigma_2 = \rho_1\rho_2, \sigma_3 = \rho_2\rho_3, \sigma_4 = \rho_3\rho_4, \sigma_5 = \rho_4\rho_5$, with defining relations

$$\begin{aligned} \sigma_1{}^3 &= \sigma_2{}^4 = \sigma_3{}^3 = \sigma_4{}^3 = \sigma_5{}^4 = (\sigma_1\sigma_2)^2 = (\sigma_2\sigma_3)^2 = (\sigma_3\sigma_4)^2 = (\sigma_4\sigma_5)^2 \\ &= (\sigma_1\sigma_2\sigma_3)^2 = (\sigma_2\sigma_3\sigma_4)^2 = (\sigma_3\sigma_4\sigma_5)^2 = (\sigma_1\sigma_2\sigma_3\sigma_4)^2 = (\sigma_2\sigma_3\sigma_4\sigma_5)^2 \\ &= (\sigma_1\sigma_2\sigma_3\sigma_4\sigma_5)^2 = 1. \end{aligned} \tag{15.3.3}$$

The generators $\sigma_1, \sigma_2, \sigma_3, \sigma_4, \sigma_5$ can be represented by linear fractional transformations $\cdot\langle S_1\rangle, \cdot\langle S_2\rangle, \cdot\langle S_3\rangle, \langle S_4\rangle, \cdot\langle S_5\rangle$, corresponding to the unit matrices

$$S_1 = \begin{pmatrix} 1 & 1 \\ -1 & 0 \end{pmatrix}, \quad S_2 = \begin{pmatrix} \mathrm{i}\eta & 0 \\ -\mathrm{i}\eta & -\eta \end{pmatrix}, \quad S_3 = \begin{pmatrix} r & 0 \\ 0 & p \end{pmatrix},$$

$$S_4 = \begin{pmatrix} s & 0 \\ 0 & s \end{pmatrix}, \quad S_5 = \begin{pmatrix} -\eta & 0 \\ 0 & -\eta \end{pmatrix},$$

where the entries of S_2, S_3, S_4, and S_5 are defined by (15.1.6) and (15.1.7).

The group [3, 4, 3, 3, 4] has an ionic subgroup $[3, 4, (3, 3)^+, 4]$, of index 2, generated by the reflections ρ_0, ρ_1, and ρ_5 and the rotations σ_3 and σ_4. The groups $[3, 4, 3, 3, 4]^+$ and $[3, 4, (3, 3)^+, 4]$ have a common subgroup $[3, 4, (3, 3)^+, 4]^+$, of index 2 in both and of index 4 in $[3, 4, 3, 3, 4]$, generated by the rotations σ_1, σ_3, σ_4, and $\sigma_{12345} = \sigma_1\sigma_2\sigma_3\sigma_4\sigma_5 = \rho_0\rho_5$. These groups all share a "trionic" subgroup $[3, 4, (3, 3)^\triangle, 4]^+$, of index 3 in $[3, 4, (3, 3)^+, 4]^+$, of index 6 in $[3, 4, 3, 3, 4]^+$ and $[3, 4, (3, 3)^+, 4]$, and of index 12 in $[3, 4, 3, 3, 4]$. This is the group generated by the four rotations σ_1, $\sigma_{34} = \sigma_3\sigma_4$, $\sigma_{43} = \sigma_4\sigma_3$, and σ_{12345}, with the defining relations

$$\begin{aligned}\sigma_1{}^3 &= \sigma_{12345}{}^2 = \sigma_{34}^2 = \sigma_{43}{}^2 = (\sigma_{34}\sigma_{43})^2 \\ &= (\sigma_1\sigma_{12345})^2 = (\sigma_{34}\sigma_{12345})^4 = (\sigma_{43}\sigma_{12345})^4 = (\sigma_{34}\sigma_{43}\sigma_{12345})^4 \\ &= (\sigma_1{}^{-1}\sigma_{34}\sigma_1\sigma_{34})^2 = (\sigma_1{}^{-1}\sigma_{43}\sigma_1\sigma_{43})^2 = (\sigma_1{}^{-1}\sigma_{34}\sigma_{43}\sigma_1\sigma_{34}\sigma_{43})^2 \\ &= (\sigma_1\sigma_{34}\sigma_{12345}\sigma_{34})^2 = (\sigma_1\sigma_{43}\sigma_{12345}\sigma_{43})^2 = 1. \end{aligned} \tag{15.3.4}$$

Since their entries are all Hamilton integers, the corresponding matrices S_1, $S_{34} = S_3S_4$, $S_{43} = S_4S_3$, and $S_{12345} = S_1S_2S_3S_4S_5$ all belong to the Hamilton special linear group $\mathrm{SL}_2(\mathbb{H}\mathsf{am}) \cong \langle A, B, M, N\rangle$. In fact, since

$$A = S_{34}S_{43}S_{12345}S_{34}S_{43}, \quad B = A^{-1}S_1,$$
$$M = S_{34}(S_{12345}S_{43})^2, \quad N = S_{34}MAS_{34}S_{43}AM,$$

the group $\mathrm{SL}_2(\mathbb{H}\mathsf{am})$ is generated by the matrices

$$S_1 = \begin{pmatrix} 1 & 1 \\ -1 & 0 \end{pmatrix}, \; S_{34} = \begin{pmatrix} \mathrm{j} & 0 \\ 0 & \mathrm{k} \end{pmatrix}, \; S_{43} = \begin{pmatrix} \mathrm{k} & 0 \\ 0 & \mathrm{i} \end{pmatrix},$$
$$S_{12345} = \begin{pmatrix} 0 & \mathrm{k} \\ \mathrm{k} & 0 \end{pmatrix}.$$

Hence the Hamilton modular group $\mathrm{PSL}_2(\mathbb{H}\mathsf{am}) \cong \langle \alpha,\ \beta,\ \mu,\ \nu\rangle$ is generated by the corresponding rotations σ_1, σ_{34}, σ_{43}, and σ_{12345}, satisfying (15.3.4). That is, $\mathrm{PSL}_2(\mathbb{H}\mathsf{am})$ is isomorphic to the group $[3, 4, (3, 3)^\triangle, 4]^+$, a subgroup of index 6 in the rotation group of the regular honeycomb {3, 4, 3, 3, 4}. That the index is 6 follows from the fact that $[3, 4, (3, 3)^\triangle, 4]^+$ is a subgroup of index 3 in the ionic subgroup $[3, 4, (3, 3)^+, 4]^+$, the group $[3, 3]^\triangle \cong \mathrm{D}_2$ generated by the half-turns σ_{34} and σ_{43} being a subgroup of index 3 in the group $[3, 3]^+ \cong \mathrm{A}_4$ generated by the rotations σ_3 and σ_4.

Corresponding to the occurrence of $[4,\ 4] \times [4,\ 4]$ as a subgroup of the Euclidean group [4, 3, 3, 4], the paracompact hyperbolic group [3, 4, 3, 3, 4] has a hypercompact subgroup [4, 4, 3, 3, 4, 4]. The latter group has two halving subgroups $[4, 4, 3, (3, \infty)^{1,1}]$, which in turn have a common halving subgroup $[(3, \infty)^{1,1,1,1}]$.

EXERCISES 15.3

1. Show that the antilinear fractional transformations $\rho_0, \rho_1, \rho_2, \rho_3, \rho_4, \rho_5$ represented by the matrices $R_0, R_1, R_2, R_3, R_4, R_5$ given in this section satisfy the relations (15.3.1) and (15.3.2) that define the Coxeter group $\bar{\mathbf{R}}_5 \cong [3,4,3,3,4]$.

2. Show that the linear fractional transformations $\sigma_1, \sigma_2, \sigma_3, \sigma_4, \sigma_5$ represented by the matrices S_1, S_2, S_3, S_4, S_5 satisfy the relations (15.3.3) that define the rotation group $\bar{\mathbf{R}}_5^+ \cong [3,4,3,3,4]^+$.

3. Show that the rotations σ_1, σ_{34}, σ_{43}, and σ_{12345} satisfy the relations (15.3.4) that define the group $[3,4,(3,3)^{\triangle},4]^+ \cong \mathrm{PSL}_2(\mathbb{H}\mathrm{am})$.

15.4 THE HURWITZ MODULAR GROUP

The Hurwitz *special linear* group $\mathrm{SL}_2(\mathbb{H}\mathrm{ur})$ consists of all 2×2 matrices over $\mathbb{H}\mathrm{ur}$ of determinant 1. The center of $\mathrm{SL}_2(\mathbb{H}\mathrm{ur})$ is the group $\mathrm{SZ}_2(\mathbb{H}\mathrm{ur}) \cong \{\pm I\}$, and the central quotient group $\mathrm{SL}_2(\mathbb{H}\mathrm{ur})/\mathrm{SZ}_2(\mathbb{H}\mathrm{ur})$ is the *Hurwitz modular group* $\mathrm{PSL}_2(\mathbb{H}\mathrm{ur})$.

The Hurwitz integers admit an unrestricted Euclidean algorithm on the norm. That is, given any Hurwitz integer H and any positive integer m, one can find a Hurwitz integer L such that $|\mathrm{H} - m\mathrm{L}|^2 < m^2$ (Dickson 1923b, pp. 148–149). Johnson & Weiss (1999b, pp. 183–184) made use of this algorithm to show that the Hurwitz special linear group $\mathrm{SL}_2(\mathbb{H}\mathrm{ur})$ is generated by the matrices

$$A = \begin{pmatrix} 0 & 1 \\ -1 & 0 \end{pmatrix}, \quad B = \begin{pmatrix} 1 & 0 \\ 1 & 1 \end{pmatrix}, \quad M = \begin{pmatrix} 1 & 0 \\ 0 & u \end{pmatrix}, \quad N = \begin{pmatrix} 1 & 0 \\ 0 & v \end{pmatrix},$$

where, as before,

$$u = \tfrac{1}{2} - \tfrac{1}{2}\mathrm{i} - \tfrac{1}{2}\mathrm{j} + \tfrac{1}{2}\mathrm{k} \quad \text{and} \quad v = \tfrac{1}{2} + \tfrac{1}{2}\mathrm{i} - \tfrac{1}{2}\mathrm{j} - \tfrac{1}{2}\mathrm{k}.$$

It follows that $\mathrm{PSL}_2(\mathbb{H}\mathrm{ur})$ is generated by the corresponding linear fractional transformations

$$\alpha = \cdot\langle A\rangle, \quad \beta = \cdot\langle B\rangle, \quad \mu = \cdot\langle M\rangle, \quad \nu = \cdot\langle N\rangle.$$

We note that $A^2 = -I$ and that $M^6 = N^6 = I$. As isometries of H^5, α is a half-turn, β is a striation, and μ and ν are rotations of period 6.

When the skew-field $\mathbb{H}$ of quaternions is taken as a four-dimensional vector space over $\mathbb{R}$, the Hurwitz integers form a four-dimensional lattice D_4. The points of D_4 are the vertices of a regular honeycomb {3, 3, 4, 3} of Euclidean 4-space E^4, whose symmetry group [3, 3, 4, 3] is generated by five reflections ρ_1, ρ_2, ρ_3, ρ_4, ρ_5, satisfying the relations

$$\begin{aligned}\rho_1{}^2 = \rho_2{}^2 = \rho_3{}^2 = \rho_4{}^2 = \rho_5{}^2 = (\rho_1\rho_2)^3 = (\rho_2\rho_3)^3 = (\rho_3\rho_4)^4 \\ = (\rho_4\rho_5)^3 = 1, \quad \rho_1 \rightleftarrows \rho_3,\ \rho_1 \rightleftarrows \rho_4,\ \rho_1 \rightleftarrows \rho_5,\ \rho_2 \rightleftarrows \rho_4, \\ \rho_2 \rightleftarrows \rho_5,\ \rho_3 \rightleftarrows \rho_5.\end{aligned} \tag{15.4.1}$$

This honeycomb is the vertex figure of a regular honeycomb {3, 3, 3, 4, 3} of hyperbolic 5-space H^5, the cellular polytopes of which are asymptotic orthoplexes (cross polytopes) {3, 3, 3, 4}.

The symmetry group $\bar{\mathbf{U}}_5 \cong [3, 3, 3, 4, 3]$ of the honeycomb {3, 3, 3, 4, 3} is generated by six reflections $\rho_0, \rho_1, \rho_2, \rho_3, \rho_4, \rho_5$, satisfying (15.4.1) and

$$\rho_0{}^2 = (\rho_0\rho_1)^3 = (\rho_0\rho_2)^2 = (\rho_0\rho_3)^2 = (\rho_0\rho_4)^2 = (\rho_0\rho_5)^2 = 1. \tag{15.4.2}$$

The combined relations (15.4.1) and (15.4.2) are indicated in the Coxeter diagram

The generators $\rho_0, \rho_1, \rho_2, \rho_3, \rho_4, \rho_5$ can be represented by antilinear fractional transformations $\odot\langle R_0\rangle, \odot\langle R_1\rangle, \odot\langle R_2\rangle, \odot\langle R_3\rangle, \odot\langle R_4\rangle, \odot\langle R_5\rangle$, defined by the unit matrices

$$R_0 = \begin{pmatrix} 0 & 1 \\ 1 & 0 \end{pmatrix}, \quad R_1 = \begin{pmatrix} -1 & 0 \\ 1 & 1 \end{pmatrix}, \quad R_2 = \begin{pmatrix} -s & 0 \\ 0 & l \end{pmatrix},$$

$$R_3 = \begin{pmatrix} -\mathrm{i} & 0 \\ 0 & -\mathrm{i} \end{pmatrix}, \quad R_4 = \begin{pmatrix} -v\eta & 0 \\ 0 & -v\eta \end{pmatrix}, \quad R_5 = \begin{pmatrix} -\mathrm{j}\eta & 0 \\ 0 & -\mathrm{j}\eta \end{pmatrix},$$

where the entries of R_2, R_4, and R_5 are defined by (15.1.6) and (15.1.7).

The direct subgroup $[3,3,3,4,3]^+$ is generated by the five rotations $\sigma_1 = \rho_0\rho_1$, $\sigma_2 = \rho_1\rho_2$, $\sigma_3 = \rho_2\rho_3$, $\sigma_4 = \rho_3\rho_4$, $\sigma_5 = \rho_4\rho_5$, with defining relations

$$\begin{aligned} \sigma_1{}^3 &= \sigma_2{}^3 = \sigma_3{}^3 = \sigma_4{}^4 = \sigma_5{}^3 = (\sigma_1\sigma_2)^2 = (\sigma_2\sigma_3)^2 = (\sigma_3\sigma_4)^2 = (\sigma_4\sigma_5)^2 \\ &= (\sigma_1\sigma_2\sigma_3)^2 = (\sigma_2\sigma_3\sigma_4)^2 = (\sigma_3\sigma_4\sigma_5)^2 = (\sigma_1\sigma_2\sigma_3\sigma_4)^2 = (\sigma_2\sigma_3\sigma_4\sigma_5)^2 \\ &= (\sigma_1\sigma_2\sigma_3\sigma_4\sigma_5)^2 = 1. \end{aligned} \tag{15.4.3}$$

The generators $\sigma_1, \sigma_2, \sigma_3, \sigma_4, \sigma_5$ can be represented by linear fractional transformations $\cdot\langle S_1\rangle, \cdot\langle S_2\rangle, \cdot\langle S_3\rangle, \langle S_4\rangle, \cdot\langle S_5\rangle$, corresponding to the unit matrices

$$S_1 = \begin{pmatrix} 1 & 1 \\ -1 & 0 \end{pmatrix}, \quad S_2 = \begin{pmatrix} s & 0 \\ -s & l \end{pmatrix}, \quad S_3 = \begin{pmatrix} p & 0 \\ 0 & -u \end{pmatrix},$$

$$S_4 = \begin{pmatrix} u\eta & 0 \\ 0 & u\eta \end{pmatrix}, \quad S_5 = \begin{pmatrix} s & 0 \\ 0 & s \end{pmatrix},$$

where the entries of S_2, S_3, S_4, and S_5 are defined by (15.1.6) and (15.1.7).

The commutator subgroup of both $[3,3,3,4,3]$ and $[3,3,3,4,3]^+$, of respective indices 4 and 2, is the ionic subgroup $[(3,3,3)^+, 4, 3^+]$ generated by σ_1, σ_2, σ_3, and σ_5, with the defining relations

$$\begin{aligned} \sigma_1^3 &= \sigma_2{}^3 = \sigma_3{}^3 = \sigma_5{}^3 = (\sigma_1\sigma_2)^2 = (\sigma_2\sigma_3)^2 = (\sigma_1\sigma_2\sigma_3)^2 \\ &= (\sigma_1\sigma_5)^2 = (\sigma_2\sigma_5)^2 = (\sigma_3{}^{-1}\sigma_5\sigma_3\sigma_5{}^{-1})^2 = 1. \end{aligned} \tag{15.4.4}$$

Since their entries are all Hurwitz integers, the corresponding matrices S_1, S_2, S_3, and S_5 all belong to the Hurwitz special linear group $\mathrm{SL}_2(\mathbb{H}\mathrm{ur}) \cong \langle A, B, M, N\rangle$. Define the matrix U by

$$U = S_3(S_5S_3{}^{-1})^4S_3{}^{-1}S_5{}^{-1} = \begin{pmatrix} s & 0 \\ 0 & l \end{pmatrix}.$$

Then it can be verified that

$$A = S_1S_2U^{-1}, \quad B = US_2{}^{-1},$$
$$M = (S_3S_5S_3S_5{}^{-1}S_3S_5S_3)^{-1},\ N = -(S_3S_5)^2.$$

Hence the group $SL_2(\mathbb{H}ur)$ is generated by the matrices

$$S_1 = \begin{pmatrix} 1 & 1 \\ -1 & 0 \end{pmatrix},\ S_2 = \begin{pmatrix} s & 0 \\ -s & l \end{pmatrix},\ S_3 = \begin{pmatrix} p & 0 \\ 0 & -u \end{pmatrix},$$
$$S_5 = \begin{pmatrix} s & 0 \\ 0 & s \end{pmatrix},$$

where $l = -vu$, $p = -uv$, and $s = l^{-1}$.

The Hurwitz modular group $PSL_2(\mathbb{H}ur) \cong \langle \alpha, \beta,\ \mu,\ \nu \rangle$ is thus generated by the corresponding rotations σ_1, σ_2, σ_3, and σ_5, satisfying (15.4.4). That is, $PSL_2(\mathbb{H}ur)$ is isomorphic to the group $[(3,3,3)^+,4,3^+]$, the commutator subgroup of the symmetry group of the regular honeycomb {3, 3, 3, 4, 3}. It can also be identified with the group $3[3^{2,1,1,1}]^+$, i.e., the rotation group $\bar{\mathbf{Q}}_5^+ \cong [3^{2,1,1,1}]^+$ extended by an automorphism group C_3 permuting three of its generators.

Since the Coxeter group $\bar{\mathbf{U}}_5 \cong [3,\ 3,\ 3,\ 4,\ 3]$ contains $\bar{\mathbf{R}}_5 \cong [3,\ 4,\ 3,\ 3,\ 4]$ as a subgroup of index 10, its rotation group $\bar{\mathbf{U}}_5^+ \cong [3,\ 3,\ 3,\ 4,\ 3]^+$ likewise contains $\bar{\mathbf{R}}_5^+ \cong [3,\ 4,\ 3,\ 3,\ 4]^+$ as a subgroup of index 10. The latter group has an ionic subgroup $[3,4,(3,3)^+,4]^+$, which, since the matrices corresponding to its generators have only Hurwitz integers as entries, is also a subgroup of $[(3,3,3)^+,4,3^+] \cong PSL_2(\mathbb{H}ur)$, and again the index is 10. Finally the group $[3,4,(3,3)^{\triangle},4]^+ \cong PSL_2(\mathbb{H}am)$, being of index 3 in $[3,4,(3,3)^+,4]^+$, is of index 30 in $[(3,3,3)^+,4,3^+]$. We have thus proved that the Hurwitz modular group $PSL_2(\mathbb{H}ur)$ contains the Hamilton modular group $PSL_2(\mathbb{H}am)$ as a subgroup of index 30.

EXERCISES 15.4

1. Show that the antilinear fractional transformations $\rho_0, \rho_1, \rho_2, \rho_3, \rho_4, \rho_5$ represented by the matrices $R_0, R_1, R_2, R_3, R_4, R_5$ given in this section satisfy the relations (15.4.1) and (15.4.2) that define the Coxeter group $\bar{\mathbf{U}}_5 \cong [3,3,3,4,3]$.

2. Show that the linear fractional transformations $\sigma_1, \sigma_2, \sigma_3, \sigma_4, \sigma_5$ represented by the matrices S_1, S_2, S_3, S_4, S_5 satisfy the relations (15.4.3) that define the rotation group $\bar{\mathbf{U}}_5^+ \cong [3,3,3,4,3]^+$.

3. Show that the rotations $\sigma_1, \sigma_2, \sigma_3$, and σ_5 satisfy the relations (15.4.4) that define the group $[(3,3,3)^+, 4, 3^+] \cong \mathrm{PSL}_2(\mathbb{H}\mathsf{ur})$.

15.5 THE HYBRID MODULAR GROUP

The hybrid *special linear* group $\mathrm{SL}_2(\mathbb{H}\mathsf{yb})$ consists of all 2×2 matrices over $\mathbb{H}\mathsf{yb}$ of determinant 1. The center of $\mathrm{SL}_2(\mathbb{H}\mathsf{yb})$ is the group $\mathrm{SZ}_2(\mathbb{H}\mathsf{yb}) \cong \{\pm I\}$, and the central quotient group $\mathrm{SL}_2(\mathbb{H}\mathsf{yb})/\mathrm{SZ}_2(\mathbb{H}\mathsf{yb})$ is the *hybrid modular group* $\mathrm{PSL}_2(\mathbb{H}\mathsf{yb})$.

Since the hybrid integers admit a Euclidean algorithm on the norm (Dickson 1923b, p. 194), the same procedure that works for the Hurwitz group $\mathrm{SL}_2(\mathbb{H}\mathsf{ur})$ can be used to show that the hybrid special linear group $\mathrm{SL}_2(\mathbb{H}\mathsf{yb})$ is generated by the matrices

$$A = \begin{pmatrix} 0 & 1 \\ -1 & 0 \end{pmatrix}, \quad B = \begin{pmatrix} 1 & 0 \\ 1 & 1 \end{pmatrix}, \quad M = \begin{pmatrix} 1 & 0 \\ 0 & \omega \end{pmatrix}, \quad N = \begin{pmatrix} 1 & 0 \\ 0 & \mathrm{j} \end{pmatrix},$$

where ω is the quaternion $-\tfrac{1}{2} + \tfrac{1}{2}\sqrt{3}\mathrm{i}$. It follows that the hybrid modular group $\mathrm{PSL}_2(\mathbb{H}\mathsf{yb})$ is generated by the linear fractional transformations

$$\alpha = \cdot\langle A\rangle, \quad \beta = \cdot\langle B\rangle, \quad \mu = \cdot\langle M\rangle, \quad \nu = \cdot\langle N\rangle.$$

We note that $A^2 = -I$ and that $M^3 = N^4 = I$. As isometries of H^5, α is a half-turn, β is a striation, and μ and ν are rotations of respective periods 3 and 4.

When the skew-field $\mathbb{H}$ of quaternions is taken as a four-dimensional vector space over $\mathbb{R}$, the hybrid integers form a four-dimensional lattice $\mathsf{A}_2 \oplus \mathsf{A}_2$. The points of the lattice are the vertices of a uniform honeycomb $\{3,6\} \times \{3,6\}$ of Euclidean 4-space E^4, whose cellular polychora are triangular double prisms $\{3\}\times\{3\}$. The reducible Coxeter group $[3,6] \times [3,6]$, which is a subgroup of index 2 in the symmetry group $[[3,6] \times [3,6]]$ of the honeycomb, is the direct product of two apeirohedral groups $[3,6]$, each generated by three reflections

$$\rho_1, \rho_2, \rho_3 \quad \text{or} \quad \rho_{\bar{1}}, \rho_{\bar{2}}, \rho_{\bar{3}}.$$

The group can be realized in hyperbolic 5-space H^5, and another reflection ρ_0 can be adjoined to generate a hypercompact Coxeter group $[6,3,3,3,3,6]$, whose fundamental region is the closure of a double prismatic pyramid with its apex on the absolute hypersphere. Defining relations for this group are implicit in the diagram

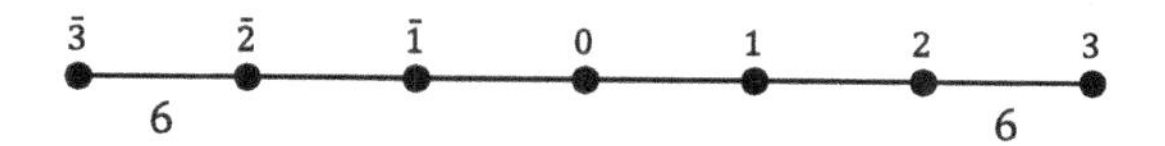

The generators ρ_0, ρ_1, ρ_2, ρ_3, $\rho_{\bar{1}}$, $\rho_{\bar{2}}$, $\rho_{\bar{3}}$ can be represented by antilinear fractional transformations $\odot\langle R_0\rangle$, $\odot\langle R_1\rangle$, $\odot\langle R_2\rangle$, $\odot\langle R_3\rangle$, $\odot\langle R_{\bar{1}}\rangle$, $\odot\langle R_{\bar{2}}\rangle$, $\odot\langle R_{\bar{3}}\rangle$, defined by the unit matrices

$$R_0 = \begin{pmatrix} 0 & 1 \\ 1 & 0 \end{pmatrix}, \; R_1 = \begin{pmatrix} -1 & 0 \\ 1 & 1 \end{pmatrix}, \; R_2 = \begin{pmatrix} -\omega & 0 \\ 0 & \omega^2 \end{pmatrix}, \; R_3 = \begin{pmatrix} -\mathrm{i} & 0 \\ 0 & -\mathrm{i} \end{pmatrix},$$

$$R_{\bar{1}} = \begin{pmatrix} \mathrm{j} & 0 \\ 1 & \mathrm{j} \end{pmatrix}, \; R_{\bar{2}} = \begin{pmatrix} \omega^2\mathrm{j} & 0 \\ 0 & \omega^2\mathrm{j} \end{pmatrix}, \; R_{\bar{3}} = \begin{pmatrix} -\mathrm{k} & 0 \\ 0 & -\mathrm{k} \end{pmatrix}.$$

The direct subgroup $[6,3,3,3,3,6]^+$ is generated by the six rotations

$$\begin{gathered} \sigma_1 = \rho_0\rho_1, \; \sigma_2 = \rho_1\rho_2, \; \sigma_3 = \rho_2\rho_3, \\ \acute{\sigma}_1 = \rho_0\rho_{\bar{1}}, \; \acute{\sigma}_2 = \rho_{\bar{1}}\rho_{\bar{2}}, \; \acute{\sigma}_3 = \rho_{\bar{2}}\rho_{\bar{3}}, \end{gathered} \tag{15.5.1}$$

each of period 3 except for σ_3 and $\acute{\sigma}_3$, which are of period 6. These generators can be represented by linear fractional transformations $\cdot\langle S_1\rangle, \cdot\langle S_2\rangle, \cdot\langle S_3\rangle, \cdot\langle \acute{S}_1\rangle, \cdot\langle \acute{S}_2\rangle, \cdot\langle \acute{S}_3\rangle$, corresponding to the unit matrices

$$S_1 = \begin{pmatrix} 1 & 1 \\ -1 & 0 \end{pmatrix}, \; S_2 = \begin{pmatrix} \omega & 0 \\ -\omega & \omega^2 \end{pmatrix}, \; S_3 = \begin{pmatrix} \omega^2\mathrm{i} & 0 \\ 0 & -\omega\mathrm{i} \end{pmatrix},$$
$$\acute{S}_1 = \begin{pmatrix} 1 & \mathrm{j} \\ \mathrm{j} & 0 \end{pmatrix}, \; \acute{S}_2 = \begin{pmatrix} \omega & 0 \\ \omega^2\mathrm{j} & \omega \end{pmatrix}, \; \acute{S}_3 = \begin{pmatrix} \omega^2\mathrm{i} & 0 \\ 0 & \omega^2\mathrm{i} \end{pmatrix}.$$

The group [6, 3, 3, 3, 3, 6] has two isomorphic halving subgroups

$$[6,3,3,3,3,6,1^+] \quad \text{and} \quad [1^+,6,3,3,3,3,6],$$

one generated by the reflections ρ_0, $\rho_{\bar{1}}$, $\rho_{\bar{2}}$, $\rho_{\bar{3}}$, ρ_1, ρ_2, and $\rho_{323} = \rho_3\rho_2\rho_3$ and the other by the reflections ρ_0, ρ_1, ρ_2, ρ_3, $\rho_{\bar{1}}$, $\rho_{\bar{2}}$, and $\rho_{\bar{3}\bar{2}\bar{3}} = \rho_{\bar{3}}\rho_{\bar{2}}\rho_{\bar{3}}$. These have a common halving subgroup of their own,

$$[1^+,6,3,3,3,3,6,1^+] \cong [3^{[3]},3,3,3^{[3]}],$$

generated by the reflections $\rho_0, \rho_1, \rho_2, \rho_{323}, \rho_{\bar{1}}, \rho_{\bar{2}}$, and $\rho_{\bar{3}\bar{2}\bar{3}}$, satisfying the relations indicated in the diagram

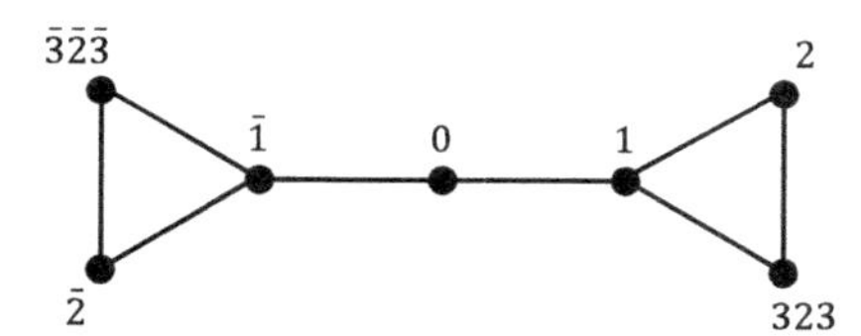

All of these above groups have a common commutator subgroup

$$[1^+,6,(3,3,3,3)^+,6,1^+] \cong [6,3,3,3,3,6]^{+3} \cong [3^{[3]},3,3,3^{[3]}]^+,$$

of index 8 in [6, 3, 3, 3, 3, 6] and of index 4 or 2 in the others, generated by the six rotations

$$\begin{aligned} &\sigma_1,\ \sigma_2,\ \sigma_{33} = {\sigma_3}^2 = (\rho_2\rho_3)^2 = \rho_2\rho_{323}, \\ &\acute{\sigma}_1,\ \acute{\sigma}_2,\ \acute{\sigma}_{33} = {\acute{\sigma}_3}^2 = (\rho_{\bar{2}}\rho_{\bar{3}})^2 = \rho_{\bar{2}}\rho_{\bar{3}\bar{2}\bar{3}}, \end{aligned} \tag{15.5.2}$$

all of period 3. Since their entries are all hybrid integers, the corresponding matrices S_1, S_2, $S_{33} = -{S_3}^2$ and $\acute{S}_1$, $\acute{S}_2$, $\acute{S}_{33} = -{\acute{S}_3}^2$ all belong to the hybrid special linear group $\mathrm{SL}_2(\mathbb{H}\mathsf{yb}) \cong \langle A, B, M, N \rangle$. The matrix group has an automorphism of period 4, conjugation by N, with

$$N^{-1}SN = \acute{S}, \quad N^{-1}\acute{S}N = \H{S}, \quad N^{-1}\H{S}N = \dddot{S}, \quad N^{-1}\dddot{S}N = S$$

for any matrix S. It can be verified that

$$A = S_1 S_2 {S_{33}}^{-1}, \quad B = S_{33} {S_2}^{-1}, \quad \text{and} \quad M = S_{33} N^{-1} {S_{33}}^{-1} N.$$

Hence the group $\mathrm{SL}_2(\mathbb{H}\mathsf{yb})$ is generated by the matrices

$$S_1 = \begin{pmatrix} 1 & 1 \\ -1 & 0 \end{pmatrix}, \; S_2 = \begin{pmatrix} \omega & 0 \\ -\omega & \omega^2 \end{pmatrix}, \; S_{33} = \begin{pmatrix} \omega & 0 \\ 0 & \omega^2 \end{pmatrix},$$
$$N = \begin{pmatrix} 1 & 0 \\ 0 & \mathrm{j} \end{pmatrix}.$$

The hybrid modular group $\mathrm{PSL}_2(\mathbb{H}\mathsf{yb}) \cong \langle \alpha, \beta, \mu, \nu \rangle$ is thus generated by the corresponding rotations σ_1, σ_2, σ_{33}, and ν, satisfying the relations (14.3.4) and

$$\nu^4 = ({\sigma_1}^{-1}\nu^{-1}\sigma_1\nu)^2 = ({\sigma_2}^{-1}\nu^{-1}\sigma_2\nu)^3 = ({\sigma_{33}}^{-1}\nu^{-1}\sigma_{33}\nu)^3 = 1. \tag{15.5.3}$$

Since the conjugate rotations $\acute{\sigma}_1$, $\acute{\sigma}_2$, and $\acute{\sigma}_{33}$ are given by

$$\acute{\sigma}_1 = \nu^{-1}\sigma_1\nu, \quad \acute{\sigma}_2 = \nu^{-1}\sigma_2\nu, \quad \text{and} \quad \acute{\sigma}_{33} = \nu^{-1}\sigma_{33}\nu, \tag{15.5.4}$$

$\mathrm{PSL}_2(\mathbb{H}\mathsf{yb})$ is isomorphic to the extended group $4[6,3,3,3,3,6]^{+3}$, the semidirect product of the commutator subgroup

$$[6,3,3,3,3,6]^{+3} \cong \langle \sigma_1, \sigma_2, \sigma_{33}, \acute{\sigma}_1, \acute{\sigma}_2, \acute{\sigma}_{33} \rangle$$

and the automorphism group C_4 generated by the quarter-turn ν.

EXERCISES 15.5

1. Show that the antilinear fractional transformations $\rho_0, \rho_1, \rho_2, \rho_3, \rho_{\bar{1}}, \rho_{\bar{2}}, \rho_{\bar{3}}$ represented by the matrices $R_0, R_1, R_2, R_3, R_{\bar{1}}, R_{\bar{2}}, R_{\bar{3}}$ given in this section satisfy the relations indicated in the diagram for the hypercompact group $[6,3,3,3,3,6]$.

2. Show that the rotations $\sigma_1, \sigma_2, \sigma_3, \acute{\sigma}_1, \acute{\sigma}_2, \acute{\sigma}_3$ are represented by the linear fractional transformations $\cdot\langle S_1\rangle, \cdot\langle S_2\rangle, \cdot\langle S_3\rangle, \cdot\langle \acute{S}_1\rangle, \cdot\langle \acute{S}_2\rangle, \cdot\langle \acute{S}_3\rangle$.

3. Assuming the relations for the Coxeter group [6, 3, 3, 3, 3, 6], show that the relations $\rho_0, \rho_1, \rho_2, \rho_{323}, \rho_{\bar{1}}, \rho_{\bar{2}}, \rho_{\bar{3}\bar{2}\bar{3}}$ satisfy the relations indicated in the diagram for the subgroup $[3^{[3]}, 3, 3, 3^{[3]}]$.

4. Show that the automorphism $S \mapsto N^{-1}SN$ takes the matrices S_1, S_2, and S_3 into the matrices $\acute{S}_1, \acute{S}_2$, and $\acute{S}_{33}$. Apply the automorphism again to obtain matrices $\H{S}_1, \H{S}_2$, and $\H{S}_{33}$ and a third time to obtain matrices $\acute{\H{S}}_1, \acute{\H{S}}_2$, and $\acute{\H{S}}_{33}$. Show that a fourth application of the automorphism yields the original matrices.

15.6 SUMMARY OF MODULAR GROUPS

Each of the discrete groups of linear fractional or antilinear fractional transformations described in this chapter and the preceding one is related to a paracompact or hypercompact Coxeter group operating in hyperbolic n-space ($2 \leq n \leq 5$). In most cases these are subgroups of the symmetry group of some regular honeycomb.

The rational modular group is the rotation group of the regular tessellation $\{3, \infty\}$ of the hyperbolic plane H^2:

$$\mathrm{PSL}_2(\mathbb{Z}) \cong [3, \infty]^+ \tag{15.6.1}$$

Three semiquadratic modular groups are likewise the rotation groups of other regular hyperbolic tessellations:

$$\begin{aligned} \mathrm{PSL}_{1+1}(\mathbb{Z}[\sqrt{2}]) &\cong [4, \infty]^+, \\ \mathrm{PSL}_{1+1}(\mathbb{Z}[\tau]) &\cong [5, \infty]^+, \\ \mathrm{PSL}_{1+1}(\mathbb{Z}[\sqrt{3}]) &\cong [6, \infty]^+. \end{aligned} \tag{15.6.2}$$

The Gaussian modular (Picard) group and the Eisenstein modular group are subgroups of the rotation groups of the regular cellulations {3, 4, 4} and {3, 3, 6} of hyperbolic 3-space:

$$\begin{aligned} \mathrm{PSL}_2(\mathbb{G}) &\cong [3,4,1^+,4]^+, \\ \mathrm{PSL}_2(\mathbb{E}) &\cong [(3,3)^+,6,1^+]. \end{aligned} \tag{15.6.3}$$

The Hamilton and Hurwitz pseudo-modular groups are subgroups of the rotation group of the regular honeycomb {3, 4, 3, 4} of hyperbolic 4-space, and the hybrid pseudo-modular group is a subgroup of a hypercompact Coxeter group also operating in H^4:

$$\begin{aligned} \mathrm{PS^*L}_2(\mathbb{H}\mathsf{am}) &\cong [3,4,3^*,4]^+, \\ \mathrm{PS^*L}_2(\mathbb{H}\mathsf{ur}) &\cong [3,4,3^+,4]^+, \\ \mathrm{PS^*L}_2(\mathbb{H}\mathsf{yb}) &\cong [\infty,3,3,3,6,1^+]^+. \end{aligned} \tag{15.6.4}$$

The Hamilton and Hurwitz modular groups are subgroups of the rotation groups of the regular honeycombs {3, 4, 3, 3, 4} and {3, 3, 3, 4, 3} of hyperbolic 5-space, and the hybrid modular group is an extension of the commutator subgroup of a hypercompact Coxeter group operating in H^5:

$$\begin{aligned} \mathrm{PSL}_2(\mathbb{H}\mathsf{am}) &\cong [3,4,(3,3)^\triangle,4]^+, \\ \mathrm{PSL}_2(\mathbb{H}\mathsf{ur}) &\cong [(3,3,3)^+,4,3^+], \\ \mathrm{PSL}_2(\mathbb{H}\mathsf{yb}) &\cong 4[1^+,6,(3,3,3,3)^+,6,1^+]. \end{aligned} \tag{15.6.5}$$

Connections between hyperbolic isometries (or Möbius transformations) and real, complex, or quaternionic linear fractional transformations have been established by numerous authors. Relevant citations include Klein (1879) Poincaré (1882), Picard (1884), Bianchi (1891, 1892), Fricke & Klein (1897) Vahlen (1902), Ahlfors (1985), Maclachlan, Waterman & Wielenberg (1989) Waterman (1993), Wilker (1993), Schulte & Weiss (1994), and Monson & Weiss (1995, 1997). More details can be found in Johnson & Weiss 1999a,b and in Feingold, Kleinschmidt & Nicolai 2009.

Having described the real, complex, and quaternionic cases, it would be natural to expect to continue with the octonions. As we shall see in the next section, there is no difficulty in defining systems of octonionic integers and interpreting them geometrically. However, since any real division algebra of more than four dimensions is nonassociative, the notions of linear fractional transformations and modular groups cannot be extended. (It may not be entirely coincidental that regular hyperbolic honeycombs exist only in spaces of two, three, four, and five dimensions.)

15.7 INTEGRAL OCTONIONS

The alternative division ring $\mathbb{O}$ of octonions is an eight-dimensional vector space over the real numbers. Systems of octonions analogous to the rational integers $\mathbb{Z}$; the Gaussian and Eisenstein integers $\mathbb{G} = \mathbb{Z}[\mathrm{i}]$ and $\mathbb{E} = \mathbb{Z}[\omega]$; and the Hamilton, Hurwitz, and hybrid integers $\mathbb{H}\mathsf{am}$, $\mathbb{H}\mathsf{ur}$, and $\mathbb{H}\mathsf{yb}$ have been investigated by L. E. Dickson (1923a,b) and others, e.g., Coxeter (1946) and Conway & Smith (2003, chap. 9).

As with complex numbers and quaternions, each octonion

$$\mathbf{x} = x_0 + x_1\mathrm{e}_1 + x_2\mathrm{e}_2 + x_3\mathrm{e}_3 + x_4\mathrm{e}_4 + x_5\mathrm{e}_5 + x_6\mathrm{e}_6 + x_7\mathrm{e}_7$$

has a *conjugate*

$$\tilde{\mathbf{x}} = x_0 - x_1\mathrm{e}_1 - x_2\mathrm{e}_2 - x_3\mathrm{e}_3 - x_4\mathrm{e}_4 - x_5\mathrm{e}_5 - x_6\mathrm{e}_6 - x_7\mathrm{e}_7,$$

in terms of which we can define its *trace* $\mathrm{tr}\,\mathbf{x} = \mathbf{x} + \tilde{\mathbf{x}} = 2x_0$ and *norm* $N(\mathbf{x}) = \mathbf{x}\tilde{\mathbf{x}} = x_0{}^2 + \cdots + x_7{}^2$, both of which are real. Traces are additive and norms multiplicative, i.e., $\mathrm{tr}\,(\mathbf{x}+\mathbf{y}) = \mathrm{tr}\,\mathbf{x} + \mathrm{tr}\,\mathbf{y}$ and $N(\mathbf{xy}) = N(\mathbf{x}) \cdot N(\mathbf{y})$. Just as every real number a satisfies a quadratic equation $x^2 - 2ax + a^2 = 0$, every octonion $\mathbf{a}$ satisfies a "rank equation" $\mathbf{x}^2 - (\mathbf{a} + \tilde{\mathbf{a}})\,\mathbf{x} + \mathbf{a}\tilde{\mathbf{a}} = 0$. (Similar definitions apply and corresponding equations hold for complex numbers and quaternions.) The nonzero octonions under multiplication form a *Moufang loop* $\mathrm{GM}(\mathbb{O})$, alternative but not associative.

A. *Cayley-Graves integers.* According to Dickson (1923b, pp. 141–142), a set of complex, quaternionic, or octonionic integers should have the following properties: (1) for each number in the set, the coefficients of its rank equation are rational integers, (2) the set is closed under subtraction and multiplication, and (3) the set contains 1. Restating the first property and imposing somewhat stricter conditions in place of the last two, consistent with the definition given in § 15.1, we shall adopt the following criteria for a *basic system* of integral elements:

(1) the trace and the norm of each element are rational integers;

(2) the elements form a subring of $\mathbb{C}$, $\mathbb{H}$, or $\mathbb{O}$ with a set of invertible units (elements of norm 1) closed under multiplication;

(3) when $\mathbb{C}$, $\mathbb{H}$, or $\mathbb{O}$ is taken as a vector space over $\mathbb{R}$, the elements are the points of a two-, four-, or eight-dimensional lattice spanned by the units.

Dickson also demands of a set of integers that it be *maximal*, not a subset of a larger set meeting the other criteria. Though a maximal system like $\mathbb{H}\mathsf{ur}$ may have desirable properties that an extendable system like $\mathbb{H}\mathsf{am}$ lacks, we do not make this a requirement: basic systems may or may not be maximal.

Up to isomorphism, there are exactly four basic systems of octonionic integers (Johnson 2013, pp. 58–59). The simplest system consists of all octonions

$$\mathbf{g} = g_0 + g_1e_1 + g_2e_2 + g_3e_3 + g_4e_4 + g_5e_5 + g_6e_6 + g_7e_7$$

whose eight components g_i are rational integers. Geometrically these are the points of an eight-dimensional lattice C_8, the vertices of a regular honeycomb $\{4, 3^6, 4\}$ of Euclidean 8-space E^8. In honor of John Graves, who discovered the octonions—or "octaves," as he called them—in 1843, and Arthur Cayley, who rediscovered them in 1845, we denote this system by $\mathbb{O}\mathsf{cg}$ and call its elements the *Cayley-Graves integers* or, following Conway & Smith (2003, p. 100), the *Gravesian octaves*. There are sixteen units, namely,

$$\pm 1, \quad \pm e_1, \quad \pm e_2, \quad \pm e_3, \quad \pm e_4, \quad \pm e_5, \quad \pm e_6, \quad \pm e_7, \qquad (15.7.1)$$

with the following nonassociative multiplication table.

	e_1	e_2	e_3	e_4	e_5	e_6	e_7
e_1	-1	e_4	e_7	$-e_2$	e_6	$-e_5$	$-e_3$
e_2	$-e_4$	-1	e_5	e_1	$-e_3$	e_7	$-e_6$
e_3	$-e_7$	$-e_5$	-1	e_6	e_2	$-e_4$	e_1
e_4	e_2	$-e_1$	$-e_6$	-1	e_7	e_3	$-e_5$
e_5	$-e_6$	e_3	$-e_2$	$-e_7$	-1	e_1	e_4
e_6	e_5	$-e_7$	e_4	$-e_3$	$-e_1$	-1	e_2
e_7	e_3	e_6	$-e_1$	e_5	$-e_4$	$-e_2$	-1

B. *Coxeter-Dickson integers.* With all their coordinates in $\mathbb{Z}$, the system $\mathbb{O}\mathsf{cg}$ of Gravesian octaves is the analogue in $\mathbb{O}$ of the commutative ring $\mathbb{G}$ of Gaussian integers in $\mathbb{C}$ and the noncommutative ring $\mathbb{H}\mathsf{am}$ of Hamilton integers in $\mathbb{H}$. As we saw in §15.1, the Hurwitz integers $\mathbb{H}\mathsf{ur}$ can be obtained from $\mathbb{H}\mathsf{am}$ by adjoining quaternions with coordinates in $\mathbb{Z} + ½$. Something similar can be done to extend the Cayley-Graves integers to a larger system.

Dickson (1923a, pp. 319–325) showed that certain sets of octonions having all eight coordinates in $\mathbb{Z}$, four in $\mathbb{Z}$ and four in $\mathbb{Z} + ½$, or all eight in $\mathbb{Z} + ½$ form a system of octonionic integers. In fact, he obtained three such systems, each containing all the octonions with all coordinates in $\mathbb{Z}$ and some of the ones with all or half their coordinates in $\mathbb{Z} + ½$. But this is not the whole story. Coxeter (1946) found that there are actually *seven* of these systems, one corresponding to each of the unit octonions $e_1, \ldots, e_7$. Each system has 240 units, consisting of the 16 Gravesian units and 224 others having coordinates of the type $(\pm½, \pm½, \pm½, \pm½, 0, 0, 0, 0)$. Of these 128 are unique to a particular system, and three sets of 32 are each shared with two other systems.

We shall denote any of these seven systems by $\mathbb{O}\mathsf{cd}$ and call its elements the *Coxeter-Dickson integers* or the *Dicksonian octaves*; the systems can be distinguished as $\mathbb{O}\mathsf{cd}(1), \ldots, \mathbb{O}\mathsf{cd}(7)$. Conway & Smith (2003, pp. 100–105) call the seven systems "octavian rings," one of which they single out as the "octavian integers." The elements of any one system are the points of a *Gosset lattice* E_8, the vertices of a uniform honeycomb 5_{21} of Euclidean 8-space E^8, the symmetry group of which is the Coxeter group $\tilde{\mathbf{E}}_8 \cong [3^{5,2,1}]$. The cellular polytopes of 5_{21} are simplexes α_8 and orthoplexes β_8, and its vertex figure is the uniform polytope 4_{21}, which has 240 vertices (Coxeter 1946, §§ 8–9).

The identity 1 and any of the Gravesian units e_i span a two-dimensional sublattice C_2 whose points are the vertices of a regular tessellation $\{4, 4\}$ of E^2. The points of each such lattice correspond to the elements of a ring $\mathbb{G}$ of Gaussian integers. Each of these rings is a subring of the system $\mathbb{O}\mathsf{cg}$ of Gravesian octaves and hence also of each of the seven systems $\mathbb{O}\mathsf{cd}(i)$ of Dicksonian octaves.

C. *Coupled Hurwitz integers.* The octonions satisfy the alternative laws $(aa)b = a(ab)$ and $(ab)b = a(bb)$ but are not generally associative. Together with the identity element 1, any of the seven sets of Gravesian units e_i, e_j, and e_k that form an associative triple, so that $(\mathrm{e}_i\mathrm{e}_j)\mathrm{e}_k = \mathrm{e}_i(\mathrm{e}_j\mathrm{e}_k)$, span a four-dimensional lattice C_4 whose points correspond to a ring $\mathbb{H}\mathsf{am}(ijk)$ of Hamilton integers. By adjoining points with coordinates in $\mathbb{Z}+½$, each such lattice can be extended to a lattice D_4 whose points correspond to a ring $\mathbb{H}\mathsf{ur}(ijk)$ of Hurwitz integers. The four complementary Gravesian units span another four-dimensional lattice C_4, orthogonal to the first, which can likewise be extended to a lattice D_4.

The two orthogonal D_4 lattices span an eight-dimensional lattice $\mathsf{D}_4 \oplus \mathsf{D}_4$ whose points are the vertices of a uniform prismatic honeycomb

$$\{3, 3, 4, 3\} \times \{3, 3, 4, 3\}$$

of E^8. The cellular polytopes of this honeycomb are hexadecachoric (16-cell) double prisms $\{3,3,4\} \times \{3,3,4\}$. As octonions, the lattice points constitute a system $\mathbb{O}\mathsf{ch}$ of *coupled Hurwitz integers* or *Hurwitzian octaves*; Conway & Smith (2003, pp. 101–103) call this a "double Hurwitzian" system. There are forty-eight units, consisting of the sixteen Gravesian units and thirty-two others having coordinates 0, i, j, and k equal to $\pm½$ and the other four equal to 0 or vice versa. Since there are seven choices for i, j, and k, there are seven such systems, namely,

$$\mathbb{O}\mathsf{ch}(124), \mathbb{O}\mathsf{ch}(235), \mathbb{O}\mathsf{ch}(346), \mathbb{O}(457), \mathbb{O}\mathsf{ch}(561), \\ \mathbb{O}\mathsf{ch}(672), \mathbb{O}\mathsf{ch}(713).$$

Each system $\mathbb{O}\mathsf{ch}(ijk)$ of Hurwitzian octaves is a subring of three systems $\mathbb{O}\mathsf{cd}(i)$, $\mathbb{O}\mathsf{cd}(j)$, and $\mathbb{O}\mathsf{cd}(k)$ of Dicksonian octaves.

The intersection of all seven systems of Dicksonian octaves (or any two systems of Hurwitzian octaves), which Conway & Smith (2003, p. 103) call the "Kleinian octaves," is not a basic system of integral octonions, since the corresponding eight-dimensional lattice is not spanned by the common units. Nor is their union, whose units are not closed under multiplication.

To bring out the connection with the complex numbers and the quaternions, we may follow Dickson and denote the Gravesian units by

$$1, \quad \mathrm{i}, \quad \mathrm{j}, \quad \mathrm{k}, \quad \mathrm{e}, \quad \mathrm{ie}, \quad \mathrm{je}, \quad \mathrm{ke} \qquad (15.7.2)$$

and their negatives. This notation can be related to the one we have been using by setting $\mathrm{i} = e_1$, $\mathrm{j} = e_2$, $\mathrm{e} = e_3$, $\mathrm{k} = e_4$, $\mathrm{je} = e_5$, $\mathrm{ke} = -e_6$, $\mathrm{ie} = e_7$. Thus i, j, and $\mathrm{k} = \mathrm{ij}$ form an associative triple. Coxeter (1946, § 5) also finds it useful to define $\mathrm{h} = ½(\mathrm{i} + \mathrm{j} + \mathrm{k} + \mathrm{e})$. A group $\bar{\mathrm{S}}\mathrm{L}(\mathbb{G}) \cong \{\pm 1, \pm \mathrm{i}\}$ of Gaussian units is generated by i. A group $\mathrm{SL}(\mathbb{H}\mathsf{am}) \cong \{\pm 1, \pm \mathrm{i}, \pm \mathrm{j}, \pm \mathrm{k}\}$ of Hamilton units is generated by i and j. The analogous Moufang loop $\mathrm{SM}(\mathbb{O}\mathsf{cg})$ of Gravesian units, of order 16, is generated by i, j, and e. The Moufang loop $\mathrm{SM}(\mathbb{O}\mathsf{cd})$

of Dicksonian units, of order 240, is generated by i, j, and h, with the lattice E_8 being spanned by the units

$$1, \quad i, \quad j, \quad k, \quad h, \quad ih, \quad jh, \quad kh. \qquad (15.7.3)$$

Let the unit octonions u, v, and $w = (uv)^{-1}$ be defined by

$$u = \tfrac{1}{2} - \tfrac{1}{2}i - \tfrac{1}{2}j + \tfrac{1}{2}k, \quad v = \tfrac{1}{2} + \tfrac{1}{2}i - \tfrac{1}{2}j - \tfrac{1}{2}k, \quad w = \tfrac{1}{2} - \tfrac{1}{2}i + \tfrac{1}{2}j - \tfrac{1}{2}k.$$

Then a group $\bar{S}L(\mathbb{E}) \cong \{\pm 1, \pm u, \pm u^2\}$ of Eisenstein units is generated by u. A group $SL(\mathbb{H}ur) \cong 2A_4$ of Hurwitz units, of order 24, is generated by u and v. When the division algebra $\mathbb{O}$ of octonions is taken as a vector space over $\mathbb{R}$, the lattice $\mathsf{D}_4 \oplus \mathsf{D}_4$ of the ring $\mathbb{O}\mathsf{ch}$ of Hurwitzian octaves is spanned by the unit octonions

$$1, \quad u, \quad v, \quad w, \quad e, \quad ue, \quad ve, \quad we. \qquad (15.7.4)$$

The Moufang loop $SM(\mathbb{O}\mathsf{ch})$ of Hurwitzian units, of order 48, is generated by u, v, and e.

D. *Compound Eisenstein integers.* When the field $\mathbb{C}$ of complex numbers is taken as a vector space over $\mathbb{R}$, the ring $\mathbb{E}$ of Eisenstein integers is a two-dimensional lattice A_2 spanned by 1 and the primitive cube root of unity $\omega = -½ + ½\sqrt{3}i$ (or $\omega^2 = -½ - ½\sqrt{3}i$). The points of the lattice are the vertices of a regular tessellation $\{3, 6\}$ of the Euclidean plane E^2. The group $\bar{S}L(\mathbb{E}) \cong \{\pm 1, \pm\omega, \pm\omega^2\}$ of Eisenstein units is generated by $-\omega$ or $-\omega^2$.

When the division ring $\mathbb{H}$ of quaternions is taken as a vector space over $\mathbb{R}$, the ring $\mathbb{H}\mathsf{yb}$ of hybrid integers is a four-dimensional lattice $\mathsf{A}_2 \oplus \mathsf{A}_2$ spanned by 1, ω, j, and ωj, where $\omega = -½ + ½\sqrt{3}i$ and $\omega j = -½j + ½\sqrt{3}k$. The points of the lattice are the vertices of a uniform prismatic honeycomb $\{3, 6\} \times \{3, 6\}$ of E^4. The cellular polychora of $\{3, 6\} \times \{3, 6\}$ are triangular double prisms $\{3\} \times \{3\}$. The group $SL(\mathbb{H}\mathsf{yb}) \cong \{\pm 1, \pm\omega, \pm\omega^2, \pm j, \pm\omega j, \pm\omega^2 j\}$ of hybrid units is generated by ω and j.

The lattice $A_2 \oplus A_2$ is the direct sum of two sublattices A_2, one spanned by 1 and ω and the other by j and ωj, the points of which represent the ring $\mathbb{E}$ of Eisenstein integers. Other pairings, e.g., 1 and j, span four sublattices C_2 whose points correspond to the ring $\mathbb{G}$ of Gaussian integers.

The quaternionic ring $\mathbb{H}$yb has an octonionic analogue, the system $\mathbb{O}$ce of *compound Eisenstein integers* or *Eisenstein octaves*, which can be realized as an eight-dimensional lattice $4A_2 = A_2 \oplus A_2 \oplus A_2 \oplus A_2$ spanned by the unit octonions

$$1, \quad \omega, \quad \mathrm{j}, \quad \omega\mathrm{j}, \quad \mathrm{e}, \quad \omega\mathrm{e}, \quad \mathrm{je}, \quad \omega\mathrm{j.e}, \qquad (15.7.5)$$

where $\omega = -\frac{1}{2} + \frac{1}{2}\sqrt{3}\mathrm{i}$. The points of the lattice are the vertices of a uniform honeycomb $\{3, 6\}^4$ of Euclidean 8-space E^8. The cellular polytopes of $\{3, 6\}^4$ are triangular quadruple prisms $\{3\}^4$. There are twenty-four units, namely, the eight listed here and their negatives and eight others involving $\omega^2 = -\frac{1}{2} - \frac{1}{2}\sqrt{3}\mathrm{i})$, i.e.,

$$\omega^2, \quad \omega^2\mathrm{j}, \quad \omega^2\mathrm{e}, \quad \omega^2\mathrm{j.e} \qquad (15.7.6)$$

and their negatives. The twenty-four compound Eisenstein units form a Moufang loop SM($\mathbb{O}$ce) generated by ω, j, and e, with the following multiplication table.

	j	e	je	ω	ωj	ωe	ωj.e	ω^2	ω^2j	ω^2e	ω^2j.e
j	-1	je	$-$e	ω^2j	$-\omega^2$	ωj.e	$-\omega$e	ωj	$-\omega$	ω^2j.e	$-\omega^2$e
e	$-$je	-1	j	ω^2e	$-\omega$j.e	$-\omega^2$	ωj	ωe	$-\omega^2$j.e	$-\omega$	ω^2j
je	e	$-$j	-1	ωj.e	ω^2e	$-\omega^2$j	$-\omega$	ω^2j.e	ωe	$-\omega$j	$-\omega^2$
ω	ωj	ωe	ω^2j.e	ω^2	ω^2j	ω^2e	je	1	j	e	ωj.e
ωj	$-\omega$	ωj.e	$-\omega^2$e	j	-1	ω^2j.e	$-$e	ω^2j	$-\omega^2$	je	$-\omega$e
ωe	$-\omega$j.e	$-\omega$	ω^2j	e	$-\omega^2$j.e	-1	j	ω^2e	$-$je	$-\omega^2$	ωj
ωj.e	ωe	$-\omega$j	$-\omega^2$	ω^2j.e	e	$-$j	-1	je	ω^2e	$-\omega^2$j	$-\omega$
ω^2	ω^2j	ω^2e	ωj.e	1	j	e	ω^2j.e	ω	ωj	ωe	je
ω^2j	$-\omega^2$	ω^2j.e	$-\omega$e	ωj	$-\omega$	je	$-\omega^2$e	j	-1	ωj.e	$-$e
ω^2e	$-\omega^2$j.e	$-\omega^2$	ωj	ωe	$-$je	$-\omega$	ω^2j	e	$-\omega$j.e	-1	j
ω^2j.e	ω^2e	$-\omega^2$j	$-\omega$	je	ωe	$-\omega$j	$-\omega^2$	ωj.e	e	$-$j	-1

The lattice $4\mathsf{A}_2$ is the direct sum (in three ways) of two sublattices $\mathsf{A}_2 \oplus \mathsf{A}_2$, e.g., one spanned by 1, ω, j, and ωj, and the other by e, ωe, je, and ωj.e, the points of which represent the ring $\mathbb{H}\mathsf{yb}$ of hybrid integers. Other choices of four octonions, such as 1, j, e, and je, span sublattices C_4 whose points correspond to the ring $\mathbb{H}\mathsf{am}$ of Hamilton integers.

In the foregoing presentation, the units j, e, and je are identified with e_2, e_3, and e_5, with the other units ω, ωj, ωe, and ωj.e defined accordingly, but we can equally well choose any three Gravesian units that form an associative triple. Thus, corresponding to the seven possible triples, there are seven isomorphic systems

$$\mathbb{O}\mathsf{ce}(124), \mathbb{O}\mathsf{ce}(235), \mathbb{O}\mathsf{ce}(346), \mathbb{O}\mathsf{ce}(457), \mathbb{O}\mathsf{ce}(561), \\ \mathbb{O}\mathsf{ce}(672), \mathbb{O}\mathsf{ce}(713).$$

If we let $g = ½(i - j + ie - je)$, then a group $\mathrm{SL}(\mathbb{H}\mathsf{am}) \cong 2\mathrm{D}_2$ of Hamilton units is generated by g and e, while a group $\mathrm{SL}(\mathbb{H}\mathsf{yb}) \cong 2\mathrm{D}_3$ of hybrid units is generated by u and g. The loop $\langle u, g, e\rangle$ is seen to be isomorphic to the Moufang loop $\mathrm{SM}(\mathbb{O}\mathsf{ce}) \cong \langle \omega, j, e\rangle$ by the correspondence $-u \leftrightarrow \omega$, $g \leftrightarrow j$, $e \leftrightarrow e$. Thus a loop of Eisenstein units is also generated by u, g, and e, and it is evident that this is a subloop of the loop $\mathrm{SM}(\mathbb{O}\mathsf{cd})$ of Dicksonian units. It follows that $\mathbb{O}\mathsf{ce}$, realized as a lattice $4\mathsf{A}_2$ spanned by the unit octonions

$$1, \quad u, \quad g, \quad ug, \quad e, \quad ue, \quad ge, \quad ug.e, \tag{15.7.7}$$

is a subsystem of $\mathbb{O}\mathsf{cd}$. Unlike the Eisenstein integers and the hybrid quaternionic integers, the compound Eisenstein integers do not constitute a maximal system.

Together with the rational integers $\mathbb{Z}$, the two systems of Gaussian and Eisenstein complex integers $\mathbb{G} = \mathbb{Z}[i]$ and $\mathbb{E} = \mathbb{Z}[\omega]$, and the three systems of integral quaternions $\mathbb{H}\mathsf{am} = \mathbb{Z}[i, j]$, $\mathbb{H}\mathsf{yb} = \mathbb{Z}[\omega, j]$, and $\mathbb{H}\mathsf{ur} = \mathbb{Z}[u, v]$, we have four basic systems of integral octonions: the Gravesian octaves $\mathbb{O}\mathsf{cg} = \mathbb{Z}[i, j, e]$, the Eisenstein octaves $\mathbb{O}\mathsf{ce} = \mathbb{Z}[\omega, j, e] = \mathbb{Z}[u, g, e]$, the Hurwitzian octaves $\mathbb{O}\mathsf{ch} = \mathbb{Z}[u, v, e]$, and the Dicksonian octaves $\mathbb{O}\mathsf{cd} = \mathbb{Z}[i, j, h]$. The ten systems belong to four families, with alternative symbols adapted from Conway & Smith (2003, p. 126). The elements of each system can be identified with the points of a lattice or the vertices of a regular or uniform honeycomb. These results are summarized in Table 15.7.

Table 15.7 *Basic Systems of Integers*

Family Symbol	Basic System	Units	Lattice	Honeycomb
$\mathcal{G}^1$	$\mathbb{Z}$	2	C_1	$\{\infty\}$
$\mathcal{G}^2$	$\mathbb{G}$	4	C_2	$\{4,4\}$
$\mathcal{E}^2$	$\mathbb{E}$	6	A_2	$\{3,6\}$
$\mathcal{G}^4$	$\mathbb{H}\mathsf{am}$	8	C_4	$\{4,3,3,4\}$
$\mathcal{E}^4$	$\mathbb{H}\mathsf{yb}$	12	$2\mathsf{A}_2$	$\{3,6\}^2$
$\mathcal{H}^4$	$\mathbb{H}\mathsf{ur}$	24	D_4	$\{3,3,4,3\}$
$\mathcal{G}^8$	$\mathbb{O}\mathsf{cg}$	16	C_8	$\{4,3^6,4\}$
$\mathcal{E}^8$	$\mathbb{O}\mathsf{ce}$	24	$4\mathsf{A}_2$	$\{3,6\}^4$
$\mathcal{H}^8$	$\mathbb{O}\mathsf{ch}$	48	$2\mathsf{D}_4$	$\{3,3,4,3\}^2$
$\mathcal{D}^8$	$\mathbb{O}\mathsf{cd}$	240	E_8	$\{3^5,3^{2,1}\}$

Six of the ten basic systems are maximal and seven of them are minimal, containing no smaller basic system of the same dimension. The lattice C_4 being a sublattice of D_4, the system $\mathbb{H}\mathsf{am}$ is a subsystem of $\mathbb{H}\mathsf{ur}$. Likewise, $\mathsf{C}_8 = 2\mathsf{C}_4$ is a sublattice of $2\mathsf{D}_4$, which is a sublattice of E_8; correspondingly, $\mathbb{O}\mathsf{cg}$ is a subsystem of $\mathbb{O}\mathsf{ch}$, which is a subsystem of $\mathbb{O}\mathsf{cd}$. Moreover, $4\mathsf{A}_2$ is a sublattice of E_8, and $\mathbb{O}\mathsf{ce}$ is a subsystem of $\mathbb{O}\mathsf{cd}$. These and other connections between the various systems are described by Conway & Smith (2003, §10.4).

EXERCISES 15.7

1. What are the common units of the seven systems of Dicksonian octaves, i.e., the units of the "Kleinian octaves"?

2. Show that the product of Hurwitzian octaves belonging to different systems need not be an octonionic integer.

3. Show that the unit octonions i, j, and h generate the 240 units of one system of Dicksonian octaves.

15.8 OCTONIONIC MODULAR LOOPS

The *Gravesian special Moufang* loop $SM_2(\mathbb{O}cg)$ of invertible 2×2 matrices with entries in $\mathbb{O}cg$ is generated by the matrices

$$A = \begin{pmatrix} 0 & 1 \\ -1 & 0 \end{pmatrix}, \; B = \begin{pmatrix} 1 & 0 \\ 1 & 1 \end{pmatrix}, \; M = \begin{pmatrix} 1 & 0 \\ 0 & \mathrm{i} \end{pmatrix}, \; N = \begin{pmatrix} 1 & 0 \\ 0 & \mathrm{j} \end{pmatrix},$$
$$P = \begin{pmatrix} 1 & 0 \\ 0 & \mathrm{e} \end{pmatrix}.$$

Identifying any such matrix with its negative, we obtain the *Gravesian modular loop* $PSM_2(\mathbb{O}cg) \cong SM_2(\mathbb{O}cg)/SZ_2(\mathbb{O}cg)$, which is generated by the elements

$$\alpha = \cdot\langle A\rangle, \; \beta = \cdot\langle B\rangle, \; \mu = \cdot\langle M\rangle, \; \nu = \cdot\langle N\rangle, \; \pi = \cdot\,\langle P\rangle.$$

Because octonionic multiplication is nonassociative, the elements of the loop $PSM_2(\mathbb{O}cg)$ cannot be treated as projective or linear fractional transformations.

Each of the other three basic systems of integral octonions defines analogous special Moufang and modular loops. For the Eisenstein octaves $\mathbb{O}ce$, the entries i and j in the matrices M and N should be replaced by ω and j or by u and g, for the Hurwitzian octaves $\mathbb{O}ch$, i and j should be replaced by u and v. For the Dicksonian octaves $\mathbb{O}cd$, e should be replaced by h in matrix P.

As with the corresponding finite loops of units, the four octonionic modular loops are interrelated. The loops $PSM_2(\mathbb{O}cg)$, $PSM_2(\mathbb{O}ce)$, and $PSM_2(\mathbb{O}ch)$ are all subloops of $PSM_2(\mathbb{O}cd)$, and $PSM_2(\mathbb{O}cg)$ is also a subloop of $PSM_2(\mathbb{O}ch)$. Moreover, the octonionic loop $PSM_2(\mathbb{O}cd)$ contains each of the real complex, or quaternionic groups $PSL_2(\mathbb{Z})$, $PSL_2(\mathbb{G})$, $PSL_2(\mathbb{E})$, $PSL_2(\mathbb{H}am)$, $PSL_2(\mathbb{H}yb)$, and $PSL_2(\mathbb{H}ur)$ as a subloop.

Feingold, Kleinschmidt & Nicolai (2009, pp. 1329–1336) have described how the Dicksonian modular loop $PSM_2(\mathbb{O}cd)$ is related to the hyperbolic Coxeter group $\bar{\mathbf{T}}_9 \cong [3^{6,2,1}]$. The modular loops $PSM_2(\mathbb{O}cg)$, $PSM_2(\mathbb{O}ce)$, and $PSM_2(\mathbb{O}ch)$ have analogous connections with hypercompact subgroups of $\bar{\mathbf{T}}_9$.

TABLES

Collected in the following pages for ready reference are lists, together with such other data as might be found useful, of real transformation groups and groups generated by reflections. An explanation of the terminology and notation used in each table, with some additional commentary, follows.

A. *Real transformation groups.* Table A lists the principal groups of collineations of real projective, affine, and areplectic n-space; circularities of real inversive n-space; and isometries and similarities of Euclidean n-space and other real metric spaces. Each of these is an extension or a quotient group of the general linear group of invertible matrices of order n, $n+1$, or $n+2$, or one of its subgroups, including the symplectic, orthogonal, and pseudo-orthogonal groups described in Chapters 5 and 6. The translation group T^n is the additive group of the vector space $\mathbb{R}^n$.

For n equal to 2, 3, 4, or 5, the projective and special projective pseudoorthogonal groups $\mathrm{PO}_{n,1}$ and $\mathrm{P^+O}_{n,1}$, which are the isometry and direct isometry groups of hyperbolic n-space or the circularity and homography groups of inversive $(n-1)$-space, can be expressed as linear or antilinear fractional transformations over $\mathbb{R}$, $\mathbb{C}$, or $\mathbb{H}$, as given by (10.4.7).

B. *Groups generated by reflections*. These are the Coxeter groups that were described in Chapters 11, 12, and 13. Table B lists all irreducible spherical and Euclidean groups and all compact and paracompact hyperbolic groups generated by reflections. The fundamental region for such a group is the closure of a simplex whose dihedral angles are submultiples of π. For a group generated by n reflections all of whose subgroups of rank $n-1$ are spherical, this is a *Möbius, Cartan*, or *Lannér* simplex according as the group is spherical, Euclidean, or hyperbolic; for a hyperbolic group some of whose subgroups are Euclidean, it is a *Koszul* simplex, with one or more absolute vertices.

Each group is identified by a numeric "Coxeter symbol" corresponding to its Coxeter diagram, as well as a literal *Cartan symbol* for spherical and Euclidean groups or a *Witt symbol* for Euclidean and hyperbolic groups. The $(n-1)$-dimensional content of the fundamental region is the *size* of the group, to be discussed further below. A spherical group has a finite *order*, and its *center* is of order 2 or 1 depending on whether or not the group contains the central inversion. The *Coxeter number h* is the period of the product of the generators. The *Schläflian* σ, the determinant of the Schläfli matrix, is positive, zero, or negative according as the group is spherical, Euclidean, or hyperbolic; for a crystallographic group σ is an integer.

Corresponding to each finite irreducible crystallographic group generated by n reflections is a simple Lie algebra with the same Cartan symbol. If h is the Coxeter number of the group, the algebra has dimension $n(h+1)$. In particular, the respective dimensions of the classical Lie algebras A_n, B_n, C_n, and D_n are $n(n+2), n(2n+1), n(2n+1)$, and $n(2n-1)$, while the exceptional Lie algebras G_2, F_4, E_6, E_7, and E_8 have dimensions $2(6+1) = 14$, $4(12+1) = 52$, $6(12+1) = 78$, $7(18+1) = 133$, and $8(30+1) = 248$.

The size of a spherical group generated by n reflections is inversely proportional to its order and can be calculated by dividing the content of the $(n-1)$-sphere S^{n-1} by the order of the group. The size of the circular group $[p]$ is the arc length π/p of the circular dion (p). The area of a Möbius triangle $(p\,q\,2)$ whose closure is the fundamental region for the full polyhedral group $[(p,q,2)] \cong [p,q]$ is equal to the spherical excess $(p^{-1}+q^{-1}-½)\pi$. In general, the size of a spherical group can be expressed in terms of the gamma function and is commensurable with a power of π.

The size of a Euclidean group is defined only up to a scalar factor, but relative sizes are of interest. When, relative to commensurable scales, the size of one group is a multiple of the size of another, the first is a subgroup of the second, the index being the ratio of the respective sizes.

The content of a simplex P in Euclidean m-space is readily calculated. Let the vertices of P have Cartesian coordinates

$$(p_i) = (p_{i1}, p_{i2}, \ldots, p_{im}), \quad 1 \leq i \leq m+1,$$

and let P be the $(m+1)\times(m+1)$ matrix with rows $((p_i), 1)$. Then the content of P is given by

$$|\mathsf{P}| = \frac{1}{m!}|\det P|. \tag{1}$$

The size of a hyperbolic group generated by n reflections is commensurable with a power of π when n is odd, i.e., when the group operates in an even-dimensional space. In particular, the area of a Lannér or Koszul triangle $(p\,q\,r)$ whose closure is the fundamental region for a group $[(p,q,r)]$ is equal to the hyperbolic defect $(1-p^{-1}-q^{-1}-r^{-1})\pi$. Values for some hyperbolic Coxeter groups have only recently been calculated (Johnson, Kellerhals, Ratcliffe & Tschantz 1999).

The volume of the fundamental region for a compact or paracompact Coxeter group in hyperbolic 3-space can be expressed in terms of functions devised by Lobachevsky (1836) and Schläfli (1860) for dealing with non-Euclidean volumes. The *Lobachevsky function* Л is defined by

$$Л(x) = -\int_0^x \log|2\sin t|\,dt, \tag{2}$$

while the *Schläfli function* S, defined by an infinite series, evaluates to four times the volume of a spherical orthoscheme with specified angles (cf. Coxeter 1935a, pp. 15–16, 22–28; Milnor 1982, pp. 17–22; Alekseevsky, Vinberg & Solodovnikov 1988, chap. 7, § 3; Ratcliffe 1994/2006, § 10.4). The Lobachevsky function is continuous, periodic with period π, and equal to zero at multiples of $\pi/2$. Applied to a hyperbolic orthoscheme with essential angles α, β, γ, the Schläfli function is imaginary, and the volume is one-fourth the value of the (real) "hyperbolic Schläfli function"

$$\begin{aligned} \mathrm{i}S(\tfrac{1}{2}\pi - \alpha, \beta, \tfrac{1}{2}\pi - \gamma) &= Л(\alpha + \delta) - Л(\alpha - \delta) \\ &\quad + Л(\tfrac{1}{2}\pi - \delta + \beta) + Л(\tfrac{1}{2}\pi - \delta - \beta) \\ &\quad + Л(\gamma + \delta) - Л(\gamma - \delta) + 2Л(\tfrac{1}{2}\pi - \delta), \qquad (3) \end{aligned}$$

where $\delta = \tan^{-1} \sqrt{\cos^2 \beta - \sin^2 \alpha \sin^2 \gamma} /(\cos \alpha \cos \gamma)$.

Other tetrahedra can be dissected into orthoschemes and the volumes of the several pieces then added together. In a few cases, the resulting expressions are too big for the table. Exact formulas for the sizes of the five compact groups whose fundamental regions are the closures of cycloschemes, as given by Cho & Kim (1999, p. 365), are:

$$\begin{aligned} \widehat{\mathbf{AB}}_3 &: Л(\tfrac{1}{6}\pi + \tfrac{1}{2}\omega) - Л(\tfrac{1}{6}\pi - \tfrac{1}{2}\omega) + Л(\tfrac{1}{8}\pi + \tfrac{1}{2}\omega) \\ &\qquad - Л(\tfrac{1}{8}\pi - \tfrac{1}{2}\omega) - \tfrac{1}{2}Л(\tfrac{1}{6}\pi + \omega) + \tfrac{1}{2}Л(\tfrac{1}{6}\pi - \omega), \\ \widehat{\mathbf{AH}}_3 &: Л(\tfrac{1}{6}\pi + \tfrac{1}{2}\omega) - Л(\tfrac{1}{6}\pi - \tfrac{1}{2}\omega) + Л(\tfrac{1}{10}\pi + \tfrac{1}{2}\omega) \\ &\qquad - Л(\tfrac{1}{10}\pi - \tfrac{1}{2}\omega) - \tfrac{1}{2}Л(\tfrac{1}{6}\pi + \omega) + \tfrac{1}{2}Л(\tfrac{1}{6}\pi - \omega), \\ \widehat{\mathbf{BB}}_3 &: 2Л(\tfrac{1}{6}\pi + \tfrac{1}{2}\omega) - 2Л(\tfrac{1}{6}\pi - \tfrac{1}{2}\omega) + \tfrac{1}{4}Л(\tfrac{1}{2}\pi - 2\omega), \qquad (4) \\ \widehat{\mathbf{BH}}_3 &: Л(\tfrac{1}{8}\pi + \tfrac{1}{2}\omega) - Л(\tfrac{1}{8}\pi - \tfrac{1}{2}\omega) + Л(\tfrac{1}{10}\pi + \tfrac{1}{2}\omega) \\ &\qquad - Л(\tfrac{1}{10}\pi - \tfrac{1}{2}\omega) - \tfrac{1}{2}Л(\tfrac{1}{6}\pi + \omega) + \tfrac{1}{2}Л(\tfrac{1}{6}\pi - \omega), \\ \widehat{\mathbf{HH}}_3 &: 2Л(\tfrac{1}{6}\pi + \tfrac{1}{2}\omega) - 2Л(\tfrac{1}{6}\pi - \tfrac{1}{2}\omega) - \tfrac{1}{2}Л(\tfrac{3}{10}\pi + \omega) + \tfrac{1}{2}Л(\tfrac{3}{10}\pi - \omega). \end{aligned}$$

The angle ω takes a different value in each formula, with $\cos\omega$ respectively equal to $\sqrt{2} - ½$, $2 - ½\sqrt{5}$, $3/4$, $1/4(3 + \sqrt{2} + \sqrt{5} - \sqrt{10})$, or $1/8(7 - \sqrt{5})$.

Size formulas for other odd-dimensional groups generally involve the Riemann zeta function or the related L-functions of Dirichlet. One five-dimensional group has no known elementary formula.

Table A *Real Transformation Groups*

Geometry	Symbol	Transformations	Groups
Projective	P^n	Projective collineation Equiprojective collineation (n odd)	$GP_n \cong PGL_{n+1}$ $SP_n \cong PSL_{n+1}$
Affine	A^n	Affine collineation (affinity) Direct affine collineation Unit affine collineation Equiaffine collineation	$GA_n \cong T^n \rtimes GL_n$ $G^+A_n \cong T^n \rtimes G^+L_n$ $\bar{S}A_n \cong T^n \rtimes \bar{S}L_n$ $SA_n \cong T^n \rtimes SL_n$
Areplectic	Ap^n	Areplexity Equiplexity } (n even)	$GAp_n \cong T^n \rtimes Gp_n$ $SAp_n \cong T^n \rtimes Sp_n$
Inversive	I^n	Circularity Homography (Möbius transformation)	$GI_n \cong PO_{n+1,1}$ $SI_n \cong P^+O_{n+1,1}$
Elliptic	eP^n	Elliptic isometry Direct elliptic isometry (n odd)	PO_{n+1} P^+O_{n+1}
Euclidean	E^n	Euclidean similarity Direct Euclidean similarity Euclidean isometry Direct Euclidean isometry	$\tilde{E}_n \cong T^n \rtimes \tilde{O}_n$ $\tilde{E}_n^+ \cong T^n \rtimes \tilde{O}_n^+$ $E_n \cong T^n \rtimes O_n$ $E_n^+ \cong T^n \rtimes O_n^+$
Hyperbolic	H^n	Hyperbolic isometry Direct hyperbolic isometry	$PO_{n,1}$ $P^+O_{n,1}$
Spherical	S^n	Spherical isometry Direct spherical isometry	O_{n+1} O_{n+1}^+
Pseudospherical	$\ddot{S}^n$	Pseudospherical isometry Direct pseudospherical isometry Positive pseudospherical isometry Equipositive pseudospherical isometry	$O_{n,1}$ $O_{n,1}^+$ $O_{n,1^+}$ $O_{n^+,1^+}$

Table B *Groups Generated by Reflections*

1. *Spherical Groups*

Group	Cartan Symbol	Size	Order	Center	h	σ^*
][	$\mathbf{A}_0$	0	1	1	1	1
[ ]	$\mathbf{A}_1$	1	2	2	2	2
[3]	$\mathbf{A}_2$	$\pi/3$	6	1	3	3
[4]	$\mathbf{B}_2$	$\pi/4$	8	2	4	2
[5]	$\mathbf{H}_2$	$\pi/5$	10	1	5	$2-\bar{\tau}^{\dagger}$
[6]	$\mathbf{G}_2$	$\pi/6$	12	2	6	1
[7]	$\mathbf{I}_2(7)$	$\pi/7$	14	1	7	$2-\xi_7{}^{\ddagger}$
[8]	$2\mathbf{B}_2$	$\pi/8$	16	2	8	$2-\sqrt{2}$
[9]	$\mathbf{I}_2(9)$	$\pi/9$	18	1	9	$2-\xi_9{}^{\ddagger}$
[10]	$2\mathbf{H}_2$	$\pi/10$	20	2	10	$2-\tau^{\dagger}$
[11]	$\mathbf{I}_2(11)$	$\pi/11$	22	1	11	$2-\xi_{11}{}^{\ddagger}$
[12]	$2\mathbf{G}_2$	$\pi/12$	24	2	12	$2-\sqrt{3}$
$[p]$, $p > 12$	$\mathbf{I}_2(p)$	π/p	$2p$	$(p,2)$	p	$4\sin^2 \pi/p$
[3, 3]	$\mathbf{A}_3$	$\pi/6$	24	1	4	4
[3, 4]	$\mathbf{B}_3$	$\pi/12$	48	2	6	2
[3, 5]	$\mathbf{H}_3$	$\pi/30$	120	2	10	$2\bar{\tau}^2$

* If the fundamental region P for a Coxeter group with n generators has angles $\alpha_{ij} = \alpha_{ji} = \pi/p_{ij}$ $(1 \le i \le j \le n)$ and A is the symmetric $n \times n$ matrix with entries $a_{ij} = -\cos\alpha_{ij}$, then $\sigma(\mathsf{P}) = \det 2A = 2^n \det A$ is the *Schläflian* of P.

† $\tau = ½(\sqrt{5}+1)$, $\bar{\tau} = ½(\sqrt{5}-1) = \tau^{-1}$.

‡ ξ_7, ξ_9, and ξ_{11} are the largest roots of the equations $x^3 + x^2 - 2x - 1 = 0$, $x^3 - 3x + 1 = 0$, and $x^5 + x^4 - 4x^3 - 3x^2 + 3x + 1 = 0$, respectively.

Group	Cartan Symbol	Size*	Order	Center	h	σ
[3, 3, 3]	$\mathbf{A}_4$	$\pi^2/60$	120	1	5	5
$[3^{1,1,1}]$	$\mathbf{D}_4$	$\pi^2/96$	192	2	6	4
[3, 3, 4]	$\mathbf{B}_4$	$\pi^2/192$	384	2	8	2
[3, 4, 3]	$\mathbf{F}_4$	$\pi^2/576$	1152	2	12	1
[3, 3, 5]	$\mathbf{H}_4$	$\pi^2/7200$	14400	2	30	$\bar{\tau}^4$
[3, 3, 3, 3]	$\mathbf{A}_5$	$\pi^2/270$	720	1	6	6
$[3^{2,1,1}]$	$\mathbf{D}_5$	$\pi^2/720$	1920	1	8	4
[3, 3, 3, 4]	$\mathbf{B}_5$	$\pi^2/1440$	3840	2	10	2
$[3^5]$	$\mathbf{A}_6$	$\pi^3/7!$	7!	1	7	7
$[3^{3,1,1}]$	$\mathbf{D}_6$	$\pi^3/32 \cdot 6!$	32·6!	2	10	4
$[3^4, 4]$	$\mathbf{B}_6$	$\pi^3/64 \cdot 6!$	64·6!	2	12	2
$[3^{2,2,1}]$	$\mathbf{E}_6$	$\pi^3/72 \cdot 6!$	72·6!	1	12	3
$[3^6]$	$\mathbf{A}_7$	$2\pi^3/15 \cdot 7!$	8!	1	8	8
$[3^{4,1,1}]$	$\mathbf{D}_7$	$\pi^3/60 \cdot 7!$	64·7!	1	12	4
$[3^5, 4]$	$\mathbf{B}_7$	$\pi^3/120 \cdot 7!$	128·7!	2	14	2
$[3^{3,2,1}]$	$\mathbf{E}_7$	$\pi^3/540 \cdot 7!$	576·7!	2	18	2
$[3^7]$	$\mathbf{A}_8$	$\pi^4/3 \cdot 9!$	9!	1	9	9
$[3^{5,1,1}]$	$\mathbf{D}_8$	$\pi^4/384 \cdot 8!$	128·8!	2	14	4
$[3^6, 4]$	$\mathbf{B}_8$	$\pi^4/768 \cdot 8!$	256·8!	2	16	2
$[3^{4,2,1}]$	$\mathbf{E}_8$	$\pi^4/51840 \cdot 8!$	17280·8!	2	30	1
$[3^{n-1}]$, $n > 8$	$\mathbf{A}_n$	$2\pi^{n/2}/\Gamma(\frac{1}{2}n)\cdot(n+1)!$	$(n+1)!$	1	$n+1$	$n+1$
$[3^{n-3,1,1}]$	$\mathbf{D}_n$	$\pi^{n/2}/2^{n-2}\Gamma(\frac{1}{2}n)\cdot n!$	$2^{n-1} \cdot n!$	$(n, 2)$	$2n-2$	4
$[3^{n-2}, 4]$	$\mathbf{B}_n$	$\pi^{n/2}/2^{n-1}\Gamma(\frac{1}{2}n) \cdot n!$	$2^n \cdot n!$	2	$2n$	2

*According as $n = 2k$ or $n = 2k+1$, the expression $\pi^{n/2}/\Gamma(\frac{1}{2}n) \cdot n!$ evaluates to

$$\frac{\pi^k}{(2k)!\,(k-1)!} \quad \text{or} \quad \frac{\pi^k}{(2k+1)!!\,(2k-1)!!\,k!}.$$

2. *Euclidean Groups*

Group	Cartan Symbol	Witt Symbol	Sizes*
$[\infty]$	$\tilde{\mathbf{A}}_1$	$\mathbf{W}_2$	1, 2, 3, ...
$[3^{[3]}]$	$\tilde{\mathbf{A}}_2$	$\mathbf{P}_3$	a^2, $2g^2$, $6g^2$
$[4, 4]$	$\tilde{\mathbf{C}}_2$	$\mathbf{R}_3$	c^2, $2c^2$
$[3, 6]$	$\tilde{\mathbf{G}}_2$	$\mathbf{V}_3$	g^2, $3g^2$
$[3^{[4]}]$	$\tilde{\mathbf{A}}_3$	$\mathbf{P}_4$	a^3, $2b^3$, $4c_3$
$[4, 3^{1,1}]$	$\tilde{\mathbf{B}}_3$	$\mathbf{S}_4$	b^3, $2c_3$
$[4, 3, 4]$	$\tilde{\mathbf{C}}_3$	$\mathbf{R}_4$	c_3, $4b^3$
$[3^{[5]}]$	$\tilde{\mathbf{A}}_4$	$\mathbf{P}_5$	a^4
$[4, 3, 3^{1,1}]$	$\tilde{\mathbf{B}}_4$	$\mathbf{S}_5$	b^4, $2c^4$, $12f^4$, $48f^4$
$[4, 3, 3, 4]$	$\tilde{\mathbf{C}}_4$	$\mathbf{R}_5$	c^4, $8b^4$, $6f^4$, $24f^4$
$[3^{1,1,1,1}]$	$\tilde{\mathbf{D}}_4$	$\mathbf{Q}_5$	d^4, $2b^4$, $4c^4$, $24f^4$, $96f^4$
$[3, 3, 4, 3]$	$\tilde{\mathbf{F}}_4$	$\mathbf{U}_5$	f^4, $4f^4$
$[3^{[6]}]$	$\tilde{\mathbf{A}}_5$	$\mathbf{P}_6$	a^5
$[4, 3, 3, 3^{1,1}]$	$\tilde{\mathbf{B}}_5$	$\mathbf{S}_6$	b^5, $2c^5$
$[4, 3, 3, 3, 4]$	$\tilde{\mathbf{C}}_5$	$\mathbf{R}_6$	c^5, $16b^5$
$[3^{1,1}, 3, 3^{1,1}]$	$\tilde{\mathbf{D}}_5$	$\mathbf{Q}_6$	d^5, $2b^5$, $4c^5$

* The sizes listed for each group indicate the number of copies of its fundamental region, or of the fundamental region for a different group, that can be amalgamated to form a similar fundamental region for an isomorphic subgroup. The variables a, b, c, etc., may take any positive integer values.

Group	Cartan Symbol	Witt Symbol	Sizes
$[3^{[7]}]$	$\tilde{\mathbf{A}}_6$	$\mathbf{P}_7$	a^6
$[4, 3^3, 3^{1,1}]$	$\tilde{\mathbf{B}}_6$	$\mathbf{S}_7$	b^6, $2c^6$
$[4, 3^4, 4]$	$\tilde{\mathbf{C}}_6$	$\mathbf{R}_7$	c^6, $32b^6$
$[3^{1,1}, 3^2, 3^{1,1}]$	$\tilde{\mathbf{D}}_6$	$\mathbf{Q}_7$	d^6, $2b^6$, $4c^6$
$[3^{2,2,2}]$	$\tilde{\mathbf{E}}_6$	$\mathbf{T}_7$	e^6
$[3^{[8]}]$	$\tilde{\mathbf{A}}_7$	$\mathbf{P}_8$	a^7, $144e^7$
$[4, 3^4, 3^{1,1}]$	$\tilde{\mathbf{B}}_7$	$\mathbf{S}_8$	b^7, $2c^7$
$[4, 3^5, 4]$	$\tilde{\mathbf{C}}_7$	$\mathbf{R}_8$	c^7, $64b^7$
$[3^{1,1}, 3^3, 3^{1,1}]$	$\tilde{\mathbf{D}}_7$	$\mathbf{Q}_8$	d^7, $2b^7$, $4c^7$
$[3^{3,3,1}]$	$\tilde{\mathbf{E}}_7$	$\mathbf{T}_8$	e^7
$[3^{[9]}]$	$\tilde{\mathbf{A}}_8$	$\mathbf{P}_9$	a^8, $5760e^8$
$[4, 3^5, 3^{1,1}]$	$\tilde{\mathbf{B}}_8$	$\mathbf{S}_9$	b^8, $2c^8$
$[4, 3^6, 4]$	$\tilde{\mathbf{C}}_8$	$\mathbf{R}_9$	c^8, $128b^8$
$[3^{1,1}, 3^4, 3^{1,1}]$	$\tilde{\mathbf{D}}_8$	$\mathbf{Q}_9$	d^8, $2b^8$, $4c^8$, $270e^8$
$[3^{5,2,1}]$	$\tilde{\mathbf{E}}_8$	$\mathbf{T}_9$	e^8
$[3^{[n]}]$, $n > 9$	$\tilde{\mathbf{A}}_{n-1}$	$\mathbf{P}_n$	a^{n-1}
$[4, 3^{n-4}, 3^{1,1}]$	$\tilde{\mathbf{B}}_{n-1}$	$\mathbf{S}_n$	b^{n-1}, $2c^{n-1}$
$[4, 3^{n-3}, 4]$	$\tilde{\mathbf{C}}_{n-1}$	$\mathbf{R}_n$	c^{n-1}, $2^{n-2}b^{n-1}$
$[3^{1,1}, 3^{n-5}, 3^{1,1}]$	$\tilde{\mathbf{D}}_{n-1}$	$\mathbf{Q}_n$	d^{n-1}, $2b^{n-1}$, $4c^{n-1}$

3. *Hyperbolic Groups*

Group	Witt Symbol	Size	$-\sigma$*
$[\pi i/\lambda]$, $\lambda > 0$	$\tilde{\mathbf{A}}_1(\lambda)$	λ	$4\sinh^2\lambda$
$[3, p]$, $p > 6$	$\bar{\mathbf{A}}_2(p)$	$\pi/6 - \pi/p$	$2(2\cos 2\pi/p - 1)$
$[4, q]$, $q > 4$	$\bar{\mathbf{B}}_2(q)$	$\pi/4 - \pi/q$	$4\cos 2\pi/q$
$[q, q]$, $q > 4$	$\bar{\mathbf{D}}_2(q)$	$\pi/2 - 2\pi/q$	$8\cos 2\pi/q$
$[p, q]$, $4 < p < q$	$\bar{\mathbf{I}}_2(p, q)$	$\pi/2 - \pi/p - \pi/q$	$4(\cos 2\pi/p + \cos 2\pi/q)$
$[(3, 3, r)]$, $r > 3$	$\hat{\mathbf{A}}_2(r)$	$\pi/3 - \pi/r$	$4(2\cos\pi/r - 1)(\cos\pi/r + 1)$
$[(3, p, q)]$, $3 < p \le q$	$\bar{\mathbf{I}}_2(3, p, q)$	$2\pi/3 - \pi/p - \pi/q$	$8(\cos^2\pi/p + \cos^2\pi/q + \cos\pi/p\cos\pi/q) - 6$
$[(p, q, r)]$, $3 < p \le q \le r$	$\bar{\mathbf{I}}_2(p, q, r)$	$\pi - \pi/p - \pi/q - \pi/r$	$8(\cos^2\pi/p + \cos^2\pi/q + \cos^2\pi/r + 2\cos\pi/p\cos\pi/q\cos\pi/r - 1)$
$[p^{[3]}]$, $p > 3$	$\hat{\mathbf{I}}_2(p)$	$\pi - 3\pi/p$	$8(2\cos\pi/p - 1)(\cos\pi/p + 1)^2$
$[3, \infty]$	$\bar{\mathbf{P}}_2$	$\pi/6$	2
$[4, \infty]$	$\bar{\mathbf{Q}}_2$	$\pi/4$	4
$[5, \infty]$	$\bar{\mathbf{HP}}_2$	$3\pi/10$	$2\tau^2$
$[6, \infty]$	$\bar{\mathbf{GP}}_2$	$\pi/3$	6
$[p, \infty]$, $p > 6$	$\bar{\mathbf{W}}_2(p)$	$\pi/2 - \pi/p$	$8\cos^2\pi/p$
$[(3, 3, \infty)]$	$\bar{\mathbf{O}}_2$	$\pi/3$	4
$[(p, q, \infty)]$, $3 \le p \le q$, $q > 3$	$\bar{\mathbf{W}}_2(p, q)$	$\pi - \pi/p - \pi/q$	$8(\cos\pi/p + \cos\pi/q)^2$
$[\infty, \infty]$	$\bar{\mathbf{N}}_2$	$\pi/2$	8
$[(3, \infty, \infty)]$	$\hat{\mathbf{P}}_2$	$2\pi/3$	18
$[(p, \infty, \infty)]$, $p > 3$	$\hat{\mathbf{W}}_2(p)$	$\pi - \pi/p$	$8(\cos\pi/p + 1)^2$
$[\infty^{[3]}]$	$\bar{\mathbf{M}}_2$	π	32

* See first note in Part 1.

Group	Witt Symbol	Size		$-\sigma$
		Formula	Value	
[4, 3, 5]	$\mathbf{B\bar{H}}_3$	$\frac{1}{4}\mathrm{i}S(\frac{1}{4}\pi,\ \frac{1}{3}\pi,\ \frac{3}{10}\pi)$	0.0358850633	$2\bar{\tau}$
[3, 5, 3]	$\bar{\mathbf{J}}_3$	$\frac{1}{4}\mathrm{i}S(\frac{1}{6}\pi,\ \frac{1}{5}\pi,\ \frac{1}{6}\pi)$	0.0390502856	$4\bar{\tau}-1$
$[5, 3^{1,1}]$	$\mathbf{D\bar{H}}_3$	$\frac{1}{2}\mathrm{i}S(\frac{1}{4}\pi,\ \frac{1}{3}\pi,\ \frac{3}{10}\pi)$	0.0717701267	$4\bar{\tau}$
$[(3^3, 4)]$	$\widehat{\mathbf{AB}}_3$	(see notes)	0.0857701820	$1+2\sqrt{2}$
[5, 3, 5]	$\bar{\mathbf{K}}_3$	$\frac{1}{4}\mathrm{i}S(\frac{3}{10}\pi,\ \frac{1}{3}\pi,\ \frac{3}{10}\pi)$	0.0933255395	$5\bar{\tau}-1$
$[(3^3, 5)]$	$\widehat{\mathbf{AH}}_3$	(see notes)	0.2052887885	$5\tau-2$
$[(3, 4)^{[2]}]$	$\widehat{\mathbf{BB}}_3$	$7^{3/2}L(2,\ 7)/96$	0.2222287320	7
[(3, 4, 3, 5)]	$\widehat{\mathbf{BH}}_3$	(see notes)	0.3586534401	$2\tau(1+2\sqrt{2})+1$
$[(3, 5)^{[2]}]$	$\widehat{\mathbf{HH}}_3$	(see notes)	0.5021308905	$7\tau-1$
[3, 3, 6]	$\bar{\mathbf{V}}_3$	$\frac{1}{8}Л(\frac{1}{3}\pi)=\frac{1}{12}Л(\frac{1}{6}\pi)$	0.0422892336	1
[3, 4, 4]	$\bar{\mathbf{R}}_3$	$\frac{1}{6}Л(\frac{1}{4}\pi)$	0.0763304662	2
$[3, 3^{[3]}]$	$\bar{\mathbf{P}}_3$	$\frac{1}{4}Л(\frac{1}{3}\pi)$	0.0845784672	3
[4, 3, 6]	$\mathbf{B\bar{V}}_3$	$\frac{5}{16}Л(\frac{1}{3}\pi)$	0.1057230840	2
$[3, 4^{1,1}]$	$\bar{\mathbf{O}}_3$	$\frac{1}{3}Л(\frac{1}{4}\pi)$	0.1526609324	4
[3, 6, 3]	$\bar{\mathbf{Y}}_3$	$\frac{1}{2}Л(\frac{1}{3}\pi)$	0.1691569344	3
[5, 3, 6]	$\mathbf{H\bar{V}}_3$	$\frac{1}{2}Л(\frac{1}{3}\pi)+\frac{1}{4}Л(\frac{11}{30}\pi)-\frac{1}{4}Л(\frac{1}{30}\pi)$	0.1715016613	τ^2

Group	Witt Symbol	Size		$-\sigma$
		Formula	Value	
$[4,\ 3^{[3]}]$	$\overline{\mathbf{BP}}_3$	$\frac{5}{8}Л(\frac{1}{3}\pi)$	0.2114461680	6
$[6,\ 3^{1,1}]$	$\overline{\mathbf{DV}}_3$	$\frac{5}{8}Л(\frac{1}{3}\pi)$	0.2114461680	4
$[4, 4, 4]$	$\bar{\mathbf{N}}_3$	$\frac{1}{2}Л(\frac{1}{4}\pi)$	0.2289913985	4
$[6, 3, 6]$	$\bar{\mathbf{Z}}_3$	$\frac{3}{4}Л(\frac{1}{3}\pi)$	0.2537354016	3
$[(3^2,\ 4^2)]$	$\widehat{\mathbf{BR}}_3$	$\frac{2}{3}Л(\frac{1}{4}\pi)$	0.3053218647	8
$[5,\ 3^{[3]}]$	$\overline{\mathbf{HP}}_3$	$Л(\frac{1}{3}\pi) + \frac{1}{2}Л(\frac{11}{30}\pi) - \frac{1}{2}Л(\frac{1}{30}\pi)$	0.3430033226	$3\tau^2$
$[(3^3,\ 6)]$	$\widehat{\mathbf{AV}}_3$	$\frac{5}{8}Л(\frac{1}{3}\pi) + \frac{1}{3}Л(\frac{1}{4}\pi)$	0.3641071004	$(1+\sqrt{3})^2$
$[3^{[]\times[]}]$	$\overline{\mathbf{DP}}_3$	$\frac{5}{4}Л(\frac{1}{3}\pi)$	0.4228923360	12
$[4^{1,1,1}]$	$\bar{\mathbf{M}}_3$	$Л(\frac{1}{4}\pi)$	0.4579827971	8
$[6,\ 3^{[3]}]$	$\overline{\mathbf{VP}}_3$	$\frac{3}{2}Л(\frac{1}{3}\pi)$	0.5074708032	9
$[(3, 4, 3, 6)]$	$\widehat{\mathbf{BV}}_3$	$Л(\frac{1}{3}\pi) + Л(\frac{5}{24}\pi) - Л(\frac{1}{24}\pi)$	0.5258402692	$(\sqrt{2}+\sqrt{3})^2$
$[(3,\ 4^3)]$	$\widehat{\mathbf{CR}}_3$	$Л(\frac{1}{4}\pi) + Л(\frac{7}{24}\pi) - Л(\frac{1}{24}\pi)$	0.5562821156	$(\sqrt{2}+2)^2$
$[(3, 5, 3, 6)]$	$\widehat{\mathbf{HV}}_3$	$Л(\frac{1}{3}\pi) + Л(\frac{11}{60}\pi) - Л(\frac{1}{60}\pi)$	0.6729858045	$(\tau+\sqrt{3})^2$
$[(3,\ 6)^{[2]}]$	$\widehat{\mathbf{VV}}_3$	$\frac{5}{2}Л(\frac{1}{3}\pi)$	0.8457846720	12
$[4^{[4]}]$	$\widehat{\mathbf{RR}}_3$	$2Л(\frac{1}{4}\pi)$	0.9159655942	16
$[3^{[3,3]}]$	$\widehat{\mathbf{PP}}_3$	$3Л(\frac{1}{3}\pi) = 2Л(\frac{1}{6}\pi)$	1.0149416064	27

$[3, 3, 3, 5]$	$\bar{\mathbf{H}}_4$	$\pi^2/10800$	0.00091385226	$2\bar{\tau}^3$
$[4, 3, 3, 5]$	$\overline{\mathbf{BH}}_4$	$17\pi^2/21600$	0.00776774420	$2\bar{\tau}$
$[5, 3, 3^{1,1}]$	$\overline{\mathbf{DH}}_4$	$17\pi^2/10800$	0.01553548841	$4\bar{\tau}$
$[5, 3, 3, 5]$	$\bar{\mathbf{K}}_4$	$13\pi^2/5400$	0.02376015874	$6\bar{\tau} - 2$
$[(3^4, 4)]$	$\widehat{\mathbf{AF}}_4$	$11\pi^2/4320$	0.02513093713	$2 + 2\sqrt{2}$
----	----	----	----	----
$[4, 3^{2,1}]$	$\bar{\mathbf{S}}_4$	$\pi^2/1440$	0.00685389195	2
$[3, 4, 3, 4]$	$\bar{\mathbf{R}}_4$	$\pi^2/864$	0.01142315324	2
$[3, 3^{[4]}]$	$\bar{\mathbf{P}}_4$	$\pi^2/720$	0.01370778389	4
$[3, 4, 3^{1,1}]$	$\bar{\mathbf{O}}_4$	$\pi^2/432$	0.02284630648	4
$[4_{\rceil} 3_{\lceil} 3, 4]$	$\bar{\mathbf{N}}_4$	$\pi^2/288$	0.03426945973	4
$[4, 3^{1,1,1}]$	$\bar{\mathbf{M}}_4$	$\pi^2/144$	0.06853891945	8
$[4, 3^{[4]}]$	$\overline{\mathbf{BP}}_4$	$\pi^2/144$	0.06853891945	8
$[(3^2, 4, 3, 4)]$	$\widehat{\mathbf{FR}}_4$	$\pi^2/108$	0.09138522594	8
$[3^{[3]\times[\,]}]$	$\overline{\mathbf{DP}}_4$	$\pi^2/72$	0.13707783890	16

Group	Witt Symbol	Size		$-\sigma$
		Formula	Value	
$[3, 3, 3, 4, 3]$	$\bar{\mathbf{U}}_5$	$7\zeta(3)/46080$	0.000182604130	1
$[4, 3, 3^{2,1}]$	$\bar{\mathbf{S}}_5$	$7\zeta(3)/15360$	0.000547812391	2
$[3, 3, 4, 3, 3]$	$\bar{\mathbf{X}}_5$	$7\zeta(3)/9216$	0.000913020651	2
$[3^{2,1,1,1}]$	$\bar{\mathbf{Q}}_5$	$7\zeta(3)/7680$	0.001095624782	4
$[3, 4, 3, 3, 4]$	$\bar{\mathbf{R}}_5$	$7\zeta(3)/4608$	0.001826041303	2
$[3, 3^{[5]}]$	$\bar{\mathbf{P}}_5$	$5^{3/2}L(3, 5)/4608$	0.002074051961	5
$[3, 4, 3, 3^{1,1}]$	$\bar{\mathbf{O}}_5$	$7\zeta(3)/2304$	0.003652082605	4
$[4, 3_{\urcorner} 3_{\ulcorner} 3, 4]$	$\bar{\mathbf{N}}_5$	$7\zeta(3)/1536$	0.005478123908	4
$[(3^5, 4)]$	$\widehat{\mathbf{AU}}_5$	———*	0.007573474422	$(1+\sqrt{2})^2$
$[4, 3, 3^{1,1,1}]$	$\bar{\mathbf{M}}_5$	$7\zeta(3)/768$	0.010956247815	8
$[3^{1,1,1,1,1}]$	$\bar{\mathbf{L}}_5$	$7\zeta(3)/384$	0.021912495631	16
$[(3^2, 4)^{[2]}]$	$\widehat{\mathbf{UR}}_5$	$7\zeta(3)/288$	0.029216660841	8
$[4, 3^2, 3^{2,1}]$	$\bar{\mathbf{S}}_6$	$\pi^3/777600$	$0.3987432701 \times 10^{-4}$	2
$[3^{1,1}, 3, 3^{2,1}]$	$\bar{\mathbf{Q}}_6$	$\pi^3/388800$	$0.7974865401 \times 10^{-4}$	4
$[3, 3^{[6]}]$	$\bar{\mathbf{P}}_6$	$13\pi^3/1360800$	$2.9620928633 \times 10^{-4}$	6

* No elementary formula available.

$[3^{3,2,2}]$	$\bar{\mathbf{T}}_7$	$3^{1/2}L(4,\,3)/860160$	$0.1892871372 \times 10^{-5}$	3
$[4,\,3^3,\,3^{2,1}]$	$\bar{\mathbf{S}}_7$	$L(4)/362880$	$0.2725266071 \times 10^{-5}$	2
$[3^{1,1},\,3^2,\,3^{2,1}]$	$\bar{\mathbf{Q}}_7$	$L(4)/181440$	$0.5450532141 \times 10^{-5}$	4
$[3,\,3^{[7]}]$	$\bar{\mathbf{P}}_7$	$7^{5/2}L(4,\,7)/3317760$	$4.1106779054 \times 10^{-5}$	7
$[3^{4,3,1}]$	$\bar{\mathbf{T}}_8$	$\pi^4/4572288000$	$0.0213042335 \times 10^{-6}$	2
$[4,\,3^4,\,3^{2,1}]$	$\bar{\mathbf{S}}_8$	$17\pi^4/9144576000$	$0.1810859845 \times 10^{-6}$	2
$[3^{1,1},\,3^3,\,3^{2,1}]$	$\bar{\mathbf{Q}}_8$	$17\pi^4/4572288000$	$0.3621719690 \times 10^{-6}$	4
$[3,\,3^{[8]}]$	$\bar{\mathbf{P}}_8$	$17\pi^4/285768000$	$5.7947515032 \times 10^{-6}$	8
$[3^{6,2,1}]$	$\bar{\mathbf{T}}_9$	$\zeta(5)/22295347200$	$0.0004650871 \times 10^{-7}$	1
$[4,\,3^5,\,3^{2,1}]$	$\bar{\mathbf{S}}_9$	$527\zeta(5)/44590694400$	$0.1225504411 \times 10^{-7}$	2
$[3^{1,1},\,3^4,\,3^{2,1}]$	$\bar{\mathbf{Q}}_9$	$527\zeta(5)/22295347200$	$0.2451008823 \times 10^{-7}$	4

NOTE: Formulas in odd-dimensional hyperbolic spaces involve the Riemann zeta function and Dirichlet L-functions, defined by

$$\zeta(s) = \sum_{n=1}^{\infty} \frac{1}{n^s} \quad \text{and} \quad L(s,\,p) = \sum_{n=1}^{\infty} \frac{\chi_p(n)}{n^s}, \text{ with } L(s) = L(s,\,4),$$

where the character $\chi_p(n) = \left(\frac{n}{p}\right)$ is $+1$ if n is a quadratic residue modulo p, -1 if n is a quadratic nonresidue, or 0 if $(n,\,p) > 1$. Some volume formulas given in terms of the Lobachevsky function Л have alternative versions derived from the identities

$$Л(\tfrac{1}{4}\pi) = \tfrac{1}{2}L(2) \quad \text{and} \quad Л(\tfrac{1}{3}\pi) = \tfrac{2}{3}Л(\tfrac{1}{6}\pi) = \tfrac{1}{4}\sqrt{3}L(2,\,3).$$

LIST OF SYMBOLS

LISTED IN THE FOLLOWING PAGES, with brief definitions, are some of the symbols used in this book to denote number systems, geometries and measures, vectors and matrices, mappings, continuous and discrete groups, polytopes and honeycombs, and a few other things. Following each symbol definition is a reference to the section or sections of the text where a fuller treatment may be found. Some widely used symbols first appear in the Preliminaries.

Many algebraic and geometric concepts involve some kind of duality, which is reflected notationally by the interchange of parentheses and brackets and by the caron (inverted circumflex) now commonly used in symbols for dualized figures or structures. One basic dualizing operation is matrix transposition: the transpose of a matrix A is denoted by $A^{\vee}$.

A parenthesized F in a symbol generally stands for either the real field $\mathbb{R}$ or the complex field $\mathbb{C}$. For a transformation group, the real field is usually to be understood if none is specified. Thus the symbols $\mathrm{T}^n(\mathbb{R})$ and $\mathrm{GL}_n(\mathbb{R})$ for the real translation and general linear groups can be abbreviated to T^n and GL_n, and O_n denotes the real orthogonal group. However, the unitary group U_n operates on a complex vector space, and the real, complex, and quaternionic symplectic groups are distinguished as Sp_n, $\mathrm{S\check{p}}_n$, and $\mathrm{S\tilde{p}}_n$.

Many transformation groups are subgroups or quotient groups of others. The direct product of two groups H and K is denoted by

$H \times K$, and the direct product of n copies of a group H by H^n. If N is a normal subgroup of a group G and $Q \cong G/N$ another subgroup of G, with $N \vee Q \cong G$ and $N \cap Q \cong 1$, then G is an extension of N by Q, expressible as a semidirect product $N \rtimes Q$. The wreath product $H \wr S_n$ is a semidirect product $H^n \rtimes S_n$.

Discrete groups generated by reflections can be represented either by Coxeter's bracket notation or by alphabetic Cartan or Witt symbols, as given in Table B. The principal finite reflection groups are listed here, along with related symbols for polytopes and honeycombs.

Number systems

$\mathbb{F}_q$	Finite field with q elements	2.1
$\mathbb{Q}$	Ordered field of rational numbers	2.1
$\mathbb{R}$	Complete ordered field of real numbers	2.1
$\mathbb{C} = \mathbb{R}_{(2)}$	Field of complex numbers	8.1
$\mathbb{H} = \mathbb{R}_{(4)}$	Division ring of quaternions	10.4
$\mathbb{Z}$	Ordered domain of rational integers	14.1
$\mathbb{G}, \mathbb{E}$	Domains of Gaussian and Eisenstein complex integers	14.2–14.3
$\mathbb{H}$am, $\mathbb{H}$ur	Rings of Hamilton and Hurwitz quaternionic integers	15.1
$\mathbb{H}$yb	Ring of hybrid quaternionic integers	15.1

Geometries

$\mathrm{P}^n, \mathbb{C}\mathrm{P}^n$	Real and complex projective n-space	1.1, 1.3
$\mathrm{A}^n, \mathbb{C}\mathrm{A}^n$	Real and complex affine n-space	1.1, 8.1
$\mathrm{I}^n, \mathbb{C}\mathrm{I}^n$	Real and complex inversive n-space	1.1, 1.3
eP^n	Elliptic n-space	1.1, 2.3
E^n	Euclidean n-space	1.1, 2.3
H^n	Hyperbolic n-space	1.1, 2.4
$\mathrm{S}^n, \ddot{\mathrm{S}}^n$	Spherical and pseudospherical n-space	1.1, 3.1–3.2
$\mathrm{I}\textsc{u}^n$	Inversive unitary n-space	1.3, 9.4
$\mathrm{P}\textsc{u}^n$	Elliptic unitary n-space	1.3, 7.1
$\mathrm{E}\textsc{u}^n$	Euclidean unitary n-space	1.3, 7.1
$\mathrm{H}\textsc{u}^n$	Hyperbolic unitary n-space	1.3, 7.1
$\mathrm{S}\textsc{u}^n, \ddot{\mathrm{S}}\textsc{u}^n$	Spherical and pseudospherical unitary n-space	1.3, 7.2

Measures

$\|XY, \check{U}\check{V}\|$	Cross ratio of points X and Y and hyperplanes $\check{U}$ and $\check{V}$	2.2
$[XY]$	Elliptic distance between points X and Y	2.3
$\|XY\|$	Euclidean distance between points X and Y	2.3
$]XY[$	Hyperbolic distance between points X and Y	2.4
$(\check{U}\check{V})$	Angle between intersecting hyperplanes $\check{U}$ and $\check{V}$	2.3
$\|\check{U}\check{V}\|$	Distance between parallel hyperplanes $\check{U}$ and $\check{V}$	2.3
$)\check{U}\check{V}($	Distance between diverging hyperplanes $\check{U}$ and $\check{V}$	2.4
$[\![XY]\!]$	Spherical distance between points X and Y	3.1
$]\!]XY[\![$	Pseudospherical distance between points X and Y	3.2

$((\breve{U}\breve{V}))$	Angle between separating great hyperspheres $\breve{U}$ and $\breve{V}$	3.1
$))\breve{U}\breve{V}(($	Distance between separated great hyperspheres $\breve{U}$ and $\breve{V}$	3.2
$\|XYZ\|$	Directed area of point triple (XYZ)	5.5

Vectors and matrices

A	Generic $m \times n$ matrix (over a ring) with $(i,j) \mapsto a_{ij}$	2.2
$A^{\check{}}$	Transpose of A, $n \times m$ matrix with $(i,j) \mapsto a_{ji}$	2.2
$\bar{A}$	Conjugate of A, $m \times n$ matrix with $(i,j) \mapsto \bar{a}_{ij}$	7.1
A^*	Antitranspose of $A, n \times m$ matrix with $(i,j) \mapsto \bar{a}_{ji}$	7.1
$D = \backslash d \backslash$	Diagonal matrix $\backslash d_{11}, \ldots, d_{nn} \backslash$	4.1
I, I_n	Identity matrix $\backslash 1, \ldots, 1 \backslash$	2.3
$\dot{I}, I_{n,1}$	Pseudo-identity matrix $\backslash 1, \ldots, 1, -1 \backslash$	2.4
J, J_{2m}	Juxtation matrix $\backslash J_2, \ldots, J_2 \backslash$, $J_2 = \begin{pmatrix} 0 & 1 \\ -1 & 1 \end{pmatrix}$	5.5
$\mathsf{F}^{m \times n}$	Vector space of $m \times n$ matrices over field F	4.1
$\mathrm{Mat}_n(\mathsf{F})$	Ring of $n \times n$ matrices over F	4.1
$\mathrm{Diag}_n(\mathsf{F})$	Commutative ring of $n \times n$ diagonal matrices over F	4.1
(x)	Row, $1 \times n$ matrix $(x_1, \ldots, x_n)$	2.2
$[u]$	Column, $n \times 1$ matrix $[u_1, \ldots, u_n] = (u_1, \ldots, u_n)^{\check{}}$	2.2
$(0), [0]$	Row or column with all entries 0	4.1
$(\delta_i), [\delta_j]$	Row or column with entries δ_{ij} (Kronecker delta)	4.2
$(\epsilon), [\epsilon]$	Row or column with all entries 1	4.3
$\mathsf{F}^n = \mathsf{F}^{1 \times n}$	Vector space of n-tuples (x) over F (rows)	2.2
$\check{\mathsf{F}}^n = \mathsf{F}^{n \times 1}$	Vector space of n-tuples $[u]$ over F (columns)	2.2
$\langle (x_i)_{i=1}^k \rangle$	Subspace spanned by k rows, $\{\sum \lambda_i (x_i) : \lambda_i \in \mathsf{F}\}$	2.2, 4.1
$\langle [u_j]_{j=1}^k \rangle$	Subspace spanned by k columns, $\{\sum [u_j] \rho_j : \rho_j \in \mathsf{F}\}$	2.2, 4.1
PF^n	Projective linear space of 1-subspaces $\langle (x) \rangle$ of F^n	4.4
$\mathsf{P}\check{\mathsf{F}}^n$	Projective linear space of 1-subspaces $\langle [u] \rangle$ of $\check{\mathsf{F}}^n$	4.4

Mappings and forms

$+(h)$	Translation $(x) \mapsto (x) + (h)$ determined by vector (h)	4.2
$\lambda \cdot$	Scaling operation $(x) \mapsto \lambda(x)$ determined by scalar λ	4.2
$\cdot A$	Linear transformation $(x) \mapsto (x)A$ determined by matrix A	4.2
$\langle \cdot A \rangle$	Projective linear transformation $\langle (x) \rangle \mapsto \langle (x)A \rangle$	4.4
$\langle \cdot A \rangle^{\check{}}$	Projective linear cotransformation $\langle (x) \rangle \mapsto \langle (x)A \rangle^{\check{}}$	4.5
$\odot A$	Antilinear transformation $(z) \mapsto (\bar{z})A$	8.1
$\langle \odot A \rangle$	Projective antilinear transformation $\langle (z) \rangle \mapsto \langle (\bar{z})A \rangle$	8.3
$\langle \cdot A \rangle^*$	Projective antilinear cotransformation $\langle (z) \rangle \mapsto \langle (z)A \rangle^*$	8.3
$\cdot \langle A \rangle$	Linear fractional transformation $z \mapsto \frac{a_{11} z + a_{21}}{a_{12} z + a_{22}}$	9.3
$\odot \langle A \rangle$	Antilinear fractional transformation $z \mapsto \frac{a_{11} \bar{z} + a_{21}}{a_{12} \bar{z} + a_{22}}$	9.3
$(x\,y)$	Bilinear form on F^n with $[(x), (y)] \mapsto (x)A(y)^{\check{}}$	2.2
$[u\,v]$	Bilinear form on $\check{\mathsf{F}}^n$ with $([u], [v]) \mapsto [u]^{\check{}} A^{-1} [v]$	2.2
$(z\,\bar{w})$	Sesquilinear form on $\mathbb{C}^n$ with $[(z), (w)] \mapsto (z)A(w)^*$	7.1
$[\bar{w}\,z]$	Sesquilinear form on $\check{\mathbb{C}}^n$ with $([w], [z]) \mapsto [w]^* A^{-1} [z]$	7.1

Linear and afffine groups

$\mathrm{T}(\mathsf{R})$	Additive group of ring R	2.1
$\mathrm{L}(\mathsf{R})$	Multiplicative semigroup of R	4.1
$\mathrm{GL}(\mathsf{K})$	Multiplicative group of division ring K (nonzero elements)	2.1

$\mathrm{T}^n(\mathsf{F})$	Translation group, additive group of vector space F^n	4.1
$\mathrm{L}_n(\mathsf{F})$	Linear semigroup (mappings $\cdot A : \mathsf{F}^n \to \mathsf{F}^n$, $A \in \mathrm{Mat}_n(\mathsf{F})$)	4.1
$\mathrm{GL}_n(\mathsf{F})$	General linear group (invertible mappings $\cdot A : \mathsf{F}^n \to \mathsf{F}^n$)	4.1
$\mathrm{G^+L}_n(\mathsf{F})$	Direct linear group (mappings $\cdot A$ with $\det A > 0$)	4.1
$\bar{\mathrm{S}}\mathrm{L}_n(\mathsf{F})$	Unit linear group (mappings $\cdot A$ with $\lvert \det A \rvert = 1$)	4.1
$\mathrm{SL}_n(\mathsf{F})$	Special linear group (mappings $\cdot A$ with $\det A = 1$)	4.1
$\mathrm{GZ}(\mathsf{F})$	General scalar group (nonzero scalar mappings $\cdot \lambda I$)	4.1
GZ^+	Positive scalar group (mappings $\cdot \lambda I : \mathbb{R}^n \to \mathbb{R}^n$ with $\lambda > 0$)	4.1
$\bar{\mathrm{S}}\mathrm{Z}(\mathsf{F})$	Unit scalar group (mappings $\cdot \lambda I$ with $\lvert\lambda\rvert = 1$)	4.1
$\mathrm{SZ}_n(\mathsf{F})$	Special scalar group (mappings $\cdot \lambda I$ with $\lambda^n = 1$)	4.4
$\mathrm{GA}_n(\mathsf{F})$	General affine group, $\mathrm{T}^n(\mathsf{F}) \rtimes \mathrm{GL}_n(\mathsf{F})$	4.2
$\mathrm{G^+A}_n(\mathsf{F})$	Direct affine group, $\mathrm{T}^n(\mathsf{F}) \rtimes \mathrm{G^+L}_n(\mathsf{F})$	4.2
$\bar{\mathrm{S}}\mathrm{A}_n(\mathsf{F})$	Unit affine group, $\mathrm{T}^n(\mathsf{F}) \rtimes \bar{\mathrm{S}}\mathrm{L}_n(\mathsf{F})$	4.2
$\mathrm{SA}_n(\mathsf{F})$	Special affine group, $\mathrm{T}^n(\mathsf{F}) \rtimes \mathrm{SL}_n(\mathsf{F})$	4.2
$\mathrm{GD}_n(\mathsf{F})$	General dilative group, $\mathrm{T}^n(\mathsf{F}) \rtimes \mathrm{GZ}(\mathsf{F})$	4.2
$\mathrm{GD}_n^+(\mathsf{F})$	Positive dilative group, $\mathrm{T}^n(\mathsf{F}) \rtimes \mathrm{GZ}^+$	4.2
$\bar{\mathrm{S}}\mathrm{D}_n(\mathsf{F})$	Unit dilative group, $\mathrm{T}^n(\mathsf{F}) \rtimes \bar{\mathrm{S}}\mathrm{Z}(\mathsf{F})$	4.2
$\mathrm{PGL}_n(\mathsf{F})$	Projective general linear group, $\mathrm{GL}_n(\mathsf{F})/\mathrm{GZ}(\mathsf{F})$	4.4
$\mathrm{PSL}_n(\mathsf{F})$	Projective special linear group, $\mathrm{SL}_n(\mathsf{F})/\mathrm{SZ}_n(\mathsf{F})$	4.4
GCom	General commutator quotient group, $\mathrm{GL}_n/\mathrm{SL}_n$	4.4
$\mathrm{G^+Com}$	Direct commutator quotient group, $\mathrm{G^+L}_n/\mathrm{SL}_n$	4.4
$\bar{\mathrm{S}}\mathrm{Com}$	Unit commutator quotient group, $\bar{\mathrm{S}}\mathrm{L}_n/\mathrm{SL}_n$	4.4
Metric groups		
O_n	Orthogonal group (mappings $\cdot P : \mathbb{R}^n \to \mathbb{R}^n$ with $PP\check{} = I$)	6.1
O_n^+	Special orthogonal group, $\mathrm{O}_n \cap \mathrm{SL}_n$	6.1
$\tilde{\mathrm{O}}_n$	Orthopetic group, $\mathrm{O}_n \times \mathrm{GZ}^+$	6.2
$\tilde{\mathrm{O}}_n^+$	Direct orthopetic group, $\tilde{\mathrm{O}}_n \cap \mathrm{G^+L}_n \cong \mathrm{O}_n^+ \times \mathrm{GZ}^+$	6.2
$\mathrm{O}_{n,1}$	Pseudo-orthogonal group (mappings $\cdot P$ with $P\ddot{P}\check{} = I$)	6.3
$\mathrm{O}_{n,1}^+$	Special pseudo-orthogonal group, $\mathrm{O}_{n,1} \cap \mathrm{SL}_{n+1}$	6.3
$\mathrm{Sp}_n, \mathrm{S\breve{p}}_n$	Symplectic group (mappings $\cdot S$ with $SJS\check{} = J$, n even)	5.5
$\mathrm{Gp}_n, \mathrm{G\breve{p}}_n$	Gyropetic group, $\mathrm{Sp}_n \times \mathrm{GZ}^+$ or $\mathrm{S\breve{p}}_n \times \mathrm{GZ}^+$	5.5
U_n	Unitary group (mappings $\cdot R : \mathbb{C}^n \to \mathbb{C}^n$ with $RR^* = I$)	10.1
SU_n	Special unitary group, $\mathrm{U}_n \cap \mathrm{SL}_n(\mathbb{C})$	10.1
$\tilde{\mathrm{U}}_n$	Zygopetic group, $\mathrm{U}_n \times \mathrm{GZ}^+$	10.2
$\widetilde{\mathrm{SU}}_n$	Proto-zygopetic group, $\mathrm{SU}_n \times \mathrm{GZ}^+$	10.2
$\mathrm{U}_{n,1}$	Pseudo-unitary group (mappings $\cdot R$ with $RR^* = I$)	10.3
$\mathrm{SU}_{n,1}$	Special pseudo-unitary group, $\mathrm{U}_{n,1} \cap \mathrm{SL}_{n+1}(\mathbb{C})$	10.3
$\mathrm{S\tilde{p}}_n, \mathrm{G\tilde{p}}_n$	Quaternionic symplectic and gyropetic groups	10.4
E_n	Full Euclidean group, $\mathrm{T}^n \rtimes \mathrm{O}_n$	6.2
E_n^+	Direct Euclidean group, $\mathrm{T}^n \rtimes \mathrm{O}_n^+$	6.2
$\tilde{\mathrm{E}}_n$	Full euclopetic group, $\mathrm{T}^n \rtimes \tilde{\mathrm{O}}_n$	6.2
$\tilde{\mathrm{E}}_n^+$	Direct euclopetic group, $\mathrm{T}^n \rtimes \tilde{\mathrm{O}}_n^+$	6.2
PO_n	Projective orthogonal group, $\mathrm{O}_n/\bar{\mathrm{S}}\mathrm{Z}$	6.1
$\mathrm{P^+O}_n$	Special projective orthogonal group, $\mathrm{O}_n^+/\bar{\mathrm{S}}\mathrm{Z}$ (n even)	6.1
$\mathrm{PO}_{n,1}$	Projective pseudo-orthogonal group, $\mathrm{O}_{n,1}/\bar{\mathrm{S}}\mathrm{Z}$	6.3
$\mathrm{P^+O}_{n,1}$	Special projective pseudo-orthogonal group, $\mathrm{O}_{n,1}^+/\bar{\mathrm{S}}\mathrm{Z}$ (n odd) or $\mathrm{O}_{n^+,1}/\bar{\mathrm{S}}\mathrm{Z}$ (n even)	6.3

PU_n	Projective unitary group, $\mathrm{U}_n/\bar{\mathrm{S}}\mathrm{Z}(\mathbb{C})$	10.1
$\mathrm{PU}_{n,1}$	Projective pseudo-unitary group, $\mathrm{U}_{n,1}/\bar{\mathrm{S}}\mathrm{Z}(\mathbb{C})$	10.1
Finite reflection groups		
$][$	Identity group, $\mathbf{A}_0 \cong \mathrm{C}_1$	11.2
$[\,]$	Bilateral group, $\mathbf{A}_1 \cong \mathrm{D}_1$	11.2
$[p]$	Full polygonal group $(\mathbf{A}_2, \mathbf{B}_2, \mathbf{H}_2, \mathbf{G}_2, \ldots)$	11.2
$[p,\ q]$	Full polyhedral group $(\mathbf{A}_3, \mathbf{B}_3, \mathbf{H}_3)$	11.4
$[p,\ q,\ r]$	Full polychoric group $(\mathbf{A}_4, \mathbf{B}_4, \mathbf{F}_4, \mathbf{H}_4)$	11.5
$[3^{n-1}]$	Anasymmetric group, $\mathbf{A}_n \cong (n+1)\cdot \mathrm{S}_n \cong \mathrm{S}_{n+1}$ $(n \geq 1)$	11.5
$[3^{n-2}, 4]$	Bisymmetric group, $\mathbf{B}_n \cong {}^2\mathrm{S}_n \cong \mathrm{C}_2 \wr \mathrm{S}_n$ $(n \geq 2)$	11.5
$[3^{n-3,1,1}]$	Demibisymmetric group, $\mathbf{D}_n \cong \tfrac{1}{2}\cdot 2\mathrm{S}_n$ $(n \geq 3)$	11.5
$[3^{n-4,2,1}]$	Exceptional (Gosset) group, $\mathbf{E}_n$ $(n = 6, 7, 8)$	11.5
Polytopes and honeycombs		
$(\,)$	Monon, 0-polytope	11.1
$\{\,\}$	Regular dion or antipodion	11.1
$\{p\}$	Regular polygon or partition	11.1
$\{p, q\}$	Regular polyhedron or tessellation	11.1
$\{p, q, r\}$	Regular polychoron or cellulation	11.5
$\mathsf{P} \vee \mathsf{Q},\ m \cdot \mathsf{P}$	Pyramid, join of polytopes P and Q or m copies of P	11.3
$\mathsf{P} + \mathsf{Q},\ n\mathsf{P}$	Fusil, sum of polytopes P and Q or n copies of P	11.3
$\mathsf{P} \times \mathsf{Q}, \mathsf{P}^n$	Prism, product of polytopes P and Q or n copies of P	11.3
" "	Prismatic honeycomb, product of Euclidean honeycombs	12.5
α_n	Regular simplex (monic pyramid), $\{3^{n-1}\} = (n+1)\cdot(\,)$	11.5
β_n	Regular orthoplex (cross polytope), $\{3^{n-2}, 4\} = n\{\,\}$	11.5
γ_n	Regular orthotope (block polytope), $\{4, 3^{n-2}\} = \{\,\}^n$	11.5
δ_n	Regular orthocomb (grid honeycomb), $\{4, 3^{n-3}, 4\} = \{\infty\}^{n-1}$	12.4
Miscellaneous		
$X \lozenge \check{U}$	Incidence: point X is on hyperplane $\check{U}$	2.1
$AB \,\not{H}\, CD$	Harmonic relation: (AB, CD) is a harmonic set	2.1
$ABC \,\not{Q}\, DEF$	Quadrangular relation: (ABC, DEF) is a quadrangular set	2.1
$AB//CD$	Separation: points A and B separate points C and D	2.1
$\mathbb{Q}(ABC)$	Net of rationality containing points A, B, C	2.1
$\mathbb{R}(ABC)$	Chain containing points A, B, C	2.1
$\cdot AB \cdot$	Line through points A and B	3.4
AB	Line segment with endpoints A and B	3.4
$\langle ABC \rangle$	Triangle with vertices A, B, C enclosing region ABC	3.4
(x, y)	Ordinary point in A^2 with abscissa x and ordinate y	5.1
$[m, b]$	Usual line in A^2 with slope m and y-intercept b	5.1
$[a]$	Vertical line in A^2 with x-intercept a	5.1
$\sigma(\mathsf{P})$	Schläflian, determinant of Schläfli matrix for simplex P	11.5
$(p\,q\,r)$	Triangle with angles $\pi/p, \pi/q, \pi/r$	12.3

BIBLIOGRAPHY

WORKS CITED IN THE TEXT or in the notes on the tables are listed here by author and original date of publication. Listings also include information regarding reprints, revisions, and translations. Names of journals have been abbreviated in accordance with the style of *Mathematical Reviews*.

L. V. AHLFORS. 1985, "Möbius transformations and Clifford numbers," in *Differential Geometry and Complex Analysis*, H. E. Rauch memorial volume, I. Chavel and H. M. Farkas, eds. (Springer-Verlag, Berlin–New York, 1985), 65–73.

D. V. ALEKSEEVSKY, E. B. VINBERG, and A. S. SOLODOVNIKOV. 1988, "Geometry of spaces of constant curvature" (Russian), in *Geometriya 2*, Itogi Nauki i Tekhniki, Ser. Sovremennye problemy matematiki, Fundamental'nye napravleniya, t. 29 (Vsesoyuz Inst. Nauchn. i Tekhn. Inform., Moscow, 1988), 5–146. Translated by V. Minachin as Part I of *Geometry II: Spaces of Constant Curvature*, E. B. Vinberg, ed., Encyclopaedia of Mathematical Sciences, Vol. 29 (Springer-Verlag, Berlin–Heidelberg–New York, 1993), 1–138. Cf. Vinberg & Shvartsman 1988.

D. V. ALEKSEEVSKY, A. M. VINOGRADOV, and V. V. LYCHAGIN. 1988, "Basic ideas and concepts of differential geometry" (Russian), in *Geometriya 1*, Itogi Nauki i Tekhniki, Ser. Sovremennye problemy matematiki, Fundamental'nye napravleniya, t. 28 (Vsesoyuz Inst. Nauchn. i Tekhn. Inform., Moscow, 1988), 5–289. Translated by E. Primrose as *Geometry I: Basic Ideas and Concepts of Differential Geometry*, R. V. Gamkrelidze, ed., Encyclopaedia of Mathematical Sciences, Vol. 28 (Springer-Verlag, Berlin–Heidelberg–New York, 1991).

E. M. ANDREEV. 1970a, "Convex polyhedra in Lobachevsky spaces" (Russian), *Mat. Sb. (N.S.)* **81** (**123**), 445–478. Translated in *Math. USSR Sb.* **10** (1970), 413–440.

1970b, "On convex polyhedra of finite volume in Lobachevsky space" (Russian), *Mat. Sb. (N.S.)* **83** (**125**), 256–260. Translated in *Math. USSR Sb.* **12** (1970), 255–259.

E. ARTIN. 1957, *Geometric Algebra*, Interscience Tracts in Pure and Applied Mathematics, No. 3. Wiley–Interscience, New York, 1957, 1988.

R. ARTZY. 1965, *Linear Geometry*. Addison-Wesley, Reading, MA, 1965.

H. ASLAKSEN. 1996, "Quaternionic determinants," *Math. Intelligencer* **18**, No. 3, 57–65.

R. BAER. 1952, *Linear Algebra and Projective Geometry*, Pure and Applied Mathematics Series, Vol. 2. Academic Press, New York, 1952; Dover, Mineola, NY, 2005.

J. C. BAEZ. 2002, "The octonions," *Bull. Amer. Math. Soc. (N.S.)* **39**, 145–205; errata, *ibid.* **42** (2005), 213.

T. F. BANCHOFF. 1988, "Torus decompositions of regular polytopes in 4-space," in *Shaping Space: A Polyhedral Approach*, M. Senechal and G. Fleck, eds. (Birkhäuser, Boston–Basel, 1988), 221–230, or *Shaping Space: Exploring Polyhedra in Nature, Art, and the Geometrical Imagination*, M. Senechal, ed. (Springer, New York–Heidelberg–Dordrecht, 2013), 257–266.

W. H. BARKER and R. E. HOWE. 2007, *Continuous Symmetry: From Euclid to Klein*. American Math. Soc., Providence, 2007.

A. F. BEARDON. 1979, "Hyperbolic polygons and Fuchsian groups," *J. London Math. Soc.* (2) **20**, 247–254.

1983, *The Geometry of Discrete Groups*, Graduate Texts in Mathematics, No. 91. Springer-Verlag, New York–Heidelberg–Berlin, 1983; 1995.

2005, *Algebra and Geometry*. Cambridge Univ. Press, Cambridge–New York, 2005.

E. BELTRAMI. 1868a, "Saggio di interpretazione della geometria non-euclidea," *Giorn. Mat.* **6**, 284–312. Reprinted in *Opere matematiche*, vol. I (Ulrico Hoepli, Milan, 1902), 374–405. Translated as "Essay on the interpretation of noneuclidean geometry" in Stillwell 1996, 7–34.

1868b,"Teoria fondamentale degli spazii di curvatura costante," *Ann. Mat. Pura Appl.* (2) **2**, 232–255. Reprinted with additional text in *Opere matematiche*, vol. I (Ulrico Hoepli, Milan, 1902), 406–429. Translated as "Fundamental theory of spaces of constant curvature" in Stillwell 1996, 41–62.

L. BIANCHI. 1891, "Geometrische Darstellung der Gruppen linearer Substitutionen mit ganzen complexen Coefficienten nebst Anwendungen auf die

Zahlentheorie," *Math. Ann.* **38**, 313–333. Reprinted in *Opere*, vol. I, pt. 1 (Edizione Cremonese, Rome, 1952), 233–258.

1892, "Sui gruppi de sostituzioni lineari con coeficienti appartenenti a corpi quadratici imaginari," *Math. Ann.* **40**, 332–412. Reprinted in *Opere*, vol. I, pt. 1 (Edizione Cremonese, Rome, 1952), 270–373.

D. Blanuša. 1955, "Über die Einbettung hyperbolischer Räume in euklidische Räume," *Monatsh. Math.* **59**, 217–229.

J. Bolyai. 1832, "Scientiam spatii absolute veram exhibens: a veritate aut falsitate Axiomatis XI Euclidei (a priori haud unquam decidenda) independentem," essay published as appendix to W. Bolyai's *Tentamen: Juventutem studiosam in elementa Matheseos purae, elementaris ac sublimioris, methodo intuitiva, evidentiaque huic propria, introducendi*, Tom. I (Reformed College, Maros-Vásárhely [Hungary], 1832), 26+ii pp., 1 plate. Translated by G. B. Halsted as *The Science Absolute of Space* (The Neomon, Austin, 1891; 4th ed., 1896) and by F. Kárteszi as *Appendix: The Theory of Space*, North-Holland Mathematics Studies, Vol. 138 (North-Holland / Elsevier, Amsterdam, and Akadémiai Kiadó, Budapest, 1987). Halsted's translation reprinted as supplement to Carslaw's translation of Bonola 1906 (Dover, New York, 1955).

R. Bonola. 1906, *La geometria non-euclidea: Esposizione storico-critica del suo sviluppo*. N. Zanichelli, Bologna, 1906. Translated by H. Liebmann as *Die nichteuklidische Geometrie: Historisch-kritische Darstellung ihrer Entwicklung*, Wissenschaft und Hypothese, Bd. 4 (Teubner, Leipzig–Berlin, 1908), and by H. S. Carslaw as *Non-Euclidean Geometry: A Critical and Historical Study of Its Development* (Open Court, Chicago, 1912, and La Salle, IL, 1938; Dover, New York, 1955). Dover edition of Carslaw's translation includes English translations (by G. B. Halsted) of Bolyai 1832 and Lobachevsky 1840.

R. H. Bruck. 1955, "Recent advances in the foundations of Euclidean plane geometry," in *Contributions to Geometry*, H. E. Slaught Memorial Paper No. 4 (*Amer. Math. Monthly* **62**, No. 7, Pt. II), 2–17.

R. P. Burn. 1985, *Groups: A Path to Geometry*. Cambridge Univ. Press, Cambridge–New York, 1985; 1987.

É. Cartan. 1927, "La Géométrie des groupes simples," *Ann. Mat. Pura Appl.* (4) **4**, 209–256. Reprinted in *Œuvres complètes*, pt. I, vol. 2 (Gauthier-Villars, Paris, 1952; 2nd ed., Centre National de la Recherche Scientifique, Paris, 1984), 793–840.

1928, "Complément au mémoire sur la géométrie des groupes simples," *Ann. Mat. Pura Appl.* (4) **5**, 253–260. Reprinted in *Œuvres complètes*, pt. I, vol. 2 (Gauthier-Villars, Paris, 1952; 2nd ed., Centre National de la Recherche Scientifique, Paris, 1984), 1003–1010.

R. W. CARTER, G. SEGAL, and I. G. MACDONALD. 1995, *Lectures on Lie Groups and Lie Algebras*, London Mathematical Society Student Texts, No. 32. Cambridge Univ. Press, Cambridge–New York, 1995.

A. CAYLEY. 1859, "A sixth memoir upon quantics," *Philos. Trans. Roy. Soc. London* **149**, 61–90. Reprinted in *Collected Mathematical Papers*, Vol. II (Cambridge Univ. Press, Cambridge, 1889), 561–592.

1872, "On the non-Euclidian geometry," *Math. Ann.* **5**, 630–634. Reprinted in *Collected Mathematical Papers*, Vol. VIII (Cambridge Univ. Press, Cambridge, 1895), 409–413.

M. CHEIN. 1969, "Recherche des graphes des matrices de Coxeter hyperboliques d'ordre $\leq$ 10," *Rev. Française Informat. Recherche Opérationnelle* **3**, Sér. R-3, 3–16.

Y. CHO and H. KIM. 1999, "On the volume formula for hyperbolic tetrahedra," *Discrete Comput. Geom.* **22**, 347–366.

W. K. CLIFFORD. 1873, "Preliminary sketch of biquaternions," *Proc. London Math. Soc.* **4**, 381–395. Reprinted in *Mathematical Papers* (Macmillan, London, 1882), 181–200.

1882, "On the classification of geometric algebras," unpublished manuscript (1876) included in *Mathematical Papers* (Macmillan, London, 1882), 397–401.

J. H. CONWAY. 1983, "The automorphism group of the 26-dimensional even unimodular Lorentzian lattice," *J. Algebra* **80**, 159–163. Cf. Conway & Sloane 1988, chap. 27.

J. H. CONWAY and N. J. A. SLOANE. 1988, *Sphere Packings, Lattices and Groups*, Grundlehren der mathematischen Wissenschaften, No. 290. Springer-Verlag, New York–Berlin–Heidelberg, 1988; 2nd ed., 1993; 3rd ed., 1998.

J. H. CONWAY and D. A. SMITH. 2003, *On Quaternions and Octonions: Their Geometry, Arithmetic, and Symmetry*. A K Peters, Natick, MA, 2003.

J. L. COOLIDGE. 1924, *The Geometry of the Complex Domain*. Clarendon Press / Oxford Univ. Press, Oxford, 1924.

H. S. M. COXETER. 1934a, "Discrete groups generated by reflections," *Annals of Math.* (2) **35**, 588–621. Reprinted in Coxeter 1995, 145–178.

1934b, "Finite groups generated by reflections, and their subgroups generated by reflections," *Proc. Cambridge Philos. Soc.* **30**, 466–482. Reprinted in Coxeter 1995, 179–195.

1935, "The functions of Schläfli and Lobatchefsky," *Quart. J. Math. Oxford Ser.* **6**, 13–29. Reprinted in Coxeter 1968, 3–20.

1940, "Regular and semi-regular polytopes, I," *Math. Z.* **46**, 380–407. Cf. Coxeter 1985, 1988. Reprinted in Coxeter 1995, 251–278.

1942, *Non-Euclidean Geometry*, Mathematical Expositions, No. 2. Univ. of Toronto Press, Toronto, 1942; 3rd ed., 1957; 6th ed., Math. Assoc. of America, Washington, DC, 1998.

1946, "Integral Cayley numbers," *Duke Math. J.* **13**, 561–578. Reprinted in Coxeter 1968, 21–39.

1948, *Regular Polytopes*. Methuen, London, 1948; Pitman, New York, 1949; 2nd ed., Macmillan, New York, and Collier-Macmillan, London, 1963, 3rd ed., Dover, New York, 1973.

1949, *The Real Projective Plane*. McGraw-Hill, New York, 1949; 2nd ed., Cambridge Univ. Press, London–Cambridge, 1955. Translated by W. Burau as *Reele Projektive Geometrie der Ebene*, Mathematische Einzelschriften, Bd. 1 (R. Oldenbourg, Munich, 1955). Cf. Coxeter 1993.

1964, "Regular compound tessellations of the hyperbolic plane," *Proc. Roy. Soc. London Ser. A* **278**, 147–167.

1966a, "The inversive plane and hyperbolic space," *Abh. Math. Sem. Univ. Hamburg* **29**, 217–242.

1966b, "Inversive distance," *Ann. Mat. Pura Appl.* (4) **71**, 73–83.

1968, *Twelve Geometric Essays*. Southern Illinois Univ. Press, Carbondale, and Feffer & Simons, London–Amsterdam, 1968. Reprinted as *The Beauty of Geometry: Twelve Essays* (Dover, Mineola, NY, 1999). Includes Coxeter 1935, 1946.

1974, *Regular Complex Polytopes*. Cambridge Univ. Press, London–New York, 1974; 2nd ed., Cambridge Univ. Press, Cambridge–New York, 1991.

1985, "Regular and semi-regular polytopes, II," *Math. Z.* **188**, 559–591. Cf. Coxeter 1940. Reprinted in Coxeter 1995, 279–311.

1988, "Regular and semi-regular polytopes, III," *Math. Z.* **200**, 3–45. Reprinted in Coxeter 1995, 313–355.

1993, *The Real Projective Plane*, 3rd ed., with an appendix for *Mathematica*. Springer-Verlag, New York–Berlin–Heidelberg, 1993.

1995, *Kaleidoscopes: Selected Writings of H. S. M. Coxeter*. Edited by F. A. Sherk, P. McMullen, A. C. Thompson, and A. I. Weiss. Wiley–Interscience, New York, 1995. Includes Coxeter 1934a, 1934b, 1940, 1985, 1988.

H. S. M. Coxeter and W. O. J. Moser. 1957, *Generators and Relations for Discrete Groups*, Ergebnisse der Mathematik und ihrer Grenzgebiete (N.F.), Bd. 14. Springer-Verlag, Berlin–Göttingen–Heidelberg, 1957; 4th ed., Springer-Verlag, Berlin–Heidelberg–New York, 1980.

H. S. M. Coxeter and G. J. Whitrow. 1950, "World-structure and non-Euclidean honeycombs," *Proc. Roy. Soc. London Ser. A* **201**, 417–437.

M. W. Davis. 2008, *The Geometry and Topology of Coxeter Groups*, London Mathematical Society Monographs, Vol. 32. Princeton Univ. Press, Princeton, NJ–Oxford, 2008.

L. E. Dickson. 1923a, "A new simple theory of hypercomplex integers," *J. Math. Pures Appl.* (9) **2**, 281–326. Reprinted in *Collected Mathematical Papers*, Vol. VI (Chelsea, Bronx, NY, 1983), 531–576.

1923b, *Algebras and Their Arithmetics*. Univ. of Chicago Press, Chicago, 1923; G. E. Stechert, New York, 1938. Translated by J. J. Burckhardt and E. Schubarth as *Algebren und ihre Zahlentheorie* (Orell Füssli, Zurich–Leipzig, 1927).

J. Dieudonné. 1943, "Les Déterminants sur un corps non-commutatif," *Bull. Soc. Math. France* **71**, 27–45.

P. Du Val. 1964, *Homographies, Quaternions, and Rotations*. Clarendon Press / Oxford Univ. Press, Oxford, 1964.

E. B. Dynkin. 1946, "Classification of simple Lie groups" (Russian), *Mat. Sb. (N.S.)* **18 (60)**, 347–352.

F. Esselmann. 1996, "The classification of compact hyperbolic Coxeter d-polytopes with $d + 2$ facets," *Comment. Math. Helv.* **71**, 229–242.

E. S. Fedorov. 1891, "Symmetry in the plane" (Russian), *Zap. Mineralog. Imper. S.-Peterburg. Obshch.* (2) **28**, 345–390 (2 plates).

A. J. Feingold, A. Kleinschmidt, and H. Nicolai. 2009, "Hyperbolic Weyl groups and the four normed division algebras," *J. Algebra* **322**, 1295–1339.

A. A. Felikson. 2002, "Coxeter decompositions of hyperbolic simplexes" (Russian), *Mat. Sb.* **193**, (12), 134–156. Translated in *Sb. Math.* **193** (2002), 1867–1888.

R. Fricke and F. Klein. 1897, *Vorlesungen über die Theorie der automorphen Funktionen*, Bd. I. Teubner, Leipzig, 1897; 2nd ed., 1926.

W. M. Goldman. 1999, *Complex Hyperbolic Geometry*, Oxford Mathematical Monographs. Clarendon Press / Oxford Univ. Press, Oxford, 1999.

T. Gosset. 1900, "On the regular and semi-regular figures in space of n dimensions," *Messenger of Math.* **29**, 43–48.

B. Grünbaum. 2009, *Configurations of Points and Lines*, Graduate Studies in Mathematics, Vol. 103. American Math. Soc., Providence, 2009.

M. Hall. 1959, *The Theory of Groups*. Macmillan, New York, 1959.

W. R. Hamilton. 1967, "Quaternion integers," unpublished manuscript (1856) printed as appendix to *Mathematical Papers*, Vol. III (Cambridge Univ. Press, London–New York, 1967), 657–665.

G. HESSENBERG. 1905, "Ueber einen geometrischen Calcül," *Acta Math.* **29**, 1–23.

D. HILBERT. 1901, "Über Flächen von konstanter Gaußscher Krümmung," *Trans. Amer. Math. Soc.* **2**, 87–99. Reprinted in *Gesammelte Abhandlungen*, Bd. II (Springer, Berlin, 1933), 437–448.

H. HOPF. 1931, "Über die Abbildungen der dreidimensionalen Sphäre auf die Kugelfläche," *Math. Ann.* **104**, 637–665. Reprinted in *Selecta* (Springer-Verlag, Berlin–Heidelberg–New York, 1964), 38–63.

J. E. HUMPHREYS. 1990, *Reflection Groups and Coxeter Groups*. Cambridge Univ. Press, Cambridge–New York, 1990; 1992.

A. HURWITZ. 1896, "Über die Zahlentheorie der Quaternionen," *Nach. Königl. Ges. Wiss. Göttingen Math.-Phys. Kl.* **1896**, 313–340. Reprinted in *Mathematische Werke*, Bd. II (Birkhäuser, Basel, 1933), 303–330.

H. C. IM HOF. 1985, "A class of hyperbolic Coxeter groups," *Expo. Math.* **3**, 179–186.

1990, "Napier cycles and hyperbolic Coxeter groups," *Bull. Soc. Math. Belgique* **42**, 523–545.

N. W. JOHNSON. 1981, "Absolute polarities and central inversions," in *The Geometric Vein: The Coxeter Festschrift*, C. Davis, B. Grünbaum, and F. A. Sherk, eds. (Springer-Verlag, New York–Heidelberg–Berlin, 1981), 443–464.

2013, "Integers," *Math. Intelligencer* **35**, No. 2, 52–59.

N. W. JOHNSON, R. KELLERHALS, J. G. RATCLIFFE, and S. T. TSCHANTZ. 1999, "The size of a hyperbolic Coxeter simplex," *Transform. Groups* **3**, 329–353.

2002, "Commensurability classes of hyperbolic Coxeter groups," *Linear Algebra Appl.* **345**, 119–147.

N. W. JOHNSON and A. IVIĆ WEISS. 1999a, "Quadratic integers and Coxeter groups," *Canad. J. Math.* **51**, 1307–1336.

1999b, "Quaternionic modular groups," *Linear Algebra Appl.* **295**, 159–189.

I. M. KAPLINSKAYA. 1974, "Discrete groups generated by reflections in the faces of simplicial prisms in Lobachevsky spaces" (Russian), *Mat. Zametki* **15**, 159–164. Translated in *Math. Notes* **15** (1974), 88–91.

R. KELLERHALS. 1989, "On the volume of hyperbolic polyhedra," *Math. Ann.* **285**, 541–569.

1991, "On Schläfli's reduction formula," *Math. Z.* **206**, 193–210.

1992, "On the volume of hyperbolic 5-orthoschemes and the trilogarithm," *Comment. Math. Helv.* **67**, 648–663.

1995, "Volumes in hyperbolic 5-space," *Geom. Funct. Anal.* **5**, 640–667.

F. KLEIN. 1871, "Ueber die sogenannte Nicht-Euklidische Geometrie," *Math. Ann.* **4**, 573–625. Reprinted in *Gesammelte mathematische Abhandlungen*, Bd. I (Springer, Berlin, 1921), 254–305. Translated as "On the so-called noneuclidean geometry" in Stillwell 1996, 69–111.

1873, "Ueber die sogenannte Nicht-Euklidische Geometrie (Zweiter Aufsatz)," *Math. Ann.* **6**, 112–145. Reprinted in *Gesammelte mathematische Abhandlungen*, Bd. I (Springer, Berlin, 1921), 311–343.

1879, "Ueber die Transformation der elliptischen Functionen und die Auflösung der Gleichungen fünften Grades," *Math. Ann.* **14**, 111–172. Reprinted in *Gesammelte mathematische Abhandlungen*, Bd. III (Springer, Berlin, 1923), 13–75.

1884, *Vorlesungen über das Ikosaeder und die Auflösung der Gleichungen vom fünften Grade*. Teubner, Leipzig, 1884. Translated by G. G. Morrice as *Lectures on the Icosahedron and the Solution of Equations of the Fifth Degree* (Trübner, London, 1888; 2nd ed., Kegan Paul, London, 1913; Dover, New York, 1956).

J. L. KOSZUL. 1968, *Lectures on Hyperbolic Coxeter Groups*. Notes by T. Ochiai. Univ. of Notre Dame, Notre Dame, IN [1968].

F. LANNÉR. 1950, "On complexes with transitive groups of automorphisms," *Medd. Lunds Univ. Mat. Sem.* **11**, 1–71.

H. LIEBMANN. 1905, *Nichteuklidsche Geometrie*. G. J. Göschen, Leipzig, 1905.

R. LIPSCHITZ. 1886, *Untersuchungen über die Summen von Quadraten*. M. Cohen, Bonn, 1886.

N. I. LOBACHEVSKY. 1829–30, "On the principles of geometry" (Russian), *Kazanskiĭ Vestnik* **1829**, II–III, 178–187; IV, 228–241; XI–XII, 227–243; *ibid.* **1830**, III–IV, 251–283; VII–VIII, 571–636. Reprinted in *Polnoe sobranie sochineniĭ*, t. I (Gosudarstvennoe Izdat. Tekhn.-Teoret. Literatury, Moscow–Leningrad, 1946), 185–261.

1835, "Imaginary geometry" (Russian), *Uchen. Zap. Imper. Kazan. Univ.* **1835**, 88 pp., 1 plate. Reprinted in *Polnoe sobranie sochineniĭ*, t. III (Gosudarstvennoe Izdat. Tekhn.-Teoret. Literatury, Moscow–Leningrad, 1951), 16–70 (1 plate). Translated by H. Liebmann in *Imaginäre Geometrie und Anwendung der imaginären Geometrie auf einige Integrale* (Teubner, Leipzig, 1904), 3–50.

1836, "Applications of the imaginary geometry to some integrals" (Russian), *Uchen. Zap. Imper. Kazan. Univ.* **1836**, 166 pp., 1 plate. Reprinted in *Polnoe sobranie sochineniĭ*, t. III (Gosudarstvennoe Izdat. Tekhn.-Teoret. Literatury, Moscow–Leningrad, 1951), 181–294 (1 plate). Translated by H. Liebmann in *Imaginäre Geometrie und Anwendung der imaginären Geometrie auf einige Integrale* (Teubner, Leipzig, 1904), 51–130.

1840, *Geometrische Untersuchungen zur Theorie der Parallellinien*. G. Fincke, Berlin, 1840; Mayer & Müller, Berlin, 1887. Translated by G. B. Halsted as *Geometrical Researches on the Theory of Parallels* (Univ. of Texas, Austin, 1891; Open Court, Chicago–London, 1914). Halsted's translation reprinted as supplement to Carslaw's translation of Bonola 1906 (Dover, New York, 1955).

C. MACLACHLAN, P. L. WATERMAN, and N. J. WIELENBERG. 1989, "Higher dimensional analogues of the modular and Picard groups," *Trans. Amer. Math. Soc.* **312**, 739–753.

W. MAGNUS. 1974, *Noneuclidean Tesselations and Their Groups*. Academic Press, New York–London, 1974.

G. E. MARTIN. 1982, *Transformation Geometry: An Introduction to Symmetry*, Undergraduate Texts in Mathematics. Springer-Verlag, New York–Berlin, 1982.

P. MCMULLEN. 1991, "The order of a finite Coxeter group," *Elem. Math.* **46**, 121–130.

P. MCMULLEN and E. SCHULTE. 2002, *Abstract Regular Polytopes*, Encyclopedia of Mathematics and Its Applications, Vol. 92. Cambridge Univ. Press, Cambridge–New York, 2002.

J. W. MILNOR. 1982, "Hyperbolic geometry: The first 150 years," *Bull. Amer. Math. Soc. (N.S.)* **6**, 9–24. Reprinted in *Collected Papers*, Vol. I (Publish or Perish, Houston, TX, 1994), 243–260.

A. F. MÖBIUS. 1855, "Die Theorie der Kreisverwandtschaft in rein geometrischer Darstellung," *Abh. Königl. Sächs. Ges. Wiss. Math.-Phys. Kl.* **2**, 529–595. Reprinted in *Gesammelte Werke*, Bd. II (Hirzel, Leipzig, 1886), 243–314.

B. R. MONSON. 1982, "The Schläflian of a crystallographic Coxeter group," *C. R. Math. Rep. Acad. Sci. Canada* **4**, 145–147.

B. R. MONSON and A. Ivić WEISS. 1995, "Polytopes related to the Picard group," *Linear Algebra Appl.* **218**, 185–204.

1997, "Eisenstein integers and related C-groups," *Geom. Dedicata* **66**, 99–117.

G. D. MOSTOW. 1968, "Quasi-conformal mappings in n-space and the rigidity of hyperbolic space forms," *Inst. Hautes Étud. Sci., Publ. Math.* **34**, 53–104.

1980, "On a remarkable class of polyhedra in complex hyperbolic space," *Pacific J. Math.* **86**, 171–276.

P. M. NEUMANN, G. A. STOY, and E. C. THOMPSON. 1994, *Groups and Geometry*. Oxford Univ. Press, Oxford–New York–Tokyo, 1994. Revised version of preliminary edition (Oxford Math. Inst., Oxford, 1980).

P. NIGGLI. 1924, "Die Flächensymmetrien homogener Diskontinuen," *Z. Kristallog. Mineralog.* **60**, 283–298. Cf. Pólya 1924.

T. H. O'Beirne. 1965, *Puzzles and Paradoxes*. Oxford Univ. Press, New York–London, 1965; Dover, New York, 1984.

A. L. Onishchik and R. Sulanke. 2006, *Projective and Cayley-Klein Geometries*, Springer Monographs in Mathematics. Springer, Berlin–Heidelberg–New York, 2006. Translation of *Projektive und Cayley-Kleinsche Geometrien* (Springer, Berlin–Heidelberg, 2006).

É. Picard. 1884, "Sur un groupe des transformations des points de l'espace situés du meme côté d'un plan," *Bull. Soc. Math. France* **12**, 43–47. Reprinted in *Œuvres*, t. I (Centre National de la Recherche Scientifique, Paris, 1978), 499–504.

H. Poincaré. 1882, "Theorie des groupes fuchsiens," *Acta Math.* **1**, 1–62. Reprinted in *Œuvres*, t. II (Gauthier-Villars, Paris, 1952), 108–168. First two sections translated as "Theory of fuchsian groups" in Stillwell 1996, 123–129.

G. Pólya. 1924, "Über die Analogie der Kristallsymmetrie in der Ebene," *Z. Kristallog. Mineralog.* **60**, 278–282. Reprinted in *Collected Papers*, Vol. IV (MIT Press, Cambridge, MA–London, 1984), 248–252. Cf. Niggli 1924.

M. N. Prokhorov. 1986, "Absence of discrete groups of reflections with a noncompact fundamental polyhedron of finite volume in a Lobachevsky space of high dimension" (Russian), *Izv. Akad. Nauk SSSR Ser. Mat.* **50**, 413–424. Translated in *Math. USSR Izv.* **28** (1987), 401–411.

J. G. Ratcliffe. 1994, *Foundations of Hyperbolic Manifolds*, Graduate Texts in Mathematics, No. 149. Springer-Verlag, New York–Berlin–Heidelberg, 1994.

2006, *Foundations of Hyperbolic Manifolds*, 2nd ed., Graduate Texts in Mathematics, No. 149. Springer, New York, 2006.

J. G. Ratcliffe and S. T. Tschantz. 1997, "Volumes of integral congruence hyperbolic manifolds," *J. Reine Angew. Math.* **488**, 55–78.

W. F. Reynolds. 1993, "Hyperbolic geometry on a hyperboloid," *Amer. Math. Monthly* **100**, 442–455.

B. Riemann. 1854, "Ueber die Hypothesen, welche der Geometrie zu Grunde liegen," *Abh. Königl. Ges. Wiss. Göttingen* **13** (1867), 133–150. Riemann's 1854 Habilitationsschrift, posthumously edited by R. Dedekind. Reprinted in *Gesammelte mathematische Werke* (Teubner, Leipzig, 1876), 254–269 or (2nd ed., 1892; Dover, New York, 1953), 272–287, and in *Gesammelte mathematische Werke, wissenschaftlicher Nachlaß und Nachträge* (Springer-Verlag, Berlin–Heidelberg–New York, and Teubner, Leipzig, 1990), 304–319. Translated by W. K. Clifford as "On the hypotheses which lie at the bases of geometry," *Nature* **8** (1873), 14–17, 36–37. Clifford's translation reprinted in his own *Mathematical Papers* (Macmillan, London, 1882), 55–71, and in *God Created the Integers: Mathematical*

Breakthroughs That Changed History, S. Hawking, ed. (Running Press, Philadelphia–London, 2005), 865–876 or (2007), 1031–1042.

G. de B. Robinson. 1940, *The Foundations of Geometry*, Mathematical Expositions, No. 1. Univ. of Toronto Press, Toronto, 1940; 4th ed., 1959.

D. Schattschneider. 1978, "The plane symmetry groups: Their recognition and notation," *Amer. Math. Monthly* **85**, 439–450.

P. Scherk. 1960, "Some concepts of conformal geometry," *Amer. Math. Monthly* **67**, 1–30.

L. Schläfli. 1858, "On the multiple integral $\int^n dx\, dy \ldots dz$, whose limits are $p_1 = a_1x+b_1y+\cdots+h_1z > 0, p_2 > 0, \ldots, p_n > 0$, and $x^2+y^2+\cdots+z^2 < 1$" (Part 1), *Quart. J. Pure Appl. Math.* **2**, 269–301. Reprinted with Schläfli 1860 in *Gesammelte mathematische Abhandlungen*, Bd. II (Birkhäuser, Basel, 1953), 219–270.

1860, "On the multiple integral $\int^n dx\, dy \ldots dz, \ldots$" (Parts 2 and 3), *Quart. J. Pure Appl. Math.* **3**, 54–68, 97–108.

1901, "Theorie der vielfachen Kontinuität," *Neue Denkschr. Schweiz. Ges. Naturwiss.* **38**, I, 1–237 (memorial volume). First full publication of early work (1850–1852), translated in part (by A. Cayley) as Schläfli 1858, 1860. Reprinted in *Gesammelte mathematische Abhandlungen*, Bd. I (Birkhäuser, Basel, 1950), 167–387.

I. J. Schoenberg. 1937, "Regular simplices and quadratic forms," *J. London Math. Soc.* **12**, 48–55. Reprinted in *Selected Papers*, Vol. I (Birkhäuser, Boston–Basel, 1988), 59–66.

P. H. Schoute. 1908, "The sections of the net of measure polytopes M_n of space Sp_n with a space Sp_{n-1} normal to a diagonal," *Konink. Akad. Wetensch. Amsterdam Proc. Sect. Sci.* **10**, 688–698.

E. Schulte and A. Ivić Weiss. 1994, "Chirality and projective linear groups," *Discrete Math.* **131**, 221–261.

H. Schwerdtfeger. 1962, *Geometry of Complex Numbers: Circle Geometry, Moebius Transformation, Non-Euclidean Geometry*, Mathematical Expositions, No. 13. Univ. of Toronto Press, Toronto, 1962; Dover, New York, 1979.

E. Snapper and R. J. Troyer. 1971, *Metric Affine Geometry*. Academic Press, New York–London, 1971; Dover, New York, 1989.

D. M. Y. Sommerville. 1914, *The Elements of Non-Euclidean Geometry*. Bell, London, 1914; Dover, New York, 1958.

A. B. Sossinsky, 2012, *Geometries*, Student Mathematical Library, Vol. 64. American Math. Soc., Providence, 2012.

P. Stäckel. 1901, "Friedrich Ludwig Wachter, ein Beitrag zur Geschichte der nichteuklidischen Geometrie," *Math. Ann.* **54**, 49–85.

K. G. Ch. von Staudt. 1847, *Geometrie der Lage*. F. Korn, Nuremberg, 1847.

1857, *Beiträge zur Geometrie der Lage*. F. Korn, Nuremberg, 1857.

E. Stiefel. 1942, "Über eine Beziehung zwischen geschlossenen Lie'schen Gruppen und diskontinuierlichen Bewegungsgruppen euklidischer Räume und ihre Anwendung auf die Aufzählung der einfachen Lie'schen Gruppen," *Comment. Math. Helv.* **14**, 350–380.

J. Stillwell. 1996, *Sources of Hyperbolic Geometry*, History of Mathematics, Vol. 10. American Math. Soc. and London Math. Soc., Providence, 1996. Includes English translations of Beltrami 1868a, 1868b; Klein 1871; Poincaré 1882.

K. Th. Vahlen. 1902, "Ueber Bewegungen und complexe Zahlen," *Math. Ann.* **55**, 585–593.

O. Veblen and J. W. Young. 1910, *Projective Geometry*, Vol. I. Ginn, Boston, 1910; 1916; Blaisdell, New York–Toronto–London, 1965.

1918, *Projective Geometry*, Vol. II. Ginn, Boston, 1918; Blaisdell, New York–Toronto–London, 1965.

E. B. Vinberg. 1984, "The absence of crystallographic groups of reflections in Lobachevsky spaces of large dimension" (Russian), *Trudy Moskov. Mat. Obshch.* **47**, 68–102, 246. Translated in *Trans. Moscow Math. Soc.* **47** (1985), 75–112.

1985, "Hyperbolic reflection groups" (Russian), *Usp. Mat. Nauk* **40**, 29–64. Translated in *Russ. Math. Surv.* **40** (1985), 31–75.

E. B. Vinberg and O. V. Shvartsman. 1988, "Discrete groups of motions of spaces of constant curvature" (Russian), in *Geometriya 2*, Itogi Nauki i Tekhniki, Ser. Sovremennye problemy matematiki, Fundamental'nye napravleniya, t. 29 (Vsesoyuz Inst. Nauchn. i Tekhn. Inform., Moscow, 1988), 147–259. Translated by V. Minachin as Part II of *Geometry II: Spaces of Constant Curvature*, E. B. Vinberg, ed., Encyclopaedia of Mathematical Sciences, Vol. 29 (Springer-Verlag, Berlin–Heidelberg–New York, 1993), 139–248. Cf. Alekseevsky, Vinberg & Solodovnikov 1988.

P. L. Waterman. 1993, "Möbius transformations in several dimensions," *Adv. in Math.* **101**, 87–113.

H. Weyl. 1925, "Theorie der Darstellung kontinuierlicher halbeinfacher Gruppen durch lineare Transformationen, I," *Math. Z.* **23**, 271–309. Reprinted with Weyl 1926a and 1926b in *Selecta* (Birkhäuser, Basel–Stuttgart, 1956), 262–366, and in *Gesammelte Abhandlungen*, Bd. II (Springer-Verlag, Berlin–Heidelberg–New York, 1968), 543–647.

1926a, "Theorie der Darstellung . . ., II," *Math. Z.* **24**, 328–376.

1926b, "Theorie der Darstellung . . ., III," *Math. Z.* **24**, 377–395; Nachtrag, *ibid.*, 789–791.

1939, *The Classical Groups: Their Invariants and Representations*, Princeton Mathematical Series, No. 1. Princeton Univ. Press, Princeton, NJ, 1939; 2nd ed., 1946; 1997.

1952, *Symmetry*. Princeton Univ. Press, Princeton, NJ, 1952; 1982; 1989.

J. B. WILKER. 1981, "Inversive geometry," in *The Geometric Vein: The Coxeter Festschrift*, C. Davis, B. Grünbaum, and F. A. Sherk, eds. (Springer-Verlag, New York–Heidelberg–Berlin, 1981), 379–442.

1993, "The quaternion formalism for Möbius groups in four or fewer dimensions," *Linear Algebra Appl.* **190**, 99–136.

E. WITT. 1941, "Spiegelungsgruppen und Aufzählung halbeinfacher Liescher Ringe," *Abh. Math. Sem. Univ. Hamburg* **14**, 289–322. Reprinted in *Collected Papers / Gesammelte Abhandlungen* (Springer-Verlag, Berlin–Heidelberg–New York, 1998), 213–246.

I. M. YAGLOM. 1955, *Geometretricheskie preobrazovaniya I: Dvizheniya i preobrazovaniya podobiya.* Gosudarstvennoe Izdat. Tekhn.-Teoret. Literatury, Moscow, 1955. Translated by A. Shields as *Geometric Transformations I* [*Isometries*], New Mathematical Library, No. 8 (Random House and L. W. Singer, New York, 1962; republished by Math. Assoc. of America, Washington, DC), and *Geometric Transformations II* [*Similarities*], New Mathematical Library, No. 21 (Random House and L. W. Singer, New York, 1968; republished by Math. Assoc. of America, Washington, DC).

1956, *Geometretricheskie preobrazovaniya II: Lineinye i krugovye preobrazovaniya.* Gosudarstvennoe Izdat. Tekhn.-Teoret. Literatury, Moscow, 1956. Translated by A. Shenitzer as *Geometric Transformations III* [*Affine and Projective Transformations*], New Mathematical Library, No. 24 (Random House and L. W. Singer, New York, 1973; republished by Math. Assoc. of America, Washington, DC), and *Geometric Transformations IV: Circular Transformations*, Anneli Lax New Mathematical Library, No. 44 (Math. Assoc. of America, Washington, DC, 2009).

INDEX